Ergodic Theorems for Group Actions

Mathematics and Its Applications

Volume 78

Ergodic Theorems for Group Actions

Informational and Thermodynamical Aspects

by

Arkady Tempelman

Department of Statistics,
Penn State University,
University Park, U.S.A.

SPRINGER-SCIENCE+BUSINESS MEDIA, B.V.

Library of Congress Cataloging-in-Publication Data

Tempel'man, A. A. (Arkadiĭ Aleksandrovich)
[Ėrgodicheskie teoremy na gruppakh. English]
Ergodic theorems for group actions : informational and thermodynamical aspects / by Arkady Tempelman.
p. cm. -- (Mathematics and its applications (Main series) ; v. 78)
Translation of: Ėrgodicheskie teoremy na gruppakh.
Includes bibliographical references and index.
ISBN 978-90-481-4155-5 ISBN 978-94-017-1460-0 (eBook)
DOI 10.1007/978-94-017-1460-0
1. Group theory. 2. Ergodic theory. I. Title. II. Series: Mathematics and its applications (Kluwer Academic Publishers) ; v. 78.
QA174.2.T4613 1992
512'.55--dc20 92-11459

ISBN 978-90-481-4155-5

Printed on acid-free paper

SERIES EDITOR'S PREFACE

'Et moi, ..., si j'avait su comment en revenir, je n'y serais point allé.'

Jules Verne

The series is divergent; therefore we may be able to do something with it.

O. Heaviside

One service mathematics has rendered the human race. It has put common sense back where it belongs, on the topmost shelf next to the dusty canister labelled 'discarded nonsense'.

Eric T. Bell

Mathematics is a tool for thought. A highly necessary tool in a world where both feedback and nonlinearities abound. Similarly, all kinds of parts of mathematics serve as tools for other parts and for other sciences.

Applying a simple rewriting rule to the quote on the right above one finds such statements as: 'One service topology has rendered mathematical physics ...'; 'One service logic has rendered computer science ...'; 'One service category theory has rendered mathematics ...'. All arguably true. And all statements obtainable this way form part of the raison d'être of this series.

This series, *Mathematics and Its Applications*, started in 1977. Now that over one hundred volumes have appeared it seems opportune to reexamine its scope. At the time I wrote

> "Growing specialization and diversification have brought a host of monographs and textbooks on increasingly specialized topics. However, the 'tree' of knowledge of mathematics and related fields does not grow only by putting forth new branches. It also happens, quite often in fact, that branches which were thought to be completely disparate are suddenly seen to be related. Further, the kind and level of sophistication of mathematics applied in various sciences has changed drastically in recent years: measure theory is used (non-trivially) in regional and theoretical economics; algebraic geometry interacts with physics; the Minkowsky lemma, coding theory and the structure of water meet one another in packing and covering theory; quantum fields, crystal defects and mathematical programming profit from homotopy theory; Lie algebras are relevant to filtering; and prediction and electrical engineering can use Stein spaces. And in addition to this there are such new emerging subdisciplines as 'experimental mathematics', 'CFD', 'completely integrable systems', 'chaos, synergetics and large-scale order', which are almost impossible to fit into the existing classification schemes. They draw upon widely different sections of mathematics."

By and large, all this still applies today. It is still true that at first sight mathematics seems rather fragmented and that to find, see, and exploit the deeper underlying interrelations more effort is needed and so are books that can help mathematicians and scientists do so. Accordingly MIA will continue to try to make such books available.

If anything, the description I gave in 1977 is now an understatement. To the examples of interaction areas one should add string theory where Riemann surfaces, algebraic geometry, modular functions, knots, quantum field theory, Kac-Moody algebras, monstrous moonshine (and more) all come together. And to the examples of things which can be usefully applied let me add the topic 'finite geometry'; a combination of words which sounds like it might not even exist, let alone be applicable. And yet it is being applied: to statistics via designs, to radar/sonar detection arrays (via finite projective planes), and to bus connections of VLSI chips (via difference sets). There seems to be no part of (so-called pure) mathematics that is not in immediate danger of being applied. And, accordingly, the applied mathematician needs to be aware of much more. Besides analysis and numerics, the traditional workhorses, he may need all kinds of combinatorics, algebra, probability, and so on.

In addition, the applied scientist needs to cope increasingly with the nonlinear world and the extra mathematical sophistication that this requires. For that is where the rewards are. Linear models are honest and a bit sad and depressing: proportional efforts and results. It is in the nonlinear world that infinitesimal inputs may result in macroscopic outputs (or vice versa). To appreciate what I am hinting at: if electronics were linear we would have no fun with transistors and computers; we would have no TV; in fact you would not be reading these lines.

There is also no safety in ignoring such outlandish things as nonstandard analysis, superspace and anticommuting integration, p-adic and ultrametric space. All three have applications in both electrical engineering and physics. Once, complex numbers were equally outlandish, but they frequently proved the shortest path between 'real' results. Similarly, the first two topics named have already provided a number of 'wormhole' paths. There is no telling where all this is leading - fortunately.

Thus the original scope of the series, which for various (sound) reasons now comprises five subseries: white (Japan), yellow (China), red (USSR), blue (Eastern Europe), and green (everything else), still applies. It has been enlarged a bit to include books treating of the tools from one subdiscipline which are used in others. Thus the series still aims at books dealing with:

- a central concept which plays an important role in several different mathematical and/or scientific specialization areas;
- new applications of the results and ideas from one area of scientific endeavour into another;
- influences which the results, problems and concepts of one field of enquiry have, and have had, on the development of another.

Quite generally, ergodic theory is concerned with statistical properties of dynamical systems. As such, the techniques and results of ergodic theory are of great importance to all who deal with dynamical systems, which means (at least) all of the sciences and engineering. For instance, in an already quite general setting one considers an operator A on a Banach space of functions and studies averages of $f, Af, A^2f, \ldots$. In this case the 'time' semigroup is $\mathbb{N} = \{0, 1, 2, \ldots\}$ and the operator A provides a linear representation of the semigroup $\mathbb{N}$ on the Banach space of functions in question.

This book is a comprehensive treatise on extensions of the classical ergodic theorems to the case of 'group time', e.g. multi-time, and (dynamical) systems defined by a representation of homogeneous random fields on groups and homogeneous spaces. In addition the results are applied to the study of information-theoretic and thermodynamical characteristics of such random fields.

Group time, e.g. periodic and multi-parameter, processes and dynamical systems and random fields are central to many considerations and applications, and thus I expect this authoritative comprehensive treatment to be most useful to a wide range of mathematicians, scientists, and engineers.

The shortest path between two truths in the real domain passes through the complex domain.

J. Hadamard

La physique ne nous donne pas seulement l'occasion de résoudre des problèmes ... elle nous fait pressentir la solution.

H. Poincaré

Never lend books, for no one ever returns them; the only books I have in my library are books that other folk have lent me.

Anatole France

The function of an expert is not to be more right than other people, but to be wrong for more sophisticated reasons.

David Butler

Bussum, March 1992

Michiel Hazewinkel

Contents

Series Editor's Preface v

Preface xi

Notations xviii

Introduction 1

§ 1. Basic ergodic and mean–value theorems 1
§ 2. Further ergodic theorems for operators 5
§ 3. Means, averaging sequences and ergodic theorems on groups 6

Chapter 1: *Means and averageable functions* 10

§ 1. Means and quasi–averageable functions 10
§ 2. Averageable functions 14
§ 3. Averageable elements in general Banach spaces 24
§ 4. Fixed point theorems 26
§ 5. Some classes of averageable functions and Banach space elements 32
§ 6. Means of positive definite functions 42
§ 7. Averageability of homogeneous random fields and of orbital functions of dynamical systems 44
Bibliographical notes 49

Chapter 2: *Ergodicity and mixing* 51

§ 1. Main definitions 51
§ 2. Ergodicity and mixing of Gaussian homogeneous random functions 62
§ 3. WM–spaces 68
§ 4. SM–spaces 76
Bibliographical notes 83

Chapter 3: *Averaging sequences. Universal ergodic theorems* 85

§ 1. General mean value theorems 85
§ 2. Averaging nets on group extensions and products . . . 88

§ 3. Universal ergodic theorems on amenable semigroups . 95
§ 4. Existence of universally averaging nets of measures on general groups 98
§ 5. Universally $O_{W(B)}$–averaging sequences 103
§ 6. Averaging with respect to contractive representations . 111
§ 7. C_a–averaging nets 111
Bibliographical notes 115

Chapter 4: *Mean ergodic theorems* 117

§ 1. Mean averaging nets: definition and existence 117
§ 2. Mean ergodic theorems on amenable semigroups . . . 127
§ 3. Mean ergodic theorems for weakly converging unitary representations and for strongly quasimixing dynamical systems 127
§ 4. Mean ergodic theorems for strongly quasimixing random fields 132
§ 5. Mean ergodic theorem on SQM–spaces 132
§ 6. Mean averaging nets on group extensions and products 133
§ 7. Construction of mean averaging sequences of sets on non–amenable groups 135
§ 8. Spectral criterion 139
§ 9. Mean ergodic theorem for homogeneous generalized random fields 144
§ 10. Mean ergodic theorem on the free group F_2 149
Bibliographical notes 153

Chapter 5: *Maximal and dominated ergodic theorems* . . 156

§ 1. Preliminaries 156
§ 2. Regular nets of sets 173
§ 3. Hardy – Littlewood type inequalities for regular nets . 183
§ 4. Hardy – Littlewood type inequalities for some non–regular and singular weights 185
§ 5. Maximal and dominated ergodic theorems: The transfer approach 193
§ 6. Maximal ergodic theorems: The dual space approach . 195
Bibliographical notes 204

Chapter 6: *Pointwise ergodic theorems* 207

§ 1. Pointwise averaging nets: definition and preliminary discussion 207
§ 2. Pointwise averaging in dense subsets of L^2: Spectral approach 210
§ 3. Pointwise averaging with regular ergodic nets of sets . 217
§ 4. Ergodic theorems with regular ergodic weights 222
§ 5. Pointwise averaging with spheres and spherical layers . 224
§ 6. Pointwise averaging with convolution weights 226
§ 7. General ratio ergodic theorems 229
§ 8. Construction of pointwise averaging sets on general connected groups 234
§ 9. Pointwise averaging nets on homogeneous spaces . . . 241
Bibliographical notes 243

Chapter 7: *Ergodic theorems for homogeneous random measures* 246

§ 1. Homogeneous and λ–homogeneous random fields . . . 246
§ 2. Local ergodic theorems. Lebesgue decomposition of homogeneous random measures 249
§ 3. "Global" ergodic theorems 253
§ 4. Homogeneous random measures on homogeneous spaces 259
§ 5. Homogeneous random signed measures 261
Bibliographical notes 262

Chapter 8: *Specific informational and thermodynamical characteristics of homogeneous random fields* 264

§ 1. Main notations and definitions 264
§ 2. Generalized ergodic theorems 268
§ 3. Specific energy 271
§ 4. Relative entropy and associated convergence theorems . 275
§ 5. Generalizations of the McMillan theorem. Specific entropy . 278
§ 6. Specific free energy 286
§ 7. Convergence theorems for Gibbsian measures. Specific relative entropy 287
§ 8. The variational principle 291

Bibliographical notes 294

Appendix 296

§ 1. Groups and semigroups 296
§ 2. Homogeneous and group–type homogeneous spaces . . 309
§ 3. Amenable semigroups and ergodic nets 318
§ 4. Positive definite functions 325
§ 5. Representations of semigroups in Banach spaces . . . 328
§ 6. Weakly almost periodic elements and functions 333
§ 7. Dynamical systems 337
§ 8. Homogeneous random functions 338
§ 9. Measurability and continuity of representations, dynamical systems and homogeneous random fields 343
§ 10. Flight–functions and FD–continuous spectrum . . . 349
§ 11. Relative entropy on random σ–algebras 353
§ 12. The Banach convergence principle 356
§ 13. Directions and nets 357
§ 14. Correspondence between "left" and "right" objects and conditions 362

References 364

Index 394

Preface

This book is a considerably extended English translation of a previous book by the author, "Ergodic theorems on groups" (Vilnius, Mokslas, 1986; in Russian).

The main work on the book was done at the Institute of Mathematics and Informatics of the Lithuanian Academy of Sciences in Vilnius; the final draft was written during my stay at North Dakota State University in Fargo, North Dakota, USA, in 1990–1991.

The book is devoted to the extension of the classical ergodic theorems to linear representations of semigroups in Banach spaces, to dynamical systems with group time and to homogeneous random fields on groups and homogeneous spaces, in particular, on $\mathbb{R}^n$, $\mathbb{Z}^n$, the Lorentz group, and Euclidean and Lobachevsky spaces. The generalized von Neumann theorem is, in its turn, treated as a special case of the general ergodic theorem for "averageable" functions which also includes the mean–value theorem for weakly almost periodic, and some other classes of functions.

Ergodicity and mixing of dynamical systems and homogeneous random fields are also studied.

We apply these results to the study of the existence, and the properties, of specific informational and thermodynamical characteristics of homogeneous random fields on amenable discrete groups.

We will usually restrict ourselves to the consideration of extensions of ergodic theorems to group and semigroup representations by isometric or contractive operators in the Hilbert space and in the spaces L^α, $1 \leq \alpha < \infty$. In the one– and multi–parameter cases ergodic theorems for more general operators on Banach spaces can be found in the excellent book "Ergodic theorems" by U.Krengel.

The book is intended for research workers, graduate students, mathematicians, and theoretical physicists.

For understanding the presented material, knowledge of the usual university courses in Probability Theory, Functional Analysis and Topology is sufficient; the necessary additional information is stated in the "Appendix".

I take this opportunuity to express my deep gratitude to my teachers R.Dobrushin, D.Raikov and A.Yaglom as well as to Ya.Sinai and V.Statulevichius for their constant friendly attention and advice. Discussions with Professors S. Gindikin, G. Ol'shansky, J. Fritz, H. Furstenberg, I. Kornfeld, M. Lin, A. Naftalevich, J. Olsen, A. Shulman and B. Weiss were very useful.

I want to express my deep gratitude to Ms. Dausa Žvirėnaitė, whose great skill, enormous patience and labour miraculously turned almost unreadable hand–written pages into a camera–ready manuscript.

I am also grateful to Prof. M. Hazewinkel, who kindly invited me to write this book for Kluwer Academic Publishers and whose practical help was considerable.

I would also like to thank Drs. J. Jaura and A. Žalys, who helped to prepare the bibliography, and to thank Rob Hoksbergen for help with the language checking.

Notations

This list does not contain the special notation introduced in § 1 of Chapt. 7 and used in that chapter.

I. Groups and semigroups

X := semigroup or group;
G := group;
$Ax^{-1} := \{y : y \in X,\ yx \in A\}, x \in X,\ A \subset X$;
$x^{-1}A := \{y : y \in X,\ xy \in A\}, x \in X,\ A \subset X$;
$\cong$:= group isomorphism;
$\mathrm{Int}_p X$:= the proper interior of X;
Y or G/K := homogeneous space with group of motions G and stationary subgroup K;
Y_q := the least full subsemigroup in X containing the carrier of a measure q;
G/Γ := factor group of G with respect to the invariant subgroup Γ;
e := identity element of a semigroup;
$\mathfrak{B}$ or $\mathfrak{B}(X)$:= σ–algebra of all measurable (resp. Borel) subsets of a measurable (resp. topological) semigroup X;
r_x (resp. l_x) := right (resp. left) translation in X by the element $x \in X$;
R_x (resp. L_x) := right (resp. left) translation corresponding to the element $x \in X$ in the function space $\Phi_B(X)$;
Aut (G) := the group of continuous automorphisms of a topological group G;
$\mathrm{Aut}\,(\Gamma|G, F) := \{\alpha_f : \alpha_f \in \mathrm{Aut}\,(\Gamma),\ \alpha_f(\gamma) = f\gamma f^{-1},\ \gamma \in \Gamma,\ f \in F\}$ if $G = \Gamma \overset{\alpha}{\odot} F$;
$A \circ \Sigma$:= semidirect product of groups A and $\Sigma \subset \mathrm{Aut}\,(A)$ (see Appendix, Subsect. 2.2);
$G = \Gamma \overset{\alpha}{\circ} F$:= right semidirect factorization of G (Γ is an invariant subgroup of G and $G/\Gamma \cong F$; see Appendix, § 2);

$G = G_1 \odot \ldots \odot G_n$:= left semidirect factorization of G (see Appendix, § 2);
$G = G_n \odot \ldots \odot G_1$:= right semidirect factorization of G (see Appendix, § 2);
$S : x \mapsto S_x$:= continuous linear representation of a semigroup X in a Banach space;
$U : g \mapsto U_g$:= continuous unitary representation of a group G;
$\Delta(\cdot)$:= modular function of a group, i.e. the left modular function of the right Haar measure (see Appendix, Subsect. 1.6).

II. Measures and functions on semigroups

$\operatorname{var} \nu$:= variation of the signed measure ν;
$\mathfrak{BM}(\mathfrak{B})$:= Banach space of all signed measures ν of bounded variation on $\mathfrak{B}$ with the norm $\|\nu\| := \operatorname{var} \nu$;
$\mathfrak{M}_+(\mathfrak{B})$:= class of all measures on $\mathfrak{B}$;
$\Delta_\mu^{(l)}(\cdot)$, $\Delta_\mu^{(r)}(\cdot)$:= left, resp. right, modular functions of a measure μ on X;
$\mathcal{P}$ or $\mathcal{P}(\mathfrak{B})$:= class of all probability measures on $\mathfrak{B}$;
$\widetilde{\mathcal{P}}$:= class of all measures in $\mathcal{P}$ with finite carrier;
MAX := class of all maximal nets;
DOM := class of all dominant nets;
$\lambda * \mu$:= convolution of λ, $\mu \in \mathcal{P}$ (see Appendix, Subsect. 1.7);
λ^{*n} := n–th convolution power of $\lambda \in \mathcal{P}$;
Φ_B or $\Phi_B(X)$:= Banach space of all bounded functions on X with values in the Banach space B;
F_B or $F_B(X)$:= subspace of all measurable B–valued functions on X;
$A_B^{(r)}$ or $A_B^{(r)}(X)$ (resp. $A_B^{(l)}$ or $A_B^{(l)}(X)$) := subspace of all right–averageable (resp. left–averageable) functions in Φ_B;
C or $C(X)$:= Banach space of all bounded continuous functions on a topological semigroup X with norm $\|\varphi\| := \sup_{x \in X} |\varphi(x)|$;
C_a or $C_a(X)$:= subspace of all functions φ in $C(X)$ for which $\lim_{x \to \infty} \varphi(x)$ (X is locally compact);
C_0 or $C_0(X)$:= subspace of all functions φ in C_a with $\lim_{x \to \infty} \varphi(x) = 0$;

$UC^{(r)}(X)$ (resp. $UC^{(l)}(X)$) := subspace of all right (resp. left) uniformly continuous functions in $C(X)$ (see Appendix, Subsect. 1.1);
AP or $AP(X)$:= subspace of all almost periodic functions in $C(X)$;
W or $W(X)$:= subspace of all weakly almost periodic functions in $C(X)$;
P or $\mathbf{P}(X)$:= set of all positive definite functions on a group X;
$\mathbf{EP}(X)$ and $\mathbf{ESP}(X)$:= sets of all elementary functions, resp. all elementary spherical, functions in **P**;
$O_{W(B)}$ or $O_{W(B)}(X)$:= subset of $\Phi_B(X)$, consisting of all orbital functions $\varphi_{S,b}(x) := S_x b$, $x \in X$, where $b \in W(B|S)$ (see part III below);
$O_H^{(c)}$:= set of all orbital functions with respect to contractive representations in a Hilbert space H;
$\mathbf{M}(f)$, $\mathbf{M}^{(r)}(f)$ and $\mathbf{M}^{(l)}(f)$:= mean value, right and left mean values, of a function f, respectively;
$\mathfrak{M}(f)$, $\mathfrak{M}^{(r)}(f)$ and $\mathfrak{M}^{(l)}(f)$:= sets of all corresponding mean values of $f \in \Phi_B$.

III. Banach and Hilbert spaces

B := Banach space;
$|b|$, $\|b\|$ or $\|b\|_B$:= norm of $b \in B$;
$[C]$ or $[C]_B$:= closure of the set $C \subset B$;
B^* := Banach space conjugate (dual) to B;
$\langle b, l\rangle$ or $l(b)$:= value of a functional $l \in B^*$ at $b \in B$;
$A(B)$ or $A(B|S)$:= subset of all elements of B that are averageable with respect to the representation S;
$W(B)$ or $W(B|S)$:= subset of all elements of B with respect to the representation S;
$B = L \oplus M$:= decomposition of B into a direct sum of subspaces L and M;
Π_L^B := projector in B onto L with respect to a given subspace M (if $B = L \oplus M$);
H or $\mathcal{H}$:= Hilbert space;
$\langle\cdot,\cdot\rangle$ or $\langle\cdot,\cdot\rangle_H$:= scalar product in H;
$\mathrm{Isom}\,(H)$:= set of all isometric linear operators in H;
Π_L^H := orthogonal projector in H onto a subspace L.

IV. Special spaces, semigroups, groups

$\mathbb{R}$ or $\mathbb{R}^1$:= group of all reals (with respect to natural addition);
$\mathbb{Z}$ or $\mathbb{Z}^1$:= group of all integers (with respect to natural addition);
$\mathbb{N}$:= set of all natural numbers;
$\mathbb{C}$ or $\mathbb{C}^1$:= set of all complex numbers;
$\mathbb{R}^m$:= linear space of vectors $a = (a_1, \ldots, a_m)$, $a_i \in \mathbb{R}$, $i = \overline{1,m}$, with the scalar product $\langle a', a'' \rangle = \sum_{i=1}^{m} a_i' a_i''$ ($m = \overline{1,\infty}$);
$\mathbb{C}^m$:= Hilbert space of all m–dimensional vectors $a = (a_1, \ldots, a_m)$, $a_i \in \mathbb{C}$, with the scalar product $\langle a', a'' \rangle = \sum_{i=1}^{m} a_i' \overline{a_i''}$ ($m = \overline{1,\infty}$);
$\mathbb{Z}^m$:= subgroup of all vectors with integer coordinates in $\mathbb{R}^m$;
$\mathbb{R}^m_+$, $\mathbb{Z}^m_+$:= semigroups of all vectors with non–negative coordinates in $\mathbb{R}^m$ and $\mathbb{Z}^m$;
$\mathbb{R}^m_{++}$, $\mathbb{Z}^m_{++}$:= semigroups of all vectors with positive coordinates in $\mathbb{R}^m$ and $\mathbb{Z}^m$;
$\mathbb{R}_+ := \mathbb{R}^1_+$; $\mathbb{R}_{++} := \mathbb{R}^1_{++}$; $\mathbb{Z}_+ := \mathbb{Z}^1_+$; $\mathbb{Z}_{++} := \mathbb{Z}^1_{++}$;
$R^{\times}_{++} := \mathbb{R}_{++}$ as a group with respect to multiplication;
$GL(m,\mathbb{R})$, $GL_0(m,\mathbb{R})$, $SL(m,\mathbb{R})$, $SO(m)$, $SO_0(m)$ are groups defined in the Appendix (Example 1.1);
E_m := m–dimensional Euclidean homogeneous space (see Appendix, Example 2.4);
L_m := m–dimensional Lobachevsky homogeneou space (see Appendix, Example 2.5);
$\mathcal{T}(T)$:= class of all transformations of a set T;
$\mathcal{T}_l$ and $\mathcal{T}_r(T) := \mathcal{T}(T)$ as a left, resp. right, semigroup, respectively (see Appendix, Subsect. 1.4);
$\mathcal{L}_l(B)$ and $\mathcal{L}_r(B)$:= left, resp. right, semigroup (with respect to multiplication) of bounded linear operators in B;
$\mathcal{L}_l(B,w)$ and $\mathcal{L}_l(B,s) := \mathcal{L}_l(B)$ as semitopological semigroup with respect to the weak, resp. strong, topology.

V. Measurable and probability spaces

$(\Omega, \mathcal{F})$:= measurable space (Ω := set, and $\mathcal{F}$:= σ–algebra of subsets

of Ω);
$(\Omega, \mathcal{F}, m)$:= σ–finite measure space;
$(\Omega, \mathcal{F}, P)$:= probability space;
End $(\Omega, \mathcal{F}, m)$:= set of endomorphisms (measure–preserving transformations) of $(\Omega, \mathcal{F}, m)$;
m–a.e. := almost everywhere with respect to the measure m;
$\mathbf{E}_P\xi$ and $\mathbf{D}_P\xi$:= expectation, resp. variance, of a random variable $\xi(\omega)$ over $(\Omega, \mathcal{F}, P)$;
$\int_A \varphi\, dm$:= integral over the set $A \in \mathcal{F}$;
$\int \varphi\, dm := \int_\Omega \varphi\, dm$;
$\mathcal{X}_A(\cdot)$:= characteristic function (indicator) of the set A;
L_B or $L_B(\Omega, \mathcal{F}, m)$:= linear manifold of all integrable B–valued simple functions on Ω;
L_B^α or $L_B^\alpha(\Omega, \mathcal{F}, m)$:= space of all B–valued m–measurable functions with the finite seminorm

$$\|f\|_{L_B^\alpha} := \left\{ \int \|f(\omega)\|_B^\alpha m(d\omega) \right\}^{1/\alpha}$$

and also its factor space with respect to the subspace $\{f : \|f\|_{L_B^\alpha} = 0\}$;

$$L(\Omega, \mathcal{F}, m) := L_{\mathbf{C}}(\Omega, \mathcal{F}, m); \quad L^\alpha(\Omega, \mathcal{F}, m) := L_{\mathbf{C}}^\alpha(\Omega, \mathcal{F}, m);$$
$$L_+^\alpha(\Omega, \mathcal{F}, m) := \{f : f \in L_{\mathbf{R}}^\alpha(\Omega, \mathcal{F}, m), \quad f \geq 0 \quad m\text{–a.e.}\};$$

vraisup $f(\omega)$ or esssup $f(w)$:= essential least upper bound of f (with respect to a measure m).

VI. Other notations and abberviations

$\Lambda \Delta M := (\Lambda \cup M) \setminus (\Lambda \cap M)$;
$|\Lambda|$ or card Λ := cardinality of the set Λ;
$\square$:= end of proof;
$[a, b] := \{x : x \in \mathbf{R}, a \leq x \leq b\}$;
$(a, b) := \{x : x \in \mathbf{R}, a < x < b\}$;
$i = \overline{m, n}$:= i ranges over all integers in $[m, n]$ $(m, n \in \mathbf{Z}_+)$;

$i = \overline{m, \infty}$:= i ranges over all integers in $[m, \infty)$ ($m \in \mathbb{Z}_+$);
App. := Appendix;
resp. := respectively;
$a := C$ or $C =: a$ denotes that the quantity (or symbol) a is defined by the expression C.

Introduction

§ 1. Basic ergodic and mean–value theorems

1.1. The Birkhoff and von Neumann ergodic theorems. This book deals with problems connected with generalizations of classical ergodic theorems for endomorphisms and flows in measure spaces; first of all with the "pointwise" Birkhoff and the "mean" von Neumann ergodic theorems. We shall briefly discuss the content and the role of these two theorems.

Theorem 1.1. **(G.D.Birkhoff)** *If T is an endomorphism (i.e. a measurable measure–preserving transformation) of a σ–finite measure space $(\Omega, \mathcal{F}, m)$, then for any function f $(f \in L^\alpha(\Omega, \mathcal{F}, m),\ \alpha \geq 1)$ the limit*

$$\lim_{n\to\infty} \frac{1}{n} \sum_{k=0}^{n-1} f(T^k\omega) =: \widehat{f}(\omega) \tag{1.1}$$

exists for m–almost all $\omega \in \Omega$, and the function $\widehat{f}$ is invariant with respect to T: $\widehat{f}(T\omega) = \widehat{f}(\omega)$ m–a.e.

If $m(\Omega) = 1$ (we write P instead of m in this case), then $\widehat{f}(\cdot) = \mathbf{E}(f|\mathfrak{I})$, where $\mathfrak{I}$ is the σ–subalgebra of all T–invariant sets in F; if T is ergodic, i.e. $\mathfrak{I} = \{\emptyset, \Omega\} \bmod \mathrm{P}$, then

$$\widehat{f}(\omega) = \mathbf{E}_P(f) := \int_\Omega f(\omega)P(d\omega),$$

i.e. the "time mean" $\widehat{f}$ coincides with the "phase mean" $\mathbf{E}_P(f)$.

Theorem 1.2. **(J. von Neumann)** *If U is an isometric linear operator in a Hilbert space H, then for any $h \in H$ the limit*

$$\lim_{n\to\infty} \frac{1}{n} \sum_{k=0}^{n-1} U^k h =: \widehat{h} \tag{1.2}$$

exists and the element $\widehat{h}$ *is invariant with respect to* U: $U\widehat{h} = \widehat{h}$. *This theorem implies that for any* $f \in L^2$ $(\Omega, \mathcal{F}, m)$ *the limit* (1.1) *also exists in the sense of the* L^2*-metric.*

Note that the mappings $k \mapsto T^k$ and $k \mapsto U^k$ are actions (homomorphisms) of the semigroup $\mathbb{Z}_+$ in the space $(\Omega, \mathcal{F}, m)$ by endomorphisms and in the space H by isometric linear operators, respectively:

$$T^k \in \mathrm{End}\,(\Omega, \mathcal{F}, \mathrm{m}), \quad \mathrm{U^k} \in \mathrm{Isom}\,(\mathrm{H})$$

and

$$T^{k_1}T^{k_2} = T^{k_1+k_2}, \quad U^{k_1}U^{k_2} = U^{k_1+k_2}, \quad k, k_1, k_2 \in \mathbb{Z}_+.$$

Both the Birkhoff and von Neumann theorems can be easily transfered to semiflows $t \mapsto T_t$ (i.e. to measurable actions of the semigroup $\mathbb{R}_+$ in $(\Omega, \mathcal{F}, m)$) and to measurable isometric representations $t \mapsto U_t$ (i.e. to measurable actions of $\mathbb{R}_+$ in H by isometric operators); of course, we have to consider the limits

$$\lim_{T\to\infty} \frac{1}{T} \int_0^T f(T_t\omega)dt \quad \text{and} \lim_{T\to\infty} \frac{1}{T} \int_0^T \mathrm{U_t h\,dt}$$

instead of (1.1) and (1.2).

In the probabilistic language Birkhoff's theorem means that for any measurable strict sense stationary random process $\xi(t)$, $t \in \mathbb{R}_+$, for which $\mathbf{E}|\xi(0)| < \infty$, with probability *1* the limit

$$\lim_{T\to\infty} \frac{1}{T} \int_0^T \xi(t)dt = \mathbf{E}[\xi(0)|\mathfrak{I}_\xi]$$

exists.

Von Neumann's theorem means that for any measurable stationary random process with $\mathbf{E}[\xi(0)]^2$ we have

$$\lim_{T\to\infty} \mathbf{E}\left(\frac{1}{T} \int_0^T \xi(t)dt - \mathbf{E}[\xi(0)|\mathfrak{I}_\xi]\right)^2 = 0.$$

These theorems allow us to justify the "consistency" of statistical inference for ergodic stationary processes in two directions. In statistical mechanics and various applications of the theory of random processes it is possible, knowing the measure P or some of its characteristics, to a priori infer some properties of the realization on a large time interval $[0, \tau]$ of an ergodic process or of an orbit of an ergodic semiflow. For example, one can predict that the realization of the random variable

$$\widehat{\xi}_\tau = \frac{1}{\tau} \int_0^\tau \xi(t)dt$$

is close to the expectation $a := \mathbf{E}_P \xi(0)$, and that the quantity

$$\nu_{\Lambda,\tau} := \tau^{-1} \int_0^\tau \mathcal{X}_\Lambda(T_t \omega)dt$$

(the relative time the trajectory $\{T_t\omega,\ t \in [0,\tau]\}$ of some point ω belongs to a set $\Lambda \subset \mathcal{F}$) is close to the probability $P(\Lambda)$. On the other hand, one can estimate unknown values of the measure $P(\Lambda)$ or some of its characteristics on the basis of an observed realization of the process $\xi(t)$, $t \in [0,\tau]$, or of the orbit $\{T_t\omega,\ t \in [0,\tau]\}$. For example, the quantity $\widehat{\xi}_\tau$ is a "consistent" estimator of the parameter $a = \mathbf{E}_P \xi(0)$, and $\nu_{\Lambda,\tau}$ is a "consistent" estimator of the probability $P(\Lambda)$: in both cases, if τ is large, Birkhoff's theorem implies that, for "typical" points $\omega \in \Omega$ the errors $\xi_\tau - a$ and $\nu_{\Lambda,\tau} - P(\Lambda)$ are small, and the von Neumann's theorem implies that the mean–square errors $(\mathbf{E}_P(\widehat{\xi}_\tau - a)^2)^{\frac{1}{2}}$ and $[\mathbf{E}_P(\nu_{\Lambda,\tau} - P(\Lambda))^2]^{\frac{1}{2}}$ are small.

1.2. The Hopf ratio ergodic theorem. Let $(\Omega, \mathcal{F}, m)$ be a σ–finite measure space and let T be an endomorphism in Ω. We denote by $\mathfrak{J}$ the σ–algebra of all T–invariant sets in $\mathcal{F}$.

Theorem 1.3.A. (E. Hopf) *Let $f \in L^1$ $(\Omega, \mathcal{F}, m)$, $g \in$ $\in L^1_+$ $(\Omega, \mathcal{F}, m)$, and let $\sum_{k=1}^\infty g(T^k\omega > 0$ m–a.e. The limit*

$$\lim_{n\to\infty} \frac{\sum_{k=1}^n f(T^k\omega)}{\sum_{k=1}^n g(T^k\omega)} =: h_{f,g}(\omega) \tag{1.3}$$

exists and is finite m–a.e. on Ω. If $\alpha > 1$ and $f/g \in L^\alpha(\Omega, \mathcal{F}, gdm)$, then (1.3) holds in the sense of $L^\alpha(\Omega, \mathcal{F}, gdm)$–convergence.

Denote $O(\omega|T) := \{\omega, T\omega, T^2\omega, \ldots\}$. An endomorphism T is called *conservative* if for almost all ω and for any $\Lambda \in \mathcal{F}$ either $O(\omega|T) \cap \Lambda = \emptyset$ or $O(\omega|T) \cap \Lambda$ is infinite. We say that T is *dissipative* if $O(\omega|T) \cap \Lambda$ is finite for almost all ω and all Λ. By the Poincaré Recurrence Theorem, T is conservative if $m(\Omega) < \infty$.

Theorem 1.3.B. (E. Hopf) *There is a unique decomposition of Ω with the following properties:*
a) $C \cap D = \emptyset$;
b) $C,\ D \in \mathfrak{I}$;
c) the restriction of T to C is conservative;
d) the restriction of T to D is dissipative.

Theorem 1.3.C. (E. Hopf) *If T is conservative, then, under the conditions of Theorem 3.A,*

$$\int_\Lambda f\,dm = \int_\Lambda h_{f,g} \cdot g\,dm, \quad \Lambda \in \mathfrak{I},$$

i.e. $h_{f,g} = \mathbf{E}_g(\frac{f}{g}|\mathfrak{I})$, where $\mathbf{E}_g(\varphi|\mathfrak{I})$ is the conditional expectation of φ with respect to the σ–algebra $\mathfrak{I}$ in the measure space $(\Omega, \mathcal{F}, gdm)$.

Theorem 1.3 reduces to the Birkhoff and von Neumann theorems when $m(\Omega) < \infty$, $g(\omega) \equiv 1$.

1.3. Bohr's mean–value theorem.

Theorem 1.4. *For every continuous almost periodic function f on $\mathbb{R}$ the following limit exists uniformly with respect to x:*

$$\lim_{T\to\infty} \frac{1}{T} \int_0^T f(t+x)dt =: \mathbf{M}(f).$$

The number $\mathbf{M}(f)$ is called the *mean value* of f. It seems to be similar to the Birkhoff and von Neumann ergodic theorems; moreover, the later investigations showed that both the von Neuman and the Bohr theorems are special cases of a general statement (see §3 below).

§ 2. Further ergodic theorems for operators

2.1. The Dunford–Schwartz ergodic theorem. Following K. Jacobs, we say that an operator S in L^1 $(\Omega, \mathcal{F}, m)$ is a *strong contraction* if $\|S\|_{L^1} \leq 1$ and S can be extended from $L^1 \cap L^\infty$ to L^∞ with $\|S\|_{L^\infty} \leq 1$. By the Riesz convexity theorem, S can be extended to all of L^α, and $\|S\|_{L^\alpha} \leq 1$, $1 \leq \alpha \leq \infty$.

Theorem 2.1. (N. Dunford–J. Schwartz) *If S is a strong contraction, then for any $f \in L^\alpha$ $(\Omega, \mathcal{F}, m)$ the limit*

$$\lim_{n\to\infty} \frac{1}{n} \sum_{k=0}^{n-1} S^{(k)} f(\omega) =: \widehat{f}(\omega) \tag{2.1}$$

exists m–a.e. and $S\widehat{f} = \widehat{f}$ m–a.e. $(\alpha \geq 1)$.

If $\alpha > 1$ or $\alpha = 1$ and $m(\Omega) < \infty$ then (1.4) *is also true in the sense of L^α–convergence.*

This theorem reduces to the Birkhoff and the von Neumann theorems when $Sf(\omega) = f(T\omega)$, where T is an endomorphism of $(\Omega, \mathcal{F}, m)$.

2.2. The Chacon–Ornstein ratio ergodic theorem. Let $(\Omega, \mathcal{F}, m)$ be a σ–finite measure space. Let S be a positive contraction in L^1 $(\Omega, \mathcal{F}, m)$. Suppose $g \in L^1_+$ $(\Omega, \mathcal{F}, m)$, $\int g\, dm = 1$ and $\sum_{k=0}^\infty S^{(k)} g(\omega) > 0$. Denote $m_g(d\omega) = g(\omega) m(d\omega)$.

Theorem 2.2.A. (S. Chacon–D. Ornstein) *If $f \in L^1_+$ $(\Omega, \mathcal{F}, m)$, then the limit*

$$\lim_{n\to\infty} \frac{\sum_{k=0}^n S^k f(\omega)}{\sum_{k=0}^n S^k g(\omega)} =: h_{f,g}(\omega) \qquad (2.2).$$

If $\alpha > 1$ and $\frac{f}{g} \in L^\alpha(\Omega, \mathcal{F}, m_g)$ the the limit (2.1) exists also in the sense of convergence in $L^\alpha(\Omega, \mathcal{F}, m_g)$ and is finite a.e. on Ω. If $\alpha > 1$ and $\frac{f}{g} \in L^\alpha(\Omega, \mathcal{F}, m_g)$, then (2.1) holds in the sense of convergence in $L^\alpha(\Omega, \mathcal{F}, m_g)$.

S is said to be conservative if for any $f \in L^1_+$ we have $\sum_{k=0}^\infty S^k f(\omega) = = \infty$ m–a.e. on the set $\{\omega : \sum_{k=0}^\infty S^k f(\omega) > 0\}$; S is dissipative if $\sum_{k=0}^\infty S^k f(\omega) < \infty$ m–a.e. for all $f \in L^1_+$.

Let $B \in \mathcal{F}$; $L(B) := \{f : f \in L^1(\Omega),\ f(\omega) = 0 \text{ on } \Omega \setminus B\}$. A set $B (\in \mathcal{F})$ is said to be S–*absorbing* if $f \in L(B)$ implies $Sf \in L(B)$; if S is conservative, then the class $\mathcal{A}$ of all S–absorbing sets is a σ–algebra.

Theorem 2.2.B. (E. Hopf) *There is a unique decomposition with the following properties:*

a) $C \cap D = \emptyset$;

b) $C \in \mathcal{A}$;

c) the restriction of S to $L^1(C)$ is conservative;

d) for all $f \in L^1_+$ we have $\sum_{k=0}^\infty S^k f(\omega) < \infty$ m–a.e. on D.

Theorem 2.2.C. (J. Neveu–S. Chacon) *Let S be conservative, $f \in L^1(\Omega)$, $g \in L^1_+(\Omega)$, and $\sum_{k=0}^\infty S^k g(\omega) > 0$. Then $\int_A f\,dm = = \int_A h_{f,g} g\,dm$, i.e. $h_{f,g} = \mathbf{E}_g(\frac{f}{g}|\mathcal{A})$, where $\mathbf{E}_g(\varphi|\mathcal{A})$ is the conditional expectation of the function φ with respect to the measure m_g.*

§ 3. Means, averaging sequences and ergodic theorems on groups

Consider a topological semigroup X, a Borel measure μ on X and a measurable linear representation S: $x \mapsto S_x$ of X in a Banach space B. In order to extend the von Neumann theorem to these objects we have to find out for which elements b in B there is at least one net of

μ–integrable sets $\{A_n,\, n \in \mathbb{N}\}$ in X such that

$$\begin{aligned} &(B)\lim_{n\in\mathbb{N}} \frac{1}{\mu(A_n)} \int_{A_n} S_x b\mu(dx) =: \widehat{b} \\ &\text{exists and} \\ &S_x\widehat{b} = \widehat{b}, \quad x \in X, \end{aligned} \tag{3.1}$$

or, more general, at least one net of Borel probability measures $\{\nu_n,\, n \in \mathbb{N}\}$ on X such that

$$\begin{aligned} &(B)\lim_{n\in\mathbb{N}} \int S_x b\nu_n(dx) = \widehat{b} \\ &\text{exists and} \\ &S_x\widehat{b} = \widehat{b}, \quad x \in X. \end{aligned} \tag{3.1$'$}$$

((3.1′) turns into (1.3) if $\nu_n(E) := \mu(A_n \cap E)/\mu(A_n)$). An element $\widehat{b}$ with property (3.1′) is called a mean of b, and the net $\{\nu_n\}$ is called an *averaging net* of b (with respect to S). If each element of the closed convex hull $K(b)$ of the orbit $\{S_x b,\, x \in X\}$ possesses a unique mean, we shall say that the element b is *averageable.*

Two problems arise here. The first one is to find extensive classes of elements which are averageable with respect to every linear representation S belonging to some family of representations; this problem is closely connected with the problem of the existence of S–fixed points in S–invariant closed convex sets (and it greatly stimulated the development of the fixed–point problem). The second one is to find (i.e. to characterize in algebraic and topologic terms) nets of measures $\{\nu_n\}$ or of sets $\{A_n\}$ which average all elements of a given set; theorems concerning the solution of this problem are called *ergodic theorems.*

A special case of ergodic theorems are *mean–value theorems* for functions $f \in C(X)$; here S is the translation representation. In the case of right translations (3.1) and (3.1′) mean, respectively, that the following limits exist, uniformly with respect to x:

$$\lim_{n\in N} \frac{1}{\mu(A_n)} \int_{A_n} f(xy)\mu(dy) \equiv: \mathbf{M}^{(r)}(f) \tag{3.2}$$

and

$$\lim_{n\in N}\int f(xy)\nu_n(dy) \equiv: \mathbf{M}^{(r)}(f). \tag{3.2'}$$

Thus we arrive at the notions of a *right mean* $\mathbf{M}^{(r)}(f)$ of a function $f \in C(X)$, of a *right-averageable function* and of a *right averaging net* of measures (or sets). The corresponding "left" notions can be introduced similary. Bohr's theorem is an example of a mean-value theorem: it asserts that the net of intervals $[0,\tau]$, $\tau \in \mathbb{R}_+$, averages all continuous almost periodic functions on $\mathbb{R}$.

Similar mean–value theorems have also been considered for functions belonging to $C_B(X)$, the space of all bounded continuous functions with values in an arbitrary Banach space B. For an element $b \in B$ the ergodic theorems (3.1) and (3.1′) are equivalent to the mean–value theorems for the "orbital function" $\varphi_{b,S}(x) := S_x b$; hence, in such a general treatment, the difference between "ergodic theorems" and "mean–value theorems" becomes a difference of points of view. In this book we use both points of view.

Note that the notion of a right mean $\mathbf{M}^{(r)}(f)$ defined by relation (3.2′) for right–averageable functions essentially differs from the notion of a right–invariant Banach mean $m^{(r)}$, which is defined as an invariant (with respect to the right translations) normed positive linear functional on the whole space $C(X)$ (although on an amenable semigroup $m^{(r)}(f) = \mathbf{M}^{(r)}(f)$ for all right–averageable functions in $C(X)$).

Of course, the wider the set of all averageable functions or elements which is covered by an ergodic theorem, the stronger this theorem is. Therefore the question naturally arises whether there exist "universally" right–averaging nets of measures and sets on X, i.e. nets which average *all* right–averageable B–valued functions and, consequently, *all* elements which are averageable (with respect to some measurable uniformly bounded linear representation of X).

The most direct generalizations of the von Neumann ergodic theorem are theorems that characterize "mean averaging" nets of measures (or sets), i.e. nets which average all elements of Hilbert spaces with respect to any isometric representation of a semigroup X. Similarly, the generalizations of the Birkhoff theorem means finding "pointwise averaging" nets of measures (or sets), i.e. nets such that for any general dynamical system $\{T_x,\ x \in X\}$ in $(\Omega,\ \mathcal{F},\ m)$ and any function $f \in L^1(\Lambda, \mathcal{F}, m)$ the limit

$$\lim_{n \in N} \int_X f(T_x \omega) \nu_n(dx) = \mathbf{E}(f | \mathfrak{J})$$

(or, respectively,

$$\lim_{n \in N} \frac{1}{\mu(A_n)} \int_{A_n} f(T_x \omega) dx = \mathbf{E}(f | \mathfrak{J}))$$

exists with probability *1*.

Various generalizations of the von Neumann and Bohr theorems were connected with the development of profound methods in functional analysis; the basic works here are due to Garret Birkhoff, L. Alaoglu–G. Birkhoff, M. Day, W. Eberlein, F. Riesz and N. Wiener. N. Wiener has extended the Birkhoff pointwise ergodic theorem to the groups $\mathbb{R}^m$ and $\mathbb{Z}^m$, $m > 1$, and thereby established the deep connection between this theorem and the Hardy–Litttlewood inequalities by means of what is now called the "transfer principle"; A. Calderón has considerably developed this approach and has extended the pointwise ergodic theorems to general groups.

General ergodic theorems are an important tool in proving "consistency" of statistical inference for homogeneous random fields. During the last two decades random field found important applications in theoretical physics. Dobrushin [1]–[3] and Lanford–Ruelle [1] (see also Ruelle [1]) have shown that random fields are adequate models of important objects considered in statistical mechanics. The similar role of random fields in quantum field theory was shown by K. Simanzik, E. Nelson, G. Glimm and A. Jaffe (see, e.g., Simon [1]). In such applications ergodic theorems are a basic argument in proving the existence of various specific physical and informational characteristics (see Chapter 8).

CHAPTER 1
MEANS AND AVERAGEABLE FUNCTIONS

§ 1. Means and Quasi–averageable Functions

In this chapter we use the following notations: X is a semigroup with identity e; B is a Banach space; Φ_B is the Banach space of all bounded functions $f\colon X \mapsto B$ with the norm $\|f\|_{\Phi_B} := \sup_{x\in X} \|f(x)\|_B$, $f \in \Phi_B$; $\widetilde{\mathcal{P}}$ is the set of all discrete probability measures with finite carriers (it is well known that $\widetilde{\mathcal{P}}$ is a semigroup with respect to convolution). We associate with any $\lambda \in \widetilde{\mathcal{P}}$ two averaging operators "with weight λ ": the right one, R_λ, and the left one, L_λ:

$$(R_\lambda f)(x) = \int f(xy)\lambda(dy) = \sum_{i=1}^{k} \alpha_i f(xx_i),$$

$$(fL_\lambda)(x) = \int f(yx)\lambda(dy) = \sum_{i=1}^{k} \alpha_i f(x_i x),$$

where $\{x_1, \ldots, x_k\} = c(\lambda)$, the carrier of λ, and $\alpha_i = \lambda(\{x_i\})$. In particular, if $\lambda = \delta_{x_0}$ is the measure concentrated at the point x_0, then the operators $R_{x_0} := R_{\delta_{x_0}}$ and $L_{x_0} := L_{\delta_{x_0}}$ are the right and the left translation in Φ_B, respectively, that is, $(R_{x_0})(x) = f(xx_0)$ and $(fL_{x_0})(x) = f(x_0 x)$. It is very easy to verify the following properties:

(i) $R_\lambda(f) = L_\lambda(f) = f$ if $f(x) \equiv b$, $b \in B$;

(ii) R_λ and L_λ are linear contractions in Φ_B, $\lambda \in \widetilde{\mathcal{P}}$;

(iii) $R_{\lambda_1} R_{\lambda_2} = R_{\lambda_1 * \lambda_2}$, $L_{\lambda_1} L_{\lambda_2} = L_{\lambda_1 * \lambda_2}$, that is, both $R : \lambda \to R_\lambda$ and $L : \lambda \to L_\lambda$ are representations of $\widetilde{\mathcal{P}}$ in Φ_B, R_λ being a left one and L_λ being a right one;

(iv) the operators R_{λ_1} and L_{λ_2} commute, i.e. $R_{\lambda_1} L_{\lambda_2} = L_{\lambda_2} R_{\lambda_1}$, $\lambda_1, \lambda_2 \in \widetilde{\mathcal{P}}$.

Definition 1.1. Let $f \in \Phi_B$. An element $a \in B$ is called an *right mean* (or, for brevity, an *r–mean*) of f if there exists a sequence of measures $\lambda_n \in \widetilde{\mathcal{P}}$ such that

$$\lim_{n\to\infty} \|R_{\lambda_n} f - a\|_{\Phi_B} = \lim_{n\to\infty} \sup_{x\in X} \| \int f(xy)\lambda_n(dy) - a\|_B = 0.$$

A sequence λ_n with this property is called an *right averaging sequence* (or, for brevity, an *r-averaging sequence*) for f and a. Similarly a *left mean* (*l–mean*) and a *left averaging sequence* (*l–averaging sequence*) are defined. We define *a mean* of f as an element of B which is both a right and a left mean of f.

We denote by $\mathfrak{M}^{(r)}(f)$, $\mathfrak{M}^{(l)}(f)$ and $\mathfrak{M}(f)$ the sets of all r–means, all l–means and all means of f, respectively; $\mathbf{M}^{(r)}(f)$, $\mathbf{M}^{(l)}(f)$, $\mathbf{M}(f)$ are particular elements of these sets; $A\widetilde{S}^{(l)}(f,a)$ and $A\widetilde{S}^{(r)}(f,a)$ are the sets of all l–averaging, respectively all r–averaging sequences, in $\widetilde{\mathcal{P}}$ for given $f \in \Phi_B$ and $a \in B$;

$$A\widetilde{S}^{(l)}(f) := \bigcup_{a\in\mathfrak{M}^{(l)}(f)} A\widetilde{S}^{(l)}(f,a);$$

$$A\widetilde{S}^{(r)}(f) := \bigcup_{a\in\mathfrak{M}^{(r)}(f)} A\widetilde{S}^{(r)}(f,a).$$

According to our definition, $\mathfrak{M}^{(l)}(f) \neq \emptyset$ iff $A\widetilde{S}^{(l)}(f) \neq \emptyset$, and $\mathfrak{M}^{(r)}(f) \neq \emptyset$ iff $A\widetilde{S}^{(r)}(f) \neq \emptyset$.

Definition 1.2. A function $f \in \Phi_B$ is called, respectively, *r–quasi–averageable, l–quasi–averageable*, or *quasi–averageable* iff $\mathfrak{M}^{(r)}(f) \neq \emptyset$, $\mathfrak{M}^{(l)}(f) \neq \emptyset$, or $\mathfrak{M}(f) \neq \emptyset$, respectively.

We shall consider some simple properties of r– and l–means and of r– and l–quasi–averageable functions; for brevity we shall formulate statements only for "right" objects; formulations for the "left" ones can be obtained simply by symmetry.

Proposition 1.1. *If f is a r-quasi-averageable function, then*
(i) $\|\mathbf{M}^{(r)}(f)\|_B \leq \|f\|_{\Phi_B}$;
(ii) $\inf_{x\in X} f(x) \leq \mathbf{M}^{(r)}(f) \leq \sup_{x\in X} f(x)$ *when* $B = \mathbb{R}$;
(iii) $\overline{\mathbf{M}^{(r)}(f)} = \mathbf{M}^{(r)}(\bar{f})$ *when* $B = \mathbb{C}$.

Proof. Statement *(i)* follows from the property *(ii)* of the operators R_λ :
$\| \mathbf{M}^{(r)}(f) \|_B = \lim_{n\to\infty} \| R_{\lambda_n} f \|_{\Phi_B} \leq \|f\|_{\Phi_B}$ if $\lambda_n \in A\widetilde{S}^{(r)}(f, \mathbf{M}^{(r)}(f))$.
Statements *(ii)* and *(iii)* can be proved similarly.

□

Proposition 1.2. *If f is a B-valued r-quasi-averageable function on X and l is a continuous linear mapping of B into a Banach space D, then the D-valued function $l \circ f$ is also r-quasi-averageable and $l(\mathfrak{M}^{(r)}(f)) \subset \mathfrak{M}^{(r)}(l \circ f)$.*

Proof. Let ε be an arbitrary positive number. Let us consider $x_1, \ldots, x_k \in X$ and $\alpha_1, \ldots, \alpha_k > 0$, $\sum_{i=1}^n \alpha_i = 1$, such that $\sup_{x\in X} \| \sum_{i=1}^k \alpha_i f(xx_i) - \mathbf{M}^{(r)}(f)\|_B \leq \varepsilon$; then $\| \sum_{i=1}^n \alpha_i l \circ f(xx_i) - l(\mathbf{M}^{(r)}(f))\|_D \leq \varepsilon\|l\|$ and, consequently, $l(\mathbf{M}^{(r)}(f)) \in \mathfrak{M}^{(r)}(l \circ f)$

□

Let us denote by $K_R(f)$ the closed convex hull in Φ_B of the set $O_R(f) := \{R_x f,\, x \in X\}$, $f \in \Phi_B$.

Proposition 1.3. *For any $f \in \Phi_B$ the following statements are true.*
(i) $\mathfrak{M}^{(r)}(f) = K_R(f) \bigcap B$*) ; *thus the set $\mathfrak{M}^{(r)}(f)$ is convex;*
(ii) *if* $\{\lambda_n\} \subset \widetilde{\mathcal{P}}$ *and* $(\Phi_B) \lim_{n\to\infty} R_{\lambda_n} f = g$ *then* $(\Phi_B) \lim_{n\to\infty} R_{\nu*\lambda_n} f = R_\nu g$ *uniformly with respect to* $\nu \in \widetilde{\mathcal{P}}$;
(iii) *the set $K_R(f)$ is invariant with respect to the operators R_ν, $\nu \in \widetilde{\mathcal{P}}$, i.e. $R_\nu(K_R(f)) \subset K_R(f)$, $\nu \in \widetilde{\mathcal{P}}$;*
(iv) *the set $\mathfrak{M}^{(r)}(f)$ is invariant with respect to R_ν, $\nu \in \widetilde{\mathcal{P}}$.*

---

*) Here and often in the sequel we identify the element $b \in B$ with the function $f(x) \equiv b$.

Proof. Statement *(i)* is a simple consequence of Definition 1.1. Statement *(ii)* follows from the obvious inequality $\| R_{\nu*\lambda_n} f - R_\nu g \|_{\Phi_B} \leq \leq \| R_{\lambda_n} f - g \|$. Statement *(iii)* is implied by *(ii)* and statement *(iv)* by *(i)* and *(iii)*.

□

The following statement is a simple consequence of the previous proposition.

Proposition 1.4. *If* $\{\lambda_n\} \in A\widetilde{S}^{(r)}(f, a)$, *then*

(i) $(\Phi_B) \lim_{n\to\infty} R_{\nu*\lambda_n} f = a$ *uniformly with respect to* $\nu \in \widetilde{\mathcal{P}}$;

(ii) for arbitrary $\nu_n \in \widetilde{\mathcal{P}}$, $n = 1, 2, \ldots$, *we have:* $\nu_n * \lambda_n \in A\widetilde{S}^{(r)}(f, a)$.

Theorem 1.5. (Godement)

(i) A function $f \in \Phi_B$ *is quasi–averageable iff it is both r– and l–quasi–averageable.*

(ii) The mean $\mathbf{M}(f)$ *of a quasi–averageable function is unique; it is also the unique l–mean and the unique r–mean of this function.*

Proof. Let $f \in \Phi_B$, $a \in \mathfrak{M}^{(r)}(f)$, $b \in \mathfrak{M}^{(l)}(f)$, $\{\lambda_n\} \in A\widetilde{S}^{(r)}(f, a)$ and $\{\nu_n\} \in A\widetilde{S}^{(l)}(f, b)$. According to Proposition 1.4, $\{\nu_n * \lambda_n\} \in \in A\widetilde{S}^{(r)}(f, a) \cap A\widetilde{S}^{(l)}(f, b)$, and, consequently, $a = b = \mathbf{M}(f)$.

□

We conclude this section by a discussion of the problem of "invariance" of the sets $\mathfrak{M}^{(r)}(f)$ and $\mathfrak{M}^{(l)}(f)$ with respect to the action of the operators R_ν and L_ν on the function f $(\nu \in \widetilde{\mathcal{P}})$.

Proposition 1.6. *For any* $f \in \Phi_B$,

$$\mathfrak{M}^{(r)}(R_\nu f) \subset \mathfrak{M}^{(r)}(f) \subset \mathfrak{M}^{(r)}(f L_\nu), \tag{1.1$_r$}$$

$$\mathfrak{M}^{(l)}(f L_\nu) \subset \mathfrak{M}^{(l)}(f) \subset \mathfrak{M}^{(l)}(R_\nu f). \tag{1.1$_l$}$$

Proof. Let $a \in \mathfrak{M}^{(r)}(f)$ and $\{\lambda_n\} \in A\widetilde{S}^{(r)}(f, a)$. We have

$$\overline{\lim_{n\to\infty}}\|R_{\lambda_n}(fL_\nu) - a\|_{\Phi_B} = \overline{\lim_{n\to\infty}}\|(R_{\lambda_n}f - a)L_\nu\|_{\Phi_B} \leq$$
$$\leq \lim_{n\to\infty}\|R_{\lambda_n}f - a\|_{\Phi_B} = 0.$$

Thus $a \in \mathfrak{M}^{(r)}(fL_\nu)$.
Now let $a \in \mathfrak{M}^{(r)}(R_\nu f)$ and $\{\lambda_n\} \in A\widetilde{S}^{(r)}(R_\nu f, a)$. Then

$$(\Phi_B)\lim_{n\to\infty} R_{\lambda_n * \nu} f = (\Phi_B)\lim_{n\to\infty} R_{\lambda_n}(R_\nu f) = a.$$

Therefore $\{\lambda_n * \nu\} \in A\widetilde{S}^{(r)}(f, a)$ and $a \in \mathfrak{M}^{(r)}(f)$. Thus the inclusions (1.1_r) are proved. The inclusions (1.1_l) can be proved similarly.

□

Proposition 1.7. *If X is a group, then $\mathfrak{M}^{(r)}(f) = \mathfrak{M}^{(r)}(fL_x) = \mathfrak{M}(R_x f)$, $x \in X$; in particular, if f is r–quasi–averageable, then fL_x is r–quasi–averageable too.*

Proof. According to Proposition 1.6 we have:

$$\mathfrak{M}^{(r)}(fL_x) \subset \mathfrak{M}^{(r)}(fL_{x^{-1}}L_x) = \mathfrak{M}^{(r)}(f)$$

and $\mathfrak{M}^{(r)}(f) = \mathfrak{M}^{(r)}(R_{x^{-1}}R_x f) \subset \mathfrak{M}^{(r)}(R_x f)$. We only have to compare these inclusions with the inclusions (1.1_r).

□

§ 2. Averageable Functions

2.1. Averageability and quasi–averageability. Proposition 1.7 guarantees only that $\mathfrak{M}^{(r)}(R_\nu f) \subset \mathfrak{M}^{(r)}(f)$, $\nu \in \widetilde{\mathcal{P}}$. We shall see that $\mathfrak{M}^{(r)}(R_\nu f)$ and $\mathfrak{M}^{(r)}(f)$, $\nu \in \widetilde{\mathcal{P}}$, coincide if the semigroup X is amenable. However, Examples 2.1 and 2.2 given below show that on a non–amenable group these sets may differ and if f is r–quasi–averageable $R_\nu f$ need not be such. To avoid such pathological situations we introduce the notion of "complete" averageability.

Definition 2.1. An r–quasi–averageable function f is called *r–averageable* if $\mathfrak{M}^{(r)}(R_\nu f) = \mathfrak{M}^{(r)}(f)$, $\nu \in \widetilde{\mathcal{P}}$; *$l$–averageable* functions are defined similarly. A quasi–averageable function $f \in \Phi_B$ is called *averageable* if $\mathfrak{M}^{(r)}(R_\nu f) = \mathfrak{M}^{(l)}(fL_\nu) = \mathfrak{M}(f)$.

We denote the sets of all B–valued l-averageable functions, r–averageable functions defined on X by $A_B^{(l)} = A_B^{(l)}(X)$, $A_B^{(r)} = A_B^{(r)}(X)$ and $A_B = A_B(X)$, respectively.

Evidently, $B \subset A_B^{(l)} \cap A_B^{(r)} \cap A_B$ and $M(b) = b, \quad b \in B$. Below we shall find the connection between the notions of averageability and l– and r–averageability.

Theorem 2.1. *Let $f \in \Phi_B$. The following statements are equivalent:*
(i) f is averageable;
(ii) f is both l– and r–averageable;
(iii) $L_\nu f$ is l–averageable and $R_\nu f$ is r–averageable, $\nu \in \widetilde{\mathcal{P}}$;
(iv) $L_\nu f$ and $R_\nu f$ are quasi–averageable, $\nu \in \widetilde{\mathcal{P}}$;
(v) $L_\nu f$ and $R_\nu f$ are averageable, $\nu \in \widetilde{\mathcal{P}}$.
Thus $A_B = A_B^{(r)} \cap A_B^{(l)}$.

Proof. Evidently, *(i)* $\Rightarrow$ *(ii)* $\Rightarrow$ *(iii)*. Further, *(iii)* $\Rightarrow$ *(iv)* according to Proposition 1.6. Using Proposition 1.6 and Theorem 1.5 once more we obtain that *(iv)* $\Rightarrow$ *(i)*. Thus statements *(i)*–*(iv)* are equivalent. Evidently, *(v)* $\Rightarrow$ *(i)*. Let *(iii)* be fulfilled and $\mu, \nu \in \widetilde{\mathcal{P}}$. Then $\mathfrak{M}^{(r)}(R_\mu R_\nu f) = \mathfrak{M}^{(r)}(R_{\mu*\nu} f) \neq \emptyset$, $\mathfrak{M}^{(l)}(L_\mu R_\nu f) = \mathfrak{M}^{(l)}(R_\nu L_\mu f) \supset$ $\supset \mathfrak{M}^{(l)}(L_\mu f) \neq \emptyset$ and therefore $R_\nu f$ also has Property *(iv)* and so it is averageable. The averageability of $L_\nu f$ can be proved similarly. Thus *(iii)* $\Leftrightarrow$ *(v)*.

□

Now we shall give two examples which illustrate different kinds of "incomplete" averageability.

Example 2.1. Let $X = \{e, a, b\}$ where e is the identity and $a^2 = a$, $b^2 = b$, $ab = b$, $ba = b$. Let us set $f(a) = 1$ and $f(e) = f(b) = 0$.

Then $\alpha f(xa) + (1-\alpha) f(xb) \equiv \alpha$, $\alpha \in [0,1]$; therefore $\mathfrak{M}^{(r)}(f) = [0,1]$ and f is r–averageable. But $\mathfrak{M}^{(r)}[f(xb)] = \{0\}$ and thus f is not averageable. Note that in this case the set $\mathfrak{M}^{(r)}(f)$ is not invariant even with respect to the translation R_b of f (according to Proposition 1.7 such a phenomenon is impossible if X is a group).

Example 2.2. Let $X = F_2$, the free group with two generators a and b. Let us denote by A_i^r (resp. B_i^r) the set of all elements (words) the irreducible notation of which finishes by a^i (resp. b^i), $i \in \mathbb{Z}\backslash\{0\}$; $A_0^r := \bigcup_{i\neq 0} B_i^r$, $B_0^r := \bigcup_{i \neq 0} A_i^r$; A_i^l and B_i^l are similar sets defined with respect to the beginning of the words. Let us consider the function $f(x) = \mathcal{X}_{A_0^r}(x)$, the indicator of the set A_0^r. According to Proposition 1.1, $\mathfrak{M}^{(r)}(f) \subset [0,1]$. We shall show that in fact $\mathfrak{M}^{(r)}(f) = [0,1]$. Clearly, $\mathcal{X}_{A_0^r}(xa^{-i}) = \mathcal{X}_{A_0^r}(x)$ and, since the sets A_i do not intersect,

$$0 \leq \left(\frac{1}{n}\right) \sum_{i=1}^{n} \mathcal{X}_{A_0^r}(xa^{-i}) \leq \left(\frac{1}{n}\right) \mathcal{X}_{\bigcup_{i=1}^{n} A_i^r}(x) \leq \frac{1}{n}$$

for all $n = 1, 2, \ldots$ and $x \in X$. Consequently, $0 \in \mathfrak{M}^{(r)}(f)$. Similarly, in view of the fact that $A_0^r \cap B_0^r = \emptyset$ and $A_0^r \cup B_0^r = X\backslash\{e\}$ we have

$$\begin{aligned} 0 \leq 1 - \left(\frac{1}{n}\right) \sum_{i=1}^{n} \mathcal{X}_{A_0^r}(xb^{-i}) &= \left(\frac{1}{n}\right) \sum_{i=1}^{n} \mathcal{X}_{B_0^r \cup \{e\}}(xb^{-i}) = \\ &= \left(\frac{1}{n}\right) \mathcal{X}_{\bigcup_{i=1}^{n} B_i^r}(x) \leq \frac{1}{n}, \end{aligned}$$

and therefore $1 \in \mathfrak{M}^{(r)}(f)$. By Proposition 1.3, $\mathfrak{M}^{(r)}(f) = [0,1]$. Let $|x|$ be the length of the word x. If $x_1, x_2, \ldots, x_n$, $x \in X$ and $|x| \geq \max_{1 \leq i \leq n} |x_i|$ then, for arbitrary $\alpha_1, \ldots, \alpha_n \geq 0$, $\sum_{i=1}^{n} \alpha_i = 1$, we have $\sum_{i=1}^{n} \alpha_i f(x_i x) = f(x)$ and therefore $\mathfrak{M}^{(l)}(f) = \emptyset$ (this can also be deduced from Theorem 1.5).

Let $\nu_1 \in \widetilde{\mathcal{P}}$ and $\nu_1(\{e\}) = \frac{1}{2}$, $\nu_1(\{b\}) = \frac{1}{2}$; evidently, $\mathcal{X}_{A_0^r}(xb) =$ $= \mathcal{X}_{A_0^r b^{-1}}(x) = \mathcal{X}_{A_0^r}(x) - \mathcal{X}_{B_{-1}^r}(x) + \mathcal{X}_{B_0^r}(x) + \mathcal{X}_{\{e\}}(x) = 1 - \mathcal{X}_{B_{-1}^r}(x)$ and $R_{\nu_1} \mathcal{X}_{A_0^r} = \left(\frac{1}{2}\right)\left(1 + \mathcal{X}_{A_0^r} - \mathcal{X}_{B_{-1}^r}\right)$. Since $B_{-1}^r \subset A_0^r$ we have

$\frac{1}{2} \leq R_{\nu_1}\mathcal{X}_{A_0^r} \leq 1$ and thus $\mathfrak{M}^{(r)}(R_{\nu_1}f) \subset [\frac{1}{2},1]$. We shall show that in fact $\mathfrak{M}^{(r)}(R_{\nu_1}f) = [\frac{1}{2},1]$. We have $\frac{1}{n}\sum_{i=1}^n R_{\nu_1}f(xa^{-i}) = = \frac{1}{2} + \frac{1}{2n}\mathcal{X}_{\bigcup_{i=1}^n A_i^r(x)} - \frac{1}{2n}\mathcal{X}_{\bigcup_{i=1}^n B_{-1}^r a^i(x)}$ and thus $\frac{1}{2} \in \mathfrak{M}^{(r)}(R_{\nu_1}f)$. Similarly,

$$1-\left(\frac{1}{n}\right)\sum_{i=1}^n R_{\nu_1}\mathcal{X}_{A_0^r}(xb^{-i}) =$$
$$=\left(\frac{1}{2n}\right)\sum_{i=1}^n(1-\mathcal{X}_{A_0^r}(xb^{-i})+\mathcal{X}_{B_{-1}^r}(xb^{-i})) =$$
$$=\left(\frac{1}{2n}\right)\sum_{i=1}^n(\mathcal{X}_{B_0^r}(xb^{-i})+\mathcal{X}_{\{e\}}(xb^{-i})+\mathcal{X}_{b_{-1}^r}(xb^{-i})) =$$
$$=\left(\frac{1}{2n}\right)\left(\mathcal{X}_{\bigcup_{i=1}^n B_i^r}+\mathcal{X}_{\bigcup_{i=1}^n B_{i-1}^r}\right) \leq \frac{3}{2n},$$

and therefore $1 \in \mathfrak{M}^{(r)}(R_{\nu_1}f)$. It follows from Proposition 1.3 that $\mathfrak{M}^{(r)}(R_{\nu_1}f) = [\frac{1}{2},1]$. Thus our function f is r–quasi–averageable but is neither quasi–averageable nor r–averageable. Now let us set $\varphi = = \mathcal{X}_{A_0^r} \cap A_0^r$. Arguing as above it is easy to show that $\mathfrak{M}(\varphi) = \{0\}$, but $\mathfrak{M}^{(l)}(L_{\nu_1}\varphi) = \mathfrak{M}^{(r)}(R_{\nu_1}\varphi) = \emptyset$ and hence φ is quasi–averageable but is neither l– nor r–averageable.

2.2. Properties of averageable functions.

Lemma 2.2. *Let $f \in A_B^{(r)}$ and $a \in \mathbf{M}^{(r)}(f)$. For any sequence $\{\lambda_n\} \subset \widetilde{\mathcal{P}}$ there exists a sequence $\{\nu_n\} \subset \widetilde{\mathcal{P}}$ such that $\{\nu_n * \lambda_n\} \subset \subset A\widetilde{S}^{(r)}(f,a)$.*

Proof. According to the definition of r–averageability, $a \in \mathfrak{M}^{(r)}(R_{\lambda_n}f)$ for any measure $\lambda_n \in \widetilde{\mathcal{P}}$. Therefore, for any positive integer n a measure $\nu_n \in \widetilde{\mathcal{P}}$ can be found such that $\|R_{\nu_n*\lambda_n}f - a\|_{\Phi_B} = = \|R_{\nu_n}R_{\lambda_n}f - a\|_{\Phi_B} \leq \frac{1}{n}$. Clearly, $\{\nu_n * \lambda_n\} \in A\widetilde{S}^{(r)}(f,a)$.

□

Lemma 2.3. *Let B_i, $i = \overline{1,k}$, be Banach spaces, let f_i, $i = \overline{1,k}$, be B_i-valued r-quasi-averageable functions, $a_i \in \mathfrak{M}^{(r)}(f_i)$, and let $f_i \in A^{(r)}_{B_i}$, $i = \overline{2,k}$, if $k \geq 2$. Then $\bigcap_{i=1}^{k} A\widetilde{S}^{(r)}(f_i, a_i) \neq \emptyset$.*

Proof. If $k = 1$ our assertion is trivial. Let us suppose that it is true for some $k \geq 1$ and let $\{\mu_n{}^{(k)}\} \in \bigcap_{i=1}^{k} A\widetilde{S}^{(r)}(f_i, a_i)$. According to Lemma 2.2 there exists a sequence $\{\nu_n\} \subset \widetilde{\mathcal{P}}$ such that $\mu_n{}^{(k+1)} :=$ $:= \nu_n * \mu_n{}^{(k)} \in A\widetilde{S}^{(r)}(f_{k+1}, a_{k+1})$. Proposition 1.4 implies that $\{\mu_n{}^{(k+1)}\} \in \bigcap_{i=1}^{k} A\widetilde{S}^{(r)}(f_i, a_i)$ and so $\{\mu_n{}^{(k+1)}\} \in \bigcap_{i=1}^{n} A\widetilde{S}^{(r)}(f_i, a_i)$. It remains to use induction.

□

Theorem 2.4. *Let $f \in A^{(r)}_B$. Then*

(i) f possesses a unique r-mean $\mathbf{M}^{(r)}(f)$;

(ii) $R_\nu f \in A^{(r)}_B$, and $\mathbf{M}^{(r)}(f) = \mathbf{M}^{(r)}(R_\nu f)$, $\nu \in \widetilde{\mathcal{P}}$.

Proof. *(i).* Let $a,b \in \mathfrak{M}^{(r)}(f)$. According to Lemma 2.3 there exists a sequence $\{\mu_n\} \in A\widetilde{S}^{(r)}(f, a) \cap A\widetilde{S}^{(r)}(f, b)$ and therefore $a = b$.

(ii). Let $\lambda, \nu \in \widetilde{\mathcal{P}}$. Then $\mathfrak{M}^{(r)}(R_\lambda R_\nu f) = \mathfrak{M}^{(r)}(R_{\lambda*\nu} f) =$ $= \mathfrak{M}^{(r)}(f) = \{\mathbf{M}^{(r)}(f)\}$.

□

Theorem 2.5. *(i) The set $A^{(r)}_B$ is a (closed) subspace of Φ_B;*

(ii) the mapping $\mathbf{M}^{(r)} : f \to \mathbf{M}^{(r)}(f)$ is a continuous linear operator from $A^{(r)}_B$ onto B and $\|\mathbf{M}^{(r)}\| = 1$;

(iii) the space $A^{(r)}_B$ is invariant with respect to the operators R_λ, $\lambda \in \widetilde{\mathcal{P}}$;

(iv) if X is a group, then the space $A^{(r)}_B$ is invariant with respect to L_λ, $\lambda \in \widetilde{\mathcal{P}}$;

(v) the set A_B is a subspace of Φ_B invariant with respect to R_λ and L_λ $\lambda \in \widetilde{\mathcal{P}}$.

Proof. Statement *(iii)* follows from Theorem 2.4. Let α and β be numbers, f,$g \in A^{(r)}_B$. According to Lemma 2.3 there exists a sequence

$\{\nu_n\} \in A\widetilde{S}^{(r)}(f) \cap A\widetilde{S}^{(r)}(g)$. Evidently,

$$(\Phi_B)\lim_{n\to\infty} R_{\nu_n}(\alpha f + \beta g) = \alpha \mathbf{M}^{(r)}(f) + \mathbf{M}^{(r)}(g),$$

that is, $\alpha\mathbf{M}^{(r)}(f) + \beta\mathbf{M}^{(r)}(\alpha f + \beta g)$. Thus $\alpha f + \beta g$ is an r–quasi–averageable function. Reasoning as in the proof of Theorem 2.4 via Lemma 2.3 (with $f_1 = \alpha f + \beta g$, $f_2 = f$, $f_3 = g$) we see that

$$\mathfrak{M}^{(r)}(\alpha f + \beta g) = \{\alpha\mathbf{M}^{(r)}(f) + \beta\mathbf{M}^{(r)}(g)\}$$

.

If $\lambda \in \widetilde{\mathcal{P}}$, then $\alpha\mathbf{M}^{(r)}(f) + \beta\mathbf{M}^{(r)}(g) = \alpha\mathbf{M}^{(r)}(R_\lambda f) + \beta\mathbf{M}^{(r)}(R_\lambda g) \in \in \mathfrak{M}^{(r)}(\alpha R_\lambda f + \beta R_\lambda g) = \mathfrak{M}^{(r)}(R_\lambda(\alpha f + \beta g))$. It follows from Proposition 1.6 that

$$\mathfrak{M}^{(r)}(R_\lambda(\alpha f + \beta g)) = \{\alpha\mathbf{M}^{(r)}(f) + \beta\mathbf{M}^{(r)}(g)\} = \mathfrak{M}^{(r)}(\alpha f + \beta g).$$

Therefore $\alpha f + \beta g \in A_B^{(r)}$ and $\mathbf{M}^{(r)}(\alpha f + \beta g) = \alpha\mathbf{M}^{(r)}(f) + \beta\mathbf{M}^{(r)}(g)$. Thus $A_B^{(r)}$ is a linear manifold in Φ_B and $\mathbf{M}^{(r)}$ is a linear operator from $A_B^{(r)}$ onto B. According to Proposition 1.1, $\|\mathbf{M}^{(r)}\| \leq 1$, and since $\mathbf{M}^{(r)}(b) = b$, $b \in B$, we have $\|\mathbf{M}^{(r)}\| = 1$. Suppose $f_n \in A_B^{(r)}$, $f \in \Phi_B$ and $(\Phi_B)\lim_{n\to\infty} f_n = f$. Denote $a := (B)\lim_{n\to\infty}\mathbf{M}^{(r)}(f_n)$ and let $\lambda_n \in \widetilde{\mathcal{P}}$ and $\|R_{\lambda_n} f_n - \mathbf{M}^{(r)}(f_n)\|_{\Phi_B} \leq \frac{1}{n}$. We have

$$\overline{\lim_{n\to\infty}} \|R_{\lambda_n} f - a\|_{\Phi_B} \leq \overline{\lim_{n\to\infty}} \|R_{\lambda_n} f - R_{\lambda_n} f_n\|_{\Phi_B} + \overline{\lim_{n\to\infty}} \|R_{\lambda_n} f_n - \\ -\mathbf{M}^{(r)}(f_n)\|_{\Phi_B} + \overline{\lim_{n\to\infty}} \|\mathbf{M}^{(r)}(f_n) - a\|_{\Phi_B} \leq \overline{\lim_{n\to\infty}} \|f - f_n\|_{\Phi_B} = 0$$

Consequently, f is an r–quasi–averageable function and $a \in \mathfrak{M}^{(r)}(f)$. Let us suppose that $b \in \mathfrak{M}^{(r)}(f)$. According to Lemma 2.3 we can consider sequences $\{\mu_n^{(k)}\} \in A\widetilde{S}^{(r)}(f,b) \cap A\widetilde{S}^{(r)}(f_k)$, $k = 1,2,\ldots$. Letting $n \to \infty$ in both sides of the inequality $\|R_{\mu_n^{(k)}} f_k - R_{\mu_n^{(k)}} f\|_{\Phi_B} \leq \leq \|f_k - f\|_{\Phi_B}$ we obtain $\|\mathbf{M}^{(r)}(f_k) - b\|_{\Phi_B} \leq \|f_k - f\|_{\Phi_B}$ $(k = 1,2,\ldots)$. This implies that $b = \lim_{k\to\infty}\mathbf{M}^{(r)}(f_k) = a$ and therefore it is the unique mean of f. Note that $R_\lambda f = (\Phi_B)\lim_{k\to\infty} R_\lambda f_k, \quad \lambda \in \widetilde{\mathcal{P}}$, and hence it is the unique mean of $R_\lambda f$. Thus $f \in A_B^{(r)}$, and $A_B^{(r)}$ is a closed subspace of Φ_B. So statements *(i)–(iii)* are proved. Let X be a group and $\nu \in \widetilde{\mathcal{P}}$, $x \in X$, $f \in A_B^{(r)}$; then, in view of Proposition 1.7, $\mathfrak{M}^{(r)}(R_\nu(fL_x)) = \mathfrak{M}^{(r)}(R_\nu f) = \mathfrak{M}^{(r)}(f) = \mathfrak{M}^{(r)}(fL_x)$ and thus $fL_x \in A_B^{(r)}$, $x \in X$; therefore, according to *(iii)*, $fL_\lambda \in A_B^{(r)}$, $\lambda \in \widetilde{\mathcal{P}}$. So statement *(iv)* is proved too. Statement *(v)* follows from statement *(iii)* and Theorem 2.1.

□

The two simple Propositions 2.6 and 2.7, stated below, help in some cases to prove the "complete" r–averageability of r–quasi–averageable functions.

Definition 2.2. The operator $\mathrm{M}^{(r)}$ introduced in Theorem 2.5 is called the *right averaging operator*. The *left averaging operator* $\mathrm{M}^{(r)}$: $A^{(l)}(B) \mapsto B$ is defined similarly.

Proposition 2.6. *Let Q be an R–invariant convex set in Φ_B and let each function in Q be r–quasi–averageable with a unique right mean. Then $Q \in A_B^{(r)}$.*

Proof. Let $f \in Q$, $\nu \in \widetilde{\mathcal{P}}$. Then the stated properties of Q imply that $R_\nu f \in Q$ and each of the sets $\mathfrak{M}^{(r)}(R_\nu f)$ and $\mathfrak{M}^{(r)}(f)$ consists of a single element and therefore $\mathfrak{M}^{(r)}(R_\nu f) = \mathfrak{M}^{(r)}(f)$.

□

Proposition 2.7. *Let X be a group and let Q be an R–invariant linear manifold of r–quasi–averageable functions in $\Phi_B(X)$ with the property $\mathfrak{M}^{(r)}(f+g) \supset \mathfrak{M}^{(r)}(f) + \mathfrak{M}^{(r)}(g)$, $f, g \in Q$. Then $Q \supset A_B^{(r)}$.*

Proof. Let $f \in Q$, $\nu \in \widetilde{\mathcal{P}}$ and $\{x_1, \ldots, x_k\} = c(\nu)$ (the carrier of ν); $\alpha_i = \nu(\{x_i\})$. The properties of Q and Propositions 1.2 and 1.7 imply:

$$\mathfrak{M}^{(r)}(R_\nu f) = \mathfrak{M}^{(r)}\left(\sum_{i=1}^{k} \alpha_i f(xx_i)\right) \supset \sum_{i=1}^{k} \alpha_i \mathfrak{M}^{(r)}[f(xx_i)] = \mathfrak{M}^{(r)}(f)$$

It remains to use Proposition 1.6.

□

2.3 Averageability and amenability: means of r–averageable functions and invariant Banach means on $\mathbf{A}_{\mathbf{B}}^{(\mathbf{r})}$.

Theorem 2.8. *If X is an r–amenable semigroup, then any r–quasi-averageable function f in $\Phi_B(X)$ is r–averageable.*

Proof. Let $a \in \mathfrak{M}^{(r)}(f)$ and $\{\lambda_n\} \subset A\widetilde{S}^{(r)}(f, a)$. Let us consider an ergodic net of normed measures $\{\nu_\alpha, \alpha \in A\}$ on X (see App., §3). It is easy to verify that $\lim_{\alpha\in A} \|\nu_\alpha * \lambda - \nu_\alpha\|_{\mathbb{B}\mathfrak{M}} = 0$, $\lambda \in \widetilde{\mathcal{P}}$. Therefore we have for any λ, $\nu \in \widetilde{\mathcal{P}}$,

$$\lim_{\alpha\in A} \|\nu_\alpha * \nu * \lambda - \nu_\alpha * \nu\|_{\mathbb{B}\mathfrak{M}} + \lim_{\alpha\in A} \|\nu_\alpha - \nu_\alpha * \nu\|_{\mathbb{B}\mathfrak{M}} = 0.$$

Using Proposition 1.4 we obtain for any $\nu \in \widetilde{\mathcal{P}}$,

$$\begin{aligned} &\lim_{\alpha\in A} \|R_{\nu_\alpha} R_\nu f - a\|_{\Phi_B} \leq \\ \leq &\|f\| \lim_{n\to\infty} \lim_{\alpha\in A} \|\nu_\alpha * \nu - \nu_\alpha * \nu * \lambda_n\|_{\mathbb{B}\mathfrak{M}} + \\ &+ \lim_{n\to\infty} \sup_{\alpha\in A} \|R_{\nu_\alpha * \nu * \lambda_n} f - a\|_{\Phi_B} = 0 \end{aligned}$$

Clearly, we can choose an increasing sequence $\{\alpha_k\} \subset A$ such that $\lim_{k\to\infty} \|R_{\alpha_k} R_\nu f - a\|_{\Phi_B} = 0$. Thus $a \in \mathfrak{M}^{(r)}(R_\nu f)$, and $\mathfrak{M}^{(r)}(f) \subset$ $\subset \mathfrak{M}^{(r)}(R_\nu f)$. It remains to use Proposition 1.6.

□

Theorem 2.9. *Let Q be an R–invariant linear manifold in $A_{\mathbb{C}}^{(r)}$. If X is Q–r–amenable (see App., §3), then the restriction of any r–invariant Banach mean $\mathfrak{M}^{(r)}$ to Q coincides with the functional $\mathbf{M}^{(r)}$.*

Proof. For any $\varepsilon \geq 0$ there exist elements $x_1, \dots, x_k \in X$ and positive numbers $\alpha_1, \dots, \alpha_k$ with $\sum_{i=1}^k \alpha_i = 1$ such that $\sup_{x \in X} |\mathbf{M}^{(r)}(f) - \sum_{i=1}^k \alpha_i f(xx_i)| \leq \varepsilon$. Let $m^{(r)}$ be an r–invariant Banach mean on Q. We have

$$|\mathbf{M}^{(r)}(f) - m^{(r)}(f)| = |m^{(r)}(f) - \sum_{i=1}^{k} \alpha_i f(xx_i))| \leq$$

$$\leq \sup_{x \in X} |\mathbf{M}^{(r)}(f) - \sum_{i=1}^{k} \alpha_i f(xx_i)| \leq \varepsilon$$

Therefore $\mathbf{M}^{(r)}(f) = m^{(r)}(f)$.

□

Corollary 2.10. *Every semigroup X is $A_{\mathbb{C}}^{(r)}$–right–amenable and the right–invariant Banach mean on $A_{\mathbb{C}}^{(r)}$ is unique: $m^{(r)}(f) = \mathbf{M}^{(r)}(f)$, $f \in A_{\mathbb{C}}^{(r)}$.*

2.4. Main ergodic decomposition. Let $Q \subset \Phi_B$. Denote by $D_Q^{(r)}$ the subspace of Φ_B generated by the set of functions $\{\Delta f_y x = f(x) - f(xy),\ f \in Q,\ y \in X\}$ and by $K(\Delta f)$ the closed convex hull of the set $\{\Delta f_y,\ y \in X\}$. It is easy to verify the following properties of the spaces $D_Q^{(r)}$:

(D_1) if $Q_1 \subset Q_2$, then $D_{Q_1}^{(r)} \subset D_{Q_2}^{(r)}$;

(D_2) if Q is a subspace of Φ_B and a subset L generates Q, then $D_L^{(r)} = D_Q^{(r)}$;

(D_3) if Q is an R–invariant subspace of Φ_B, then $D_Q^{(r)} \subseteq Q$.

If A is a Banach space, then, as usual, the notation $B \oplus C$ means that B and C are subspaces of A, $A = B + C := \{a : a = b + c, b \in B, c \in C\}$ and $B \cap C = \{0\}$; in this case, for any $a \in A$ the decomposition $a = b + c$ where $b \in B$, $c \in C$ is unique; the operator Π_B^A: $a \mapsto b$ is the projector in A onto the space B with respect to C. If Q is a subspace of Φ_B, then $B_Q := B \cap Q$ is the subspace of constants in Q.

Theorem 2.11. *Let Q be an R–invariant subspace of $A_B^{(r)}$; $Q^0 :=$ $:= \{f : f \in Q, \mathbf{M}^{(r)}(f) = 0\}$. Then*

(i) $D_Q^{(r)} = Q^0$;

(ii) $D_Q^{(r)}$ *is R–invariant;*

(iii) $Q = B_Q \oplus D_Q^{(r)}$;

(iv) $\Pi_{B_Q}^Q f = \mathbf{M}^{(r)}(f)(\in K(f)), \quad f \in Q$;

(v) $\Pi_{D_Q^{(r)}}^Q f \in K(\Delta f), \quad f \in Q$.

Proof. It is easy to verify that Q^0 is an R–invariant subspace of Q. Now, $Q = B_Q \oplus Q^0$ since $B_Q \cap Q^0 = \{0\}$ and any function $f(\in Q)$ admits the decomposition $f = \mathbf{M}^{(r)}(f) + (f - \mathbf{M}^{(r)}(f))$ where $\mathbf{M}^{(r)}(f) \in$ $\in B_Q$, according to the definition of rl–mean, and $f - \mathbf{M}^{(r)}(f) \in Q^0$. We have $\Pi_{B_Q}^Q = \mathbf{M}^{(r)}(f)$ and $\Pi_{Q^0}^Q = f - \mathbf{M}^{(r)}(f)$.

Let $f \in Q$, $\varepsilon \geq 0$, and let λ be an element of $\widetilde{\mathcal{P}}$ such that

$$\|R_\lambda f - \mathbf{M}^{(r)} f\|_{\Phi_B} \leq \varepsilon;$$

$c(\lambda) =: \{y_1, \ldots, y_k\}$, $\alpha_i := \lambda(\{y_i\})$, $i = \overline{1,k}$. Then

$$\begin{aligned} &\|\Pi_{Q^0}^Q f(x) - \sum_{i=1}^{k} \alpha_i (f(x) - f(xy_i))\|_{\Phi_B} = \\ =&\|(f(x) - \mathbf{M}^{(r)}(f)) - (f(x) - \sum_{i=1}^{k} \alpha_i f(xy_i))\|_{\Phi_B} = \\ =&\|R_\lambda f - \mathbf{M}^{(r)}(f)\|_{\Phi_B} \leq \varepsilon \end{aligned} \tag{2.1}$$

Since ε is an arbitrary positive number and f is an arbitrary function in Q, we have $Q^0 \subseteq D_Q^{(r)}$. On the other hand, the continuity of the rl–mean $\mathbf{M}^{(r)}$ implies that $D_Q^{(r)} \subset Q^0$. Thus we have shown that $D_Q^{(r)} = Q^0$ and therefore *(i)–(iv)* are valid. Statement *(v)* also follows from (2.1).

□

§ 3. Averageable Elements in General Banach Spaces

In this section we shall transfer the notions and results of § 1 and § 2 concerning averageability with respect to the right and the left translation in the Banach spaces Φ_B to general Banach spaces.

Let X be a semigroup and let B be a Banach space. We consider a uniformly bounded left representation S: $x \mapsto S_x$ of X in B. Let $b \in B$. The set $O(b) := \{S_x b, x \in X\}$ is called *the orbit* of b with respect to S. We also denote by $K(b)$ the closed convex hull of the orbit $O(b)$ and by I the set of all S–invariant elements ("S–fixed points") in B, that is, $a \in I$ iff $S_x a = a, \quad x \in X$. It is easy to verify that

(i) $K(b)$ is a S–invariant set, that is, $S_x(K(b)) \subset K(b)$, $x \in X$;

(ii) I is a (closed) subspace of B.

For $\nu \in \widetilde{\mathcal{P}}$ we denote $S_\nu = \int S_x b\nu(dx)$ (the Bochner integral). We recall that $\widetilde{\mathcal{P}}$ is a semigroup with respect to composition of measures; the mapping $\nu \mapsto S_\nu$ is a representation of $\widetilde{\mathcal{P}}$ in B.

Definition 3.1. An element $a \in B$ is called a *mean of b with respect to the representation* S if for some sequence $\{\nu_n\}$ in $\widetilde{\mathcal{P}}$ we have $\lim_{n\to\infty} S_{\nu_n} b = a$. Let us denote by $\mathfrak{M}(b|S)$ or $\mathfrak{M}(b)$ the set of all means of b (with respect to S). An element $b(\in B)$ is said to be *quasi-averageable with respect to S* (or for short *S–quasi–averageable*) if $\mathfrak{M}(b|S) \neq \emptyset$. A quasi–averageable element b is said to be *averageable with respect to S* (or simply *S–averageable*) if $\mathfrak{M}(S_\nu b) = \mathfrak{M}(b)$, $\nu \in \widetilde{\mathcal{P}}$.

Evidently, any function $f(\in \Phi_B(X))$ is r–(quasi)–averageable iff it is an R–(quasi)–averageable element in Φ_B. On the other hand, to any element $b(\in B)$ corresponds its B–valued *orbital function* (with respect to S), $\varphi_b(x) := S_x b$, $x \in X$. Let us denote by τ the mapping $b \mapsto \varphi_b$. It is easy to verify that τ is a linear mapping from B onto a subspace $\tau(B)$ of Φ_B. Moreover, if $S_e = I$ for some $e \in X$, $\|b\|_B = \|\varphi_b(e)\| \leq \sup_{x\in X} \|S_x b\|_B \leq \|b\| \sup_{x\in X} \|S_x\|$ and thus τ and τ^{-1} are continuous linear mappings. If, as usual, we identify constant functions in Φ_B with elements in B, we can write $\tau(I) = I$ and the restriction of τ to I can be considered as the identity mapping. And, certainly, $\tau(K(b)) = (K(\varphi_b)$, $b \in B$. According to Proposition 1.2, $\mathfrak{M}^{(r)}(\varphi_b) = \mathfrak{M}(b)$. Note also that $\tau(S_x b) = R_x \varphi_b = R_x \tau(b)$, $x \in X$, and hence the actions of X in B by S and in $\tau(B)(\subset \Phi_B)$ by R are equivalent. It follows from what was said above that the function $\varphi_b(\in \Phi_B)$ is r–(quasi)–averageable iff the element $b(\in B)$ is S–(quasi)–averageable. Now it is clear how to transfer all the results of §§ 1 and 2 to the general problem connected with averaging with respect to a uniformly bounded linear left representation of X in a Banach space B. We shall state the generalized results in the following propositions.

Proposition 3.1. $\mathfrak{M}(b) = I \cap K(b)$.

Proposition 3.2. *If* $a \in \mathfrak{M}(b)$, *then* $\|a\| \leq \|b\| \sup_{x \in X} \|S_x\|$.

Proposition 3.3. *If* $b(\in B)$ *is averageable, then it possesses only one mean* $\mathbf{M}(b)$, *and* $\mathbf{M}(S_x b) = b$, $x \in X$.

Proposition 3.4. *The averageable elements form an* S*–invariant subspace of* B.

We shall denote this subspace by $A(B|S)$ or simply by $A(B)$.

Proposition 3.5. *If* X *is right–amenable, then any* S*–quasi–averageable element is* S*–averageable.*

Proposition 3.6. *Let* Q *be an* S*–invariant linear manifold in* B. *If all elements of* Q *are quasi–averageable and each of them possesses a unique mean, then* $Q \subset A(B)$.

Proposition 3.7. (i) $A(B) = I \oplus D$, *where* D *is the subspace of* B *generated by the elements* $b = c - S_x c$, $c \in B$, $x \in X$;

(ii) *the subspaces* I *and* D *are* S*–invariant;*

(iii) $D = \{b : b \in A(B),\ \mathbf{M}(b) = 0\}$.

Proposition 3.8. *(i)* $\Pi_I^{a(B)} = \mathbf{M}(b)$, $b \in A(B)$ *and thus* $S_x\mathbf{M}(b) = = \mathbf{M}(b)$, $x \in X$, $b \in B$;

(ii) $\Pi_D^{A(B)} b \in D_b$ *where* D_b *is the subspace of* B *generated by the set* $\{e = b - S_x b,\ x \in X\}$, $b \in A(B)$.

Proposition 3.9. *If* Q *is an* S*-invariant subspace of* $A(B)$*, then* $\Pi_D^{A(B)} Q = D \cap Q =: D_Q$, $\Pi_I^{A(B)} Q = I \cap Q =: I_Q$ *and* $Q = I_Q \oplus D_Q$.

Proposition 3.10.
$\|\Pi_I^{A(B)}\| \le \sup_{x\in X} \|S_x\|$; $\|\Pi_D^{A(B)}\| \le 1 + \sup_{x\in X} \|S_x\|$.

Proposition 3.11. *If the representation* S *is irreducible and* $S \not\equiv \mathbb{I}$ *(the identity operator), then* $A(B) = D$ *and thus* $\mathbf{M}(b) = 0$, $b \in A(B)$; *if* $S \equiv \mathbb{I}$, *then* $A(B) = B = I$ *and* $\mathbf{M}(b) = b$, $b \in B$.

According to Proposition 3.8, the operator $\mathbf{M}$: $b \mapsto \mathbf{M}(b)$ defined on $A(B)$ coincides with the projector $\Pi_I^{A(B)}$; it will be called the *averaging operator* on $A(B)$ with respect to S or simply the S–averaging operator in $A(B)$.

§ 4. Fixed Point Theorems

Let X be a semigroup, B a Banach space, and S: $x \longrightarrow S_x$ a uniformly bounded representation of X in B. It follows from Proposition 3.1 that quasi–averageability of an element $b(\in B)$ is equivalent to the existence of an S–invariant (or "fixed") point in the S–invariant closed convex set $K(b)$. In this section we will discuss the problem of existence of such points in invariant closed convex sets.

We recall that I denotes the subspace of all S–invariant elements of B.

4.1. Contractive representations in uniformly convex spaces. First of all let us recall the notion of uniform convexity.

Definition 4.1. A Banach space B is said to be *uniformly convex* if for any ε $(0 < \varepsilon < 2)$ there exists a number $\delta(\varepsilon) > 0$ such that $\|a+b\| < 2-\delta(\varepsilon)$ whenever $a, b \in B$, $\|a\| = \|b\| = 1$ and $\|a-b\| \geq \varepsilon$.

D. P. Milman has shown that any uniformly convex space is reflexive (cf.,e.g. Yosida [1], or Diestel [1]).

Example 4.1. A Hilbert space is uniformly convex. This follows from the equality $\|x+y\|^2 + \|x-y\|^2 = 2(\|x\|^2 + \|y\|^2)$.

Example 4.2. **(Clarkson)** The spaces L^α and l^α, $1 < \alpha < \infty$, are uniformly convex (cf. Bourbaki [1], Diestel [1]).

The spaces L^1 and l^1 are not reflexive and thus they are not uniformly convex.

Lemma 4.1. **(Alaoglu and Garret Birkhoff)** *If F is a closed convex set in a uniformly convex Banach space B, then there exists one and only one point b_0 in F such that $\|b_0\| = \inf_{b\in F} \|b\|$.*

Proof. Existence. Let us set $M := \inf_{b\in F} \|b\|$. Let $\{b_n,\ n = \overline{1,\infty}\} \subset F$ and $\lim_{n\to\infty} \|b_n\| = M$. Suppose that the sequence $\{b_n\}$ diverges. Then the sequence $\frac{b_n}{\|b_n\|}$ diverges too, and there exist a number $0 < \varepsilon < 2$ and a sequence of positive integers $\{K_n\}$ such that $\|\frac{b_n}{\|b_n\|} - \frac{b_{n+K_n}}{\|b_{n+K_n}\|}\| \geq \varepsilon$, $n = 1, 2, \ldots$. Due to uniform convexity, $(\frac{1}{2})\, \|\frac{b_n}{\|b_n\|} + \frac{b_{n+K_n}}{\|b_{n+K_n}\|}\| < 1 - \delta(\varepsilon)/2$ and thus $(\frac{1}{2})\|b_n + b_{n+K_n}\| < M$ for sufficiently large n. This is impossible since $(\frac{1}{2})\,(b_n + b_{n+K_n}) \in F$. Thus $\lim_{n\to\infty} b_n =: b_0$ exists, $b_0 \in F$ and $\|b_0\| = M$.

Unicity. Let $b_0, b_1 \in F$ and $\|b_0\| = \|b_1\| = M := \inf_{b\in F} \|b\|$. Then $\|M^{-1}b_0 + M^{-1}b_1\| < 2 - \delta(\frac{1}{M}\|b_0 - b_1\|)$ and hence $\|\frac{1}{2}(b_0 + b_1)\| < M$. This is impossible since $\frac{1}{2}(b_0 + b_1) \in F$.

□

Theorem 4.2. **(Alaoglu and Garret Birkhoff)** *If $S\colon x \to S_x$ is a contractive representation of a semigroup in a uniformly convex Banach space B and F is an S-invariant closed convex set in B, then* *the element b_0 defined in Lemma 4.1 is S–invariant; thus $I \cap F \neq \emptyset$.*

Proof. Let b_0 be the least–norm point in F. Then $S_x b_0 \in F$, $x \in X$, and $\|S_x b_0\| \leq \|b_0\|$ (since $\|S_x\| \leq 1$). It follows from Lemma 4.1 that $S_x b_0 = b_0$, that is, $b_0 \in I \cap F$.

□

4.2. Bilaterally uniformly bounded representations. The following statement is an important step in the study of fixed points and the averageability of orbital functions of representations in general Banach spaces.

Lemma 4.3. **(Ryll–Nardzewski)** *Let $(X, \mathfrak{B})$ be a measurable semigroup, q a probability measure on $\mathfrak{B}$ and S a bilaterally uniformly bounded representation of X in B. We suppose that S is measurable with respect to all convolution powers q^n of q $(n = 1, 2, \ldots)$. If $\int S_x b q(dx) = b$ for some $b \in W(B, S)$, then $S_x b = b$ for q–almost all $x \in X$.*

Proof. Let $(\Omega, \mathcal{F}, P) := \prod_{i=1}^{\infty} (\Omega_i, \mathcal{F}_i, P_i)$ where $(\Omega_i, \mathcal{F}_i, P_i) = (X, \mathfrak{B}, q)$, $i = \overline{1, \infty}$; $\xi_i(\omega) := x_i$ if $\omega = (x_1, x_2, \ldots)$; $\widetilde{\mathcal{F}}^{(n)} = \mathcal{F}_1 \otimes \cdots \otimes \mathcal{F}_n$, and let $\mathcal{F}^{(n)}$ be the σ–subalgebra of $\mathcal{F}$ consisting of all the sets of the form $\widetilde{\Lambda} \times \Omega_{n+1} \times \Omega_{n+2} \times \cdots$ where $\widetilde{\Lambda} \in \widetilde{\mathcal{F}}^{(n)}$. Let $P^{(n)}$ be the restriction of P to $\mathcal{F}^{(n)}$, and set $\widetilde{P}^{(n)} = P_1 \times \cdots \times P_n$. Then $\beta_n := S_{\xi_1 \ldots \xi_n} b$ is a B–valued random variable over $(\Omega, \mathcal{F}, P)$ and, since $\|\beta_n\| \leq \|b\|$ a.s., the Bochner integrals $\int_\Lambda S_{\xi_1 \cdots \xi_n} b \, \widetilde{P}^{(n)}$, $\Lambda \in \mathcal{F}^{(n-1)}$, exist. Using the well–known theorem on the permutability of the Bochner integral and a bounded operator in B (cf. Dunford and Schwartz [1], Theorem 3.2.19) as well as Fubini's theorem, we obtain that for any $\Lambda \in \mathcal{F}^{(n-1)}$,

$$\int_\Lambda \beta_n \, dP = \int_{\widetilde{\Lambda}} S_{\xi_1 \cdots \xi_{n-1}} \left(\int_{\Omega_n} S_{\xi_n} b \, dP_n \right) d\widetilde{P}^{(n-1)} =$$

$$= \int_{\widetilde{\Lambda}} S_{\xi_1 \cdots \xi_{(n-1)}} \left(\int S_x b \, dq \right) d\widetilde{P}^{(n-1)} = \int_\Lambda \beta_{n-1} \, dP$$

Thus β_n is a B–valued martingale with respect to the σ–algebras

$\mathcal{F}^{(n)}$. Note that $\|\beta_n\| \leq \sup_{x\in X} \|S_x\|\|b\|$ and that the β_n belong to the weakly compact set $[O(b)]_B$. According to the B–valued martingale convergence theorem (cf. Chattergi [1,2], Buldygin [1]*) (L_B^1) $\lim_{n\to\infty} \beta_n$ exists. Consequently, $(L_B^1) \lim_{n\to\infty}(\beta_n - \beta_{n-1}) = 0$, that is,

$(L_B^1) \lim_{n\to\infty} S_{\xi_1\cdots\xi_{n-1}}(S_{\xi_n}b - b) = 0$. And since

$$\|S_{\xi_n}b - b\|_B \leq c^{-1}\|S_{\xi_1\cdots\xi_{n-1}}(S_{\xi_n}b - b)\|_B$$

a.s. (here $c := \inf\{\frac{\|S_x b'\|}{\|b'\|},\ x \in X,\ b' \in B\} > 0$) we also have $\lim_{n\to\infty} \int_\Omega \|S_{\xi_n}b - b\|_B \, dP = 0$. According to the definition of the X–valued random elements ξ_n we have $\int_\Omega \|S_{\xi_n}b - b\|dP = \int_X \|S_x b - b\| \, dq = 0$, i.e. $S_x b = b$ for q–almost all $x \in X$.

□

Note 4.1. If the representation s is continuous, the hypotheses of Lemma 4.3 imply that $s_x b = b$ for all $x \in c(q)$ (the carrier of q), and hence for all $x \in Y_q$, the smallest l-full semigroup containing $c(q)$. M. Lin [2] has shown that if x is commutative, q is regular, and s is bounded, this conclusion is true also for any $b \in B$. Derriennic [1] has given an example which shows that this can be false if S is amenable, but not commutative.

Lemma 4.4. *Let $(\Omega,\ \mathfrak{B},\ P)$ be a probability space and $f \in L_B^1\ (X,\ \mathfrak{B},\ P)$. Then $\int f\, dP \in K(f)$, the closed convex hull of the set $f(X) \subset B$.*

Proof. Let us fix an arbitrary $\varepsilon \in (0, 1)$. There exists a partitioning $X = \cup A_i$ ($A_i \in \mathfrak{B}$, $A_i \cap A_j = \emptyset$ if $i \neq j$) such that for all $x_i \in A_i$ we have $\sum_{i=1}^{\infty} \|f(x_i)\|P(A_i) =: M < \infty$ and $\|\int f(x)P(dx) - \sum_{i=1}^{\infty} f(x_i)P(A_i)\| < \frac{\varepsilon}{4}$ (cf. e.g., Hille and Phillips [1], Subsect.

---

*) In Buldygin [1] it is supposed that B is separable. The general case can be easily reduced to this one since P–measurability of β_n implies the existence of a separable subspace $B_0 \subset B$ such that $P\{\beta_n \in B_0\} = 1$, $n = \overline{1,\infty}$ (see, e.g., Dunford and Schwartz [1]).

3.7). Let $\sum_{i=n+1}^{\infty}\|f(x_i)\|P(A_i) < \frac{\varepsilon}{4}$ and $\sum_{i=n+1}^{\infty}P(A_i) < \frac{\varepsilon}{3M}$; $\alpha_i :=$ $:= P(A_i)/\sum_{k=1}^{n}P(A_k)$. Evidently, $\alpha_i > 0$, $i = \overline{1,n}$, and $\sum_{i=1}^{n}\alpha_i = 1$. Now $\|\sum_{i=1}^{n}\alpha_i f(x_i) - \sum_{i=1}^{n}f(x_i)P(A_i)\| \le (\varepsilon/3)(1-\varepsilon/3)^{-1} < \frac{\varepsilon}{2}$. Thus $\|\int f\,dP - \sum\limits_{i=1}^{n}\alpha_i f(x_i)\|_{\varepsilon}$.

□

Lemma 4.5. *Let $(X, \mathfrak{B})$, S and q be the same as in Lemma 4.3, and set $\lambda_n^{(q)} = \frac{1}{n}\sum_{k=1}^{n}q^k$ where q^n is the k–th convolution power of the measure q. Then $\lim\limits_{n\to\infty}\int S_x b\lambda_n^{(q)}(dx) =: b_0$ exists for any $b \in W(B,S)$ and $S_x b_0 = b_0$ for q–almost all $x \in X$.*

Proof. Let us consider the linear operator T: $Tb = \int S_x bq(dx)$ in B. Evidently, $\|T\| \le \sup(\|S_x\|, x \in X)$. Now

$$T^2 b = \int S_x\left(\int S_y bq(dy)\right)q(dx) = \iint S_x S_y bq(dx)q(dy) =$$
$$= \iint S_{xy} bq(dx)q(dy) = \int S_x bq^2(dx).$$

We obtain by induction $T^k b = \int S_x bq^k(dx)$, and thus $\frac{1}{n}\sum_{i=1}^{n}T^k b =$ $= \int S_x b\lambda_n^{(q)}(dx)$. According to Lemma 4.4, $T^k b \in \underline{K}(b)$ and, consequently, for any b in $W(B,S)$ the set $\{T^k b,\ k = \overline{1,\infty}\}$ is relatively weakly compact. According to the ergodic theorem for bounded operators in Banach spaces (cf. e.g., Dunford and Schwartz [1], Theorem 8.5.1) $\lim_{n\to\infty}\frac{1}{n}\sum_{k=1}^{n}T^k b =: b_0$ exists and b_0 is T–invariant. It remains to use Lemma 4.3.

□

The main result of this subsection is the following theorem.

Theorem 4.6. **(Ryll–Nardzewski)** *Let S be a bilaterally uniformly bounded representation of X in B and let F be an S–invariant weakly compact convex set in B. Then $F \cap I \neq \emptyset$.*

Proof. Let us denote by $\mathcal{K}$ the class of all finite subsets in X. Let $M := \{x_1, \ldots, x_l\} \in \mathcal{K}$ and let us consider the probability measure

$q_M : q_M(\{x_i\}) = \frac{1}{l}$, $i = \overline{1,l}$, and an element $b \in F$. According to Lemma 4.5, the element $b_0 := \lim_{n\to\infty} \frac{1}{n} \sum_{k=1}^{n} \int S_x b q^k(dx)$ is invariant with respect to all operators S_x, $x \in M$. Let I_M be the set of all elements in F which are invariant with respect to S_x, $x \in M$. We have seen that $I_M \neq \emptyset$ for any $M \in \mathcal{K}$. It is easy to verify that the sets I_M, $M \in \mathcal{K}$, are weakly compact, and $\bigcap_{i=1}^{k} I_{M_i} = I_{\bigcup_{i=1}^{k} M_i} \neq \emptyset$ if $M_1, \ldots, M_k \in \mathcal{K}$. Since F is weakly compact this implies that $I = I_x = \bigcap_{M \in \mathcal{K}} I_M \neq \emptyset$.

□

4.3. Uniformly bounded representations of amenable semigroups.

Theorem 4.7. **(Day)** *Let S be a uniformly bounded left representation of a left–amenable semigroup X in a Banach space B; let F be an S–invariant weakly compact convex set in B. Then $F \cap I \neq \emptyset$.*

Proof. Let $b \in F$, $\{\nu_n, n \in N\}$ be some weakly left–ergodic net of measures in $\widetilde{\mathcal{P}}(X)$ (see App. (3. G)). Then $\int_F S_x b \nu_n(dx) \in F$, $n \in N$. Since F is weakly compact there are an element b_0 in F and a subnet $\{\nu_m, m \in M\}$ such that the weak limit $(W) \lim_{m\in M} \int S_x b \nu_m(dx) = b_0$ (cf., e.g., Kantorovich and Akilov [1], Theorem 1.2.3). Since the net $\{\nu_n\}$ is weakly left–ergodic we have for any $y \in X$, $l \in B^*$,

$$\begin{aligned}
\langle S_y b_0, l\rangle &= \langle b_0, S_y^* l\rangle = \lim_{m\in M} \langle \int S_x b \nu_m(dx), S_y^* l\rangle = \\
&= \lim_{m\in M} \int \langle S_x b, S_y^* l\rangle \nu_m(dx) = \lim_{m\in M} \int \langle S_{yx} b, l\rangle \nu_m(dx) = \\
&= \lim_{m\in M} \int \langle S_x b, l\rangle \nu_m(dx) = \lim_{m\in M} \langle \int S_x b \nu_m(dx), l\rangle = \langle b_0, l\rangle.
\end{aligned}$$

Consequently, $S_y b_0 = b_0$, $y \in X$.

□

§ 5. Some Classes of Averageable Functions and Banach Space Elements

5.1. Theorems on quasi–averageability. In view of Proposition 3.1 and App. (6.C), Theorems 4.2, 4.6 and 4.7 can be reformulated as follows.

Let X be an arbitrary semigroup.

Theorem 5.1. *Every element of a uniformly convex Banach space B is quasi–averageable with respect to any contractive representation of X in B.*

Theorem 5.2. *If S is a bilaterally uniformly bounded representation of X in a Banach space B, then any b in $W(B \mid S)$ is S-quasi–averageable.*

Theorem 5.3. *If S is a bounded left representation of a left–amenable semigroup X in a Banach space B, then any b in $W(B \mid S)$ is S-quasi–averageable.*

5.2. Averageability of weakly almost periodic functions.

Theorem 5.4. **(Ryll–Nardzewski)** *Every weakly almost periodic function on a group X is averageable.*

Theorem 5.5. **(Eberlein–Rosen)** *Every weakly almost periodic function on an amenable semigroup X is averageable.*

Example 2.1 shows the essence of the amenability condition in Theorem 5.5.

We shall give a combined proof of both theorems. Let us recall that $W(\Phi_{\mathbb{C}})$ denotes the subspace of all weakly almost periodic functions in $\Phi_{\mathbb{C}}$.

Proof. The representation R: $x \mapsto R_x$ of X in $\Phi_{\mathbb{C}}$ is uniformly bounded and is, of course, bilaterally uniformly bounded if X is a group. According to Theorems 5.2 and 5.3, any $f \in W(\Phi_{\mathbf{C}})$ is r–quasi–averageable. Similarly, considering the representation L: $x \mapsto L_x$, we see that any $f \in W(\Phi_{\mathbf{C}})$ is l–quasi–averageable, too, and according to Theorem 1.5 its mean is unique. Since $W(\Phi_{\mathbf{C}})$ is a subspace in $\Phi_{\mathbf{C}}$, Proposition 2.6 implies that all functions in $W(\Phi_{\mathbf{C}})$ are l–and r–averageable and thus, in view of Theorem 2.1, they are averageable.

□

5.3. Averageability in Banach spaces with respect to contractive representations. From Theorem 5.1 and Proposition 3.5 the following theorem is immediate.

Theorem 5.6. *If S is a contractive left representation of a right–amenable semigroup X in a uniformly convex Banach space B, then*

(i) $A(B) = B$;

(ii) $\mathbf{M}(b)$ is the least–norm element in $K(b)$.

Now let us turn to the non–amenable case. Here additional restrictions on B are necessary.

Definition 5.1. A Banach space B is said to be *smooth* if for any $a, b \in B$ the function $\varphi_{a,b}(t) = \| a + tb \|$ is differentiable at $t = 0$.

We denote by $\tau_a(b) := \varphi'_{a,b}(0)$.

Example 5.1. A Hilbert space is smooth, since $\|a + tb\|^2 = \|a\|^2 + $ $+ 2t\mathrm{Re}\,(a, b) + t^2\|b\|^2$.

Example 5.2. The spaces $L^\alpha(\Omega, \mathcal{F}, m)$ and l^α, $1 < \alpha < \infty$, are smooth (cf. Diestel [1]). However the spaces L^1 and l^1 are not.

Definition 5.2. Let B be a real Banach space. A hyperplane $\Gamma_{\alpha,f} :=$ $:= \{b : f(b) = \alpha\}$, $f \in B^*$, $\alpha \in \mathcal{R}$ is said to be *tangential* to the sphere $S_r := \{b\colon \|b\| = r\}$ at the point $a \in S_r$ if $a \in \Gamma_{\alpha,f}$ and $\|b\| \geq r$, $b \in \Gamma_{\alpha,f}$.

It is well-known (cf. Dunford and Schwartz [1], §5.9) that if B is a smooth Banach space, then $\tau_a(\cdot)$ is a linear functional on B, $a \in B$; the hyperplane $\{b : \tau_a(b) = r\}$ is the single tangential hyperplane to S_r at the point a if $a \in S_r$.

Theorem 5.7. **(Aloaglu and Garret Birkhoff)** *If S is a contractive representation of a semigroup X in a smooth uniformly convex Banach space B, then*

(i) $A(B) = B$;

(ii) $M(b)$ is the least-norm element in $K(b)$.

Proof. Taking into account Theorem 5.1 and Proposition 3.6 it is sufficient to prove that $|K(a) \cap I| \leq 1$, $a \in B$. Let us suppose that $K(a) \cap I$ contains two different points, b and c. We set $b_1 := a - b$, $c_1 := a - c$ and $d := c - b = b_1 - c_1$. Evidently, $K(b_1) = K(a) - b$, $K(c_1) = K(a) - c$, $0 \in K(b_1) \cap K(c_1)$ and $d \in I$. Hence, for any $\varepsilon > 0$ there is a measure $\nu \in \widetilde{\mathcal{P}}$ such that $\|S_\nu b_1\| \leq \varepsilon$. For any $t \in \mathbb{R}$ we have $\|d + tb_1\| \geq \|S_\nu(d + tb_1)\| = \|d + tS_\nu b_1\| \geq \|d\| - t\varepsilon$ and, since ε was chosen arbitrarily, $\|d+tb_1\| \geq \|d\|$. Similarly, $\|d+tc_1\| \geq \|d\|$. If b_1 and c_1 are scalar multiples, that is, $b_1 = \alpha c_1$, for some $\alpha \in \mathbb{C}$, $\alpha \neq 0$, $\alpha \neq 1$, then $a = (1 - \alpha)^{-1}(b - \alpha c)$ and $a \in I$, $K(a) \cap I = K(a) = \{a\}$ and $b = c = a$. If b_1 and c_1 are not scalar multiples, then $\{\omega : \omega = d + tb_1, t \in \mathbb{R}\}$ and $\{\omega\colon \omega = d + tc_1, t \in \mathbb{R}\}$ are two different tangential straight lines to the circle $S_{\|d\|}$ at the point d in the two-dimensional real subspace Π generated by the elements b_1 and c_1. This means that Π is a non-smooth space and therefore B cannot be smooth, either. Thus we arrive at a contradiction to the condition of the theorem.

□

Now we shall give an example which will show the essence of the conditions imposed on X, B and S in Theorems 5.6 and 5.7.

Example 5.3. **(Alaoglu and Garret Birkhoff)** Let $B_1 = \mathbb{R}^2$ with the norm $\|(\xi, \eta)\|_1 = |\xi| + |\eta| + \frac{1}{100}(\xi^2 + \eta^2)^{\frac{1}{2}}$. This space is uniformly convex but not smooth: the unit sphere has "angles" at the points (1,0), (0,1), (-1,0), and (0,-1). Let us consider in B_1 the operators T_1: $(\xi,\eta) \mapsto (\xi,0)$ and T_2: $(\xi,\eta) \mapsto (\xi + \frac{1}{20}\eta, 0)$. It is not difficult to verify that $\|T_1\| = \|T_2\| = 1$ and $T_1^2 = T_1$, $T_2^2 = T_2$, $T_2T_1 = T_1$, $T_1T_2 = T_2$; T_1 and T_2 form an l–amenable but not an r–amenable contraction semigroup. If $b = (0,1)$ then $T_1 b = (0,0)$ and $T_2 b = (1/20,0)$. But $T_i(0,0)=(0,0)$ and $T_i(1/20,0)=(1/20,0)$, and thus $(0,0)$, $(1/20,0) \in K(b) \cap I$. The element b is quasi–averageable with respect to $\{T_1, T_2\}$, but not averageable. This shows the essence of the r–amenability condition in Theorem 5.6.

Now let us consider the space $B_2 = \mathbb{R}^2$ with the Euclidean norm $\|(\xi,\eta)\|_2 = (\xi^2 + \eta^2)^{\frac{1}{2}}$; it is uniformly convex and smooth. The operators T_1 and T_2 are bounded in this space too, but T_2 is not a contraction. Thus, the condition that S is a contractive representation cannot be replaced in Theorems 5.6 and 5.7 by the weaker condition of uniform boundedness (we shall see below that this can be done if X is a group or an amenable semigroup).

5.4. Averageability with respect to uniformly bounded representations. Theorem 5.3 and Proposition 3.5 imply the following statement.

Theorem 5.8. **(Day)** *If S is a uniformly bounded representation of an amenable semigroup X in a Banach space B, then $W(B|S) \subset$ $\subset A(B|S)$. Consequently, if B is reflexive, then $A(B|S) = B$.*

If X is a group, Theorems 5.6–5.8 are contained in the following statement.

Theorem 5.9. **(Ryll–Nardzewski)** *If S is a uniformly bounded representation of a group X in a Banach space B, then $W(B|S) \subset$ $\subset A(B|S)$; if B is reflexive, then $A(B|S) = B$.*

Proof. According to Theorem 5.2, any element of $W(B|S)$ is quasi–averageable and, in view of Proposition 3.6, it remains to show that its mean is unique. Suppose that c, $d \in \mathfrak{M}(b)$ and $c \neq d$. Let us consider a functional $l \in B$ such that $\langle c, l\rangle \neq \langle d, d\rangle$. According to Proposition 1.2, $\langle c, l\rangle$ and $\langle d, l\rangle$ are means of the function $\varphi(x) = \langle S_x b, l\rangle$, which is weakly almost periodic (cf. App. (6.E)); this is a contradiction to Theorems 5.4 and 2.4.

□

Example 5.3 shows that Theorem 5.9 does not hold if X is a non-amenable *semi*group.

5.5. Averageability in the spaces H and L^α_B, $\alpha > 1$. First of all we shall state two simple consequences of Theorems 5.7 and 5.9. Let $(\Omega, \mathcal{F}, m)$ be a measure space.

Corollary 5.10. *If B is either $L^\alpha(\Omega, \mathcal{F}, m)$, $1 < \alpha < m$, or a Hilbert space and S is a contractive representation of a semigroup in B, then $A(B|S) = B$; for any $b \in B$ the mean $\mathbf{M}(b)$ coincides with the unique least-norm element in $K(b)$.*

Corollary 5.11. *If S is the same as in Corollary 5.10 and is a uniformly bounded representation of a group X in B, then $A(B|S) = B$.*

If $B = H$, a Hilbert space, then Proposition 3.7 can be stated in the following more precise form.

Theorem 5.12. *If S is an isometric representation of a semigroup in the Hilbert space H, then the ergodic decomposition with respect to S: $H = I \oplus D$ is orthogonal (that is, I and D are orthogonal complements to each other).*

Proof. Let $a \in I$ and $b = c - S_y c$, $c \in H$. Then $\langle a, b\rangle = \langle a, c\rangle - \langle a, S_y c\rangle = \langle a, c\rangle - \langle S_y a, S_y c\rangle = 0$. Since D is generated by the elements of the form $b = c - S_y c$, $c \in H$, the theorem is proved.

□

In Subsection 5.6 this statement will be transfered to “compatible” isometric representations in L^α, $1 < \alpha < \infty$.

5.6. κ–equivalent representations. Consistent representations in L^α. Let us denote by $B^{(1)}$ and $B^{(2)}$ Banach spaces, by $A^{(i)}$ a dense linear manifold in $B^{(i)}$, $i = 1, 2$, by κ a one–to–one linear mapping from $A^{(1)}$ onto $A^{(2)}$, and by $T_{(i)}$ a bounded linear operator in $B^{(i)}$, $i = 1, 2$.

Definition 5.3. We say that two operators $T^{(1)}$ and $T^{(2)}$ are κ–*equivalent* (notation: $T^{(1)} \overset{\kappa}{\sim} T^{(2)}$) if:

(i) $T^i(A^{(i)} \subseteq A^{(i)}$, $i = 1, 2$;

(ii) $T^{(1)} = \kappa^{-1} T^{(2)} \kappa$.

Evidently, κ–equivalence is a symmetric relation: $T^{(1)} \overset{\kappa}{\sim} T^{(2)}$ means the same as $T^{(2)} \overset{\kappa^{-1}}{\sim} T^{(1)}$.

If $S^{(1)}$: $x \mapsto S_x^{(1)}$ is a representation of a semigroup X in $B^{(1)}$ and $S_x^{(1)} \overset{\kappa}{\sim} S_x^{(2)}$, $x \in X$, then $S^{(2)}$: $x \mapsto S_x^{(2)}$ is a representation of X in $B^{(2)}$. In this case we shall say that the representations $S^{(1)}$ and $S^{(2)}$ are κ–*equivalent* and write: $S^{(1)} \overset{\kappa}{\sim} S^{(2)}$.

We recall that a mapping κ defined as above is said to be *closed* if the relations $b_n^{(i)} \in A^{(i)}$, $b_n^{(2)} = \kappa(b_n^{(1)})$ and $\lim_{n\to\infty} b_n^{(i)} = b^{(i)}$, $i = 1, 2$, imply that $b_n^{(i)} \in A^{(i)}$, $i = 1, 2$, and $\kappa(b^{(1)} = b^{(2)}$. Evidently, if x is closed, then x^{-1} is closed too.

Theorem 5.13. *Let $S^{(1)} \overset{\kappa}{\sim} S^{(2)}$ and let κ be closed. Then*

(i) $I^{(1)} \cap A^{(1)} \overset{\kappa}{\longleftrightarrow} I^{(2)} \cap A^{(2)}$;

(ii) if $b \in A^{(1)}$ and $A(B^{(i)}) = B^{(i)}$, $i = 1, 2$, then $\mathbf{M}(b) \in A^{(1)}$, $\mathbf{M}(\kappa b) \in A^{(2)}$ *and* $\mathbf{M}(\kappa b) = \kappa \mathbf{M}(b)$;

(iii) if $A(B^{(i)}) = B^{(i)}$, $i = 1, 2$, then $D^{(1)} \cap A^{(1)} \overset{\kappa}{\longleftrightarrow} D^{(2)} \cap A^{(2)}$;

(iv) if $A(B^{(i)}) = B^{(i)}$, $i = 1, 2$, then $\Pi_{I^{(1)}}^{B^{(1)}} \overset{\kappa}{\sim} \Pi_{I^{(2)}}^{B^{(2)}}$ and $\Pi_{D^{(1)}}^{B^{(2)}} \overset{\kappa}{\sim} \Pi_{D^{(2)}}^{B^{(2)}}$;

(v) if $A^{(1)} = B^{(1)}$ and $A(B^{(1)}) = B^{(1)}$, then $A(B^{(2)}) = B^{(2)}$.

Proof. *(i)*. If $b \in I^{(1)} \cap A^{(1)}$, then $S_x^{(2)} \kappa b = \kappa S_x^{(1)} b = \kappa b$, i.e. $\kappa b \in I^{(2)} \cap A^{(2)}$, and hence $\kappa(I^{(1)}) \subset I^{(2)} \cap A^{(2)}$. Changing the roles of the indexes 1 and 2 we obtain $\kappa^{-1}(I^{(2)} \cap A^{(2)}) \subset I^{(1)} \cap A^{(1)}$. Thus statement *(i)* is proved.

(ii). According to Lemma 2.3 there exists a sequence of measures $\{\lambda_n\} \subset \widetilde{\mathcal{P}}$ such that $\lim_{n\to\infty} S^{(1)}_{\lambda_n} b = \mathbf{M}(b)$ and $\lim_{n\to\infty} S^{(1)}_{\lambda_n} \kappa b = \mathbf{M}(\kappa b)$. We have $\mathbf{M}(b) \in A^{(1)}$, $\mathbf{M}(\kappa b) \in A^{(2)}$ and $\mathbf{M}(\kappa b) = \kappa \mathbf{M}(b)$.

(iii). This statement follows from *(ii)*, since $\mathbf{M}(b) = 0$ for some $b \in A^{(1)}$ is equivalent to $\mathbf{M}(\kappa b) = \kappa \mathbf{M}(b) = 0$.

(iv). The first assertion is a simple consequence of statement *(ii)* and the second one follows from the equality $\Pi^{B^{(i)}}_{D^{(i)}} = I - \Pi^{B^{(i)}}_{I^{(2)}}$.

(v). First of all, according to the Closed Graph Theorem, the mapping κ is continuous. Therefore $\kappa\mathbf{M}(b) \in \mathfrak{M}(\kappa b)$, $b \in B^{(1)}$, and, consequently, all elements in $A^{(2)}$ are quasi–averageable. If $c \in \mathfrak{M}(\kappa b)$, then, by Lemma 2.3, there exists a sequence $\{\lambda_n\} \subset \widetilde{\mathcal{P}}$ such that $\lim_{n\to\infty} S^{(1)}_{\lambda_n} b = \mathbf{M}(b)$ and $\lim_{n\to\infty} \kappa S^{(1)}_{\lambda_n} b = \lim_{n\to\infty} S^{(2)}_{\lambda_n} \kappa b = c$. Since κ is continuous it follows from these relations that $c = \kappa(\mathbf{M}(b))$, and thus $\kappa(\mathbf{M}(b))$ is the unique mean of κb with respect to $S^{(2)}$. It remains to use Proposition 3.6.

□

The notion of consistency of representations introduced below is a special case of the notion of κ–equivalence.

Definition 5.4. Let $A \subset [0, \infty)$, let B be a Banach space and let $(\Omega, \mathcal{F}, m)$ be a measure space. We shall say that a family of bounded linear operators $\{S^\alpha,\ \alpha \in A\}$ which act in the corresponding spaces $L^\alpha_B = L^\alpha_B(\Omega, \mathcal{F}, m)$ is *consistent* if for any $\alpha_1, \alpha_2 \in A$ we have $S^{\alpha_1} f = S^{\alpha_2} f$, $f \in L^{\alpha_1}_B \cap L^{\alpha_2}_B$, or, in other words, if the operators S^{α_1} and S^{α_2} are equivalent with respect to the identity mapping from the set $L^{\alpha_1}_B \cap L^{\alpha_2}_B$ onto itself. A family $\{S^\alpha\}$ of representations of a semigroup X in the Banach spaces L^α_B is said to be *consistent* if for any x the family of operators $\{S^\alpha_x,\ \alpha \in A\}$ is consistent.

Lemma 5.14. *The identity mapping κ from the set $L^{\alpha_1}_B \cap L^{\alpha_2}_B$ onto itself, considered as a linear mapping from $L^{\alpha_1}_B$ into $L^{\alpha_2}_B$, is closed,* $\alpha_1, \alpha_2 \in [1, \infty)$.

Proof. $L^{\alpha_i}_B$–convergence of $\{f_n\}$ to $f^{(i)}$ implies the existence of a subsequence $\{f_{n_k}\}$ converging to $f^{(i)}$ almost everywhere, $(i = 1, 2)$; therefore $f^{(1)} = f^{(2)} \in L_B^{\alpha_1} \cap L_B^{\alpha_1}$.

□

From Theorem 5.13 and Lemma 5.14, taking $A_1 = A_2 = L_B^1 \cap L_B^2$, we easily obtain the following statement. Let S^α be a representation in L^α and $A(L^\alpha|S) = L^\alpha$. We denote by I^α and D^α the spaces I and D which figure in Proposition 3.7.

Theorem 5.15. *Let S^{α_i}, $i = 1, 2$, be consistent representations in $L_B^{\alpha_i}$, and $A(L_B^{\alpha_i}|S) = L_B^{\alpha_i}$. Then*

(i) $I_B^{\alpha_1} \cap L_B^{\alpha_2} = I_B^{\alpha_2} \cap L_B^{\alpha_1} = I_B^{\alpha_1} \cap I_B^{\alpha_2}$;

(ii) $D_B^{\alpha_1} \cap L_B^{\alpha_2} = D_B^{\alpha_2} \cap L_B^{\alpha_1} = D_B^{\alpha_1} \cap D_B^{\alpha_2}$;

(iii) the pairs of operators $\Pi_{I_B^{\alpha_i}}^{L_B^{\alpha_i}}$, $i = 1, 2$, and $\Pi_{D_B^{\alpha_i}}^{L_B^{\alpha_i}}$, $i = 1, 2$, are consistent;

(iv) $\Pi_{I_B^{\alpha_i}}^{L_B^{\alpha_i}}(L_B^{\alpha_1} \cap L_B^{\alpha_2}) = I_B^{\alpha_1} \cap I_B^{\alpha_2}$ and $\Pi_{D_B^{\alpha_i}}^{L_B^{\alpha_i}}(L_B^{\alpha_1} \cap L_B^{\alpha_2}) = D_B^{\alpha_1} \cap D_B^{\alpha_2}$;

(v) the set $I_B^{\alpha_1} \cap I_B^{\alpha_2}$ is dense in $I_B^{\alpha_1}$ and in $I_B^{\alpha_2}$; the set $D_B^{\alpha_1} \cap D_B^{\alpha_2}$ is dense in $D_B^{\alpha_1}$ and in $D_B^{\alpha_2}$.

Let α, $\beta \geq 1$ be conjugate numbers (i.e. $\frac{1}{\alpha} + \frac{1}{\beta} = 1$). Then $\langle f, g \rangle = \int fg \, dm$ is a continuous bilinear functional on $L^\alpha \times L^\beta$. Recall that the annihilator M^1 of a subspace M of L^β is the subspace $M^\perp = \{f : f \in L^\alpha, \langle f, g \rangle = 0, \forall g \in M\}$.

Theorem 5.16. *Let α and β be conjugate numbers, α, $\beta \geq 1$, and let S^α and S^β be representations of X in L^α and in L^β which are consistent with an isometric representation S^2 in L^2, and let $A(L^\alpha) = L^\alpha$, $A(L^\beta) = L^\beta$. Then $I^\alpha = (D^\beta)^\perp$ and $D^\alpha = (I^\beta)^\perp$.*

Proof. According to Theorem 5.12, $\langle f, g \rangle = 0$, $f \in D^2$, $g \in I^2$. Therefore, in view of statement (v) of Theorem 5.15, $I^\alpha \subset (D^\beta)^\perp$ and $D^\alpha \subset (I^\beta)^\perp$. Since $L^\beta = I^\beta \oplus D^\beta$ we have (cf. Dunford and Schwartz [1], § 6.9). On the other hand, $L^\alpha + I^\alpha \oplus D^\alpha$. Therefore if either $I^\alpha \neq (D^\beta)^\perp$ or $D^\alpha \neq (I^\beta)^\perp$, then $(D^\beta)^\perp \oplus (I^\beta)^\perp$ would be a proper subset in L^α.

□

Corollary 5.17. *Let X be a group, let α and β be conjugate numbers, $1 < \alpha, \beta < \infty$, and let S^α and S^β be consistent isometric representations of X in L^α and L^β. Then $I^\alpha = (D^\beta)^\perp$ and $D^\alpha = (I^\beta)^\perp$.*

Proof. We can extend the operators S_x^α and S_x^β (from $L^\alpha \cap L^\beta \cap L^2$ into L^2) to continuous operators S_x^2 in L^2. According to the Riesz convexity theorem (cf., e.g., Dunford and Schwartz [1], Corollary 6.10.12), $\|S_x^2\| \leq 1$. Clearly, $S^2 \colon x \mapsto S_x^2$ is a representation of X in L^2 which is consistent with S^α and S^β and, since X is a group, this representation is unitary. It remains to use Theorem 5.16.

□

5.7. Averageability with respect to α–contractive representations in $\mathbf{L^1}$. Let $(\Omega, \mathcal{F}, m)$ be a measure space.

Definition 5.5. Let $1 < \alpha \leq \infty$. A representation S in L_B^1 will be called *α–contractive* if it is a contractive representation and is consistent with a contractive representation $S^{(\alpha)}$ in L_B^α.

By the Riesz convexity theorem, S^α is consistent with contractive representations S^β in L_B^β, $1 < \beta < \alpha$, as well. The ∞–contractive representations in S^1 are also called *strongly contractive* representations. For example, the isometric representations U^α in L^α associated with a dynamical system $\{T_x,\ x \in X\}$ are consistent, and hence U^1 is strongly contractive. Corollary 5.10 and Theorem 5.13 imply the following statement.

Theorem 5.18. *Let $m(\Omega) < \infty$ and let S be an α–contractive representation (for some $\alpha > 1$) of a semigroup X in $L^1(\Omega, \mathcal{F}, m)$. Then $A(L^1) = L^1$.*

The following example shows that without the assumption $m(\Omega) < < \infty$ Theorem 5.18 is not valid.

Example 5.5. Let $\Omega = \mathbb{R}$, $X = \mathbb{Z}$, and let S be Lebesque measure on $\mathbb{R}$, $S_x f(\omega) = f(\omega - x)$, $x \in \mathbb{R}$. Evidently, S is a strong contractive representation of $\mathbb{R}$ in L_1. It is easy to see that $I^1 = \{0\}$. If $f(\omega) = g(\omega) - g(\omega - x)$, $g \in L^1$, $x \in \mathbb{Z}$, then $\int_{\infty}^{\infty} f(\omega)\, d\omega = 0$. Since the functional $l(f) := \int_{\infty}^{\infty} f(\omega)\, dx$ is continuous on L^1, we have $\int_{\infty}^{\infty} f(\omega)\, dx = 0$ for all $f \in D^1$. Therefore $D^1 \neq L^1$. But, according to Proposition 3.7, $A(L^1) = I^1 \oplus D^1 = D^1$, and thus $A(L^1) \neq L^1$.

In view of Theorem 5.18, Definition 3.1 gives the unique mean $M(f|U^1)$ with respect to an α–contractive representation if $\alpha = 1$ and $m(\Lambda) < \infty$. However, Example 5.5 shows that this does not work if $\alpha = 1$ and $m(\Lambda) = \infty$. In spite of this we can give a reasonable definition of the mean $\mathbf{M}^{(1)}(f|U^1)$ for $\varphi \in L^1$ in the case $m(\Lambda) = \infty$ too. Indeed, we set $\mathbf{M}^{(1)}(f|S) = \mathbf{M}^{(\beta)}(f|S)$, $1 < \beta < 0$, for any $f \in L^0(\Lambda, \mathcal{F}, m)$, the linear manifold of all m–integrable simple functions on Λ. Since for any $\varphi \in L^0$ the function $\beta \mapsto \|f\|_\beta$ is continuous on $[1, \infty]$ (see, e.g., Bourbaki [1], Proposition 4.6.4) the linear operator $\varphi \mapsto \mathbf{M}^{(1)}(f)$, $\varphi \in L^0$, is of norm $\|\mathbf{M}^{(1)}\|_{L^1} \leq 1$ and therefore we can extend it to a linear operator on L^1 with $\|\mathbf{M}^{(1)}\|_{L^1} \leq 1$. Thus we can complement Definition 3.1 in the following way.

Definition 5.6. Let $m(\Lambda) = \infty$ and let $s\colon x \to S_x$ be a contractive representation of X in $L^1(\Lambda, \mathcal{F}, m)$ of degree $\alpha > 1$; the operator $\mathbf{M}^{(1)}$: $L^1 \to I^{(1)}$ defined above is called the *averaging operator* (with respect to S) and for any $\varphi \in L^1$ the function $\mathbf{M}^{(1)}(f|S) = \mathbf{M}^{(1)}(f)(\in L^1)$ is called the *mean of φ with respect to the representation S*.

Of course, the family of operators $\{\mathbf{M}^{(\beta)}, 1 \leq \beta \leq \alpha\}$ is consistent.

Now we can also consider the subspace $\widetilde{D}^{(1)} := \{f : f \in L^1, \mathbf{M}^{(1)}(f) = 0\}$; then $L^1 = I^{(1)} \oplus \widetilde{D}^1$, and $\mathbf{M}^{(1)} = \Pi_{I^1}^{L^1}$, the projector onto I^1 with respect to $\widetilde{D}^1$.

It is evident that the mean $\mathbf{M}^{(1)}(\cdot|S)$ has the following properties (compare with Propositions 3.1–3.3 and 3.11):

1) $\mathbf{M}^{(1)}(S_x f) = \mathbf{M}(f)$, $x \in X$, $f \in L^1$;

2) $\|\mathbf{M}^{(1)}(f)\|_{L^1} \leq \|f\|_{L^1}$, $\varphi \in L^1$;

3) $\mathbf{M}^{(1)}(f) \in I^1$, $\varphi \in L^1$;

4) if $S \not\equiv I$ and is irreducible, then $\mathbf{M}^{(1)}(f) = 0$, $f \in L^1$.

5.8. Means of $\mathbf{C_a}$–functions. Let X be a non–compact locally compact group. Since $C_a(X) \subset W(X)$ (cf. App., § 6), we have $C_a(X) \subset A(X)$. The functional m: $m(\varphi) = \lim_{x\to\infty} \varphi(x)$ is a right– and left–invariant Banach mean on $C_a(X)$, and therefore, according to Corollary 2.10, $\mathbf{M}(\varphi) = \lim_{x\to\infty} \varphi(x)$. Thus the following result is proved.

Theorem 5.19. *$C_a(X) \subset A(X)$ and* $\mathbf{M}(\varphi) = \lim\limits_{x\to\infty} \varphi(x)$, $\varphi \in C_a(X)$.

5.9. Averaging with respect to weakly convergent representations. Let X be as in Subsect.5.8.

Theorem 5.20. *Let S be a uniformly bounded representation of X in a Banach space B and let the weak limit $(w) \lim\limits_{x\to\infty} S_x =: S_\infty$ exist. Then $A(B|S) = B$ and $S_\infty = \Pi_I^B = \mathbf{M}$ (the S–averaging operator in B).*

Proof. According to Mazur's theorem (cf. Dunford and Schwartz [1], Theorem 5.3.13), for any $b \in B$ we have $(w) \lim\limits_{x\to\infty} S_x b =: b_\infty \in K(b)$. It is also clear that $S_y b_\infty = (w) \lim\limits_{x\to\infty} S_y S_x b = (w) \lim\limits_{x\to\infty} S_x b = b_\infty$ and thus $b_\infty \in K(b) \cap I = \mathfrak{M}(b)$, $b \in B$. Using Theorem 5.19 and arguing as in the proof of Theorem 5.9, we show that b_∞ is the single point of $\mathfrak{M}(b)$. Now it remains to use Proposition 3.6.

□

§ 6. Means of Positive Definite Functions

Let X be a group, and let $\mathbf{P}(X)$ be the set of all positive definite functions ("P–functions") on X (see App., § 5). According to App. (5.B) and App. (6.F), $\mathbf{P}(X) \subset W(X)$ and hence $\mathbf{P}(X) \subset A(X)$. In this section we shall state some special properties of means of P–functions. First, Proposition 1.2 implies the following fact.

Theorem 6.1. **(Godement)** *If U is an unitary representation of X in a Hilbert space H and $\varphi(x) = \langle U_x h_1, h_2\rangle$, $h_1, h_2 \in H$, then* $\mathbf{M}(\varphi) = \langle \mathbf{M}(h_1), h_2\rangle$.

Theorem 6.2. **(Godement)** *Let $\varphi \in \mathbf{P}(X)$ and $\varphi(x) = \langle U_x h, h\rangle$ where U is the associated unitary representation (cf. App., § 5). Then*

(i) $\mathbf{M}(\varphi) = \|\mathbf{M}(h)\|^2$;

(ii) $\mathbf{M}(\varphi) = \min\{\sum_{i,k=1}^{n} \alpha_i\alpha_k\varphi(x_i^{-1}x_k),\ \alpha_i \geq 0,\ \sum_{i=1}^{n}\alpha_i = 1,$
$n = \overline{1,\infty}\}$.

Proof. In view of Theorem 6.1, $\mathbf{M}(\varphi) = (\mathbf{M}(h), h) = (\Pi_I^H h, h) = (\Pi_I^H h, \Pi_I^H h) = \|\mathbf{M}(h)\|^2$. Further, by Corollary 5.10

$$\mathbf{M}(\varphi) = \|\mathbf{M}(h)\|^2 = \min(\|f\|^2,\ f \in K(h)) = \min \|\sum_{i=1}^{n} \alpha_i U_{x_i} h\|^2 =$$

$$= \min\{\sum_{i,k=1}^{n} \alpha_i\alpha_k\langle U_{x_k}h,\ U_{x_i}h\rangle\} = \min\{\sum_{i=1}^{n} \alpha_i\alpha_j\langle U_{x_i}^{-1}h,\ h\rangle\} =$$

$$= \min\{\sum_{i,k=1}^{n} \alpha_i\alpha_k\varphi(x_i^{-1}x_k)\}$$

□

Theorem 6.3. *Any $\varphi (\in \mathbf{P}(X))$ can be represented in the form $\varphi = \mathbf{M}(\varphi) + \varphi_0$ where $\varphi_0 \in \mathbf{P}(X)$ and $\mathbf{M}(\varphi_0) = 0$.*

Proof. $\varphi(x) = \langle U_x h,\ h\rangle$ where U is the associated unitary representation, $h \in H$. According to Theorem 5.12, $h = \mathbf{M}(h) + h_0$ where $h_0 \in D$ and $h_0 \perp \mathbf{M}(h)$. Since $U_x h_0 \in D$ (see Proposition 3.7), $U_x h_0 \perp \mathbf{M}(h)$, and therefore $\varphi(x) = \langle U_x(\mathbf{M}(h) + h_0\rangle = \|\mathbf{M}(h)\|^2 + \langle U_x h_0,\ h_0\rangle = \mathbf{M}(\varphi) + \varphi_0(x)$. By App. (3.A) $\varphi_0(x) = \mathbf{P}(X)$.

□

Theorem 6.4. *If φ is an elementary P-function and $\varphi(x) \not\equiv 1$, then* $\mathbf{M}(\varphi) = 0$.

Proof. This is a simple consequence of Proposition 3.11 and of Theorem 6.2 *(i)*.

□

§ 7. Averageability of Homogeneous Random Fields and of Orbital Functions of Dynamical Systems

7.1. The mean and the background of a wide–sense homogeneous random field. Let $\mathcal{H}$ be a Hilbert space and $\mathbf{X}$ a semigroup with identity e. A wide–sense l–homogeneous $\mathcal{H}$–valued random field $\xi(\mathbf{x})$ on $\mathbf{X}$ is the orbital function of the element $\xi(e)$ with respect to the isometric translation (left) representation $\mathbf{U}$ of $\mathbf{X}$ in the Hilbert space $H_\xi \subset L^2_{\mathcal{H}}\ (\Omega,\, \mathcal{F},\, P)$ (cf. App., § 4).

Let us denote by I_ξ the subspace of H_ξ formed by all U–invariant elements in H_ξ and by K_ξ the closed convex hull of the set $\{\xi(x),\ x \in X\} \subset H_\xi$.

It is easy to verify that $\mathbf{E}\gamma = \mathbf{E}\xi(e)$, $\gamma \in K_\xi$, and that $x \mapsto {}_{x_0}\xi(x) :=$ $:= \xi(x_0 x)$ is also a wide–sense l–homogeneous random field, $x_0 \in X$.

After was said above, Proposition 1.6 and Corollary 5.10 imply the following statement.

Theorem 7.1. *A wide–sense l–homogeneous random field $\xi(x)$, $x \in X$, is an r–averageable H–valued function on X. Its mean possesses the following properties:*

(i) $\mathbf{E} \parallel \mathbf{M}^{(\mathbf{r})}(\xi) \parallel^{\mathbf{2}} \kappa < \infty$;

(ii) $\mathbf{E}\mathbf{M}^{(r)}(\xi) = \mathbf{E}\xi(e)$;

(iii) $\mathbf{M}^{(r)}(\xi) \in H_\xi$;

(iv) $\mathbf{M}^{(r)}({}_{x_1}\xi_{x_2}) = \mathbf{M}^{(r)}(\xi)$ *where* ${}_{x_1}\xi_{x_2}(x) = \xi(x_1 x\, x_2)$, $x_1, x_2 \in X$;

(v) $\mathbf{M}^{(r)}(\xi) = \Pi^{H_\xi}_{I_x i}\xi(x)$ *a.s.*, $x \in X$.

Note 7.1. Yaglom [1,2] has shown that any wide–sense l–homogeneous random field on a separable locally compact group G of type *1* admits the following spectral representation:

$$\xi(g) = \int_{\widetilde{G}} \mathrm{tr}[U_g(\lambda)Z(d\lambda)] \tag{7.1}$$

where $\widetilde{G}$ is the "dual object" of G, i.e. the set of all equivalence classes of irreducible unitary representations of G, $U(\lambda)$ is some representation of the class $l \in \widetilde{G}$ which acts in the Hilbert space $H(\lambda)$, $Z(A)$ is a countably additive function defined on some σ–algebra D of subsets of $\widetilde{G}$; $Z(A)$, $A \in D$, are random linear operators in the spaces $H(\lambda)$, and

$$\begin{aligned} &\mathbf{E}\langle Z(A_1)h_1,\, Z(A_2)h_2\rangle = \langle F(A_1 \cap A_2)h_1,\, h_2\rangle, \\ &h_1,\, h_2 \in H(\lambda), \quad A_1,\, A_2 \in D \end{aligned} \tag{7.2}$$

where F is an additive function on D whose values are positive definite Hermitian operators. Let $U_g(\lambda_I)$ be the identity representation, $U_g(\lambda_I) \equiv I$, in the one–dimensional space $H(\lambda_I)$. It follows from (7.1) and (7.2) that for all $g \in G$,

$$\mathbf{E}[(\xi(g)) - Z(\{\lambda_I\})]\overline{Z(\{\lambda_I\})} = 0. \tag{7.3}$$

It is easy to see that $Z(\{\lambda_I\}) \in I_\xi$. We deduce from (7.3) and statement *(v)* of Theorem 7.1 that $\mathbf{M}^{(r)}(\xi) = Z(\{\lambda_I\})$ a.s..

Theorem 7.2. *Let X be a group, $\xi(\cdot)$ a wide–sense l–homogeneous random field of X, and let $R(\cdot)$ be the correlation function of $\xi(\cdot)$. Then*

$$\mathbf{M}(R) = \mathbf{E} \parallel \mathbf{M}^{(r)}(\xi) \parallel^2 \tag{7.4}$$

This is simply a reformulation of Godement's formula (Theorem 6.2).

Definition 7.1. We shall say that the *subspace I_ξ is trivial* if it consists of non–random elements (scalars) of H_ξ.

Definition 7.2. The random value $\eta = \mathbf{M}(\xi) - \mathbf{E}(e)$ is called the *background* of the field $\xi(\cdot)$. If $\eta = 0$ a.s., we shall say that $\xi(\cdot)$ is *background–free*.

Evidently, $\mathbf{E}\eta = 0$.

From Theorem 7.1 the following statement can be easily deduced.

Theorem 7.3. *Any wide-sense homogeneous random field $\xi(\cdot)$ can be represented in the form $\xi(x) = \eta + \zeta(x)$ where η is the background of $\xi(\cdot)$, $\zeta(\cdot)$ is a background-free wide-sense homogeneous random field and for any $x \in X$ the random variables η and $\zeta(x)$ are orthogonal, i.e. $\mathbf{E}\langle\eta(x), \zeta(x)\rangle_{\mathcal{H}} = 0$. If the field $\xi(\cdot)$ is Gaussian, then η and $\zeta(\cdot)$ are Gaussian and independent of each other.*

Now let $Y = G/K$ be a l-homogeneous space and let π be the natural mapping of G onto Y. If $\xi(\cdot)$ is a homogeneous random field on Y, then $\tilde{\xi}(g) := \xi(\pi(g))$ is an l-homogeneous random field on G. We can define the *mean* $\mathbf{M}^{(r)}(\xi) := \mathbf{M}^{(r)}(\tilde{\xi})$ and the *background* $\eta := := \mathbf{M}(\xi) - \mathbf{E}\xi(y)$ of $\xi(\cdot)$; if $\eta = 0$ a.s. we say that $\xi(\cdot)$ is *background-free*. As before, $\xi(y) = \zeta(y) + \eta$. $\zeta(\cdot)$ is homogeneous and for any $y \in Y$ the random variables $\zeta(y)$ and η are orthogonal.

7.2 Averageability of orbital functions of dynamical systems. Let X be a semigroup; let $T = \{T_x,\ x \in X\}$ be a (left) dynamical system in the phase space $(\Omega, \mathcal{F}, m)$, and let B be a Banach space.

Theorem 7.4. *Let $f \in L^\alpha_B\ (\Omega, \mathcal{F}, m)$. If $\alpha > 1$ or $\alpha = 1$ and $m(\Omega) < \infty$; then the L^α_B-valued orbital function $\varphi_f(x) = f(T_x\Omega)$ is l-averageable; $\mathbf{M}^{(l)}(\varphi_f) = \mathbf{M}(f|U^\alpha)$, the mean of $f \in L^\alpha_B$ with respect to the associated translation representation U^α in L^α_B.*

Proof. Let us consider the right representations U^α and $\widetilde{U}^\alpha$ in L^α_B and L^α, resp., associated with T. Of course, the representations $\widetilde{U}^\alpha$ are consistent. Therefore Corollary 5.10 and Theorem 5.18 imply that the theorem is valid if $B = \mathbb{C}$ and, as we have shown in § 3, $\mathbf{M}^l(\varphi_f) = = \mathbf{M}(f)$, $f \in L^\alpha$. Let $U^\alpha_\lambda g := \int U^\alpha_x g\lambda(dx)$, and $\widetilde{U}^\alpha_\lambda f := \int \widetilde{U}^\alpha_x f\lambda(dx)$, $g \in L^\alpha_B$, $f \in L^\alpha$, $\lambda \in \widetilde{\mathcal{P}}$. Evidently, $U^\alpha_\lambda(bf(\cdot)) = b\widetilde{U}^\alpha_\lambda(f)$ and therefore $\mathfrak{M}(bf|U^\alpha) = b\mathfrak{M}(f|\widetilde{U}^\alpha) = \{b\mathbf{M}(f|\widetilde{U}^\alpha)\}$. We also have $\mathfrak{M}(U^\alpha_\lambda(bf)) =$

$= \mathfrak{M}\{b\widetilde{U}^{\alpha}_{\lambda}(f)\} = b\mathfrak{M}(\widetilde{U}^{\lambda}f) = b\mathfrak{M}(f) = \mathfrak{M}(bf)$ and thus the element bf is averageable with respect to U^{α}. Since all elements of this form generate the space L^{α}_{B} our statement follows from Proposition 3.4.

□

Let us state some simple but important properties of the means $\mathbf{M}(f|U^{\alpha})$.

Theorem 7.5. *Under the conditions of the previous theorem:*

(i) $\mathbf{M}(f) \in I^{\alpha}_{B}$, *the subspace of* T*-(and* U^{α}*-)invariant elements in* L^{α}_{B}, $f \in L^{\alpha}_{B}$;

(ii) if $f \in L^{\alpha_1}_{B} \cap L^{\alpha_2}_{B}$, *then* $\mathbf{M}(f|U^{\alpha_1}) = \mathbf{M}(f|U^{\alpha_2})$;

(iii) $\int_{\Lambda} f\,dm = \int_{\Lambda} \mathbf{M}(f)\,dm$ *if* $\Lambda \in \mathfrak{I}$, $m(\Lambda) < \infty$, $f \in L^{1}_{B}$; *in particular, if* $(\Omega, \mathcal{F}, m)$ *is a probability space then* $\mathbf{M}(f) = \mathbf{E}(f|\mathfrak{I})$, *the conditional expectation of the random variable* f *with respect to the* σ*-algebra* $\mathfrak{I}$.

Proof. Statement *(i)* follows from the definition of $\mathbf{M}(f)$. Statement *(ii)* is a simple consequence of the consistency of the representations U^{α_1} and U^{α_2}. Let us note that if $f \in L^{\alpha}_{B}$ and $\{\lambda_n\} \in A\widetilde{S}^{(l)}(\varphi_f)$, then uniformly with respect to $y \in X$,

$$(L^{\alpha}_{B}) \lim_{n\to\infty} \int_{X} f(T_{xy}\omega)\lambda_n(dx) = \mathbf{M}(f) \tag{7.5}$$

Let us suppose that $m(\Omega) < \infty$. Using Fubini's theorem we obtain

$$\int_{\Omega} \left(\int_{X} f(T_x\omega)\lambda_n(dx) \right) m(d\omega) = \int_{\Omega} f(\omega)m(d\omega).$$

From this equality and from (7.5) (with $\alpha = 1$, $y = e$) statement *(iii)* with $\Lambda = \Omega$ follows at once. In order to prove the general case we only have to apply the obtained result to the space $(\Lambda, \mathcal{F}, m_{\Lambda})$ with $m_{\Lambda}(K) = m_{\Lambda}(\Lambda \cap K)$, $k \in \mathcal{F}$.

Statement *(ii)* of this theorem and Definition 5.6 enable us to write $\mathbf{M}(f|T)$ or $\mathbf{M}(f)$ instead of $\mathbf{M}(f|U^{\alpha})$, $f \in L^{\alpha}$, $1 \leq \alpha < \infty$. The mean $\mathbf{M}(f|T)$ is often called the *temporal mean* of the function f with respect to the dynamical system T. If $f \in L^{1}_{B}$ and $m(\Lambda) < \infty$, we can also consider the *phase mean* $\mathbf{E}(f) := \frac{1}{m(\Lambda)} \int_\Lambda f(\omega) m(d\omega)$ of f. The connection between these means, stated in the following simple corollary of Theorem 7.5, is basic in many applications of ergodic theory (see also Lemma 2.1.1 below).

□

Corollary 7.6. *If the σ-algebra $\mathfrak{I}$ corresponding to the dynamical system T is trivial ($\mathfrak{I} = \{\Phi, \Omega\}$), that is, if T is ergodic or metrically transitive (see Definition 2.1.1 below), then $\mathbf{M}(f|T)$ is constant m-almost everywhere: if $m(\Omega) = \infty$ then $\mathbf{M}(f|T) = 0$, and if $m(\Omega) < \infty$ then $\mathbf{M}(f|T) = \mathbf{E}(f)$, $f \in L_B^\alpha$, $\alpha \geq 1$.*

Let us denote by D_B^α the subspace of L^α generated by the functions

$$f(\omega) = g(\omega) - g(T_x\omega), \quad g \in L_B^\alpha. \tag{7.6}$$

Proposition 3.7 and Theorem 7.4 imply the following statement.

Theorem 7.7. *If $\alpha > 1$ or $\alpha = 1$ and $m(\Omega) = \infty$, then*

$$L_B^\alpha = I_B^\alpha \oplus D_B^\alpha; \tag{7.7}$$

in other words, any $f \in L_B^\alpha$ can be represented in the form

$$f = \mathbf{M}(f) + f_1 \qquad \text{where} \qquad f_1 \in D_B^\alpha. \tag{7.8}$$

7.3. Strict-sense homogeneous random fields. The results of the previous subsection can be rephrased in probabilistic language in the following way.

Theorem 7.8. *If $\xi(x)$, $x \in X$, is a strict-sense r-homogeneous B-valued random field on a semigroup X and $\mathbf{E} \parallel \xi(e) \parallel_B^\alpha < \infty$ ($\alpha \geq 1$), then the L_B^α-valued function $\xi(\cdot)$ is l-averageable. For any $x \in X$ $\mathbf{M}^{(l)}(\xi(\cdot)) = \mathbf{E}(\xi(x)|\mathfrak{I}_\xi)$ with probability 1 (the conditional expectation with respect to the σ-algebra of the invariant sets $\mathfrak{I}_\xi$); in particular, if ξ is ergodic, then $\mathbf{M}^{(l)}(\xi(\cdot)) = \mathbf{E}\xi(x)$ a.s., $x \in X$.*

BIBLIOGRAPHICAL NOTES

1. The notion of an "ergodic element", which coincides with the notion of an averageable element (with respect to some representation of a semigroup) was introduced in Alaoglu, Birkhoff [1] (1939), [2] (1940); notions of an "ergodic function", which are close to the notion of an averageable function were introduced in Godement [1] (1948) and Maak [1] (1950), [2] (1954) (see also Jacobs [1] (1960)).

2. The averageability of almost periodic functions on groups was proved in von Neumann [2] (1934). The averageability of matrix coefficients of unitary representations of groups (including positively definite functions) was proved in Godement [1] (1948); Theorems 4.6 and 5.4 were proved in Ryll-Nardzewski [1] (1962), [2] (1966). Theorem 5.5 was proved in Eberlein [1] (1949) for commutative semigroups and in Rosen [1] (1954), who extended it to general amenable semigroups.

3. The averageability of elements of a Hilbert space with respect to a contractive representation of a semigroup was proved in G.Birkhoff [1], [2] (1939) where the "least-norm" approach was used in the proof of Theorem 5.7. Lemmas 4.2 and 5.7 as well as Example 5.3 are contained in Alaoglu and Birkhoff [2] (1940).

4. Lemma 4.3 and Theorems 4.6, 5.7 and 5.9 are proved in Ryll-Nardzewski [1] (1962), [2] (1966) where a different method of proof of the fixed point Theorem 4.6 was used; the idea of the proof presented in § 4 is also due to Ryll-Nardzewski [2] (1966); a geometrical proof of this theorem was given in Namioka, Asplund [1] (1967) (see also Greenleaf [1] (1969)).

In the case when B is a Hilbert space, Theorem 5.9 was proved earlier in Jacobs [2] (1954).

Theorems 4.7 and 5.8 were proved in Day [4] (1961) and Day [3] (1957), respectively. It should be noted that in Ryll-Nardzewski [2] and Day [4] more general versions of Theorems 5.9 and 5.8, related to representations in locally convex linear topological spaces, are proved.

5. Decompositions of spaces of averageable elements of the form $B = I \oplus D$ (see, e.g., Theorem 5.12 and Proposition 3.7) were studied in Riesz [1] (1938), Riesz – Sz.-Nagy [1] (1943), Day [1] (1942), [2] (1950), Jacobs [1] (1960) and by other authors.

Corollary 5.17 on the mutual annihilation of the spaces of D^α and I^β, $\alpha^{-1} + \beta^{-1} = 1$, is contained in Theorem 1.2.3 in Jacobs [1] (1960); other results of Subsect. 5.6 are contained in Tempelman [14] (1975).

6. Theorems 6.1 and 6.2 were proved in Godement [1] (1948).

7. § 7 is contained in Tempelman [14] (1975); in the case when X is a group and $\alpha = 1$, Theorem 7.4 was proved in Aribaud [1] (1970).

CHAPTER 2
ERGODICITY AND MIXING

§1. Main Definitions

In the sequel X is a semigroup, $T = \{T_x,\ x \in X\}$ is a left dynamical system in a phase space $(\Omega,\ \mathcal{F},\ m)$, U will denote the isometric (right) representation of X in L^2 $(\Omega,\ \mathcal{F},\ m)$ associated with T. $\mathfrak{I}$ is the σ–subalgebra of all T–invariant sets $\Lambda \in \mathcal{F}$, and I_B^α is the subspace of all T–invariant functions $f \in L_B^\alpha$, B being an arbitrary Banach space. If $f \in L_B^\alpha(\Omega,\ \mathcal{F},\ m)$, $\mathbf{M}(f) = \mathbf{M}(f|T)$ is the mean of the function f with respect to T (see Subsect. 1.7.2). Let P be a probability measure on $\mathcal{F}$: $\mathbf{E}(f) = \int_\Omega f\,dP$, $f \in L^1(\Omega,\ \mathcal{F},\ m)$.

Along with T we consider the associated isometric (unitary if X is a group) representation U of X in $L^2(\Omega, \mathcal{F}, m)$ and the matrix coefficients $R_{f,g}(x) := \mathbf{E}[f(T_x\omega)g(\omega)] = \langle Uf, g\rangle_{L^2}$, $f, g \in L^2(\Omega, \mathcal{F}, m)$.

In this section we shall define some important properties of dynamical systems and homogeneous random fields which are closely connected with the mean–value problem studied in the previous chapter and with the ergodic theorems which will be studied below. Some of these properties, namely ergodicity and weak and strong mixing, are well–known in the case $X = \mathbb{R}^m$ or $X = \mathbb{Z}^m$, $m \geq 1$ (see, e.g., Cornfeld, Sinai and Fomin [1]).

1.1. Ergodicity. Corollary 7.6 above states that if the σ–algebra $\mathfrak{I}$ is trivial, then for any $f \in L_B^\alpha$ the temporal mean $\mathbf{M}(f|T)$ coincides with the phase mean $E(f)$. In this subsection we shall study this important property more explicitly; the following simple auxiliary statement complements and makes more precise the statement of Corollary 1.7.6.

Lemma 1.1. *The following properties of T are equivalent:*

(E_1) *$\mathfrak{I}$ is trivial, i.e. either $m(\Lambda) = 0$ or $m(\Omega \backslash \Lambda) = 0$ for any*

$\Lambda \in \mathfrak{I}$;

(E_2) *any* $\mathfrak{I}$*-measurable function* $f(\omega)$, $\omega \in \Omega$ *is constant* m*–a.e.;*

(E_3) $I_B^\alpha = B \cap L_B^\alpha$ *(the subspace of all constant functions in* L_B^α*) for some* B *and* α, $1 \le \alpha < \infty$;

(E_4) $I_B^\alpha = B \cap L_B^\alpha$ *for any* α, $1 \le \alpha < \infty$, *and* B;

(E_5) $\mathbf{M}(f|T)$ *is a.e. constant,* $f \in L_B^\alpha$, *for some* B, *and* α, $1 \le \alpha < < \infty$, *or* $\alpha = 1$ *if* $M(\Omega) < \infty$;

(E_6) $\mathbf{M}(f|T)$ *is a.e. constant for any* B *and any* $f \in L_B^\alpha$, $1 \le \alpha < < \infty$;

$$(E_7) \qquad \mathbf{M}(f|T) \overset{\text{a.e.}}{=} \begin{cases} 0 & \text{if} \quad m(\Omega) = \infty, \\ \frac{1}{m(\Omega)} \int\limits_\Omega f\, dm & \text{if} \quad m(\Omega) < \infty, \end{cases}$$

for any $f \in L_B^\alpha$, $\alpha \ge 1$.

If $m = P$, *a probabilty measure, then Properties* (E_1)–(E_7) *are equivalent to each of the following properties:*

(E_8) $\mathbf{M}(f|T) = \mathbf{E}(f)$, $f \in L^1$;

(E_9) $\mathbf{E}(f|\mathfrak{I}) = \mathbf{E}f$, $P(\Lambda|\mathfrak{I}) = P(\Lambda)$ *a.s.,* $\Lambda \in \mathcal{F}$, $f \in L_B^1$;

(E_{10}) $\mathbf{M}[P(T_x\Lambda_1 \cap \Lambda_2)] = P(\Lambda_1)P(\Lambda_2)$, Λ_1, $\Lambda_2 \in \mathcal{F}$;

(E_{11}) $\mathbf{M}(R_{f,g}(\cdot)) = \mathbf{E}(f)\mathbf{E}(g)$, $f, g \in L^2$.

Proof. It is obvious that $(E_1) \Rightarrow (E_2) \Rightarrow (E_4) \Rightarrow (E_3) \Rightarrow (E_1)$. By Theorem 1.7.5, $(E_4) \Rightarrow (E_6) \Rightarrow (E_7) \Rightarrow (E_6) \Rightarrow (E_5) \Rightarrow (E_3)$; if $m = P$, then $(E_8) \Leftrightarrow (E_7)$, $(E_9) \Leftrightarrow (E_8)$. By Godement's formula (Theorem 1.5.1), $\mathbf{M}[R_{f,g}(x)] = \mathbf{M}\langle fU_x, g\rangle I_{L^2} = \langle \Pi_{I^2}^{L^2} f, g\rangle = = \langle \Pi_{I^2}^{L^2} f,\ \Pi_{I^2}^{L^2} g\rangle_{L^2} = \langle \mathbf{M}(f|T), \mathbf{M}(g|T)\rangle_{L^2}$ $f, g \in L^2$, and thus $(E_8) \Rightarrow (E_{11})$. It is obvious that $(E_{11}) \Rightarrow (E_{10})$. ($E_{10}$) with $\Lambda_1 = = \Lambda_2 = \Lambda \in I$ implies that $P(\Lambda) = [P(\Lambda)]^2$, and hence $P(\Lambda) = 0$ or 1, i.e. $(E_{10}) \Rightarrow (E_1)$.

□

Definition 1.1. A dynamical system T is said to be *ergodic, or metrically transitive*, if it has the equivalent properties (E_1)–(E_{11}) stated in Lemma 1.1.

1.2. Strong mixing and strong quasi–mixing. Let $m = P$ be a probability measure, and X a non–compact, locally compact topological group. We suppose that T is completely measurable.

Along with the usual limit $\lim_{x\to\infty}\varphi(x)$ and the "Cesaro limit" $\mathbf{M}(\varphi)$ we consider the "strong Cesaro limit" $(sC)\lim\varphi(x)$ defined as follows: $a := (sC)\lim\varphi(x)$ if $|\varphi(\cdot) - a| \in A(X)$ and $\mathbf{M}|\varphi(\cdot) - a| = 0$. Theorem 1.5.19 shows that if the ordinary limit $\lim_{x\to\infty}\varphi(x) = a$, then $(sC)\lim\varphi(x) = a$ too. Certainly, if $(sC)\lim\varphi(x) = a$ then $a = \mathbf{M}(\varphi)$.

Consider the U–matrix coefficient $R_{f,g}(x) := \mathbf{E}[f(T_x\omega)\cdot g(\omega)] = \langle fU_x, g\rangle_{L^2}$. Since $R_{f,g} \in W \subset A(X)$, we can study the existence and properties of $(sC)\lim R_{f,g}(x)$ as well as of $\lim_{x\to\infty} R_{f,g}(x)$.

The following statement will be useful in both cases.

Let $F(f,g) := \mathbf{M}(|R_{f,g}(\cdot) - \mathbf{E}[\mathbf{E}(f|\mathfrak{I})\mathbf{E}(g|\mathfrak{I})]$, $f, g \in L^2$.

Lemma 1.2. *The mapping $(f, g) \mapsto F(f, g)$ is a continuous mapping from $L^2 \times L^2$ onto R_+.*

Proof. It is easy to check that

$$|F(\varphi_1,\psi_1) - F(\varphi_2,\psi_2)| \le F(\varphi_1 - \varphi_2, \psi_1) + F(\varphi_2, \psi_1 - \psi_2),$$

and the Cauchy inequality implies that $|F(\varphi,\psi)| \le 2\|\varphi\|_{L^2}\|\psi\|_{L^2}$. Our statement follows immediately from these two inequalities.

□

Lemma 1.3. *The following properties of T are equivalent:*

(SM_1) $\lim\limits_{x\to\infty} P(T_x\Lambda_1 \cap \Lambda_2)$ *exists,* $\Lambda_1,\ \Lambda_2 \in \mathcal{F}$;

(SM_2) $\lim\limits_{x\to\infty} P(T_x\Lambda_1 \cap \Lambda_2) = \mathbf{E}[P(\Lambda_1|\mathfrak{I})P(\Lambda_2|\mathfrak{I})]$, $\Lambda_1,\ \Lambda_2 \in \mathcal{F}$;

(SM_3) $\lim\limits_{x\to\infty} \int\limits_{\Omega} f(T_x\omega)g(\omega)P(d\omega)$ *exists,* $f, g \in L^2\ (\Omega,\mathcal{F},P)$;

(SM_4) $\lim\limits_{x\to\infty} \int\limits_{\Omega} f(T_x\omega)g(\omega)P(d\omega) = \mathbf{E}[\mathbf{E}(f|\mathfrak{I})\mathbf{E}(g|\mathfrak{I})]$,
$f, g \in L^2\ \Omega,\mathcal{F},\mathcal{P})$;

(SM_5) $\lim\limits_{x\to\infty} \int_M P(T_x\Lambda_1 \cap \Lambda_2|\mathfrak{I})dP = \int\limits_M P(\Lambda_1|\mathfrak{I})P(\Lambda_2|\mathfrak{I})\,dP$, Λ_1, $\Lambda_2 \in \mathcal{F}$, $M \in \mathfrak{I}$;

(SM_6) $\lim\limits_{x\to\infty} P(T_x\Lambda_1\cap\Lambda_2|\mathfrak{I}) = P(\Lambda_1|\mathfrak{I})P(\Lambda_2|\mathfrak{I})$ *in the weak topology of* $L^1(\Omega,\ \mathfrak{I},\ P)$;

(SM_7) *The weak limit* $(w)\ \lim_{x\to\infty} U_x(=\Pi_{I^2}^{L^2})$ *exists.*

If the σ*-algebra* $\mathfrak{I}$ *is generated by a finite or countable partitioning,* $\Omega = \bigcup_{i=1}^{n} M_i$, $n \in \{1,2,\ldots,\infty\}$ *and the restriction of* T *to* M_i *is ergodic, then Properties* (SM_1)*-*(SM_7) *are equivalent to the following property:*

(SM_8) $\lim_{x\to\infty} P(T_x\Lambda_1 \cap \Lambda_2|M_i) = P(\Lambda_i|M_i)P(\Lambda_2|M_i)$, $i=\overline{1,n}$, Λ_1, $\Lambda_2 \in \mathcal{F}$; *this means that,in the terms of Definition 1.3 below,* T *is strongly mixing on each* T*-ergodic component of* Ω.

Proof. The implications $(SM_2)\Rightarrow(SM_1)$ and $(SM_4)\Rightarrow(SM_3)$ are trivial. Property (SM_1) is a special case of Property (SM_3) with $f=\mathcal{X}_{\Lambda_1}$, $g=\mathcal{X}_{\Lambda_2}$ and Property (SM_2) is a special case of Property (SM_5) when $M=\Omega$. Since the family of functions $\mathcal{X}_\Lambda$, $\Lambda\in\mathcal{F}$, generates the space $L^2(\Omega,\ \mathcal{F},\ P)$, it follows from Lemma 1.2 that $(SM_2)\Rightarrow(SM_4)$. By virtue of Theorems 1.5.20 and 1.7.5,

$$\begin{aligned}\lim_{x\to\infty}\int f(T_x\omega)g(\omega)P(d\omega) &= \lim_{x\to\infty}\langle fU_x,g\rangle_{L_2} = \mathbf{M}\langle fU_x,g\rangle_{L_2} = \\ &=\langle\Pi_{I^2}^{L^2}f,g\rangle_{L_2} = \langle(\Pi_{I^2}^{L^2})^2 f,g\rangle = \langle\Pi_{I^2}^{L^2}f,\Pi_{I^2}^{L^2}g\rangle = \\ &=\mathbf{E}\big[\mathbf{E}(f|\mathfrak{I})\mathbf{E}(g|\mathfrak{I})\big],\quad f,g\in L^2;\end{aligned}$$

hence $(SM_3)\Leftrightarrow(SM_4)$ and therefore $(SM_1)\Rightarrow(SM_2)$. Obviously, $(SM_3)\Leftrightarrow(SM_7)$. The implication $(SM_2)\Rightarrow(SM_5)$ follows from the equalities:

$$\begin{aligned}\int_M P(T_x\Lambda_1\cap\Lambda_2|\mathfrak{I})dP &= P(T_x\Lambda_1\cap\Lambda_2\cap M) = \\ &=P(T_x(\Lambda_1\cap M)\cap(\Lambda_2\cap M))\end{aligned}$$

and

$$\mathbf{E}[P(\Lambda_1\cap M|\mathfrak{I})P(\Lambda_2\cap M|\mathfrak{I})] = \int_M P(\Lambda_1|\mathfrak{I})P(\Lambda_2|\mathfrak{I})dP,$$

where Λ_1, $\Lambda_2 \in \mathcal{F}$, $M \in \mathfrak{J}$. It follows from a well-known weak convergence criterion (cf., e.g., Neveu [1], Corollary 2 of Proposition 4.2.2) that $(SM_5)\Leftrightarrow(SM_6)$. And, finally, it is evident that under the stated condition (SM_8) is equivalent to (SM_6).

□

Definition 1.2. A dynamical system T is said to be *strongly quasi-mixing*, or to *possess property* (SM), if it possesses the equivalent properties (SM_1)–(SM_7).

Lemma 1.4. *The following properties of* T *are equivalent:*

(SM_1) $\lim_{x\to\infty} P(T_x\Lambda_1 \cap \Lambda_2) = P(\Lambda_1)P(\Lambda_2)$, $\Lambda_1, \Lambda_2 \in \mathcal{F}$;

(SM_2) $\lim_{x\to\infty} \int f(T_x\omega)g(\omega)P(d\omega) = \int f\,dP \int g\,dP$, $f, g \in L^2$;

(SM_3) T *is strongly quasi-mixing and ergodic.*

Proof. The implications $(SM_1)\Rightarrow (SM_2)\Rightarrow (SM_1)$ are obvious. It is also obvious that $(SM_1)\Rightarrow (E_{10})$ and $(SM_1) + (E_9)\Rightarrow (SM_2)$.

□

Definition 1.3. If a dynamical system T possesses the (equivalent to each other) properties (SM_1)–(SM_3), it is said to be *strongly mixing*, or *to possess property (SM).*

1.3. Weak mixing and weak quasi-mixing. If in Properties (SM_1)–(SM_5), (SM_8), (SM_1), (SM_2) we replace the ordinary limit "lim" by the strong Cesaro limit "(sC) lim" we obtain new, weaker versions of these properties, which we shall denote, respectively, by (WM_1)–(WM_5), (WM_8), (WM_1) and (WM_2).

Lemma 1.5. *Properties* (WM_1)*—*(WM_5) *are pairwise equivalent.*

This can be proved similarly as the pairwise equivalence of (SM_1)–(SM_5) in Lemma 1.4.

Let us consider the probability space $(\widetilde{\Omega}, \widetilde{\mathcal{F}}, \widetilde{P}) = (\Omega \times \Omega, \mathcal{F} \times \mathcal{F}, P\times P)$ and the dynamical system $\widetilde{T} = \{\widetilde{T}_x, x \in X\}$ where $\widetilde{T}_x(\omega_1, \omega_2) = (T_x\omega_1, T_x\omega_2)$, $(\omega_1, \omega_2) \in \Omega$.

Definition 1.4. A dynamical system T with Properties (WM_1)–(WM_2) (resp. (WM_1)–(WM_2)), is said to be *weakly mixing* (resp. *weakly quasi–mixing*) or *to possess Property* (WM) (resp. *Property* (WM)).

Theorem 1.6. **(Dye–Bergelson–Rosenblatt)** *Each of the properties* (WM_1) *and* (WM_2) *of* T *is equivalent to any of the following properties:*

(WM) T *is weakly mixing;*
(WM_3) T *is weakly quasi–mixing and ergodic;*
(WM_4) T *has FD–continuous spectrum;*
(WM_5) $\widetilde{T}$ *is ergodic.*

Proof. The pairwise equivalence of Properties (WM_1), (WM_2) and (WM_3) can be proved similar to the pairwise equivalence of (SM_1), (SM_2) and (SM_3) in Lemma 1.4. Recall that $\mathbf{M} \mid R_{f,g} - a \mid = 0$ is equivalent to $\mathbf{M} \mid R_{f,g} - a \mid^2 = 0$ (see App., (6.M)). This implies that $(WM_2)\Leftrightarrow(WM_4)$. Let

$$
\begin{aligned}
&f(\widetilde{\omega}) = f(\omega_1)\overline{f(\omega_2)}, \quad \widetilde{g}(\widetilde{\omega}) = g(\omega_1)\overline{g(\omega_2)},\\
&\text{if} \quad \widetilde{\omega} = (\omega_1, \omega_2), \quad f, g \in L^2(\Omega, \mathcal{F}, m).
\end{aligned}
\tag{1.1}
$$

It is easy to check that the equality $\mathbf{M}|R_{f,g} - \int f\,dP \cdot \int g\,dP|^2 = 0$ is equivalent to the equality $\mathbf{M}(\int \widetilde{f}(\widetilde{T}_x\widetilde{\omega})\widetilde{g}(\widetilde{\omega})\widetilde{P}(d\widetilde{\omega})) = \int \widetilde{f}\,dP \int \widetilde{g}\,d\widetilde{P}$. Since the functions $\widetilde{f}$ and $\widetilde{g}$ of the form (1.1) generate the space $L^2(\widetilde{\Omega}, \widetilde{\mathcal{F}}, \widetilde{P})$, we see that all Properties (WM_1)–(WM_4) are equivalent to (E_{11}) for $\widetilde{T}$ and hence to Property (WM_5).

□

It is easy to prove the following statement.

Proposition 1.7. *The properties* (E), (SM), (SM), (WM) *and* (WM) *are invariant with respect to the equivalence relation of dynamical systems.*

1.4. Relative mixing and quasi–mixing. Consider a homogeneous space $Y = G/K$ where G is a non–compact locally compact topological group and K is a compact subgroup of G. Let $T = \{T_g,\ g \in G\}$ be a dynamical system in $(\Omega, \mathcal{F}, P)$. Denote by $\mathfrak{J}_K$ the σ–subalgebra of all K–invariant subsets in $\mathcal{F}$: $\mathfrak{J}_K = \{\Lambda : \Lambda \in \mathcal{F},\ T_k\Lambda = \Lambda\ (P-\text{mod } 0),\ k \in K\}$. Let $L^2\ (\Omega,\ \mathfrak{J}_K,\ P)$ denote the subspace of all K-invariant (i.e. $\mathfrak{J}_K$–measurable) functions in $L^2\ (\Omega,\ \mathcal{F},\ P)$. Let U: $g \mapsto U_g$ be the representation associated with T. We denote by $H^{(K)}$ the least U–invariant subspace of $L^2\ (\Omega,\ \mathcal{F},\ P)$ containing $L^2\ (\Omega,\ \mathfrak{J}_K,\ P)$, and let $U^{(K)}$ be its restriction to $H^{(K)}$. It is obvious that U is a K–class *1* representation.

Definition 1.5. We say that a dynamical system T is *G/K–relatively strongly* (or *weakly*) *(quasi)–mixing* if the corresponding (quasi)-mixing property introduced above (see Definitions 1.2–1.5) holds for sets $\Lambda_1,\ \Lambda_2 \in \mathfrak{J}_K$ and functions $f \in L^2\ (\Omega,\ \mathfrak{J}_K,\ P)$.

Note 1.1. If $K = \{e\}$, $Y = G$, any G/K–relative (quasi)–mixing property coincides with the corresponding "absolute" property.

Note 1.2. G/K–relative strong quasi–mixing of T is equivalent to any of the following properties:

a) (SM_3) and (SM_4) hold for all functions $f, g \in H^{(K)}$;

b) (SM_3) and (SM_4) hold for all functions $f \in L^2\ (\Omega,\ \mathfrak{J}_K,\ P)$ and $g \in L^2\ (\Omega,\ \mathcal{F},\ P)$;

c) $(W)\ \lim_{g\to\infty} U_g$ exists (and is equal to $\Pi^{H^{(K)}}_{L^2(\Omega,\mathfrak{J},P)}$);

G/K–relative strong mixing is equivalent to any of the following properties:

a′) (SM_2) holds for every $f, g \in H^{(K)}$;

b′) (SM_2) holds for all $f \in L^2\ (\Omega,\ \mathfrak{J}_K,\ P)$ and $g \in L^2\ (\Omega,\ \mathcal{F},\ P)$;

c′) $(W)\ \lim_{g\to\infty} U_g = 0$ on $H^{(K)} \ominus E$ where E is the subspace of constants;

d′) T is G/K–relatively strongly quasi–mixing and ergodic.

Note 1.3. Suppose that K is a normal subgroup of G. Then the σ–algebra $\mathfrak{J}_K$ is T–invariant: if $\Lambda \in \mathfrak{J}_K$, then $T_g\Lambda \in \mathfrak{J}_K$, $g \in G$. Indeed, let $k \in K$, $k_1 := g^{-1}kg$, $\Lambda \in \mathfrak{J}_K$; then $T_kT_g\Lambda = T_{kg}\Lambda = T_{gk_1}\Lambda =$

$= T_g T_{k_1} \Lambda = T_g \Lambda$. Under some restrictions on the space $(\Omega, \mathcal{F}, P)$ and the dynamical system T, there exists in $\mathfrak{I}_K$ a system Ω_K of disjoint sets such that:

a) any $\Lambda \in \mathfrak{I}_K$ is a union of sets belonging to Ω_K;

b) if $M \in \Omega_K$, then $T_g M \in \Omega_K$

(for example, if $K = \{k_1, \ldots, k_n\}$, Ω is the set of all functions, constant on cosets of K, and the T_g are translations in Ω: $(T_x\omega)(y) = \omega(yx)$, then the sets $\Lambda_{a_1,\ldots,a_n} = \{\omega : \omega(k_1) = a_1, \ldots, \omega(k_n) = a_n\}$, $a_i \in \in R$, form this partition of Ω). $\mathfrak{I}_K$ can be considered as a σ–algebra $\mathcal{F}_K$ of sets in Ω_K. Consider the restriction P_K of P to $\mathcal{F}_K$. Then $(\Omega_K, \mathcal{F}_K, P_K)$ is a probability space and the transformations $T_g^{(K)}$: $T_g^{(K)}(M) = \bigcup_{\omega \in M} T_g\omega$, $M \in \Omega_K$, form a dynamical system in this space. It is obvious that $T_{g_1}^{(K)} = T_{g_2}^{(K)}$ if $g_1, g_2^{-1} \in K$. Therefore we can consider $g \mapsto T_g$ as an action of the factor–group G/K; this is the *factor system* of the dynamical system T with repsect to K. It is obvious that G/K–relative strong or weak (quasi)–mixing of T is equivalent to the "absolute" (quasi)–mixing of $T^{(K)}$ in the corresponding sense.

We give an example of a dynamical system which is G/K–relatively strongly mixing but is not "absolutely" weakly–quasimixing.

Example 1.1. Let $G = \mathbb{Z} \times Z_2$ where $Z_2 = \{0, 1\}$ with addition mod 2. We consider in Z_2 the algebra $\mathcal{F}' := \{\{0\}, \{1\}, \{1, 2\}\}$ and the measure $P' = \frac{1}{2}\delta_0' + \frac{1}{2}\delta_1$ on $\mathcal{F}'$ where δ_i is the probability measure concentrated at the point $i \in Z_2$. Let $(\widetilde{\Lambda}, \widetilde{\mathcal{F}}, \widetilde{P}) = (\Omega, \mathcal{F}, P) \times (Z_2, \mathcal{F}', P')$. Consider a measure preserving invertible transformation T in $(\Omega, \mathcal{F}, P)$ such that $T_n = T^n$, $n \in \mathbb{Z}$, is a strongly mixing dynamical system, and now consider the following dynamical system $\widetilde{T}_g$, $g \in G$, in $(\widetilde{\Omega}, \widetilde{\mathcal{F}}, \widetilde{P})$: $T_g(\omega, i) = (T^n\omega, i + \alpha(\text{mod } 2))$ if $g = (n, \alpha, n \in \mathbb{Z}$, $\alpha \in Z_2$, $(\omega, i) \in \widetilde{\Omega}$. We have $\mathfrak{I}_{Z_2} = \{\widetilde{\Lambda} : \widetilde{\Lambda} \in \widetilde{\mathcal{F}}, \widetilde{\Lambda} = \Lambda \times Z_2, \Lambda \in \mathcal{F}\}$. It is obvious that $\widetilde{T}$ is G/Z_2–relatively strongly mixing. But $\widetilde{T}$ has not even the weak quasi–mixing property: $\widetilde{T}_g(\Omega \times \{0\} = \Omega \times \{0\}$ if $g = (n, 0)$ and $\widetilde{T}_g(\Omega \times \{0\}) = \Omega \times \{1\}$ if $g = (n, 1)$; thus $\widetilde{P}(T_g(\Omega \times \{0\}) \cap (\Omega \times \{0\})) = 1$ if $g = (n, 0)$ and $= 0$ if $g = (n, 1)$; hence the strong Cesaro limit (sC) $\lim P(T_g(\Omega \times \{0\}) \cap (\Omega \times \{0\}))$ does not exist.

1.5. Ergodicity and mixing of homogeneous random functions. Let Y be a G–space, $(\Omega, \mathcal{F}, P)$ a probability space and $\xi(x)$ a homogeneous random function on Y with values in a Hilbert space $\mathcal{H}$.

Definition 1.6. $\xi(\cdot)$ is said to be *ergodic* , or *metric transitive* (briefly, *to have Property (E)*), if it is covariant with respect to some dynamical system with this property.

Definition 1.7. Let G be a non–compact locally compact, topological group; $\xi(\cdot)$ is said to be *(G/K–relatively) weakly mixing, weakly quasi–mixing, strongly mixing* , or *strongly quasi–mixing* if it is covariant with respect to a dynamical system with the corresponding (G/K–relative) property; we shall denote these properties by (G/K–) WM, WM, SM and SM, respectively. We shall also say these are (M)–properties.

Proposition 1.8. *If $\xi(\cdot)$ has one of the Properties (E)–(W), then any image of it also has this property.*

The proof is trivial.

1.6. Mixing and quasi–mixing in the wide sense. The notions of (E) and (M) properties of homogeneous random functions introduced in subsection 1.4 were connected with the action of the translation (or some other) dynamical system T in $(\Omega, \mathcal{F}, \mathcal{P})$ or with the action of the unitary representation associated with T in the space $L^2(\Omega, \mathcal{F}, \mathcal{P})$. Now we shall consider the similar properties of a wide–sense homogeneous random function $\xi(\cdot)$ which are connected with the action of the translation unitary representation U in the subspace H_ξ of $L^2_{\mathcal{H}}(\Omega, \mathcal{F}, P)$ generated by the random variables $\xi(t)$, $t \in T$ (see App., § 8). Instead of the subspace I^2 of L^2 we shall consider the subspace $I_\xi = I^2_{\mathcal{H}} \cap H_\xi \subset H_\xi$, and instead of the conditional expectation $\mathbf{E}(f|\mathfrak{I}) = \Pi^{L^2}_{I^2} f$, $f \in L^2$, the projection $\widehat{\mathbf{E}}(\gamma|I_\xi) = \Pi^{H_\xi}_{I_\xi}\gamma$, $\gamma \in H_\xi$. Recall that $\widehat{\mathbf{E}}[\xi(t)|I_\xi] = \mathbf{M}^{(r)}(\xi(\cdot))$, $t \in T$, and the random function $\eta(t) = \mathbf{M}^{(r)}\xi(t) - \mathbf{E}\xi(t)$, $t \in T$, is the background of $\xi(\cdot)$. Let $m(t) := \overline{E\xi(t)}$ $R(s,t) := \mathbf{E}\langle\xi(\cdot), \xi(t)\rangle$, $R_0(s,t) := \mathbf{E}\langle(\xi(s) - -m(s)), (\xi(t) - m(t)\rangle = R(s,t) - \langle m(s), m(t)\rangle$. If $T = G$ is a group then $R(s,t) = R(t^{-1}s)$, $R_0(s,t) = R_0(t^{-1}s)$.

Now we are able to define "H_ξ–versions" of the (E)–(W) properties. First, one can define "H_ξ*–ergodic*" random functions: $\mathfrak{I}_\xi$ is trivial, i.e. $\mathfrak{I}_\xi = H_\xi \cap \mathcal{H}$; the following proposition states other equivalent properties.

Proposition 1.9. *The following properties of a wide–sense r–homogeneous field $\xi(\cdot)$ are equivalent:*

(i) I_ξ is trivial;

(ii) $\mathbf{M}^{(r)}(\xi)$ is a non–random element of H_ξ;

(iii) $\xi(\cdot)$ is background–free.

If $Y \equiv G$ is a group, the properties (i)–(iii) are equivalent to each of the properties:

(iv) $\mathbf{M}^{(r)}(\xi) = \mathbf{E}\xi(e)$;

(v) $\mathbf{M}(R) = \|\mathbf{E}\xi(e)\|^2$;

(vi) $\mathbf{M}(R_0) = 0$.

The proof is elementary.

Now we shall define H_ξ–versions of mixing and quasi–mixing.

Definition 1.8. A wide–sense homogeneous random function $\xi(\cdot)$ is said to be *strongly quasi–mixing in the wide sense* if it has the following equivalent properties:

(i) there exists $R(\infty s,t) := \lim\limits_{x\to\infty} R(xs,t) = \mathbf{E}\langle\widehat{\mathbf{E}}\xi(s|I_\xi),\ \widehat{\mathbf{E}}\xi(t|I_\xi)\rangle$, $s,t \in T$;

(ii) there exists $R_0(\infty s,t) := \lim_{x\to\infty} R(xs,t) = \mathbf{E}\langle\widehat{\mathbf{E}}(\xi(s)|I_\xi), \widehat{\mathbf{E}}(\xi(t)|I)\rangle - \langle m(s), m(t)\rangle$;

(iii) there exists $\lim_{x\to\infty} \mathbf{E}\langle U_x\gamma, \zeta\rangle = \mathbf{E}\langle\widehat{\mathbf{E}}(\gamma|\mathfrak{I}_\xi)\widehat{\mathbf{E}}(\zeta|I_\xi)\rangle$, γ, $\xi \in H_\xi$;

Definition 1.9. A wide–sense homogeneous random function $\xi(\cdot)$ is said to be *strongly mixing in the wide sense* if it has the following equivalent properties:

(i) $R(\infty\, s,t) = m(s)m(t)$;

(ii) $R_0(\infty\, s,t) = 0$;

(iii) $\lim_{x\to\infty} \mathbf{E}\langle U_x\gamma, \zeta\rangle = \langle\mathbf{E}(\gamma), \mathbf{E}(\zeta)\rangle$, γ, $\zeta \in H_\xi$;

(iv) $\xi(\cdot)$ is strongly quasi–mixing in the wide sense and I_ξ is trivial.

The reader can easily state the definition of the property of "*weakly quasi–mixing in the wide sense*" by substituting "(sC) lim" for "lim". We shall also consider the following stronger property of "weakly mixing in the wide sense".

Definition 1.10. A wide–sense homogeneous random function $\xi(\cdot)$ is said to be *weakly mixing in the wide sense* if it has the following pairwise equivalent properties:

(i) $(sC)\ \lim_x R(xs,t) = m(s)m(t)$;

(ii) $(sC)\ \lim_x R_0(xs,t) = 0$;

(iii) $(sC)\ \lim_x \mathbf{E}\langle U_x\gamma,\zeta\rangle_{\mathcal{H}} = \langle \mathbf{E}(\gamma),\mathbf{E}(\zeta)\rangle$, γ, $\zeta \in H_\xi$;

(iv) the random function $\xi_0(t) := \xi(t) - m(t)$ has FD–continuous spectrum and I_ξ is trivial.

The proof of the equivalence of the properties stated in each of Definitions 1.8–1.10 is elementary, as are the proofs of the following three statements.

Proposition 1.10. *A wide–sense homogeneous random function is strongly (resp. weakly) quasi–mixing in the wide sense if its background–free component is strongly (resp. weakly) mixing in the wide sense.*

Proposition 1.11. *If a homogeneous random function $\xi(\cdot)$ with $E\|\xi(t)\|^2 < \infty$ possesses an (E)–(M) property in the sense of Subsection 1.4, then it possesses the appropriate property in the wide sense too.*

Proposition 1.12. *If $\xi(\cdot)$ has one of properties (E)–(W) in the wide–sense, then any its image has this property, too.*

The real meaning of the term "in the wide sense" introduced in Definitions 1.8–1.10 will be cleared up in the next section: in the Gaussian case the mixing properties in the wide sense coincide with the appropriate "strict" properties.

§ 2. Ergodicity and Mixing of Gaussian Homogeneous Random Functions

In this section we shall establish criteria of ergodicity and mixing of Gaussian homogeneous random functions which generalize the well-known criteria due to Grenander, Maruyama and Fomin concerning Gaussian stationary random processes and sequences. Let G be a group and let T be a G–space.

Along with a random function $\xi(y)$ we consider the "centered" function $\xi_0(y) := \xi(y) - \mathbf{E}\xi(y)$.

Theorem 2.1. *If $\xi(y)$, $y \in Y$, is a Gaussian homogeneous random function, the following properties are equivalent:*
(i) $\xi(\cdot)$ is ergodic;
(ii) $\xi(\cdot)$ is weakly mixing in the wide sense;
(iii) $\xi(\cdot)$ is weakly mixing;
(iv) $\xi_0(\cdot)$ has FD-continuous spectrum.

Proof. A complex random function $\xi(y) = \xi_1(y) + i\xi_2(y)$ is Gaussian, homogeneous or possesses one of the properties *(i)–(iv)* iff the real-valued random function $\eta(l) = \xi_\alpha(y)$, $l \in (y, \alpha) \in Y \times \{1, 2\}$, possesses the appropriate property; the case with the pair of random functions $\xi(y)$ and $\xi_0(y) := \xi(y) - \mathbf{E}\xi(y)$ is similar. Therefore we shall restrict ourselves to the consideration of the case of real functions with $\mathbf{E}\xi(y) = 0$. By virtue of Proposition 1.8 and 1.12, we may also set $(\Omega, \mathcal{F}) = (\mathbb{R}^Y, \mathfrak{B}^Y)$, where $\mathfrak{B}$ is the σ–algebra of Borel sets in $\mathbb{R}$.

1° *(ii)* $\Rightarrow$ *(i)*. Let T_g be the translation dynamical system in the space $(\mathbb{R}^Y, \mathfrak{B}^Y, P)$. Suppose that $\Lambda \in \mathbb{R}^Y$ is a T–invariant set. Since Λ is $\mathbf{B}^Y$–measurable for any $\varepsilon > 0$, there exists a finite–dimensional cylindrical set $\mathfrak{I} = \{\omega : (\omega(y_1), \dots, \omega(y_n)) \in A\} \in \mathbb{R}^Y$, $A \in \mathfrak{B}^n$, such that $P(\mathfrak{I} \Delta \Lambda) < \varepsilon$ (see, e.g., Halmos [1], § 13) and thus

$$P(\mathfrak{I}) < P(\Lambda) + \varepsilon, \quad P(\Lambda \cap \bar{\mathfrak{I}}) < \varepsilon. \tag{2.1}$$

The random vector $\omega = (\omega(y_1), \dots, \omega(y_n))$ can, in general, be degenerate. Let E be the m–dimensional subspace of $\mathbb{R}^n$ $(0 \le m \le n)$

such that $P\{\omega \in E\} = 1$ and $P\{\omega \in K\} < 1$ for any subspace $K \subset E$. Suppose first that $m > 0$ and let Q be a non–degenerate linear operator in $\mathbb{R}^m$ which maps E onto the subspace $F := \{(x_1, \dots, x_m) : x_{m+1} = \dots = x_n = 0\} \subset \mathbb{R}^n$ and let $(\eta_1, \dots, \eta_m, 0, \dots, 0) := Q\omega$. The vector $\eta = (\eta_1, \dots, \eta_m)$ is a non–degenerate Gaussian vector. Denote $B := Q(A) \cap F = Q(A \cap E)$. Evidently, $\mathfrak{I} = \{\omega : \eta \in B\}$ and $\eta_i \in H_\xi$, $i = 1, \dots, m$.

Let U be the translation representation of G in H_ξ. Consider the vector $(\eta, U_g\eta) := (\eta_1, \dots, \eta_m, U_g\eta_1, \dots, U_g\eta_m)$. The covariance matrix of this vector

$$L_g = \begin{pmatrix} M & N_g \\ N_g^* & M \end{pmatrix},$$

where M is the covariance matrix of the vector η (and of the vectors $U_g\eta$, $g \in G$), $N_g = \|E(\eta_i \cdot U_g\eta_j\|_{i,j=1}^m$; N_g^* is the conjugate matrix. If $\xi(\cdot)$ is weakly mixing in the wide sense, we have

$$\mathbf{M}\Big[\sum_{i,j=1}^m \mid \mathbf{E}(\eta_i \cdot U_g\eta_j) \mid\Big] = \sum_{i,j=1}^m \mathbf{M}[\| \mathbf{E}(\eta_i \cdot U_g\eta_j) \|] =$$

$$= \sum_{i,j=1}^m \mathbf{M}[\| \langle U_g\eta_j, \eta_i\rangle_{H_\xi} \|] = 0.$$

This implies the existence of a sequence $\{g_k\} \subset G$ such that $\lim_{k\to\infty} \sum_{i,j=1}^m \mathbf{E}(\eta_i \cdot U_{g_{\bar{k}}}\eta_j) = 0$. Thus $\lim_{k\to\infty} N_{g_{\bar{k}}} = 0$ and

$$\lim_{k\to\infty} L_{g_k} = \begin{pmatrix} M & 0 \\ 0 & M \end{pmatrix}. \tag{2.2}$$

Put $\mathbf{x}' := (x_1, \dots, x_m)$, $\mathbf{x}'' := (x_{m+1}, \dots, x_{2m})$, $\mathbf{x} = (x_1, \dots, x_{2m})$, $\Pi_g(\mathbf{x}) := (L_g^{-1}\mathbf{x}, \mathbf{x})$. Because of (2.2), $\Pi_{g_k}(\mathbf{x}) \to \Pi(\mathbf{x}) := (M^{-1}\mathbf{x}', \mathbf{x}') + {}$ $+ (M^{-1}\mathbf{x}'', \mathbf{x}'')$ and hence, for sufficiently small $\lambda > 1$, $k > K_\lambda$ and $\mathbf{x} \neq 0$ we have $\Pi_{g_k}(\mathbf{x}) \geq \langle M^{-1}\mathbf{x}', \mathbf{x}'\rangle + \langle M^{-1}\mathbf{x}'', \mathbf{x}''\rangle - \lambda \mid \mathbf{x} \mid^2 \geq \lambda\|\mathbf{x}\|^2$, $\exp\{-\Pi_{g_k}(\mathbf{x})) \leq \exp -\lambda \mid \mathbf{x} \mid^2\} \in L^1(\mathbb{R}^{2m})$.

Consider the sets $\mathfrak{I}_g := T_g\mathfrak{I} = \{\omega : (U_g\eta_1, \dots, U_g\eta_m) \in B\}$. Using the Lebesgue dominated convergence theorem we obtain:

$$\lim_{k\to\infty} P(\mathfrak{I}\cap\mathfrak{I}_{g_k}) = \lim_{k\to\infty} P\{(\eta_1,\ldots,\eta_m)\in B,$$
$$(U_{g_k}\eta_1,\ldots,U_{g_k}\eta_m)\in B\} =$$
$$= \lim_{k\to\infty} \frac{1}{(2n)^m \mid L_{g_k}\mid^{\frac{1}{2}}} \int_{B\times B} \exp\{-(1/2)\Pi_{g_k}(x)\}\, dx =$$
$$= \frac{1}{(2n)^m \mid M\mid} \int_B \exp\{1(1/2)\langle M^{-1}x', x'\rangle\} dx' \cdot$$
$$\cdot \int_B \exp\{1(1/2)\langle M^{-1}x'', x''\rangle\} dx'' = (P(\mathfrak{I}))^2.$$

Put $\bar{\mathfrak{I}} = \Omega\backslash\mathfrak{I}$. Since Λ is T–invariant

$$P(\Lambda\cap\bar{\mathfrak{I}}_g) = P(\Lambda\cap T_g\bar{\mathfrak{I}}) = P(T_g\Lambda\cap T_g\bar{\mathfrak{I}}) =$$
$$= P(T_g(\Lambda\cap\bar{\mathfrak{I}})) = P(\Lambda\cap\bar{\mathfrak{I}}) \tag{2.3}$$

Let k be such that $\mid (P(\mathfrak{I}))^2 - P(\mathfrak{I}\cap\mathfrak{I}_{g_k})\mid < \varepsilon$. By virtue of (2.3) and (2.1),

$$(P(\Lambda)+\varepsilon)^2 > (P(\mathfrak{I}))^2 > P(\mathfrak{I}\cap\mathfrak{I}_{g_k}) - \varepsilon \geq P(\Lambda\cap\mathfrak{I}\cap\mathfrak{I}_{g_k}) - \varepsilon >$$
$$> P(\Lambda) - P(\Lambda\cap\bar{\mathfrak{I}}) - P(\Lambda\cap\bar{\mathfrak{I}}_{g_k}) - \varepsilon > P(\Lambda) - 3\varepsilon.$$

In the case when $m = 0$ we have $P\{\omega = 0\} = 1$ and, consequently, $P(\mathcal{J}) = 0$ or 1; therefore $(P(\Lambda)+\varepsilon)^2 > P(\mathfrak{I})^2 = P(\mathfrak{I}) > P(\Lambda) - \varepsilon$. Thus, for any $m \geq 0$ we have $(P(\Lambda)+\varepsilon)^2 > P(\Lambda) - 3\varepsilon$; since $\varepsilon > 0$ is chosen arbitrarily, we obtain $(P(\Lambda))^2 \geq P(\Lambda)$ and therefore $P(\Lambda) = 0$ or $P(\Lambda) = 1$. Thus we have proved that if *(ii)* is fulfilled, $\xi(\cdot)$ is ergodic.

2°. *(i)*⇒*(ii)*. Let $\xi(y)$ be ergodic. Fix some elements s, $y \in Y$ and consider the random fields

$$\eta_1(g) = \xi^2(gs) - E\xi^2(s) + \xi^2(gy) - E\xi^2(y)$$

and

$$\eta_2(g) = \xi^2(gs) - E\xi^2(s) - \xi^2(gy) + E\xi^2(y).$$

Both these fields are covariant with respect to the translation dynamical system T in $(\Omega, \mathcal{F}, m)$ and thus they are homogeneous and ergodic. Consequently, the invariant subspaces I_{η_1} and I_{η_2} in H_{η_1} and H_{η_2} are trivial. By virtue of Proposition 1.12, the covariance functions of η_1 and η_2 have the following property:

$$\mathbf{M}[R_{\eta_1}(g)] = \mathbf{M}[R_{\eta_2}(g)] = 0. \tag{2.4}$$

Now, using the well–known relation between the second– and fourth–order moments of Gaussian functions we obtain:

$$\begin{aligned}
&[\mathbf{E}\xi(gs)\xi(t)]^2 + [\mathbf{E}\xi(gt)\xi(s)]^2 = \\
&=(1/2)\mathbf{E}\{[\xi^2(gs) - \mathbf{E}\xi^2(s)][\xi^2(t) - \mathbf{E}\xi^2(t)]\}+ \\
&+(1/2)\mathbf{E}\{[\xi^2(gt) - \mathbf{E}\xi^2(t)][\xi^2(s) + \mathbf{E}\xi^2(s)]\} = \\
&=(1/4)R_{\eta_1}(g) - (1/4)R_{\eta_2}(g).
\end{aligned}$$

These equalities and (2.4) imply Property *(ii)*.

3°. Properties *(i)* and *(ii)* imply Property *(iii)*.

As in Subsection 1.3 we consider the product probability space $(\widetilde{\Omega}, \widetilde{\mathcal{F}}, \widetilde{P}) = (\Omega\times\Omega, \mathcal{F}\times\mathcal{F}, P\times P)$ along with the main probability space $(\Omega, \mathcal{F}, P) = (\mathbb{R}^T, \mathcal{B}^T, P)$. Put $\widetilde{\omega}(t) := \{\omega_1(t), \omega_2(t)\} \in \widetilde{\omega}_0$ and consider the coordinate random function $\widetilde{\xi}(y, \widetilde{\omega}) = \widetilde{\omega}(x) = (\xi_1(t, \widetilde{\omega}), \xi_2(y, \widetilde{\omega}))$, where $\xi_i(y, \widetilde{\omega}) = \xi_i(y, \omega_i) = \omega_i(x)$, $i = 1, 2$, are independent copies of our function $\xi(\cdot)$. Property *(ii)* of $\xi(\cdot)$ implies that

$$\mathbf{M} \mid \mathbf{E}[\xi_i(gy)\xi_i(s_0)] \mid= 0, \qquad i = 1, 2,$$

and since $\xi_1(\cdot)$ and $\xi_2(\cdot)$ are independent we also have $\mathbf{E}\xi_i(gy_0)\xi_j(s_0) = = 0$, if $i \neq j$. Thus the Gaussian random function $\widehat{\xi}(y, i)$, $(y, i) \in \in Y \times \{1, 2\}$, is weakly mixing in the wide sense and therefore, as was proved above, it is ergodic. This means that $\widetilde{\xi}(\cdot)$ is ergodic. According to Lemma 1.6, the random function $\xi(\cdot)$ is weakly mixing.

4° *(iii)* $\Rightarrow$ *(ii)* by Proposition 1.11.

5° *(iii)* $\Leftrightarrow$ *(iv)* by Lemma 1.6.

Theorem 2.1 implies (with the aid of App. (10.*F*)–(10.*I*)) some obvious corollaries which give ergodicity — weak mixing criteria for some classes of homogeneous Gaussian random functions directly in terms of the spectral measure, generalizing the well-known criteria of ergodicity of homogeneous Gaussian random processes and sequences due to Fomin, Grenander and Maruyama.

□

Corollary 2.2. *A continuous n-dimensional Gaussian strongly homogeneous random field $\xi(\cdot)$ on a separable locally compact group is weakly mixing (ergodic) iff each spectral operator measure $F_0(d\lambda)$ of the field $\xi_0(\cdot)$ has no finite-dimensional atoms.*

Corollary 2.3. *A continuous Gaussian homogeneous random field $\xi(\cdot)$ on a commutative locally compact group is weakly mixing (ergodic) iff the spectral measure of the field $\xi_0(\cdot)$ has no atoms.*

Corollary 2.4. *A Gaussian homogeneous generalized random field $\xi(\varphi)$, $\varphi \in K$, is weakly mixing (ergodic) iff the spectral measure of the field $\xi_0(\varphi)$ has no atoms.*

Corollary 2.5. *Any non-constant continuous Gaussian homogeneous random field on a compact group is not ergodic.*

Theorem 2.6. *Let G be a non-compact, locally compact group. A continuous Gaussian homogeneous random function $\xi(\cdot)$ on a G-space is strongly mixing iff it is strongly mixing in the wide sense.*

Proof. The necessity of the condition is evident. Its sufficiency can be proved by the arguments used in part 1 of the proof of Theorem 2.1; instead of (2.2) we have here:

$$\lim_{g\to\infty} L_g = \begin{pmatrix} M & 0 \\ 0 & M \end{pmatrix}.$$

□

Corollary 2.7. *A continuous Gaussian homogeneous random field on a non-compact, locally compact group is strongly mixing iff its covariance function vanishes at infinity.*

Proof. This is an evident consequence of Definition 1.9 and Theorem 2.6.

□

Theorem 2.8. *Let G be a non-compact, locally compact group and Y a homogeneous G-space. The following properties of a continuous Gaussian homogeneous random field $\xi(\cdot)$ on a homogeneous G-space Y are equivalent:*

(i) $\xi(\cdot)$ is strongly (resp. weakly) quasi-mixing;

(ii) $\xi(\cdot)$ is strongly (resp. weakly) quasi-mixing in the wide sense;

(iii) the background-free component of $\xi(\cdot)$ is strongly (resp. weakly) mixing.

Proof. 1°. *(i)* ⇒ *(ii)*; this is evident.

2°. *(ii)* ⇒ *(iii)*; this is a simple consequence of Proposition 1.10 and Theorems 2.1 and 2.6.

3°. *(iii)* ⇒ *(i)*. We shall only consider the "strong" case, as the "weak" one can be proved by substituting "(sC) lim" for "$\lim_{g\to\infty}$". First let us note that, evidently, strong mixing (property (SM_2) in Lemma 1.4) implies

(SM_2') for any Hilbert space $\mathcal{H}$,

$$\lim_{g\to\infty}\int\langle af(T_x\omega), bg(\omega)\rangle_{\mathcal{H}}dP = \langle a,b\rangle\mathbf{E}(f)\cdot\mathbf{E}(g),$$
$$a,b\in\mathcal{H},\quad f,g\in L^2.$$

Now consider the decomposition $\xi(t) = \zeta(t)+\eta$ where $\zeta(\cdot)$ is the background-free component and η is the background of $\xi(\cdot)$; recall that $\zeta(\cdot)$ and η are independent. Consider also the coordinate images $\widetilde{\zeta}(\cdot)$ and $\widetilde{\eta}$ of $\zeta(\cdot)$ and η and the spaces $(\widetilde{\Omega}^\zeta,\widetilde{\mathcal{F}}^\zeta,\widetilde{P}^\zeta)$ and $(\widetilde{\Omega}^\eta,\widetilde{\mathcal{F}}^\eta,\widetilde{P}^\eta)$ (see App., Ex. 8.1). The mapping r from $(\Omega,\mathcal{F},P)$ into $(\Omega^*,\mathcal{F}^*,P^*) :=$ $:= (\widetilde{\Omega}^\zeta\times\widetilde{\Omega}^\eta,\widetilde{\mathcal{F}}^\zeta\times\widetilde{\mathcal{F}}^\eta,\widetilde{P}^\zeta\times\widetilde{P}^\eta)$ given by the formula $\omega\stackrel{r}{\longrightarrow}(\zeta(\cdot,\omega),\eta(\omega))$ is a ξ-admissible reduction of Ω. Set $\zeta^*(y,\omega^*) := \zeta(y)$, $\eta^*(y,\omega^*) := \eta$ and $\xi^*(y,\omega^*) := \zeta^*(y)+\eta^*$ for $\omega^* = (\zeta,\eta)\in\Omega^*$. Evidently, $\zeta^*(\cdot)$, η^* and $\xi^*(\cdot)$ are the r-images of $\zeta(\cdot)$, η and $\xi(\cdot)$, resp. Let $T_g^*\omega^* :=$ $:= (T_g^{(\zeta)}\zeta,\eta)$ if $\omega^* = (\zeta(\cdot),\eta)$ (here $T_g^{(\zeta)}$ is the translation dynamical system of $\zeta(\cdot)$); clearly, $\{T_g^*,\ g \in G\}$ is a dynamical system in $(\Omega^*, \mathcal{F}^*, P^*)$ and $\xi(\cdot)$ is covariant with respect to it. Let $f_1, g_1 \in L^2(\widetilde{\Omega}^\zeta, \widetilde{\mathcal{F}}^\zeta, \widetilde{P}^\zeta)$, $f_2, g_2 \in L^2(\widetilde{\Omega}^\eta, \widetilde{\mathcal{F}}^\eta, \widetilde{P}^\eta)$, $f(\zeta,\eta) := f_1(\zeta)f_2(\eta)$, $g(\zeta,\eta) := g_2(\zeta)g_2(\eta)$. If $\zeta(\cdot)$ is strongly mixing, we have:

$$\lim_{g\to\infty} \int_{\Omega^*} f(T_g^*\omega^*)g(\omega^*)P^*(d\omega^*) = \lim_{g\to\infty} \int_{\Omega^\zeta} f_1(T_g^{(\zeta)}\zeta)g_1(\zeta)\widetilde{P}^{(\zeta)}(d\zeta) =$$

$$= \int_{\Omega^\eta} f_2(\eta)g_2(\eta)\widetilde{P}^{(\eta)}(d\eta) = \int f_1 d\widetilde{P}^\zeta \int_{\widetilde{\Omega}^\zeta} g_2\, d\widetilde{P}^\zeta \int_{\widetilde{\Omega}^\eta} f_2 g_2\, d\widetilde{P}^\eta.$$

Since the functions f and g of the considered form generate the space $L^2(\Omega^*, \mathcal{F}^*, P^*)$ we find with the aid of Lemma 1.2 that the dynamical system $\{T_g^*,\ g \in G\}$ is strongly quasi–mixing. Since the random field $\xi^*(\cdot)$ is covariant with respect to this dynamical system, it is strongly quasi–mixing, according to Definition 1.7. Now it only remains to use Proposition 1.8.

□

§ 3. WM–Spaces

3.1. Definition and ergodic properties. Let G be a locally compact group and let K be a compact subgroup of G.

Lemma 3.1. *The following properties of the homogeneous space $X = G/K$ are equivalent:*

(i) any unitary class–1 (with respect to K) representation of G which does not contain the identity representation has FD–continuous spectrum;

(ii) any non–identity unitary class-1 representation of G is infinite–dimensional;

(iii) any non–identity irreducible unitary class–1 representation of G is infinite–dimensional;

(iv) any spherical positive definite function R on G has the following form: $R(g) = R_0(g) + \mathbf{M}(R)$ where R_0 is the flight component of R;

(v) any coefficient $\varphi(g) = \langle U_g h_1, h_2 \rangle$ of a class–1 unitary representation $g \to U_g$ has the following form: $\varphi(g) = \varphi_0(g) + \mathbf{M}(\varphi)$ where φ_0 is a flight–function;

(vi) any non–identity elementary spherical positive definite function is a flight function.

If $X = G$, $K = \{e\}$, the words "spherical" and "class–1" have to be omitted.

Proof. Evidently, *(ii)* $\Leftrightarrow$ *(iii)* and *(v)* $\Rightarrow$ *(vi)*. The implications *(i)* $\Leftrightarrow$ *(iii)*, *(i)*$\Leftrightarrow$*(iv)*, *(iii)*$\Leftrightarrow$*(vi)* follow from App. (10.B)–(10.E). Finally, *(iv)*$\Rightarrow$*(v)*, since φ is a linear combination of spherical positive definite functions.

□

Definition 3.1. A homogeneous space G/K with Properties *(i)*–*(v)* stated in Lemma 3.1 is called a *weakly mixing space* or, briefly, a WM–space. If $X = G$, $K = \{e\}$ and Properties *(i)*–*(v)* are fulfilled (in this case the words "class–*1*" and "spherical" can be omitted), G is called a *WM–group*.

The following theorem, which is a simple consequence of the results of § 1 and § 2, explains this "ergodic" terminology and shows the role of WM–spaces in ergodic theory.

Theorem 3.2. *The following properties of a space G/K are equivalent:*

(i) G/K is a WM–space;

(ii) any Gaussian homogeneous continuous background–free random field on G/K is weakly mixing;

(iii) any strict–sense homogeneous continuous random field on G/K is G/K–relatively weakly quasi–mixing;

(iv) any completely measurable dynamical system $\{T_g,\ g \in G\}$ in a probability phase space is G/K–relatively weakly quasi–mixing;

(v) any wide sense homogeneous continuous background–free random

*field on G/K is weakly mixing in the wide sense**).

Proof. Let U be the unitary representation of G in L^2 $(\Omega, \mathcal{F}, P)$ associated to the dynamical system T and let $H^{(K)}$ be the least U–invariant subspace in L^2 $(\Omega, \mathcal{F}, P)$ containing L^2 $(\Omega, \mathfrak{J}_K, P)$. The restriction $U^{(K)}$ of U to $H^{(K)}$ is a class–*1* unitary representation of G. It is clear that T is weak quasi–mixing iff U has the property *(v)* of Lemma 3.1. Hence *(i)*⇒*(iv)*. By Definition 1.7, *(iv)*⇔*(iii)* and, by Theorem 2.1, *(i)*⇒*(ii)*. Any positive definite function $R(g)$ is the correlation function of a Gaussian random field on G, and if $R(g)$ is a spherical **P**–function then the field is constant on the cosets gK. Thus, *(ii)*⇒*(i)*. It is obvious that *(iii)*⇒*(ii)*, *(v)*⇔*(ii)* and *(i)*⇒*(v)*.

□

3.2. Some connections between WM–spaces. Auxiliary statements. We shall state three simple lemmas, which sometimes make easier to check whether or not a space G/K has the WM–property.

Lemma 3.3. *Let K_1 and K_2 be compact subgroups of G and $K_1 \subset \subset K_2$. If G/K_1 is a WM–space, then G/K_2 is a WM–space.*

Proof. Any class–*1* representation of G with respect to K_2 is a class–*1* representation with respect to K_1. It only remains to recall Property *(iii)* of WM–spaces.

□

The following two lemmas can also be proved easily.

Lemma 3.4. *A group $G = G_1 \otimes \cdots \otimes G_n$ is a WM–group iff all G_i, $i = \overline{1,m}$, are WM–groups.*

Lemma 3.5. *If G is a WM–group, then any factor group $G/F \neq \{e\}$ is also a WM–group.*

---

*) If Y=G, K=$\{e\}$, G/K–relative weak quasi–mixing is equivalent to "absolute" weak quasi–mixing.

3.3. Classes of groups without the WM–property. The definition of WM–spaces and Lemma 3.5 imply that the following important classes of groups do not possess the WM–property:

— compact groups (because all irreducible continuous unitary representations are finite–dimensional);

— commutative locally compact groups (because all irreducible unitary representations are one–dimensional);

— solvable locally compact groups (because there exists a commutative factor group $G/\Phi \neq \{e\}$);

— non–perfect locally compact groups (because $G/[G,G]$ is commutative);

— semi–direct product groups of the form $A \odot K$ where K is a compact group (because $G/A \cong K$).

In the following subsection we shall find some important classes of WM–spaces and WM–groups.

3.4. Intrinsic characterization and examples of WM–spaces. Definition 3.1 characterizes the WM–spaces in terms of unitary representations and spherical functions. In this subsection we shall find a characterization of these spaces in "intrinsic" algebraic and topological terms; this will enable us to find wide classes of WM–spaces.

Lemma 3.6. *The kernel Φ of a non–identity class–1 representation U of $\mathcal{G}$ (with respect to K) in a Hilbert space H is a non–transitive normal subgroup.*

Proof. Assume the contrary. Then any $g \in G$ can be represented in the form $g = \varphi k$, $\varphi \in \Phi$, $k \in K$. If $h \in I_K$, i.e. $U_k h = h$, $k \in K$, then $U_g h = U_\varphi U_k h = U_\varphi h = h$ and therefore the subspace I_K is U–invariant and the subrepresentation U' in I_K is the identity representation. Since U is a class–*1* representation, this implies that $I_K = H$ and thus U is the identity representation; this contradicts the condition of the lemma.

□

Lemma 3.7. $X = G/K$ *is a WM–space iff every large (with respect* *to K) factor group G/Φ is non–representable in a compact group.*

Proof. Sufficiency. Suppose that the condition of the lemma is fulfilled but there exists a non–identity unitary class–*1* representation U of G in a m–dimensional Hilbert space H ($m < \infty$). Let Φ be the kernel of U. By Lemma 3.6, G/Φ is a large factor group. It admits a one–to–one continuous homomorphism into the compact group $U(m, \mathbb{C})$ of all m–dimensional unitary matrices, this contradicts the condition of the lemma.

Necessity. Suppose that there exists a large factor group G/Φ admitting a homomorphism κ in a dense subset of a compact group $F \neq \{e\}$. Let j be the natural homomorphism of G onto G/Φ and $\widetilde{K} := \kappa \circ j(K)$; clearly, $\kappa \circ j(\Phi) = \{e_F\}$. Since K is compact and $K\Phi \neq G$, $\widetilde{K}$ is a closed subgroup of F and $\widetilde{K} \neq F$. Let U be a non–identity finite–dimensional class–*1* (with respect to $\widetilde{K}$) unitary representation of F (since F is compact the existence of such a representation follows from App., Example 5.2). Then $g \to U_{\kappa \circ j(g)}$ is a non–identity finite–dimensional class–*1* representation of G and thus G/K is not a WM–space.

□

Theorem 3.8. *In order that a space G/K be a WM–space it is necessary (and, if G is connected, sufficient, too) that the following condition be fulfilled:*

(W) all large factor groups G/Φ are non–compact.

Proof. Necessity is implied by Lemma 3.7. By virtue of this lemma, to prove sufficiency we only have to show that if G is connected and there exists a large factor group G/Φ representable in a compact group, then there also exists a compact large factor group G/Φ. Since G is connected, G/Φ is also connected, and by the Freudental–Weyl theorem (cf., App. (6.O)) $G/\Phi \cong \mathbb{R}^m \times \Gamma =: G'$ where Γ is a compact group. If $m = 0$, $G/\Phi \cong \Gamma$ and G/Φ is compact. Let $m \neq 0$. The natural homomorphism $G \overset{\kappa}{\mapsto} G'$ maps K onto a compact subgroup K' of G'. The group $F' := \mathbb{Z}^m \times \Gamma$ is a normal subgroup of G' and G'/F' is compact. Since there are no compact proper subgroups in $\mathbb{R}^m$, we have $K' \subset \Gamma$. Therefore $K'F' \neq G'$. $F := \kappa^{-1}(F')$ is a normal subgroup of G, $G/F \cong G'/F'$ (cf.,e.g. Pontryagin [1], § 20, (D)) and $KF \neq G$. Thus G/F is a compact large factor group.

□

Below we state two simple consequences of Theorem 3.8.

Corollary 3.9. *In order that G be a WM–group it is necessary and, if G is connected, sufficient that all factor groups $G/\Phi \neq \{e\}$ be non–compact.*

Corollary 3.10. *If G/K is a non–compact simple homogeneous space and G is connected, then G/K is a WM–space. In particular, any non–compact simple group is a WM–group.*

Example 3.1. The Euclidean homogeneous space $E_m = (\mathbb{R}^m \odot \odot SO(m))/SO(m)$, $m \geq 2$, is a WM–space; this is implied by Corollary 3.9 and App., Example 2.7. Note that neither the motion group $\mathbb{R}^m \odot SO(m)$ nor its subgroup $\mathbb{R}^m$ are WM–groups (see Subsect. 3.3). Thus, any "homogeneous and isotropic", that is homogeneous with respect to translations and rotations, Gaussian random field is weakly quasi–mixing (and mixing if it is background–free); but these properties may certainly be absent if the field is "non–isotropic" (i.e. non–homogeneous with respect to the rotation group).

Example 3.2. The group $SO_0(m,d)$, $m \geq 2$, is a WM–group. So, by Lemma 3.3, the Lobachevsky space $L_m = SO_0(m,1)/SO(m)$, $m \geq 2$, is a WM–space.

App. (2.I) and Corollary 3.10 imply the following generalization of the result stated in the Example 3.1.

Theorem 3.11. *If K is a connected γ–irreducible group of automorphisms of a connected group A, then the homogeneous space $(A \odot K)/K$ is a WM–space.*

Lemma 3.12. *(i) Let F be a discrete normal subgroup of a connected group G and let G/F be a WM–group. Then G is also a WM–group.*

(ii) Let F be a central normal subgroup of a perfect group G and let G/F be a WM–group. Then G is also a WM–group.

Proof. *(i)* Let Φ be a proper normal subgroup of G. Assume that $F\Phi = G$; then $G/\Phi \cong F/F \cap \Phi \neq \{e\}$ (cf., Pontryagin [1], § 20, (G)), and Φ is a proper closed–open set. This is impossible since G is supposed to be connected. Thus $F\Phi \neq G$. Put $\Phi_1 := \kappa(\Phi)$, the image of Φ with respect to the natural homomorphism κ: $G \longmapsto G/F$, and $G_1 := G/F$. We have $G/\Phi F \cong G_1/\Phi_1 \neq \{e\}$ and $G/\Phi F \cong$ $\cong (G/\Phi)/(\Phi F/\Phi)$. Since G_1 is a WM–group, G_1/Φ_1 is non–compact and thus G/Φ is non–compact. It remains to refer to Corollary 3.9.

(ii) Let Φ be a proper normal subgroup of G. Assume that $F\Phi = G$. Then $[G, G] \subset \Phi$, which is impossible since G is supposed to be perfect. Thus $F\Phi \neq G$. Now we can repeat word for word the remaining part of the proof of statement *(i)*.

□

Lemmas 3.4, 3.5, 3.12, and Subsect. 3.3 imply the following statement.

Theorem 3.13. *A connected semi–simple group is a WM–group iff it is of non–compact type.*

This theorem enables us to generalize the result stated in Example 3.2 as follows.

Corollary 3.14. *Any Riemannian symmetric space of non–compact type is a WM–space.*

Theorem 3.15. *Let $G = A \odot \Sigma$ be a connected group, where Σ is a semisimple group of non–compact type which is γ–irreducible on A. Then G is a WM–group.*

Proof. Let $G = A' \overset{\alpha}{\odot} \Sigma'$ and N be a proper normal subgroup of G. By App. (2.*I*), N is a transitive normal subgroup, i.e. $N\Sigma' = G$. Therefore $N \cap \Sigma'$ is a normal subgroup of Σ' and $G/N = N\Sigma'/N \cong \Sigma'/\Sigma' \cap N$ (cf., e.g., Pontryagin [1], § 20, G)). Thus the group G/N is non–compact. It remains to apply Theorem 3.8.

□

Example 3.3. The m–dimensional Poincaré group $P(m) := R^m \odot \odot SO(m-1, 1)$, $m \geq 3$, is a WM–group.

The following theorem shows that amenability and the "WM–property" of groups are opposite properties.

Theorem 3.16. *In a connected amenable group there are no subgroups which are WM–groups.*

Proof. Since any subgroup of an amenable group is amenable (cf. Greenleaf [1], Theorem 2.3.2), it is sufficient to prove that any connected amenable group G is not a WM–group. If G is solvable it is not a WM–group (cf. Subsect. 3.3). If G is non–solvable and R is its radical, then by App.(3.F) G/R is compact; Lemma 3.5 implies that G is not a WM–group in this case either.

□

All WM–spaces considered above were of one of the two types of homogeneous spaces:

type 1. homogeneous spaces with a WM–motion group;

type 2. group–type homogeneous WM–spaces.

Theorem 3.17. *If K is a simple group and G/K is a WM–space, then G/K is either of type 1 or of type 2.*

Proof. Assume that G/K is a WM–space, but G is not a WM–group. Then there exists a normal subgroup $\Phi \neq G$ such that G/Φ is representable in a compact group. By Lemma 3.6, Φ is a K–transitive normal subgroup, i.e. $\Phi K = G$. Therefore (cf. Pontryagin [1], § 20, G)) $G/\Phi \cong K/K \cap \Phi$. Since K is a simple group and $G \neq \Phi$, we have $K \cap \Phi = \{e\}$ and thus $G/\Phi \cong K$. Consequently, $G \cong \Phi \odot K$ (see App., Subsect. 2.2).

□

Note that the connectedness condition of G in all theorems stated above is essential. If we consider a stronger topology in G, e.g. the discrete topology, the WM–property can disappear (of course, if we consider a weaker topology this property remains valid). We conclude this section by two theorems concerning discrete groups.

Theorem 3.18. (von Neumann and Wigner) *If for any $g \in G$ and any $m = \overline{1, \infty}$ there exists an integer k_m divisible by m and an $f_m \in G$ such that $g^{k_m} = f_m g f_m^{-1}$, then G is a WM–group.*

Theorem 3.19. (Hewitt and Ross) *Let H be an infinite subgroup of G with the following properties:*

(i) in H there exists a finite family of sets $\{e\}$, $H_1, \dots, H_n$ which are minimal invariant sets with respect to the inner automorphisms of G;

(ii) the minimal normal subgroup of H containing H_i is H, $i = 1, \dots, n$;

(iii) the minimal normal subgroup of G containing H is G itself.

Then G is a WM–group.

The proofs of Theorems 3.18 and 3.19 are presented in Hewitt and Ross [1].

§ 4. SM–Spaces

4.1. The definition of SM–spaces. Let G be a non–compact separable connected locally compact group, and K be a compact subgroup of G.

Lemma 4.1. *The following properties of the homogeneous space $X = G/K$ are equivalent:*

(i) for any unitary class–1 (with respect to K) representation U of G in a Hilbert space H the weak limit $(w)\lim_{g\to\infty} U_g$ exists (by Theorem 1.5.20 this limit is equal to $\mathrm{M} = \Pi_{\mathfrak{J}}^{H}$); in other words, any matrix coefficient $\varphi(\cdot) := \langle U.h, f\rangle \in C_a$.

(ii) any non-identity irreducible unitary class-1 representation weakly converges to the operator 0*;*

(iii) any SP-function (spherical positive definite function) R_g, $g \in G$, *has the following form:* $R(g) = R_0(g) + M(R)$ *where* R_0 *is a SP-function and* $R_0 \in C_0(G)$.

(iv) any SP-function $R \in C_a(G)$*;*

(v) any coefficient $\varphi(g) = \langle U_g, h_1, h_2 \rangle$ *of a class-1 unitary representation* $g \mapsto U_g$ *is in* $C_a(G)$*;*

(vi) any non-identity elementary SP-function $\varphi \in C_0(G)$.

Proof. Evidently, *(i)* $\Rightarrow$ *(ii)*, *(i)* $\Rightarrow$ *(iii)* $\Leftrightarrow$ *(iv)* $\Rightarrow$ *(vi)* and *(v)*$\Rightarrow$*(iv)*. The implication *(iv)* $\Rightarrow$ *(v)* can be proved similar to the implication *(iv)* $\Rightarrow$ *(v)* in Lemma 3.1. Finally, the implication *(vi)* $\Rightarrow$ *(iv)* follows from the spectral decomposition theorem for SP-functions (cf. App. (5.F)) and the Lebesgue dominated convergence theorem.

□

Definition 4.1. A homogeneous space G/K with the properties *(i)*–*(v)* stated in Lemma 3.1 is said to be a *strongly mixing homogeneous space* or, briefly, a *SM-space*. If $X = G$, $K = \{e\}$ and the properties *(i)*–*(v)* are fulfilled (in this case the words "class *1*" and "spherical" can be omitted), G is called a *SM-group*.

It is easy to verify that the "SM-analogues" of Lemmas 3.3 and 3.5 hold.

Any SM-space is a WM-space.

4.2. Ergodic properties of SM-spaces. Lemma 1.3, Definition 1.9, Proposition 1.10, and Theorem 2.8 imply the following theorem.

Theorem 4.2. *The following properties of a space* G/K *are equivalent:*

(i) G/K *is a SM-space;*

(ii) any Gaussian homogeneous continuous background-free random field on G/K *is strongly mixing;*

(iii) a strict sense homogeneous continuous random field on G/K *is* G/K*-relatively strongly quasi-mixing;*

(iv) any measurable dynamical system $\{T_g,\ g \in G\}$ in a probability measure space is G/K–relatively strongly quasi–mixing;

(v) any wide sense homogeneous continuous random field on G/K is strongly quasi–mixing in the wide sense;

(vi) any wide sense homogeneous continuous background–free random field on G/K is strongly mixing in the wide sense).*

The proof is similar to the proof of Theorem 3.2.

Corollary 4.3. *If G is a SM–group, then any measurable dynamical system $T = \{T_g,\ g \in G\}$ in a probability measure space is strongly quasi–mixing; moreover, if T is ergodic then it is strongly mixing.*

Corollary 4.4. *If G is a SM–group, F is a non–compact subgroup of G and $\{T_g,\ g \in G\}$ is an ergodic measurable dynamical system, then the dynamical system $\{T_f,\ f \in F\}$ is strongly mixing.*

Example 4.1. From the point of view of classical ergodic theory, the case when F is a one–parameter subgroup (i.e. $F = \{f(t),\ t \in R\}$ and $f(t_1 + t_2) = f(t_1)f(t_2)$) is especially interesting; in this case $S = \{s_t := T_{f(t)},\ t \in T\}$ is a flow. Let us consider the space Φ of all unit tangent vectors of a smooth surface A of constant negative curvature; it can be identified with the homogeneous space $Y = G/L$, where $G := SL(2)/\{\pm 1\}$ and L is a discrete subgroup of G. This enables us to consider Φ as a left homogeneous G–space. Suppose that on Φ a G–invariant probability Borel measure P exists and define the dynamical system T: $T_g\varphi = g\varphi,\ \varphi \in \Phi,\ g \in G$. If we set $F = \{\left(\begin{smallmatrix} e^t & 0 \\ 0 & e^{-t} \end{smallmatrix}\right),\quad t \in \mathbb{R}\}$ then S is the geodesic flow: $S_t\varphi$ is the position of the vector $\varphi \in \Phi$ at the moment t if φ moves with unit velocity along the geodesic line which is defined by φ (cf., e.g., Cornfeld, Sinai, Fomin [1], Ch.2, § 2, Subsect. 4). The condition of Corollary 4.4 is fulfilled: G is a simple group, therefore (as we shall see below) it is a SWQ–group and, of course, the dynamical system T acts on Φ transitively. By

---

*) If $Y=G$, $K=\{e\}$, G/K–relative strong quasi–mixing is equivalent to "absolute" strong quasi–mixing.

Corollary 4.4, the geodesic flow is mixing. If we set $f = O_t^+ := \binom{1\ t}{0\ 1}$ or $F = O_t^- := \binom{1\ 0}{t\ 1}$, we obtain the so called oricyclic flows in Φ; they are also mixing.

4.3. Characterization and examples of SM–spaces. The class of SM–spaces (SM–groups) is essentially smaller than the class of WM–spaces (WM–groups): Proposition 4.5 below shows that the Condition (W) of Theorem 3.8, which characterizes the WM–spaces, has to be replaced here by the following considerably stronger condition:

(S) any non–transitive (with respect to K) normal subgroup of G is compact.

Definition 4.2. A homogeneous space G/K with a non–compact separable motion group satisfying Condition (S) will be called an *S–space*. In particular, a group G will be called an S–group if it is non–compact, separable and all its proper normal subgroups are compact.

It is easy to verify that for S–spaces the analogs of Lemmas 3.3 and 3.5 are true. Therefore, if G is an S–group, any homogeneous space G/K is an S–space. As in § 3 we shall consider two types of S–spaces:

type 1: homogeneous spaces whose motion groups are S–groups.

type 2: group–type homogeneous S–spaces.

The analogue of Theorem 3.17 is valid: *if K is a simple group and G/K is a S–space, then it is either of type 1 or of type 2.*

The aim of this subsection is to study the connection between the notions of SM–space and S–space. In one direction the solution of this problem is rather simple.

Proposition 4.5. *Any SM–space is an S–space.*

Proof. Assume that G contains a non–compact non–transitive normal subgroup N. Let κ be the natural homomorphism $G \longmapsto G_1 := G/N$; $K_1 := \kappa(K)$. Since N is non–transitive, $K_1 \neq G_1$ and since κ is continuous, K_1 is compact. Let $U^{(1)}$ be a non–identity unitary K_1–class–*1* representation of G_1 in H (see App., Ex. 4.2); of course, we may suppose that the $U^{(1)}$ averaging operator $\mathbf{M}^{(1)} = \Pi^H_{I^{(1)}} = \{0\}$, where $I^{(1)}$ is the subspace of $U^{(1)}$–invariant vectors. Set $U_g := U^{(1)}_{\kappa(g)}$. $U : g \longrightarrow U_g$ is a unitary K–class–*1* representation of G in H. Clearly, $I = I^{(1)}$ and $\mathbf{M} = \Pi^H_I = 0$. N belongs to the kernel of U and therefore $\varphi_h(g) := \langle U_g h,\ h\rangle = \|h\|^2 \neq 0$ if $g \in N$, $h \neq 0$. Since N is non-compact, U_g does not converge weakly to $\mathbf{M} = 0$ and thus G/K is not a SM–space. We have arrived at a contradiction.

□

Conjecture. Any S–space G/K with a connected motion group is an SM–space.

We shall see that this conjecture is true for important and large classes of homogeneous spaces. Let us begin with the type–*1* spaces with Property (S); certainly, in this case our problem at once reduces to the study of S–groups.

Theorem 4.6. *Let G be a connected Lie group. The following three properties are equivalent:*

(i) G is an SM–group;

(ii) G is an S–group;

(iii) G is a non–compact almost simple group with finite center.

Proof. *1*. *(i)* ⇒ *(ii)* by Proposition 4.5.

2. *(ii)* ⇒ *(iii)*. Let G be an S–group. Consider the Levi decomposition of its universal covering group G': $G' = R' \odot \Sigma'$ (see App. (1.D)). Denote by π the universal covering mapping $G' \longmapsto G$ of G; $R = \pi(R')$; $\Sigma = \pi(\Sigma')$; Φ_R and Φ_G are the kernels of π in the groups R' and G' resp. It is easy to see that Σ is almost simple. R is a normal subgroup of G and therefore compact; since R is, moreover, solvable, it is commutative and therefore isomorphic to the torus T^m, $(m \geq 0)$ (cf., e.g., Bourbaki [2], ch. 7, § 2; $T^\circ = \{e\}$). Since R' is the universal covering group of R, we have $R' = \mathbb{R}^m$ (cf. Pontryagin [1], Ex. 8.6); $\alpha_{\sigma'}$, the restriction to R' of the inner automorphism of G' corresponding to the element $\sigma' \in \Sigma'$, is a linear operator in R'. It is clear that $\Phi_R = \Phi_G \cap R'$. Φ_R is a discrete normal subgroup and therefore it belongs to the center of G'. Consequently $\alpha_{\sigma'}(g) = g$ for any $g \in \Phi_R$ and thus for any $g \in R'_1 \subset R'$, where R'_1 is the subspace of R' generated by Φ_R. Since $R'/R'_1 \cong (R'/\Phi_R)/(R'_1/\Phi_R)$ and the group $R'/\Phi_R (= R)$ is compact, R'/R'_1 is compact. Therefore $R' = R'_1$, and thus $G' = R' \otimes \Sigma'$. Consequently, Σ is a normal subgroup of G. Since $G = R\Sigma$ and G is non–compact and R is compact, we conclude that Σ is non–compact; since G is an S–group, $G = \Sigma$. Thus G is almost simple; its center is a discrete normal subgroup and therefore it is finite.

3. (iii) ⇒ (i). This implication was proved by Howe and Moore [1] (see also Zimmer [1]). The proof is based on methods from Lie group theory which are beyond the scope of this book; therefore we omit it; the reader will find it in the above mentioned work of Howe and Moore.

□

Note 4.1. Schmidt [1] has shown that any connected separable locally compact group G is an SM–group iff G is perfect, the center $Z(G)$ is compact and $G/Z(G)$ is a non–compact, connected, simple Lie group with trivial center.

Corollary 4.7. *Any irreducible Riemannian symmetric homogeneous space of non–compact type is an SM–space.*

Proof. The motion group of such a space possesses Property *(iii)* of Theorem 4.6 (cf., e.g., Helgason [1], ch. 8, § 5) and thus it is an SM–group.

□

Example 4.2. The Lobachevsky space $L_m = SO_\circ(m, 1)/SO(m)$ is an SM–space.

Now we shall turn to the S–spaces of type 2. Reasoning as in the proof of App., (2. I), it is easy to show that if $G = A \odot K = A' \odot K'$, then Property (S) follows from the following property:

(S') A does not contain proper K–invariant non–compact subgroups.

If A is commutative then the properties (S) and (S') are equivalent: in this case any proper K–invariant subgroup Γ of A corresponds to a subgroup Γ' of A' which is a non–transitive (in G/K) normal subgroup of G.

The following theorem is an important tool in the investigation of S–spaces of type *2*.

Theorem 4.8. **(Bagget and Taylor)** *Let $G = A \odot K$. Suppose G, A, and K are connected, A is non–compact and commutative. Denote by U an irreducible unitary representation of G, by L its kernel, and by κ the natural homomorphism of G onto G/L. The matrix coefficients of the representation $f \longmapsto U_{\kappa(f)}$ of the factor group G/L vanish at infinity.*

We omit the proof; it can be found in Bagget and Taylor [1].

Theorem 4.9. *Let G be as in Theorem 4.8. Condition (S) (or (S')) is necessary and sufficient in order that G/K be an SM–space.*

Proof. In view of Proposition 4.5 we only have to prove the sufficiency of (S). By Lemma 3.6, the kernel of a non–identity irreducible class–*1* representation is a non–transitive normal subgroup of G. If (S) is fulfilled, L is compact. It remains to use Theorem 4.8.

□

Example 4.3. The Euclidean space $E_m = (R^m \odot SO(m))/SO(m)$, $m \geq 2$, is an SM–space.

Howe and Moore [1] have proved a statement analogous to Theorem 4.8 for any connected algebraic group over a field of characteristic zero. The arguments used in the proof of the previous theorem convince us that the following general statement is valid.

Theorem 4.10. *If G is a connected non–compact algebraic group over a field of characteristic zero, then G/K is an SM–space iff it is an S–space.*

BIBLIOGRAPHICAL NOTES

1. The notion of "weak mixing" for dynamical systems on amenable topological semigroups was defined in Dye [1] (1965), who also proved Theorem 1.6 for such semigroups. In Bergelson, Rosenblatt [1] weak mixing is defined and Theorem 1.6 was proved for general topological groups among other results.

2. The spectral criterion of ergodicity of Gaussian stationary processes was established in Fomin [1] (1949), Grenander [1] (1950) and Maruyama [1] (1949). In its general form, Theorem 2.1 is contained in Tempelman [2] (1961), [1] (1962), [3] (1970); for commutative locally compact groups this result (Corollary 2. 3) was independently established in Birkhoff, Kampe de Feriet [1] (1962) and was repeatedly rediscovered (see, e.g., Blum, Eisenberg [1] (1973), Eisenberg [1] (1972)). The stated proof of Theorem 2.1 is based on a method due to Grenander [1].

3. WM-groups ("ergodic groups") were characterized in Tempelman [10] (1973), [14] (1975); the results concerning general WM-spaces ("ergodic spaces") were announced in Tempelman [23] (1982); the proofs are published in Tempelman [24] (1986). Theorem 3.18 is due to von Neumann and Wigner [1] (1940). Theorem 3.19 is contained in Hewitt, Ross [1] (this book also contains proof of Theorem 3.18). The proofs of the other results of §3 were given in Tempelman [24] (1986).

4. The definition of the SM-spaces and the most of the results of §4 can be found in Tempelman [23] (1982) and Tempelman [24] (1986). The implication 3)⇒1) in Theorem 4.6 was proved in Zimmer [2] (1977). Corollary 4.4 is close to Theorem 5.3 in Zimmer [2]. Lemma 3.8 is due to Bagget, Taylor [1] (1978); Theorem 4.10 is a simple consequence of Theorem 6.1 in Howe and Moore [1] (1979). The theorem stated in Note 4.1 was proved in Schmidt [1] (1984).

5. Example 4.1 shows possible applications of Corollary 4.4. Mixing of geodesic flows on surfaces of constant negative curvature was established by Hopf [2] (1939) (Hopf's proof is given, e.g., in Cornfeld, Sinai, Fomin [1] (Ch. 4, §4)).

6. The properties of "ergodicity", "mixing" and "quasi–mixing" of dynamical systems and homogeneous random fields, established in Theorems 3.2, 4.2 and Corollaries 4.3 and 4.4, are determined by properties of unitary representations on WM– and SM–groups and spaces. Gelfand and Fomin [1] (1952) and Parasyuk [1] (1953) used methods of the theory of group representation in order to investigate ergodic properties of geodesic and oricyclic flows on surfaces of constant negative curvature. Later this approach was widely used in studies of various types of dynamical systems by Auslender, Green [1] (1966), Green [1] (1966), Mautner [1] (1957), Moore [1] (1966), [2] (1970), and Auslender, Green, Hahn [1] (1963).

7. Mackey [1] (1964) has proved that every ergodic dynamical systems with a time group G, acting in a standard Borel space, with a fixed FD-spectrum is equivalent to the translation dynamical system in a compact homogeneous space.

CHAPTER 3
AVERAGING SEQUENCES. UNIVERSAL ERGODIC THEOREMS

§ 1. General Mean Value Theorems

Let $(X, \mathfrak{B})$ be a measurable semigroup; $\mathfrak{B}\mathcal{M}\ (\mathfrak{B})$ the Banach space of all signed measures of bounded variation on $\mathfrak{B}$ with norm $\|\nu\| = \operatorname{var}\nu$; $\mathcal{P}(\mathfrak{B})$ the set of all probability measures on $\mathfrak{B}$; $\widetilde{\mathcal{P}}$ the set of all probability measures ν on X whose carriers $c(\nu)$ are finite sets; and let F_B be the subspace in Φ_B consisting of the bounded measurable functions on X.

Definition 1.1. Let Q be a class of measurable r–quasi–averageable functions on X; a net of measures $\{\nu_n,\ n \in N\}$ is said to be a *Q–r–averaging net* (or to *r–average the class Q*) if all $\nu_n \in \mathcal{P}$ and $\lim\limits_{n\in N} \sup\limits_{y\in X} \| \int f(yx)\nu_n(dx) - M^{(r)}(f)\|_B = 0,\ f \in Q$. Let us associate with each Banach space B a class of B–valued r–quasi–averageable functions Q_B; we say that $\{\nu_n\}$ is a *universally Q_B–r–averaging net* if it is Q_B–r-averaging for any B.

Sometimes it is convenient to use the following "vector" terminology instead of the "functional" one given in Definition 1.1.

Definition 1.2. Let S: $x \to S_x$ be a uniformly bounded representation in a Banach space B; let Q be a class of quasi–averageable (with respect to S) elements of B. A net of measures $\{\nu_n, n \in N\}$ *averages the class Q (with respect to S)*, if $\nu_n \in \mathcal{P}$, $n \in N$, and $\lim_{n\in N} \int S_x b\nu_n(dx) = \mathbf{M}^{(r)}(b)$, $b \in B$.

Both languages, the "functional" and the "vector" one, are equivalent. On the one hand, if $\{\nu_n\}$ averages some $b \in B$ with respect to a representation S, this means that it r–averages the class of B–valued orbital functions $\varphi_{S,b}(x) = S_x b$; on the other hand, r–averaging of $f \in F_B$ is equivalent to averaging it as an element of the Banach space F_B with respect to the left representation $x \to R_x$ (see also § 1.2).

In particular, $\{\nu_n\}$ is a universally $A_B^{(r)} \cap F_B$–averaging net iff $\lim_{n \in N} \int S_x b \nu_n(dx) = M^{(r)}(b)$, $b \in A(B|A)$ for any measurable uniformly bounded left representation in any Banach space B.

If μ is a measure on $\mathfrak{B}$ and $A \in \mathfrak{B}$, $\mu(A) < \infty$, we can consider the measure ν_A "uniformly" distributed on A with respect to the basic measure μ: $\nu_{A,\mu}(c) = \mu(A \cap C)/\mu(A)$. Then for any $f \in F_B$ we have $\int_x f\, d\nu_{A,\mu} = \frac{1}{\mu(A)} \int_A f\, d\mu$, i.e. in this case $\int_X f\, d\nu_{A,\mu}$ is the arithmetic mean of f on A with respect to μ.

Definition 1.3. A net of sets $\{A_n,\ n \in N\}$ is called a *Q–r–averaging net* (a *universally Q_B–r–averaging net*) if $A_n \in \mathfrak{B}$, $0 < \mu(A_n) < \infty$ and the net $\{\nu_{A_n,\mu},\ n \in N\}$ possesses the corresponding property.

Averaging nets on a homogeneous space $Y = G/K$ are defined similarly.

Note 1.1. Let π be the natural mapping of G onto Y; let $Q \in \mathcal{F}_B(Y)$ be a class of quasi–averageable functions on Y; $\widetilde{Q} = \{\varphi\colon \varphi(g) = f(x)\}$ if $g \in \pi^{-1}(f)$. Clearly $\{\nu_n\}$ is Q–averaging iff the net of complete preimages $\{\widetilde{\nu}_n\}$ is $\widetilde{Q}$–l–averaging.

Mean value theorems, or *ergodic theorems*, for a class of r–quasi–averageable functions or elements of a Banach space are propositions which state conditions (formulated in terms which do not assume knowledge of the mean values) under which a net $\{\nu_n,\ n \in N\} \subset \mathcal{P}$ is Q–averaging; the term "mean value theorem" is more used in the "functional" approach, while the term "ergodic theorem" is more preferable in the "vector" approach. Mean value (ergodic) theorems concerning $A_B \cap F_B$–universally averaging nets will be called *universal mean value (ergodic) theorems*.

Now we shall give a version of such a universal theorem.

Let $R_x^* \nu$ ($x \in X$, $\nu \in \mathcal{P}$) be the measure defined by $R_x^*\nu(A) :=$ $:= \nu(Ax^{-1})$, $A \in \mathfrak{B}$, $\nu_{n,x}^{(r)} = \nu_n - R_x^* \nu_n$. Recall that $F_B(X, \mathfrak{B})$ is the space of all bounded $\mathfrak{B}$–measurable B–valued functions on X.

Definition 1.4. Let $Q \subset F_B(X, \mathfrak{B})$; a net of measures $\{\nu_n,\ n \in N\}$ is said to be *Q–r–ergodic* if all $\nu_n \in \mathcal{P}(\mathfrak{B})$ and for any f from Q and each $x_0 \in X$ we have $(B)\ \lim_{n\in N} \int f(yx)\nu^{(r)}_{n,x_0}(dx) = 0$ uniformly with respect to $y \in Y$.

Theorem 1.1. *Let Q be a set of r–quasi–averageable functions in $F_B(X, \mathfrak{B})$. Then*

(i) any Q–r–ergodic net of measures is Q–r–averaging;

(ii) if there exists a Q–r–ergodic net, then any f in Q possesses a single right mean value; if, moreover, Q is an invariant (with respect to the right translations R_x), $x \in X$, convex set in F_B, then $Q \subset A^{(r)}_B \cap F_B$;

(iii) if any function in Q possesses a single right mean value and Q is invariant with respect to the right translations R_x, $x \in X$, then any Q–r–averaging net of measures is Q–r–ergodic.

Proof. Let $\{\nu_n,\ n \in N\}$ be a Q–r–ergodic net of measures, let $f \in Q$, $\mathbf{M}^{(r)}(f) \in \mathfrak{M}^{(r)}(f)$, $\varepsilon > 0$, and let the elements $x_1, \ldots, x_k$ of X and positive numbers $\alpha_1, \ldots, \alpha_k$ with $\sum_{i=1}^k \alpha_i = 1$ be such that $\sup\limits_{x\in X} \|\sum_{i=1}^k \alpha_i f(xx_i) - M^{(r)}(f)\|_B \leq \varepsilon$. We have:

$$\lim_{n\in N} \sup_{y\in X} \|\int f(yx)\nu_n(dx) - \mathbf{M}^{(r)}(f)\|_B \leq$$

$$\leq \overline{\lim_{n\in N}} \sup_{y\in X} \|\int f(yx)\nu_n(dx) - \sum_{i=1}^k \alpha_i \int f(yxx_i)\nu_n(dx)\|_B +$$

$$+ \lim_{n\in N} \sup_{y\in X} \|\int (\sum_{i=1}^k \alpha_i f(yxx_i) - \mathbf{M}^{(r)}(f))\nu_n(dx)\|_B \leq$$

$$\leq \overline{\lim_{n\in N}} \sup_{x\in X} \sum_{i=1}^k \alpha_i \|\int f(yx)\nu^{(r)}_{n,x_i}(dx)\|_B + \varepsilon = \varepsilon,$$

and since ε is chosen arbitrarily we have $\lim_{n\in N} \int f(yx)\nu_n(dx) =$

$= \mathbf{M}^{(r)}(f)$ uniformly with respect to y. Thus statement *(i)* is proved. Statement *(ii)* follows from *(i)* and Proposition 1.2.6. Let the conditions of statement *(iii)* be fulfilled and let $\{\nu_n,\ n \in N\}$ be a Q–r–averaging net of measures. Then for any $x_0 \in X$ and any $f \in Q$, uniformly with respect to $y \in X$,

$$\lim_{n\in N} \int f(yx)\nu_{n,x_0}^{(r)}(dx) = \lim_{n\in N} \int f(yx)\nu_n(dx) -$$

$$- \lim_{n\in N} \int f(yxx_0)\nu_n(dx) = \mathbf{M}^{(r)}(f(y)) - M^{(r)}(f(yx_0)) = 0$$

(the last equality is justified by Proposition 1.1.7).

□

Corollary 1.2 . *Let Q be an R–invariant subset of $A_B^{(r)} \cap F_B$. Then a net $\{\nu_n,\ n \in N\}$ is Q–r–averaging iff it is Q–r–ergodic.*

The following statement shows how to construct new averaging nets using a given one.

Corollary 1.3. *Let $\{\nu_n,\ n \in N\}$ be a Q–r–averaging net, $\{\nu_n',\ n \in \in N\} \in \mathcal{P}$ and* $\lim_{n\in N} \operatorname{var}(\nu_n - \nu_n') = 0$. *Then $\{\nu_n'\}$ is Q–r–averaging.*

Proof. Our statement follows at once from the following inequality: for any $f \in F_B(X)$,

$$\sup_z \| \int f(zx)\nu_n(dx) - \int f(zx)\nu_n'(dx)\| \leq \|f\|_{F_B} \|\nu_n - \nu_n'\|_{\mathfrak{B}\mathfrak{M}}.$$

□

§ 2. Averaging Nets on Group Extensions and Products

2.1. Factor representations: definition. Let G be a group and Γ a normal subgroup; $F := G/F$; and S a left representation of G

in a Banach space B. We denote by S_Γ the restriction of S to Γ; $I(\Gamma) := \{b : b \in B,\ S_\gamma b = b,\ \gamma \in \Gamma\}$.

Lemma 2.1. *(i) The subspace $I(\Gamma)$ is S–invariant.*

(ii) The subrepresentation V' of S in $I(\Gamma)$ is constant on the left cosets with respect to Γ; i.e. $S_{g\gamma}b = S_g b$ if $g \in G$, $\gamma \in \Gamma$, $b \in I(\Gamma)$.

(iii) The mapping $V\colon f \to V_f$, where $V_f = V'_g$ if $g \in f$, is a representation of F in $I(\Gamma)$.

(iv) If G is a semitopological (resp. semimeasurable) group and S is a continuous (resp. measurable) representation, then V is continuous (resp. measurable).

(v) Let $I(\Gamma) \neq \{0\}$. If S is uniformly bounded (resp. isometric, unitary), then V has the corresponding property.

Proof. *(i)* If $b \in I(\Gamma)$, $\gamma \in \Gamma$, $g \in G$, and $\gamma' = g^{-1}\gamma g$, then $S_\gamma S_g b = S_g S_{\gamma'} b = S_g b$ and thus $S_g b \in I(\Gamma)$.

(ii) For any $b \in I(\Gamma)$, $\gamma \in \Gamma$, $g \in G$ we have: $V'_{g\gamma}b = V'_g V'_\gamma b = V'_g b$.

(iii) If $f_i = \pi(g_i)$, $i = 1, 2$, we have: $V_{f_1}V_{f_2} = V'_{g_1}V'_{g_2} = V'_{g_1 g_2} = V_{f_1 f_2}$.

Statements *(iv)*–*(v)* are obvious.

□

The representation V defined in the previous Lemma is called the *factor representation* of F (*with respect to* Γ) generated by the representation S.

2.2. Representations, their restrictions and factor representations: connection between ergodic decompositions and means. According to Proposition 1.2.4, the space of all S–averageable elements of B is S–invariant. Therefore we can assume, without loss of generality, that $B = A(B|S)$. We shall consider the main ergodic decomposition $B = I \oplus D$ with respect to S, $A(B|S_\Gamma) = I(\Gamma) \oplus D(F)$ with respect to V and $I(\Gamma) = I_V \oplus D_V$ with respect to V (see Ch. 1, Subsect. 1.3). Below we consider some relations between these decompositions.

Proposition 2.2 . *(i) The subspaces $A(B|S_\Gamma)$, $I(\Gamma)$ and $D(\Gamma)$ are S–invariant.*

(ii) $I_V = I$, and D_V is S–invariant.

(iii) $A(I(\Gamma)|V) = I(\Gamma)$.

(iv) $D \cap A(B|S_\Gamma) = D_V \oplus D(\Gamma)$.

(v) $A(B|S_\Gamma) = (D \cap A(B|S_\Gamma)) \oplus I = D_V \oplus D(\Gamma) \oplus I$.

(vi) Let $\Pi_I^{A(B|S_\Gamma)}$ be the projector in $A(B|S_\Gamma)$ onto I with respect to $D \cap A(B|S_\Gamma)$, $\Pi_{I(\Gamma)}^{A(B|S_\Gamma)}$ the projector onto I_Γ with respect to D_{S_Γ} and $\Pi_I^{I(F)}$ the projector in $I(\Gamma)$ onto $I_V = I$ with respect to D_V. Then $\Pi_I^{A(B|S_\Gamma)} = \Pi_I^{I(\Gamma}\Pi_{I(\Gamma)}^{A(B|S_\Gamma)}$; in other words, $\mathbf{M}_S(b) = \mathbf{M}_V[\mathbf{M}_{S_\Gamma}(b)]$, $b \in A(B|S_\Gamma)$.

(vii) If B is a Hilbert space and S is a unitary representation, then the subspaces which appear in decompositions (iv) and (v) and in the ergodic decomposition $A(B|S_\Gamma) = I(\Gamma) \oplus D(F)$ are orthogonal to each other; hence all the projectors which appear in (vi) are orthogonal.

Proof. *(i)* For any $b \in A(B|S_\Gamma)$ there exist elements $\gamma_i^{(n)} \in \Gamma$ and numbers $\alpha_i^{(n)} \geq 0$ $(i = \overline{1, k_n})$ such that $\sum_{i=1}^{k_n} \alpha_i^{(n)} = 1$ and $\lim_{n\to\infty} \sum_{i=1}^{k_n} \alpha_i^{(n)} S_{\gamma_i^{(n)}} b = \mathbf{M}_{S_\Gamma} b$ $(= \Pi_{I(\Gamma)} b)$. Let $g \in G$ and $\gamma_i^{(n)'} = g\gamma_i^{(n)} g^{-1}$.

Then the $\lim_{n\to\infty} \sum_{i=1}^{k_n} \alpha_i^{(n)} S_{\gamma_i^{(n)'}} S_g = \lim_{n\to\infty} \sum_{i=1}^{k_n} \alpha_i^{(n)} S_g S_{\gamma_i^{(n)}} b =$ $= \lim_{n\to\infty} S_g(\sum_{i=1}^{k_n} \alpha_i^{(n)} S_{\gamma_i^{(n)}} b) = S_g \mathbf{M}_{S_\Gamma}(b)$ exists. Thus $S_g \mathbf{M} S_\Gamma(b) \in$ $\in \mathfrak{M}_{S_\Gamma}(S_g b)$.

Similarly we can show that if $a \in \mathfrak{M}_{S_\Gamma}(S_g b)$, then $S_{g^{-1}} a = \mathbf{M}_{S_\Gamma}(b)$, i.e. $\mathfrak{M}_{S_\Gamma}(S_g b) = \{S_g \mathbf{M}_{S_\Gamma}(b)\}$. Proposition 1.2.6 implies that $S_g(A(B|S_\Gamma)) = A(B|S_\Gamma)$, $g \in G$. The S–invariance of $I(\Gamma)$ was proved in Lemma 2.1. Equality $S_g \mathbf{M}_{S_\Gamma}(b) = \mathbf{M}_{S_\Gamma}(S_g b)$ and statement *(i)* of Theorem 1.2.11 involve the S–invariance of $D(\Gamma)$.

(ii) Of course, $I \subset I(\Gamma)$, $I_V \subset I(\Gamma)$. If $b \in I(\Gamma)$, then $V_f b = b$ is equivalent to $S_g b = b$, $g \in f$. Therefore $I_V = I$.

(iii) Since $I(\Gamma)$ is S–invariant, we have for any $b \in I_{S_\Gamma}$: $K(b|S) =$ $= K(b|V)$ and thus $I \cap K(b|S) = I_V \cap K(b|V)$. Consequently, this statement is implied by Propositions 1.1.3 and Theorem 1.2.4.

(iv)–(v) We have the following ergodic decomposition of $I(\Gamma)$ with respect to V: $I(\Gamma) = I \oplus D_V$; hence $A(B|S_\Gamma) = D(\Gamma) \oplus D_V \oplus I$. Compare this decomposition with the decomposition $B = D \oplus I$; since the definition of $D(\Gamma)$, D_V and D implies $D(\Gamma) \oplus D_V \subset D \cap A(B|S_\Gamma)$, we obtain $\Pi^{A(B|S_\Gamma)}_{D\cap A(B|S_\Gamma} = \Pi^{A(B|S_\Gamma)}_{D(\Gamma)\oplus D_V}$, and the statement is proved.

(vi) This is implied by *(v)*.

(vii) This is evident by virtue of Theorem 1.4.10.

□

Note 2.1. If $A(B|S_\Gamma) = B$, statements *(iv)-(vi)* acquire a simpler form:

(iv') $D = D_V \oplus D(\Gamma)$.

(v') $B = D_V \oplus D(\Gamma) \oplus I$.

(vi') $\Pi^B_I = \Pi^{I(\Gamma)}_I \Pi^B_{I(\Gamma)}$, i.e. $\mathbf{M}_S(b) = \mathbf{M}_V[M_{S_\Gamma}(b)]$, $b \in B$.

2.3. Construction of averaging sequences on group extensions. Let G be a σ–compact, locally compact topological group and $\mu^{(l)}_G$ its left Haar measure. Let S be a uniformly bounded $B(G)$–measurable representation of G. We preserve the notations of the previous subsections. Since the subspace $A(B|S) \cap A(B|S_\Gamma)$ is S–invariant we can, without loss of generality, restrict our study to the case $B = A(B|S) = A(B|S_\Gamma)$. For any $b \in B$ and any Borel measure μ we denote by $AN(b|S, \mu)$ the class of all nets $\{G_n, n \in N\}$ such that $\{G_n\} \in \mathfrak{B}(G)$, $0 < \mu(G_n) < \infty$ and $\lim_{n\in N} \frac{1}{\mu(G_n)} \int_{G_n} S_g b\mu(dg) = \mathbf{M}(b|S)$; π is the natural homomorphism $G \mapsto F$.

Theorem 2.3. *Let* $\{\Gamma_n\} \in AN(b|S_\Gamma, \mu^{(l)}_\Gamma)$ *and* $\{F_n\} \in AN(\Pi^B_{I(\Gamma)} b|V, \mu^{(l)}_F)$. *Let* $\widetilde{F}_n$ *be a cross–section in* G *over* F_n *such that the set* $G_n := \widetilde{F}_n \Gamma \in \mathfrak{B}(\Gamma)$, $n \in N$. *Then* $\{G_n\} \in AN(b|S, \mu^{(l)}_G)$.

Proof. Put $S^{(n)} b := \frac{1}{\mu^{(l)}_G(G_n)} \int_{G_n} S_g b \mu^{(l)}_G(dg)$; similarly we introduce $S^{(n)}_\Gamma b$ and $V^{(n)} b$; $\widetilde{\mathfrak{B}}_n := \{\widetilde{A} : \widetilde{A} \in \widetilde{F}_n, \pi(\widetilde{A}) \in \mathfrak{B}(F)\}$; $\widetilde{\mu}_{\widetilde{F}_n}(\widetilde{A}) = \mu_F(\pi(\widetilde{A}))$, $\widetilde{A} \in \widetilde{\mathfrak{B}}_n$; $\widetilde{S}^{(n)} = \frac{1}{\mu_\Gamma(F_n)} \int_{\widetilde{F}_n} S_g \widetilde{\mu}_{\widetilde{F}_n}(dg)$. We have

$$S^{(n)}b = \frac{1}{\mu_G^{(l)}(G_n)} \int_F \left(\int_\Gamma \mathcal{X}_{\widetilde{F}_n\Gamma_n}(f\gamma) S_{f\gamma} \mu_\Gamma(d\gamma) \right) \mu_F(df) =$$

$$= \frac{1}{\mu_G^{(l)}(G_n)} \int_{F_n} \left(\int_\Gamma \mathcal{X}_{\widetilde{F}_n\Gamma_n}(\widetilde{f}\gamma) S_{\widetilde{f}\gamma} \mu_\Gamma(d\gamma) \right) \mu_{\widetilde{F}_n}(d\widetilde{f}) =$$

$$= \frac{1}{\mu_G^{(l)}(G_n)} \int_{\widetilde{F}_n} S_{\widetilde{f}} \left(\int_{\Gamma_n} S_g b \mu_\Gamma^{(l)}(d\gamma) \right) \widetilde{\mu}_{F_n}(d\widetilde{f}) = V^{(n)} S_\Gamma b. \quad (2.1)$$

Since $\widetilde{S}^{(n)} = V^{(n)}$ in $I(\Gamma)$, we have

$$\|S^{(n)}b - \Pi_I^B b\| \le \|\widetilde{V}^{(n)} S_\Gamma^{(n)} b - \widetilde{S}^{(n)} \Pi_{I(\Gamma)} b\| +$$
$$+ \|V^{(n)} \Pi_{I(\Gamma)}^B b - \Pi_I^{I(\Gamma)} \Pi_{I(\Gamma)}^B b\| \le$$
$$\le \sup_{g \in G} \|S_g\| \cdot \|S_\Gamma^{(n)} b - \Pi_{I(\Gamma)}^B b\| +$$
$$+ \|V^{(n)} \Pi_{I(\Gamma)}^B b - \Pi_I^{I(\Gamma)} \Pi_{I(\Gamma)}^B b\|,$$

and the hypothesis of the theorem imply that $\lim_{n \in N} S^{(n)} b = \Pi_I^B b$.

□

Theorem 2.4. *Let Γ be a central normal subgroup and $\{F_n\} \in AN(\Pi_{I(\Gamma)}^B b | V, \mu_F^{(r)})$, $\{\Gamma_n\} \in AN(b | S_\Gamma, \mu_\Gamma)$. Let $\widetilde{F}_n$ be a cross-section over F_n such that $G_n := \widetilde{F}_n \Gamma$ is measurable. Then $\{G_n\} \in AN(b | S, \mu_G^{(r)})$.*

Proof. Since Γ belongs to the center of G, equalities (2.1) remain true if we replace $\mu_G^{(l)}$ by $\mu_G^{(l)}$ and $\mu_F^{(l)}$ by $\mu_F^{(r)}$. It is only left to repeat the concluding part of the proof of Theorem 2.3.

□

Theorem 2.5. *Let Γ be a compact normal subgroup of G;*

(i) if $\{F_n\} \in AN(\Pi_{I(\Gamma)} b | V, \mu_F^{(l)})$ and $G_n := \{\pi^{-1}(F_n)\}$, then $\{G_n\} \in AN(b | S, \mu_G^{(l)})$;

(ii) if $\{F_n\} \in AN(\Pi_{I(\Gamma)} b | V, \mu_F^{(r)})$, then $G_n \in AN(b | S, \mu_G^{(r)})$.

Proof. Statement *(i)* is a simple consequence of Theorem 2.3, since for any cross–section $\widetilde{F}_n$ we have $\widetilde{F}_n\Gamma = \pi^{-1}(F_n)$. Let us prove statement *(ii)*. We have $\int_\Gamma S_{\gamma g} b\mu_\Gamma(d\gamma) = \int_\Gamma S_\gamma S_g b\mu_\Gamma(d\gamma) = \Pi^B_{I(\Gamma)} S_g(b) = S_g \Pi^B_{I(\Gamma)} b = V_{\pi(g)} \Pi^B_{I(\Gamma)}(b)$. Therefore

$$\int_{G_n} S_g b\mu_g^{(r)}(dg) = \int_{F_n}\left(\left(\int_\Gamma S_{\gamma g} b\mu_\Gamma(d\gamma)\right)\mu_F^{(r)}(d\pi(g)\right) = \int_{F_n} V_f \Pi^B_{I(\Gamma)} b\mu_F^{(r)}(df).$$

Since $\mu_G^{(r)}(G_n) = \mu_F^{(r)}(F_n)$ this implies the statement.

□

Let us put $AN(B|S,\mu) = \bigcap_{b\in B} AN(b|S,\mu)$.

Theorem 2.6. *Let $G = F \odot \Gamma$ be a left semidirect decomposition of G and S_F the restriction of the representation S to F.*

(i) If $\{F_n\} \in AN(B|S_\Gamma,\mu_\Gamma^{(l)})$ and $\{\Gamma_n\} \in AN(B|S_\Gamma,\mu_\Gamma^{(l)})$ then $G_n = \{F_n\cdot\Gamma_n\} \in AN(B|S,\mu_G^{(l)})$ and $\Pi^B_I = \Pi^B_{I(F)}\Pi^B_{I(\Gamma)}$.

(ii) If $\{F_n\} \in AN(B|S_F,\mu_F^{(r)})$ and $\{\Gamma_N\} \in AN(B|S_\Gamma,\mu_\Gamma^{(r)})$ then $\{G'_n\} := \{\Gamma_n F_n\} \in AN(B|S,\mu_G^{(r)})$.

(iii) $\Pi^B_I = \Pi_{I(F)}\Pi^B_{I(\Gamma)} = \Pi^B_{I(\Gamma)}\Pi^B_{I(F)}$.

Proof. Statement *(i)* is an obvious consequence of Theorem 2.3 and Note 2.1, since V is a subrepresentation of S_F in $I(\Gamma)$.

(iii) Let $\{F_n\}$, $\{\Gamma\}$ and $\{G_n\}$ be as in *(i)*; consider the linear operators $S^{(n)} = \frac{1}{\mu_g(G_n)}\int_{G_n} S_g\mu_G^{(l)}(dg)$, $S_F^{(n)} = \frac{1}{\mu_F^{(l)}(F_n)}\int_{F_n} S_f\mu_F^{(l)}(df)$, $S_\Gamma^{(n)} = \frac{1}{\mu_\Gamma^{(l)}(\Gamma_n)}\int_{\Gamma_n} S_g\mu_G^{(l)}\ (d\gamma)$. It is easy to check that $S_G^{(n)} = S_F^{(n)}S_\Gamma^{(n)}$; we have by *(i)*, $\Pi^B_I = \lim_{n\in N} S_G^{(n)} = \lim_{n\in N} S_F^{(n)}S_\Gamma^{(n)} = \Pi^B_{I(F)}\Pi^B_{I(\Gamma)}$. Let us consider the inverse right representations S^{-1}: $g \to S_g^{-1}$. It is easy to verify that $\{F_n^{-1}\} \in AN(B|S^{-1},\mu_F^{(r)})$ and $\{\Gamma_n^{-1}\} \in AN(B|S_\Gamma^{-1},\mu_\Gamma^{(r)})$, and it is obvious that the ergodic decompositions S of B with respect to S and S^{-1} coincide. Therefore, by symmetry, $\Pi^B_I = \Pi^B_{I(\Gamma)}\Pi^B_{I(F)}$.

(ii) Let us use the introduced notations $S_F^{(n)}$ and $S_\Gamma^{(n)}$ for operators with $\mu_F^{(r)}$ and $\mu_\Gamma^{(r)}$ instead of $\mu_F^{(l)}$ and $\mu_\Gamma^{(l)}$ and the notation $S_g^{(n)}$ with $G_n' = \Gamma_n F_n$ instead of G_n and $\mu_G^{(r)}$ instead of $\mu_G^{(l)}$. Then $S_G^{(n)} = = S_\Gamma^{(n)} S_F^{(n)}$, and $\lim_{n\in N} S_G^{(n)} = \Pi^B_{I(\Gamma)}\Pi^B_{I(F)} = \Pi^B_I$ by virtue of statement *(iii)*.

We conclude this subsection by a simple statement which is in some sense converse to Theorem 2.3 and will be useful in the sequel.

□

Lemma 2.7. *Suppose that Γ is a normal subgroup of G, $F = G/\Gamma$; and let $\{F_n\}$ and $\{\Gamma_n\}$, be nets of measurable sets in F and Γ, resp. Let $V\colon f \to V_f$ be a representation of F in B; consider the trivial extension of V onto G: $S_g = V_f$ if $g \in f$. Let $\widetilde{F}_n$ be a cross-section in G over F_n, $n \in N$.*

(i) If $\{G_n\} := \{\widetilde{F}_n\Gamma\} \in AN(B|S,\mu_G^{(l)})$, then $\{F_n\} \in AN(B|V,\mu_F^{(l)})$.

(ii) If $\{G_n'\} := \{\Gamma_n\widetilde{F}_n\} \in AN(B|S,\mu_G^{(r)})$, then $\{F_n\} \in AN(B|V,\mu_F^{(r)})$.

Proof. *(i)* As in the proof of Theorem 2.3 we have $S^{(n)} = \widetilde{V}^{(n)}S^{(n)}$. In the present case $\widetilde{V}^{(n)} = V^{(n)}$ and $S_\Gamma^{(n)} \equiv I$. Thus $S^{(n)} = V^{(n)}$, and $\lim_{n\in N} V^{(n)} = \Pi^B_{I_S} = \Pi^B_{I_V}$. Statement *(ii)* can be proved similarly.

□

2.4. Construction of averaging nets on multiple products of subgroups. Let G be decomposed into a left semidirect product of subgroups: $G = G^{(1)}\overset{l}{\odot}G^{(2)}\cdots\overset{l}{\odot}G^{(m)}$, i.e. $G = G^{(1)}\cdots G^{(m)}$, where $G^{(i)}\cap G^{(j)} = \{e\}$, $i \neq j$, and $\Gamma^{(i)} := G^{(i)}\cdots G^{(m)}$ is a normal subgroup of $\Gamma^{(i-1)}$, $i = 1,\ldots,m$ (this implies $\Gamma^{(i-1)}/\Gamma^{(i)} \cong G^{(i-1)}$). Let $S^{(i)}$ be the restriction of S to $G^{(i)}$.

Theorem 2.8. *(i) If $\{G_n^{(i)}\} \in AN(B|S^{(i)},\mu^{(l)}_{G^{(i)}})$, $i = \overline{1,m}$, and $G_n := G_n^{(1)}\cdots G_n^{(m)}$; then $\{G_n\} \in AN(B|S,\mu_G^l)$.*

(ii) If $\{G_n^{(i)}\} \in AN(B|S^{(i)},\mu^{(r)}_{G^{(i)}})$ and $G_n' = G_n^{(m)}\cdots G_n^{(1)}$, then $\{G_n'\} \in AN(B|S,\ \mu_G^{(r)})$.

Proof. In case $m = 2$ these statements coincide with statements *(i)* and *(ii)* of Theorem 2.6. The general case follows by induction with respect to m.

Now we shall suppose that G is decomposed into an almost direct product of subgroups $G^{(1)}, \ldots, G^{(m)}$, i.e. that $G = G^{(1)} \cdots G^{(m)}$ where $G^{(i)}$ are subgroups of G with the following properties:

a) G^i and G^j commute, i.e. $g^{(i)}g^{(j)} = g^{(j)}g^{(i)}$ if $g^{(i)} \in G^{(i)}$, $g^{(j)} \in G^{(j)}$, $(i \neq j)$;

b) the subgroups $G^{(i)} \cap G^{(j)}$ are compact $(i \neq j,\ i,\ j = 1, \ldots, m)$.

□

Theorem 2.9. *(i) If* $\{G_n^{(i)}\} \in A(B|S^{(i)}, \mu_G^{(l)})$, $G_n := G_n^{(1)}, \ldots, G_n^{(m)}$, *then* $\{G_n\} = A(B|S, \mu_G^{(l)})$.

(ii) If $\{G_n^{(i)}\} \in A(B|S^{(i)}, \mu_G^{(r)})$, *then* $\{G_n\} \in A(B|S, \mu_G^{(r)})$.

Proof. *(i)* Let $m = 2$. Put $S_n^{(i)} = \frac{1}{\mu_G^{(l)}(G_n^{(i)})} \int_{G_n^{(i)}} S_g^{(i)} \mu_{G^{(i)}}^{(l)}(dg)$, $i = 1, 2$, and $S_n = \frac{1}{\mu_G^{(l)}(G_n)} \int_{G_n} S_g \mu_G^{(l)}(dg)$. By App. (1.L), we have $S_n = S_n^{(1)} S_n^{(2)}$ and hence $\lim_{n \in N} = \Pi^B_{I(G^{(1)})} \Pi^B_{I(G^{(2)})}$. Since $G_n^{(1)}$ and $G_n^{(2)}$ commute, we also have: $S_n = S_n^{(2)} S_n^{(1)}$ and thus $\lim_{n \in N} S_n = \Pi^B_{I(G^{(1)})} \Pi^B_{I(G^{(2)})} = \Pi^B_{I(G^{(2)})} \Pi^B_{I(G^{(1)})} = \Pi^B_{I(G^{(1)}) \cap I(G^{(2)})} = \Pi^B_I$, and $\{G_n\} \in AN(B|S, \mu_G^{(l)})$. If $m > 2$ we have to use induction with respect to m.

Statement *(ii)* can be proved similarly.

□

§ 3. Universal Ergodic Theorems on Amenable Semigroups

In this section we shall consider universally $A_B^{(r)} \cap F_B$–r–averaging nets on r–amenable measurable semigroups. Recall that, according to our definitions, $\{\nu_n,\ n \in N\}$ is such a net iff the following two equivalent statements are true.

(i) **The universal mean value theorem:** $\lim_{n\in N} \int f(yx)\nu_n(dx) = \mathbf{M}^{(r)}(f)$ *uniformly with respect to* $y \in X$ *for any* $f \in A_B^{(r)} \cap F_B$ *and for any Banach space* B.

(ii) **The universal ergodic theorem:** $\lim_{n\in N} \int S_x b\nu_n(dx) = \mathbf{M}(b|S)$, $b \in B$, *for any Banach space* B *and any uniformly bounded measurable left representation of* X *in* B.

To prove *(i)* and *(ii)* means to find *universally* $A_B^{(r)} \cap F_B$*–r–averaging nets*. Below we show a method for finding such nets.

3.1. Averaging by ergodic nets. Theorem 1.1 and Corollary 1.2 lead to the problem of finding nets which are $A_B^{(r)} \cap F_B$–r–ergodic for any B. It follows from the inequality $\sup_{z\in X} \| \int f(zy)\nu_{n,x}^{(r)}(dy)\| \le \le \|f\|_{F_B} \|\nu_{n,x}^{(r)}\|_{\mathfrak{B}\mathcal{M}}$ that any r–ergodic net $\{\nu_n\}$ (see App., § 3) is such a net. Thus we obtain the following statement.

Theorem 3.1. *Any* r*–ergodic net of measures* $\{\nu_n\}$ *is a universally* $A_B^{(r)} \cap F_B$*–r–averaging net.*

We recall that such nets exist on any r–amenable measurable semigroup and that on any amenable measurable group G there exist r–ergodic (with respect to the right Haar measure) nets of compact sets $\{A_n\}$, i.e. nets with the property

$$\lim_{n\to\infty} \frac{\mu^{(r)}(A_n \Delta A_n g)}{\mu^r(A_n)} = 0, \qquad g \in G$$

(see App., § 3).

Corollary 3.2. *Let* G *be an amenable locally compact topological group. Any* r*–ergodic (Følner) net of sets* $\{A_n\}$ *is a universally* $A_B^{(r)} \cap F_B$*–r–averaging net with respect to the right Haar measure* $\mu_G^{(r)}$.

This corollary and App. 3.1 and 3.2 imply the following two universal ergodic theorems on $\mathbb{R}^m$.

Let $r(A)$ be the upper bound of radii of balls which are contained in $A \in \mathbb{R}^m$.

Corollary 3.3. *Let $\{A_n,\ n \in N\}$ be a net of bounded convex sets in $\mathbb{R}^m$ with $r(A_n) \to \infty$. Then $\{A_n,\ n \in N\}$ is a universally $A_B^{(r)} \cap F$-averaging net on $\mathbb{R}^m$, $m \geq 1$.*

Example 3.1. It is obvious that in $\mathbb{R}^2$ the sequence of strips $A_n = \{x\colon x = (x_1, x_2),\ 0 \leq x_1 \leq n,\ 0 \leq x_2 \leq 1,\ n = 1, 2, \dots$. It is obvious that $\{A_n\}$ does not average the one-dimensional representation $(x_1, x_2) \mapsto \mapsto e^{ix_2}$. Thus, in Corollary 3.3 the condition $r(A_n) \to \infty$ can not be omitted.

Corollary 3.4. *On $\mathbb{R}^m$ any net of sets $\{A_n,\ n \in N\}$, which are similar to a measurable set A with Lebesgue measure $0 < l(A) < \infty$, is a mean averaging net iff the similarity coefficients $\lambda(A_n : A) \xrightarrow{n \in N} \infty$.*

3.2. Construction of averaging sequences of sets using canonical semidirect factorization. Let G be an amenable locally compact connected topological group. If $G = G^{(0)} \odot \cdots \odot G^{(m)}$ is a canonical semidirect decomposition of G (see App., § 3) and $\varphi^{(i)}$ is the isomorphism $\mathbb{R} \to G^{(i)}$, we set $G_n^{(i)} := \varphi([-n, n])$ or $\varphi([0, n])$, $1 \leq i \leq m$; $G_n := G^{(0)} \cdot G_n^{(1)} \cdot \dots \cdot G_n^{(m)}$, $\widetilde{G}_n := G_n^{(m)} \dots G_n^{(1)} G^{(0)}$.

We denote by $O_{W(B)}$ the class of all orbital functions $\varphi(x) = S_x b$ where $b \in W(B|S)$ and S runs over the class of all uniformly bounded representations.

Evidently, any sequence $\{G_n^{(i)}\}$ is ergodic, and thus Corollary 3.2 and Theorem 2.8 imply the following statement.

Theorem 3.5. *If $G = G^{(0)} \odot \cdots \odot G^{(m)}$ is a canonical semidirect factorization, then the sequence $\{G_n\}$ is universally $Q_{W(B)}$-r-averaging with respect to the left Haar measure $\mu_G^{(l)}$ and $\{\widetilde{G}_n\}$ is such a sequence with respect to $\mu_G^{(r)}$.*

This solves the problem of constructing universally $O_{W(B)}$–r–averaging sequences on an amenable group if G is a simply connected Lie group or if its maximal compact normal subgroup $K(G)$ is trivial (see App., (3.G)). In order to deal with the general case we recall that if $G' = G/K(G)$, then $K(G') = \{e\}$. Theorem 2.5 shows how to make the last step.

Theorem 3.6. *If π is the natural homomorphism $G \mapsto G'$ and $\{G'_n\}$ and $\{\widetilde{G}'_n\}$ are the sequences in G' constructed in Theorem 3.3, then $\{G_n\} := \{\pi^{-1}(G'_n)\}$ and $\{\widetilde{G}_n\} := \{\pi^{-1}(\widetilde{G}'_n)\}$ are universally $Q_{W(B)}$–r–averaging sequences on G: $\{G_n\}$ with respect to $\mu_B^{(l)}$ and $\{\widetilde{G}_n\}$ with respect to $\mu_B^{(r)}$.*

Example 3.2. In $\mathbb{R} \odot \mathbb{R}^{\times}_{++}$, the affine group of the reals (see App., § 1 and § 2), the rectangles $[-a_n, a_n] \times [\sigma_n^{-1}, \sigma_n]$ with $a_n \to \infty$, $\sigma_n \to \infty$ form a universally $Q_{W(B)}$–r–averaging sequence with respect to $\mu^{(r)}$. Recall that such a sequence is r–ergodic only under the additional condition $a_n/\sigma_n \to \infty$ (see App., Example 3.3). In the isomorphic left semidirect product $\mathbb{R}^{\times}_{+} \odot \mathbb{R}$ the sequence $\{[\sigma_n^{-1}, \sigma_n] \times [-a_n, a_n]\}$ is a universally $Q_{W(B)}$–r–averaging sequence with respect to $\mu_B^{(l)}$.

The formulations of "symmetric" l–versions of statements 3.1–3.6 are evident.

§ 4. Existence of Universally Averaging Nets of Measures on General Groups

In the previous section it was proved that ergodic nets are universally averaging; at least in principle this solves the problem on averaging all averageable functions on amenable semigroups. On non-amenable groups and semigroups, even weakly ergodic nets of measures do not exist. Nevertheless, the following statement is true.

Theorem 4.1. *On any semigroup X there exist universally $A^{(r)}(B)$–r–averaging nets of measures $\{\nu_n,\ n \in N\}$ in $\widetilde{\mathcal{P}}$.*

Proof. We put $\alpha := \text{card}\, X$. Let us consider a function $f \in A_B^{(r)}$ and the subspace B_f in $A_B^{(r)}$ generated by the set $\{R_x f,\, x \in X\}$. It is evident that:

a) the weight of the Banach space B_f does not exceed α;

b) B_f is r–invariant.

Thus, any function $f \in A_B^{(r)}$ can be considered as an element of the space B_f and the orbital function with respect to R: $x \mapsto R_x$ is an element of $A_{B_f}^{(r)}$. Now we choose a representative of each class of mutually isometric isomorphic Banach spaces with weights not exceeding α; we denote the set of all these representatives by β_α.

From what was said above it follows that any net $\{\nu_n,\, n \in N\}$ that averages the set of Banach spaces $\{A_B^{(r)},\, B \in \beta_\alpha\}$ is universally $A_B^{(r)}$–r–averaging. We put $\mathcal{L}$ the set of families $\mathcal{D} = \{D_B,\, B \in \beta_\alpha\}$ where D_B is a linear operator in $A_B^{(r)}$. We turn $\mathcal{L}$ into a locally convex linear topological space by setting $\{D_B^{(1)},\, B \in \beta_\alpha\} + \{D_B^{(2)}, B \in \beta_\alpha\} = \{D_B^{(1)} + D_B^{(2)}\}$, $\lambda\{D_B,\, B \in \beta_\alpha\} = \{\lambda D_B,\, B \in \beta_\alpha\}$ and by introducing in $\mathcal{L}$ the topology of strong convergence of operators D_B in $A_B^{(r)}$ for each $B \in \beta_\alpha$. In other words, all sets $U^{\delta_1,\ldots,\delta_k}_{f_1,\ldots,f_k,B_1,\ldots,B_k} = \{\mathcal{D} = \{D_B,\, B \in \beta_\alpha\} : \|D_{B_i} f_i\|_{\Phi_{B_i}} < \delta_i,\ i = 1, \ldots, k\}$, where $f_i \in B_i$, $k = 1, 2, \ldots$, form a fundamental system of neighborhoods of zero in $\mathcal{L}$. Consider an arbitrary real functional $l \in \mathcal{L}^*$. Since it is continuous for any $\varepsilon > 0$ we can find a neighborhood $U^{\delta_1,\ldots,\delta_k}_{f_1,\ldots,f_k,B_1,\ldots,B_k}$ such that $|l(\mathcal{D})| < \varepsilon$ for $\mathcal{D} \in U^{\delta_1,\ldots,\delta_k}_{f_1,\ldots,f_k,B_1,\ldots,B_k}$. By Lemma 1.2.3, such a measure $\lambda \in \widetilde{\mathcal{P}}$ can be chosen so that

$$\|\mathbf{M}^{(r)}(f_i) - R_\lambda(f_i)\|_{\Phi_{B_i}} < \delta_i, \qquad i = 1, \ldots, k,$$

and, consequently, $|l(\mathbf{M}(r) - l(R_\lambda)| < \varepsilon$. This implies that

$$l(\mathbf{M}^{(r)}) \leq l(R_\lambda) + \varepsilon \leq \sup_{x \in X} l(R_x) + \varepsilon.$$

Since $\varepsilon > 0$ is chosen arbitrarily,

$$l(\mathbf{M}^{(r)}) \leq \sup_{x \in X} l(R_x). \tag{3.1}$$

Evidently, any convex linear combination of operators R_x, $x \in X$, can be written in the form: $R_\nu = \int R_x \nu(dx)$ where $\nu \in \widetilde{\mathcal{P}}$. Therefore, by the approximation principle (cf. Edwards [1], Theorem 2.3.1) it follows from (3.1) that in $\widetilde{\mathcal{P}}$ there exists a net $\{\nu_n,\ n \in N\}$ such that for any $B \in \beta_\alpha$ we have $\lim_{n\in N} \int R_x \nu_n(dx) = \mathbf{M}^{(r)}$ in the sense of strong operator convergence in $A_B^{(r)}$. Thus $\{\nu_n\}$ is a universally $A_B^{(r)}$-r–averaging net. Corollary 1.3 shows that $\{\nu_n\}$ is not the only net with this property.

□

Lemma 4.2. *On any semigroup there exist universally $A_B^{(r)}$–r–averaging nets of measures $\{\nu_n\} \subset \widetilde{\mathcal{P}}$ with rational values.*

Proof. Let $\{\lambda_n,\ n \in N\}$ be an $A_B^{(r)}$–r–averaging net in $\widetilde{\mathcal{P}}$. Put $c(\lambda_n) =: \{x_1^{(n)}, \dots, x_k^{(n)}\}$ and $\lambda_n(\{x_i^{(n)}\}) = \alpha_i^{(n)}$. We choose natural numbers m_n, $n \in N$, and non–negative integers $l_i^{(n)}$ $(i = 1, \dots, k)$ such that *1)* $\lim_{n\in N} \frac{m_n}{k_n} = \infty$; *2)* $l_i^{(n)} \le m_n$; *3)* $|\alpha_i^{(n)} - l_i^{(n)}/m_n| < 1/m_n$. Then

$$\Big|\sum_{i=1}^{k_n} l_i^{(n)} - m_n\Big| = m_n\Big|\sum_{i=1}^{k_n} l_i^{(n)}/m_n - 1\Big| \le$$

$$\le m_n \sum_{i=1}^{k_n} \big|l_i^{(n)}/m_n - \alpha_i^{(n)}\big| < k_n,$$

and therefore there exist non–negative integers $d_i^{(n)}$ such that $|l_i^{(n)} - d_i^{(n)}| \le 1$ and $\sum_{i=1}^{k_n} d_i^{(n)} = m_n$. Consider, the measures ν_n, $n \in N$, where $c(\nu_n) = c(\lambda_n)$ and $\nu_n(x_i^{(n)}) = \frac{d_i^{(n)}}{m_n}$; clearly $\nu_n \in \widetilde{\mathcal{P}}$ and $\mathrm{var}(\nu_n - \lambda_n) \le \max_{1\le i\le k_n} |\frac{d_i^{(n)}}{m_n} - \alpha_i^{(n)}| \le \frac{2k_n}{m_n}$. By Corollary 1.3, $\{\nu_n\}$ is an $A_B^{(r)}$–r–averaging net.

Now let X be a topological group. We let $\widetilde{\widetilde{\mathcal{P}}}$ be the class of all uniformly distributed measures in $\widetilde{\mathcal{P}}$; in other words, the notation $\nu \in \widetilde{\widetilde{\mathcal{P}}}$ means that $\nu(\{x\}) = 1/|c(\nu_n)|$, $x \in c(\nu_n)$. ($c(\nu)$ is the cardinality of

$c(\nu)$); $UC_B^{(l)} = UC_B^{(l)}(X)$ is the set of all bounded l–uniformly continuous functions on X.

□

Lemma 4.3. *On any non–discrete topological group X there exists a universally $A_B^{(r)} \cap UC_B^{(l)}$–$r$–averaging net of measures $\{\nu_n,\ n \in \in N\} \in \widetilde{\widetilde{\mathcal{P}}}$ such that 1) $|c(\nu_n)| \uparrow \infty$; 2) $c(\nu_n)$ is contained in a given dense subset $Q \subset X$.*

Proof. Let $\{V_m,\ m \in M\}$ be a decreasing net of neighbourhoods of the identity such that $\bigcap_{m\in M} V_m = \{e\}$ and let $\{\lambda_k,\ k \in K\}$ be a net with the properties defined in Lemma 4.2. Consider the direction $N := M \times K$ with the order $(m_1, k_1) \prec (m_2, k_2)$ iff $m_1 \prec m_2$, $k_1 \prec k_2$ and set $V_n := V_m$, $\lambda_n := \lambda_k$ if $n = (m, k)$. It is clear that the nets $\{V_n\}$ and $\{\nu_n\}$ possess the same properties as $\{V_m\}$ and $\{\lambda_k\}$. We put $c(\lambda_n) =: \{x_1^{(n)}, \ldots, x_{k_n}^{(n)}\}$, $\lambda_n(\{x_i^{(n)}\}) =: d_i^{(n)}/m_n$. It is evident that the $d_i^{(n)'}s$ and $m_n's$ can be chosen such that $m_n \uparrow \infty$. In any set $x_i^{(n)} V_n$ we choose $d_i^{(n)}$ arbitrary points: $y_{i,1}^{(n)}, \ldots, y_{i,d_2^{(n)}} \in Q$ $(i = \overline{1, k_n})$ in such a way that $y_{i_1,j_1}^{(n)} \neq y_{i_2,j_2}^{(n)}$ if $i_1 \neq i_2$, $j_1 \neq j_2$. Consider measures ν_n with carriers $c(\nu_n) = \bigcup_{i=1}^{k_n} \{y_{i_1,1}, \ldots, y_{i,d_i^{(n)}}\}$. Evidently, $|c(\nu_n)| = = \sum_{i=1}^{k_n} d_i^{(n)} = m_n$ and thus $|c(\nu_n)| \uparrow \infty$. If $f \in A_B^{(r)} \cap C_B^{(l)}$, then

$$\|\int f(zx)\nu_n(dx) - \mathbf{M}^{(r)}(f)\|_{\Phi_B} \leq$$

$$\leq \|(1/|c(\nu_n)|) \sum_{x\in c(\nu_n)} f(zx) - \sum_{i=1}^{k_n} (d_i^{(n)}/m_n) f(zx_i^{(n)})\|_{\Phi_B} +$$

$$+ \|\int f(zx)\lambda_n(dx) - \mathbf{M}^{(r)}(f)\|_{\Phi_B} =$$

$$=\|(1/m_n)\sum_{i=1}^{k_n}\sum_{j=1}^{d_i^{(n)}} f(zy_{ij}^{(n)}) - \sum_{i=1}^{k_n}(d_i^{(n)}/m_n)f(zx_i^{(n)})\|_{\Phi_B}+$$

$$+\|\int f(zx)\lambda_n(dx) - \mathbf{M}^{(r)}(f)\|_{\Phi_B} \leq \sup_{\substack{x,z\in X\\ y\in xV_m}} \|f(zy)-f(zx)\|_B+$$

$$+\|\int f(zx)\lambda_n(dx) - \mathbf{M}^{(r)}(f)\|_{\Phi_B}.$$

From these inequalities it follows at once that $\{\nu_n\}$ l–averages the function f. Since $f \in A_B^{(r)} \cap UC_B^{(l)}$ is chosen arbitrarily, the statement is proved.

□

Theorem 4.4. *On any non–discrete locally compact group there exist universally* $A_B^{(r)} \cap UC_B^{(l)}$*–r–averaging (with respect to the left Haar measure) nets of compact Baire sets.*

Proof. Let $\{\nu_n,\ n \in N\}$ be a net of measures with the properties stated in Lemma 4.3; $\{x_1^{(n)}, \ldots, x_{k_n}^{(n)}\} := c(\nu_n)$. Let us consider a net $\{V_n,\ n \in N\}$ of open neighborhoods of the identity e with the following properties: $V_n \downarrow \{e\}$, and if $i \neq j$ then $(x_i^{(n)})^{-1}x_j^{(n)} \notin V_n$. In any V_n we consider an open neighborhood U_n of e such that $U_nU_n^{-1} \in V_n$, $n \in N$; in U_n we can choose a compact Baire neighborhood W_n' (see, e.g., Halmos [1], § 50, Theorem 4).

Let $W_n := W_n' \cap (W_n')^{-1}$; $A := \bigcup_{i=1}^{k_n} x_i^{(n)} W_n$. It is easy to show that $x_i^{(n)}W_n \cap x_j^{(n)}W_n = \emptyset$ if $i \neq j$. Therefore, for any Banach space B and any function $f \in A_B^{(r)} \cap UC_B^{(l)}$ we have

$$\|\frac{1}{\mu(A_n)}\int_{A_n} f(yx)\mu(dx) - \mathbf{M}^{(r)}(f)\| \leq$$

$$\leq \|\frac{1}{k_n}\sum_{i=1}^{k_n} f(yx_i^{(n)}) - \mathbf{M}^{(r)}(f)\| +$$

$$+\|\frac{1}{\mu(A_n)}\int_{A_n} f(yx)\mu(dx) - \frac{1}{k_n}\sum_{i=1}^{k_n} f(yx_i^{(n)})\| \leq$$

$$\leq \|\frac{1}{k_n}\sum_{i=1}^{k_n} f(yx_i^{(n)}) - \mathbf{M}^{(r)}(f)\| +$$

$$+\frac{1}{\mu(A_n)}\|\sum_{i=1}^{k_n}\int_{x_i^{(n)}W_n}(f(yx) - f(yx_i^{(n)}))\mu(dx)\| \leq$$

$$\leq \|\frac{1}{k_n}\sum_{i=1}^{k_n} f(yx_i^{(n)}) - \mathbf{M}^{(r)}(f)\| +$$

$$+\sup\{\|f(yx) - f(yz)\|, \quad y, z \in X, \quad x \in zU_n\}.$$

These relations imply the statement.

□

§ 5. Universally $O_{W(B)}$–Averaging Sequences

In this section X is a fixed separable topological semigroup; B is the Borel σ–algebra in X; B is a Banach space; and $O_{W(B)}$ is the class of all orbital functions of the form $\varphi_{S,b}(x) = S_x b$ where S runs over the class of all bilaterally uniformly bounded continuous measurable left representations of X in B and b runs over $W(B, S)$ for any fixed S. If X is a group, then, according to Theorem 1.4.7, $O_{W(B)} \in A_B^{(r)}$ and from § 4 the existence of universally $O_{W(B)}$–r-averaging nets of sets and measures follows. Now we shall consider the question of the existence of $O_{W(B)}$-r-universally averaging sequences of sets and measures.

5.1. Averaging by means of convolutions. In this subsection a general method for constructing of $O_{W(B)}$–r–averaging sequences of measures is presented which is suitable, for example, for all separable groups.

Let $q \in \mathcal{P}(B)$; let Y_q be the r–complete subsemigroup of X generated by the carrier $c(q)$ of q; $\lambda_n^{(q)} := \frac{1}{n}\sum_{k=1}^{n} q^k$ where q^k is the k–th power of the measure q with respect to convolution.

Theorem 5.1. *Let $S\colon x \mapsto S_x$ be some bilaterally uniformly bounded l–representation of X in a Banach space B; $b \in W(B,S)$. Then in B the limit $\lim_{n\to\infty} \int S_x b\lambda_n^{(q)}(dx)$ exists if $Y_q = X$, this limit coincides with $\mathbf{M}(b)$.*

Proof. According to Lemma 1.3.5, $\lim\limits_{n\to\infty} \int S_x b\lambda_n^{(q)}(dx) =: c$ exists. Let W be the set of all $x \in X$ for which $S_x c = c$. According to the same Lemma 1.3.5, $q(W) = 1$. Since S is continuous, the set W is closed and consequently $c(q) \subset W$. Moreover, $W \cdot W = W$ and if $x \in W$, $y \in Wx^{(-1)}$ then $T_y c = T_y T_x c = T_{yx} c = c$, i.e. $y \in W$. Thus W is an r-complete subsemigroup of X and $Y_q \subset W$. If $Y_q = X$, then $W = X$ and hence $c \in I$, the set of all S–invariant vectors. Lemma 1.3.4 implies that $c \in K(b)$ and therefore $c \in \mathfrak{M}(b)$.

□

Corollary 5.2. *If X is a group and $Y_q = X$, then the sequence $\lambda_n^{(q)}$ averages all continuous weakly almost periodic functions on X.*

Corollary 5.3. *If X is a group, $Y_q = X$ and S is a bilaterally uniformly bounded continuous measurable representation in a reflexive Banach space B, then for any $b \in B$ we have* $\lim\limits_{n\to\infty} \int S_x b\lambda_n^{(q)}(dx) = \mathbf{M}(b)$.

5.2. Existence of $O_{W(B)}$–r–averaging sequences of sets. Let X be a separable non–discrete locally compact group. In this subsection it will be shown that on X there exist increasing universally $O_{W(B)}$–r–averaging (with respect to the counting measure) sequences of finite sets and increasing universally $O_{W(B)}$–r–averaging (with respect to the left Haar measure) sequences of compact sets.

Lemma 5.4. *On X there exist universally $O_{W(B)}$–r–averaging sequences of measures $\{\nu_n\} \subset \widetilde{\mathcal{P}}$.*

Proof. Let Q be a probability measure on X with a countable and dense carrier $c(q)$; evidently $Y_q = X$. Set $\lambda_n := \frac{1}{n}\sum_{i=1}^n q^k$, $n = 1, 2, \ldots$, $c(\lambda_n) =: (x_1^{(n)}, \ldots, x_n^{(n)})$ and $k_n = \min(i : i \geq 2, \sum_{j=1}^{i-1} \lambda_n(x_j^{(n)}) > 1 - \frac{1}{n})$. Consider new measures $\nu_n : \nu_n(\{x_i^{(n)}\} = = \lambda_n(\{x_i^{(n)}\})$ if $1 \leq i \leq k_n$, $\nu_n(\{\lambda_n^{(n)}\}) = 1 - \sum_{j=1}^{k_n} \lambda_n(x_j^{(n)})$, and $c(\nu_n) = \{x_1^{(n)}, \ldots, x_{k_n}^{(n)}\}$. It is clear that $\nu_n \in \widetilde{\mathcal{P}}$ and $\| \int S_x b \nu_n(dx) - - \int S_x b \lambda_n(dx)\| \leq \frac{2}{n} \sup_{x \in X} \|S_x\| \|b\|$. According to Theorem 3.5.1, $\{\lambda_n\}$ is a uniformly $O_{W(B)}$–r–averaging sequence of measures; therefore $\{\nu_n\}$ also has this property.

□

Lemma 5.4 is a "sequential" analog of Theorem 2.4.1. Analogs of Lemmas 2.4.2 and 2.4.3 (with $N = 1, 2, \ldots$) can be proved similarly. We omit their formulations and proofs.

Theorem 5.5. *There exist universally $O_{W(B)}$–r–averaging increasing sequences of finite sets $\{A_n\}$.*

Proof. The conclusion of the theorem, except the monotonicity requirement, coincides with the "sequential" analogue of Lemma 2.4.2 mentioned above. So let $\{A'_k\}$ be a universally $O_{W(B)}$–r–averaging sequence of finite sets with $|A'_k| \uparrow \infty$. According to Proposition 1.1.4, these properties are shared by any sequence $\{y^{(k)} A'_k\}$ where $y^{(k)} \in X$. We shall use this fact in order to construct a sequence $\{A''_n\}$ of disjoint sets with these properties. Let $A''_1 := A'_1$ and, if the sets $A''_1, \ldots, A''_k$ are constructed, let $A''_{k+1} := z^{(k+1)} A_{k+1}$ where $z^{(k+1)} \notin \bigcup_{i=1}^k A''_i (A'_{k+1})^{-1}$; evidently, $A''_1, A''_2, \ldots$, are pairwise disjoint. We set $A_n := \bigcup_{k=1}^n A''_k$. Since $\sum_{k=1}^\infty |A_n| = \infty$, we have for any B, any uniformly bounded continuous representation $x \to S_x$ in B and any $b \in W(B|S)$:

$$\lim_{n\to\infty}|A_n|^{-1}\sum_{x\in A_n}S_xb=\lim_{n\to\infty}(\sum_{k=1}^{n}|A''_k|)^{-1}\sum_{k=1}^{n}\sum_{x\in A''_n}S_xb=$$
$$=\lim_{n\to\infty}|A''_n|^{-1}\sum_{x\in A''_n}S_xb=\mathbf{M}(b).$$

□

Lemma 5.6. *Let X be a separable locally compact group, non-discrete and non-compact, let μ_l be the left Haar measure on X. There exists a universally $O_{W(B)}$–r–averaging sequence of finite sets $\{A_n\}$, $A_n = = \{x_1^{(n)},\dots,x_{k_n}^{(n)}\}\subset X$, and sequences of compact sets $\{W_n\}$ and of open neighborhoods $\{V_n\}$ with the following properties:*
(i) $V_n\downarrow\{e\}$ and $e\in W_n\subset V_n$;
(ii) $|A_n|\mu(W_n)\geq 1$;
(iii) for any n the sets $x_k^{(n)}W_n$, $k=1,\dots,k_n$, are disjoint;
(iv) the sets A_nW_n, $n=1, 2,\dots$ are disjoint.

Proof. Let $\{A'_n\}$ be a universally $O_{W(B)}$–r–averaging sequence of finite sets in X, $A'_n=\{x_1^{(n)},\dots,x_{k_n}^{(n)}\}$. Let us consider an arbitrary sequence $\{V_n\}$ of relatively compact open neighbourhoods of the identity e with the following properties: $V_n\downarrow\{e\}$ and $(x_i^{(n)})^{-1}x_j^{(n)}\notin V_n$ if $x_i^{(n)}\neq x_j^{(n)}$. In each V_n we choose a symmetric compact neighborhood $W_n$of e such that $W_nW_n\subset V_n$. Let $a_n:=[\frac{1}{\mu(W_n)}]$; we choose a_n points $y_1^{(n)},\dots,y_{a_n}^{(n)}\in X$ in the following way. The point $y_1^{(n)}$ is chosen arbitrarily. Suppose we have chosen some points $y_1^{(n)},\dots,y_k^{(n)}$ $(1\leq k\leq a_n)$. If $k=a_n$ our aim is achieved; if $k<a_n$, we choose $y_{k+1}^{(n)}$ so that

$$y_{k+1}^{(n)}\notin\bigcup_{r\leq k}y_r^{(n)}\bigcup_{s,f=1}^{k_n}x_s^{(n)}V_n(x_t^{(n)})^{-1}\tag{5.2}$$

(this is possible since the group X is supposed to be non-compact). We shall show now that the sets $y_i^{(n)}x_s^{(n)}$ and $y_j^{(n)}x_t^{(n)}W_n$ are disjoint if $x_s^{(n)} \neq x_t^{(n)}$ or $y_i^{(n)} \neq y_j^{(n)}$. Indeed, let

$$y_i^{(n)} x_s^{(n)} W_n \cap y_j^{(n)} x_t^{(n)} W_n \neq \emptyset.$$

Then, if $y_i^{(n)} = y_j^{(n)}$ and $x_s^{(n)} \neq x_t^{(n)}$, we obtain from (4.3) $(x_t^{(n)})^{-1} x_s^{(n)} \subset W_n W_n \subset V_n$, which contradicts Property (5.1) of the neighborhoods V_n; if $y_i^{(n)} \neq y_j^{(n)}$ and, say, $i > j$, then it follows from (5.3) that $y_i^{(n)} \in y_j^{(n)} x_t^{(n)} W_n W_n (x_s^{(n)})^{-1}$, which contradicts (4.2). Thus the sets $y_i^{(n)} x_s^{(n)} W_n$ and $y_j^{(n)} x_t^{(n)} W_n$ are disjoint. Consider the measure λ_n with $c(\nu_n) = \{y_1^{(n)}, \ldots, y_{a_n}^{(n)}\}$ and $\nu_n(\{y_i^{(n)}\}) = a_n^{-1}$ $(i = \overline{1, a_n})$, and the measure ν_n' with $c(\nu_n') = A_n'$ and $\nu_n'(\{x_j\}) = \frac{1}{k_n}$, $j = \overline{1, k_n}$. Let $\nu_n'' := \lambda_n * \nu_n'$. By Proposition 1.1.4, $\{\nu_n''\}$ is a universally $O_{W(B)}$–uniformly averaging sequence. Evidently, $c(\nu_n'') = \{y_i^{(n)} x_j^{(n)}, 1 \leq i \leq$ $\leq a_n, 1 \leq j \leq k_n\}$ and every point z of $c(\nu_n'')$ admits a unique representation of the form $z = yx$, $y \in c(\lambda_n)$, $x \in c(\nu_n')$; consequently $\nu_n''(\{y_i^{(n)} x_j^{(n)}\}) = \lambda_n(y_i^{(n)}) \nu_n'(x_j^{(n)})$. This implies that the sequences $\{A_n'' := c(\nu_n''), n = \overline{1, \infty}\}$ and $\{W_n\}$ have Properties *(i)–(iii)*. Finally, we shall construct a sequence $\{A_n\}$ with all the Properties *(i)–(iv)*. Set $A_1 = A_1''$. Suppose $A_1, \ldots, A_k$ have been constructed and choose $z^{(k+1)} \notin \bigcup_{i=1}^k A_i W_i W_{k+1} (A_{k+1}'')^{-1}$.

Set $A_{k+1} = z^{(k+1)} A_{k+1}''$; it is obvious that the sets $A_{k+1} W_{k+1}$ and $A_i W_i$, $i = \overline{1, k}$, are disjoint. Thus the sets A_n possess property *(iv)*; properties *(i)–(iii)* are preserved.

□

Theorem 5.7. *On any non-discrete separable locally compact group there exist universally $O_{W(B)}$–r–averaging (with the weight $\mu = \mu^{(l)}$, and left Haar measure) increasing sequences of compact sets.*

Proof. If X is compact, then the sequence $A_n \equiv X$ possesses the required properties. Let X be non–compact. Consider the sequences of sets $\{W_n\}$, $\{V_n\}$ and $\{A_n\}$ constructed in Lemma 3.5.6.
Let $D_n := \bigcup_{i=1}^{k_n} x_i^{(n)} W_n$ where $\{x_1^{(n)}, \ldots, x_{k_n}^{(n)}\} = A_n'$, $n = \overline{1, \infty}$. Since the sets $x_i^{(n)} W_n$ are disjoint, $\mu_l(D_n) = |A_n'| \mu_l(W_n) \geq 1$; hence for any

$f \in O_{W(B)}$,

$$\Big\|\frac{1}{\mu(D_n)}\int_{D_n} f(yx)\mu_l(dx) - \frac{1}{|A'_n|}\sum_{x\in A'_n} f(yx)\Big\|_{\Phi_B} \leq$$

$$\leq \Big\|\frac{1}{\mu(D_n)}\sum_{i=1}^{k_n}\int_{x_i^{(n)}W_n}(f(yx) - f(yx_i^{(n)})\mu_l(dx)\Big\|_{\Phi_B} \leq$$

$$\leq \sup\{\|f(z) - f(x)\|_{\Phi_B}; \quad z, x : z^{-1} \in V_n\}.$$

Consequently, $\{D_n\}$ averages f. We set $A_n := \bigcup_{k=1}^n D_k$; the sets A_n are clearly compact and the sequence $\{A_n\}$ is increasing. Since the sets D_n, $n = \overline{1,\infty}$, are disjoint, $\sum_{k=1}^\infty \mu(D_k) = \infty$ and therefore for any $f \in O_{W(B)}$ the following limit exists uniformly with respect to $y \in B$:

$$\lim_{n\in N}\frac{1}{\mu(A_n)}\int_{A_n} f(yx)dx = \lim_{n\to\infty}\sum_{k=1}^n\int_{D_k} f(yx)\mu(dx)\Big/\sum_{k=1}^n \mu(D_k) =$$

$$= \lim_{k\to\infty}\frac{1}{\mu(D_k)}\int_{D_k} f(yx)\mu(dx) = \mathbf{M}^{(r)}(f).$$

□

Theorem 5.8. *If X is a σ–compact locally compact group and is not completely disconnected, then there exist universally $O_{W(B)}$–r–averaging (with respect to μ_l) increasing sequences of compact sets on X.*

Proof. Since X is σ–compact and locally compact, in any open neighbourhood V of e there exist a compact subgroup Γ such that X/Γ is separable (see, e.g., Hewitt, Ross [1], Ch. 2, Th. 8.7).

We have assumed that X is not completely disconnected; therefore there exist a Γ with the mentioned properties which is not open (see, e.g., Pontryagin, § 22, E)), so that X/Γ is not discrete. If $\{A_n\}$ is a universally $O_{W(B)}$–r–averaging sequence of compact sets in X/Γ (see Theorem 5.7) and if r denotes the natural homomorphism $X \mapsto X/\Gamma$, then $\{\pi^{-1}(A_n)\}$ is, by Theorem 2.5, the required sequence in X.

□

Corollary 5.9 . *On any connected locally compact group there exist universally $O_{W(B)}$–r–averaging (with the weight $\mu^{(l)}$) increasing sequences of compact sets.*

Of course, "left" analogues of Theorems 5.5, 5.7, 5.8, and Corollary 5.9 (in which left averaging of functions of $O_{W(B)}$ with respect to right representations are considered) are valid.

5.4. W– and AP–averaging nets of sets. Since $AP(X) \subset \subset W(X) = O_(W(C(X))$, any universally $O_{W(B)}$–r–averaging net of measures or sets is also an AP– and W–averaging net. Therefore the following three statements are implied by the results of this chapter.

Theorem 5.10. *Any r–ergodic net of sets on an amenable locally compact group r–averages the class W.*

Theorem 5.11. *On any non–discrete locally compact group X there exist W–r–averaging (with the weight measure $\mu^{(l)}$) nets of compact Baire sets.*

Theorem 5.12. *On any non-discrete separable locally compact group and on any connected locally compact group there exist W–r–averaging (with the weight measure $\mu^{(l)}$) sequences of compact sets.*

What is the situation on non–amenable discrete groups, for example, on free groups? The following theorem partially answers this question in the case of the class AP.

Theorem 5.13. (Davis) *On any locally compact group X there exist AP–averaging (with the weight measure $\mu^{(l)}$) nets of compact Baire sets.*

Proof. In view of Theorem 3.4.10, it is sufficient to consider the case when X is discrete. Let $\widetilde{X}^{(a)}$ be the Bohr compactification of X (see App., Subsect. 6.3); let κ be the canonical mapping of X into $\widetilde{X}^{(a)}$. Then $\kappa(X)$ is dense in $\widetilde{X}^{(a)}$. If $\widetilde{X}^{(a)}$ is discrete, then it is finite and therefore X is also finite; in this case $\{A_n \equiv X\}$ is an AP–averaging "net"; $\mathbf{M}(f) = |X|^{-1}\sum_{x\in X} f(x)$, $f \in AP$. If $\widetilde{X}^{(a)}$ is non–discrete, then, by Lemma 3.3, in $\widetilde{X}^{(a)}$ there exists an $AP(\widetilde{X}^{(a)})$–r–averaging net of finite sets $\{\widetilde{A}_n, n \in N\} \in \kappa(X)$. Let us take $A_n \in \kappa^{-1}(\widetilde{A}_n)$, $(A_n \cap \kappa^{-1}(x)) = 1$, $x \in \widetilde{A}_n$. It is clear that $\{A_n, n \in N\}$ is an AP–r–averaging net on G.

□

Example 5.1. (Greenleaf) Let $X = F_2$, the free group with two generators a and b. Let us consider the sets:

$E_n = \{x_1 \cdots x_n : x_i = a \text{ or } b\}$,

$F_n = \{x_1 \cdot \ldots \cdot x_k, 1 \le k \le n, x_i = a \text{ or } b\} = \bigcup_{j=1}^n E_j$,

$R_n = \{x_1^{i_1} \ldots x_n^{i_n}, x_i = a \text{ or } b, i_j = \pm 1\}$, $R_0 = \{e\}$,

$S_n = \{x_1^{i_1} \cdots x_k^{i_k}, 0 \le k \le n, x_j = a \text{ or } b, i_j = \pm 1\} = \bigcup_{j=0}^n R_j$.

The sequences $\{R_n \cup aR_n\}$, $\{R_n \cup bR_n\}$, $\{S_n \cup aS_n\}$ and $\{S_n \cup bS_n\}$ r–average the class $AP(F_2)$, but $\{E_n\}$, $\{R_n\}$, $\{S_n\}$ and $\{F_n\}$ do not. Proofs can be found in Greenleaf [4].

Example 5.2. (Gorbis–Tempelman) Let $X = F_k$, the free group with generators $x_1, \ldots, x_k$, $k < \infty$; $A_n := \{x : x \in F_k;\ x = \prod_{i=a_n}^{b_n} x_{\bar{i}}^{j_i}, S_i^{(n)} \le j_i \le t_i^{(n)} - 1\}$ where $i \equiv \bar{i} \,(\mathrm{mod}\, k)$, $1 \le \bar{i} \le k$. If $b_n - a_n \to \infty$ and $\min\{t_i^{(n)} - s_i^{(n)} : a_n \le i \le b_n\}/(b_n - a_n) \to \to \infty$ then the sequence $\{A_n\}$ averages the class $AP(F_k)$. The proof is contained in Gorbis, Tempelman [1].

Now we turn to the space $W(X)$. Here we can try to argue as in the proof of Theorem 5.13, considering the weak periodic compactification $\widetilde{X}^{(w)}$ instead of $\widetilde{X}^{(a)}$. However, $\widetilde{X}^{(w)}$ is not a group, it is a semitopological semigroup. Therefore we only obtain a weaker version of Lemma 4.3: if $\widetilde{X}^{(w)}$ is non–discrete, then there exist nets of finite sets $\{\widetilde{A}_n, n \in N\}$ in $\widetilde{X}$ such that $\lim_{n\in N} \frac{1}{|\widetilde{A}_n|}\sum_{\widetilde{x}\in\widetilde{A}_n} f(\widetilde{x}) = \mathrm{M}(f)$, $f \in C(\widetilde{X}^{(w)})$. This gives the following weakened version of Theorem 5.13.

Theorem 5.14. (**Davis**) *On any locally compact group there exist nets of Baire compact sets $\{A_n,\ n \in N\}$ such that* $\lim\limits_{n\in N} \frac{1}{\mu(A_n)} \int_{A_n} f(x)\mu(dx) = M(f),\ f \in W.$

§ 6. Averaging with Respect to Contractive Representations

Let X be a topological semigroup and let S be a continuous measurable contractive linear representation of H in H. By Corollary 1.4.9, $A(H|S) = H$. Let $q \in \mathcal{P}(\mathfrak{B})$ and let Y_q be the r–complete subsemigroup of X generated by the carrier $C(q)$ of the measure q; $\lambda_n :=$ $:= \frac{1}{n}\sum_{k=1}^{n} q^k$. $I_A := \{h : h \in H,\ S_x h = h,\ x \in A\}$.

Theorem 6.1. *For any $h \in H$ the following limit exists:*

$$\lim_{n\to\infty} \int S_x h \lambda_n^q(dx) =: \widehat{h}, \qquad \widehat{h} \in I_{Y_q} \cap K(h|S).$$

Therefore if $Y_q = X$, then $\widehat{h} = \mathbf{M}(h|S)$.

The proof is similar to that of Theorem 1.5.1. Instead of Lemmas 1.3.3–1.3.5 their analogs for contractive representations are to be used. The analog of Lemma 1.3.3 is elementary. We have to prove that if $\int S_x h q(dx) = h$ then $S_x h = h$ for q–almost all $x \in X$. Indeed, if $\int S_x h q(dx) = h$, then $\int |\langle S_x h, h\rangle| q(dx) \geq |\int \langle S_x h, h\rangle q(dx)| = \|h\|^2$. But, on the other hand, $|\langle S_x h, h\rangle| \leq \|h\|^2$, $x \in X$, and $\int (\langle S_x h, h\rangle -$ $-\|h\|^2) q(dx) = 0$. These inequalities imply that $S_x h = h$ q–almost everywhere. The analogs of Lemmas 1.3.4 and 1.3.5 can be proved almost word–for–word as the original lemmas.

§ 7. C_a–Averaging Nets

7.1. C_a–averaging nets on groups and semigroups. Let X be a non–compact, locally compact semigroup. We recall that C_a or

$C_a(X)$ denotes the Banach space of all continuous functions f on X such that $\lim_{x\to\infty} f(x)$ exists, $\|f\| = \sup_{x\in X} f(x)$. $C_0 = C_0(X)$ is the subspace of $f \in C_a$ with $\lim_{x\to\infty} f(x) = 0$. By Theorem 1.4.17, $C_a(X) \subset A(X)$. Let $\mathcal{K}$ be the class of all compact subsets of X.

We shall consider the following property of a net $\{\nu_n,\ n \in N\} \subset \subset \mathcal{P}(\mathfrak{B}_X)$:

$(A_\circ)$ $\lim\limits_{n\in N} \sup\limits_{x\in X} \nu_n(x^{-1}M) = 0, \quad M \in \mathcal{K}.$

Speaking more precisely, this is the "right" version of Property $(A_\circ)$; as we shall see, nets with this property can be used (and, actually, have to be used) for right averaging of all functions $f \in C_a$. If we are concerned with l–averaging of C_a, we have to use the "left" version of $(A_\circ)$, in which the sets $x^{-1}M$ have to be replaced by Mx^{-1}. For brevity reasons, here we consider only r–averaging and the stated r–version of $(A_\circ)$.

In Examples 7.1–7.3 we shall show how nets of "uniformly" distributed measures with Property $(A_\circ)$ can be constructed.

Example 7.1. Let μ be a left semi–invariant measure on X such that $\mu(M) < \infty$, $M \in \mathcal{K}$; let $\{A_n,\ n \in N\}$ be a net of Borel sets with the following property

$(A'_\circ)$ $0 < \mu(A_n) < \infty$, $n \in N$, and $\lim\limits_{n\in N} \mu(A_n) = \infty$.

Let $\nu_{A_n}(C) := \mu(A_n \cap C)/\mu(A_n)$. The net $\{\nu_n,\ n \in N\}$ has the Property $(A_\circ)$. Indeed, for any $M \in \mathcal{K}$ we have $\sup_{x\in X} \nu_n(x^{-1}M) \leq \leq \sup_{x\in X}[\mu(x^{-1}M)/\mu(A_n)] = \mu(M)/\mu(A_n)$. If X is a group and there exist an $M_0 \in \mathcal{K}$ with $\mu(M_0) > 0$ and elements x_n, $n \in N$, such that $M_0 \subset x_nA_n$, then $(A_\circ)$ implies $(A'_\circ)$; this follows from the relations: $\sup_{x\in X} \nu_n(x^{-1}M_0) \geq \mu(M_0 \cap x_nA_n)/\mu(A_n) = \mu(M_0)/\mu(A_n)$.

In order to construct discrete uniformly distributed measures with property $(A_\circ)$ we shall introduce the notions of "lattice" and "quasi–lattice".

Definition 7.1. Suppose X is a group, V is a bounded neighbourhood of the identity e, and m is a natural number. A set $D \subset X$ will be called an (m, V)–*quasi–lattice* if $|D \cap xV| \leq m$, $x \in D$. D is said to be a *quasi–lattice* if it is an (m, V)–quasi–lattice for some m and V; D is a *lattice* if it is a $(1, V)$–quasi–lattice for some V.

Of course, any subset of a (quasi–) lattice is a (quasi–) lattice.

Example 7.2. Any discrete subgroup D of a group X is a lattice. Otherwise for any neighborhood of the identity V there would exist a pair of elements x_V, $y_V \in D$ such that $x_V^{-1} y_V \in V$. This contradicts the discreteness of D.

Example 7.3. Let X be a group and let D be a lattice in X. If a net $\{A_n, n \in N\} \subset D$ has the property

$(A_\circ'')$ $|A_n| < \infty$, $n \in N$, and $\lim_{n \in N} |A_n| = \infty$,

then the net of measures $\lambda_{A_n}(C) = |A_n \cap C|/|A_n|$, $C \in \mathfrak{B}$, $n \in N$, has Property $(A_\circ)$. Indeed, let D be an (m, V)–lattice and $M \in \mathcal{K}$. Consider a symmetric neighborhood of the identity W such that $W \cdot W \subset V$. From the covering $\{xW, x \in M\}$ of M we select a finite subcovering $\{y_i W, i = \overline{1, k}\}$. It is clear that the family $\{x y_i W, i = \overline{1, k}\}$ is a covering of xM, $x \in X$. If z', $z'' \in x y_i W$, then $z'(z'')^{-1} \in W \cdot W \subset V$ and $z' \in z''V$. Hence, $|D \cap x y_i W| \leq m$, $i = \overline{1, k}$, and $|D \cap xM| \leq mk$, $x \in X$. This implies our statement. Let us also note that a $(1, V)$–lattice $D = \{x_i, i = \overline{1, \infty}\}$ can be selected from any unbounded set C: we choose x_1 arbitrarily and then, if elements $x_1, \ldots, x_n$ are chosen, we take an arbitrary $x_{n+1} \in C \setminus \bigcup_{i=1}^n x_i V V^{-1}$. It is easy to verify that $\{x_i, i = \overline{1, \infty}\}$ is a $(1, V)$–lattice.

Theorem 7.1. *A net $\{\nu_n, n \in N\} \subset \mathcal{P}(\mathfrak{B})$ r–averages the class C_a iff it has Property $(A_\circ)$.*

Proof. Sufficiency. It suffices to show that

$$\lim_{n \in N} \sup_{x \in X} \left| \int_X f(xy) \nu_n(dy) \right| = 0, \quad f \in C_0.$$

For any $\varepsilon > 0$ and $f \in C_0$ there exists an a $M_\varepsilon \in \mathcal{K}$ such that $|f(y)| < \frac{1}{2}\varepsilon$ if $y \in X \setminus M_\varepsilon$; since $\{\nu_n\}$ has Property $(A_\circ)$, there also exists an $n_\varepsilon \in N$ such that $\sup_{x \in X} \nu_n(x^{-1} M_\varepsilon) < \frac{\varepsilon}{2\|f\|}$ if $n > n_\varepsilon$. Note that if $y \in x^{-1} M_\varepsilon$, then $xy \neq M_\varepsilon$. If $n > n_\varepsilon$, then for any $x \in X$,

$$\left|\int_X f(xy)\nu_n(dy)\right| \le \int_{x^{-1}M_\varepsilon} |f(xy)|\nu_n(dy)+$$
$$+\int_{X\setminus x^{-1}M_\varepsilon} |f(xy)|\nu_n(dy) \le \|f\|\nu_n(x^{-1}M_\varepsilon)+\frac{1}{2}\varepsilon \le \varepsilon.$$

Necessity. Let $X_\infty := X \cup \{\infty\}$, the one–point compactification of X; let $M \in \mathcal{K}$. Since X_∞ is normal, there exists a function $f \in C_0$ such that $f(x) = 1$, $x \in M$, and $0 \le f(x) \le 1$, $x \in X$. If $\{\nu_n,\, n \in N\}$ r–averages the class C_a, then $\lim_{n\in N} \sup_{x\in X} \int_X f(xy)\nu_n(dy) = 0$. This implies that $\lim_{n\in N} \sup_{x\in X} \nu_n(x^{-1}M) = 0$.

□

Corollary 7.2. *Any net of Borel sets with the Property* $(A'_\circ)$ *r–averages* C_a *with the weight measure* μ.

Corollary 7.3. *In a group, any net of finite subsets* A_n *of a quasi–lattice r– and l–averages* C_a *if only* $|A_n| \to \infty$.

7.2. $\mathbf{C_a}$–averaging nets on homogeneous spaces. Let $Y = G/K$ be a non–compact, locally compact homogeneous space, and let $\mathfrak{B}$ be the σ–algebra of Borel sets in Y. As in Subsection 6.1 we shall consider nets $\{\nu_n,\, n \in N\} \in \mathcal{P}(\mathfrak{B})$ with the property:

$(A_\circ)$ $\lim\limits_{n\in N} \sup\limits_{g\in G} \nu_n(gM) = 0$ for any compact set $M \subset Y$.

We make two simple remarks concerning some relations between properties of some objects on Y and their preimages in G under the natural mapping $\pi : G \mapsto Y$.

Note 7.1. Since the complete preimage $\widetilde{M}$ in G of a compact set $M \subset Y$ is compact (see Weyl [1], Ch. 1, § 3) and $g\widetilde{M} = \widetilde{gM}$, it is clear that a net $\{\nu_n,\, n \in N\}$ $(\in \mathcal{P}(\mathfrak{B}))$ has Property $(A_\circ)$ iff the net of certain prototypes $\{\widetilde{\nu}_n,\, n \in N\}$ on G has this property.

Note 7.2. If $\varphi \in C_a(Y)$, then $\widetilde{\varphi} := \varphi \circ \pi \in C_a(G)$ and $\lim\limits_{x\to\infty} \varphi(x) = \lim\limits_{g\to\infty} \widetilde{\varphi}(g)$ and $\int \widetilde{\varphi}(g\widetilde{g})\widetilde{\nu}(d\widetilde{g}) = \int \varphi(gx)\nu_n(dx)$, $g \in G$.

These notes allow to transfer at once all the results of subsection 7.1 to homogeneous spaces. For the convenience of quotation we state the "Y–versions" of these results.

Example 7.1′. Let μ be the invariant measure on Y; let $\{A_n, n \in N\} \subset \mathfrak{B}$, $0 < \mu(A_n) < \infty$, where $\nu_{A_n}(C) = \mu(A_n \cap C)/\mu(A_n)$ satisfies condition $(A_\circ)$.

Definition 7.1′. We get the notions of (m, W)*–quasi–lattice* and *lattice* in Y if we replace in Definition 7.1 Y by X and let V be a bounded spherically symmetric neighbourhood of the K–fixed point y_0.

Note 7.3. Let V be a bounded spherically symmetric neighbourhood of y_0 and let $\widetilde{V}$ be its complete preimage in G. If some $\widetilde{D}\,(\subset G)$ is a (m, W)–quasi–lattice, then $\pi(\widetilde{D})$ is a (m, W)–quasi–lattice. If $D\,(\subset Y)$ is an (m, W)–quasi–lattice, then any cross–section $\widetilde{D}$ in G over D (see App., Subsect. 1.8) is an (m, W)–quasi–lattice.

Example 7.2′. If $\{A_n, n \in N\}$ is a net of finite subsets of a quasi–lattice D and $\lim_{n\in N} |A_n| = \infty$, then the net $\{\lambda_{A_n}, n \in N\}$, where $\lambda_{A_n}(C) := (A_n \cap C)/|A_n|$, satisfies condition $(A_\circ)$. For any given V we can extract from every unbounded set $C \subset Y$ a $(1, V)$–lattice $\{y_1, y_2, \ldots\}$. This is a simple consequence of Notes 7.1, 7.3 and the assertions stated in Example 7.2.

Theorem 7.1′. *A net $\{\nu_n, n \in N\} \in \mathcal{P}(\mathfrak{B})$ averages the class $C_a(Y)$ iff it satisfies condition $(A_\circ)$.*

BIBLIOGRAPHICAL NOTES

1. The presentation of the primary results of this chapter follows Tempelman [19] (1981).

2. Theorem 1.1 is close to results in Day [2] (1950) and Eberlein [1] (1949); in the form presented it is contained in Tempelman [2] (1961).

3. Averaging with nets of measures which are simultaneously left– and right–ergodic was considered in Alaoglu, Birkhoff [2] (1940), Day [1] (1942), [2] (1950), Peck [1] (1951), and Eberlein [1] (1949). Boclé [1] (1958), [3] (1960) and Michel [1] (1967–1968) considered such nets in connection with problems of desintegration of measures and the existence of measures which are invariant with respect to groups of transformations of a measurable space.

The mean ergodic theorem with ergodic (Følner) nets of sets was proved in Tempelman [2] (1961), [4] (1962) and was rediscovered several times; the mean–value theorem (with ergodic nets) for almost periodic functions on groups was proved independently in Hewitt, Ross [1] (1963) and in Tempelman [2] (1961), [4] (1962); this theorem was proved in Lyubarsky [1] (1946) and Struble [1] (1953) for more restricted classes of nets of sets. Theorem 2.1 on the universal averaging property of ergodic nets is contained in Tempelman [2] (1961), [14] (1975), [19] (1976).

4. Theorems 3.1 and 3.4, on the existence of universal averaging nets of measures and of sets, were proved in Tempelman [14] (1975) (see also Tempelman [19] (1976)).

5. Theorems 4.10 and 4.11 were proved in Davis [1] (1967) and [2] (1974). The results stated in Example 4.1 are due to Greenleaf [2] (1973). The result of Example 4.2 was proved in Gorbis and Tempelman [1] (1988).

6. The results of § 2 generalize some results connected with the mean ergodic theorem on group extensions contained Chapter 4, § 6, in Tempelman [24] (1986).

CHAPTER 4
MEAN ERGODIC THEOREMS

§ 1. Mean Averaging Nets: Definition and Existence

1.1. Mean averaging nets on semigroups. Let X be a topological semigroup, $\mathfrak{B}$ the σ–algebra of Borel sets in X, and $\{\nu_n,\ n \in N\}$ a net of Borel probability measures.

Lemma 1.1. *Let $x \mapsto U_x$ be a measurable unitary representation of a group X in a Hilbert space H, $\varphi_{h,h'}(x) := \langle U_x h, h' \rangle$, $h, h' \in X$, and let $\{\nu_n,\ n \in N\} \subset \mathcal{P}(\mathfrak{B})$.*

(i) If for some $h \in H$ we have $\lim_{n \in N} \int \varphi_{h,h}(yx)\nu_n(dx) = \mathbf{M}(\varphi_{h,h})$ uniformly with respect to $y \in X$, then $\lim_{n \in N} \int U_x h \nu_n(dx) = \mathbf{M}(h)$.

(ii) Converse: if for some $h \in H$ we have $\lim_{n \in N} \int U_x h \nu_n(dx)$, then $\lim_{n \in N} \int \varphi_{h,h'}(yx)\nu_n(dx) = \mathbf{M}(\varphi_{h,h'})$ uniformly with respect to $y \in X$, $h, h' \in H$.

Proof. *(i)* Since $\|\mathbf{M}(h)\|^2 = \mathbf{M}[\langle U_x h, h\rangle]$ (see Theorem 1.5.2), the statement is implied by the following relations:

$$\Big\| \int_X U_x h \nu_n(dx) - \mathbf{M}(h) \Big\|^2 =$$

$$= \int_X \int_X \langle U_{y^{-1}x} h, h \rangle \nu_n(dx)\nu_n(dy) -$$

$$-2 \int_X \langle U_x h, \mathbf{M}(h) \rangle \nu_n(dx) + \|\mathbf{M}(h)\|^2 =$$

$$= \int_X \left[\int_X \varphi_{h,h}(y^{-1}x)\nu_n(dx) - \|\mathbf{M}(h)\|^2 \right] \nu_n(dy).$$

For any $h, h' \in H$, $y \in X$,

$$\begin{aligned}
&\left| \int \varphi_{h,h'}(yx)\nu_n(dx) - \mathbf{M}\varphi_{h,h'} \right| = \\
=&\left| \int \langle U_{yx}h, h'\rangle \nu_n(dx) - \langle \mathbf{M}h, h'\rangle \right| = \\
=&\left| \int \langle U_x h, U_{y^{-1}}h'\rangle \nu_n(dx) - \langle \mathbf{M}h, U_{y^{-1}}h'\rangle \right| \leq \\
\leq& \left\| \int_X U_x h \nu_n(dx) - \mathbf{M}h \right\| \, \|h'\|.
\end{aligned}$$

This implies our statement.

□

Lemma 1.2. *Suppose B runs through the class of all Banach spaces and $\mathcal{H}$ and H run through the class of all Hilbert spaces; $1 \leq \alpha < \infty$. The following properties of a net $\{\nu_n,\ n \in N\} \subset \mathcal{P}(\mathfrak{B})$ are equivalent:*

a) *for any measurable isometric representation $x \mapsto U_x$ in a Hilbert space H and for any $h \in H$ the following limit exists:*

$$\lim_{n\in N} \int_X U_x h \nu_n(dx) = \mathbf{M}(h); \tag{1.1}$$

$b_{\mathcal{H}}$) *for any measurable wide-sense l-homogeneous $\mathcal{H}$-valued random field $\xi(\cdot)$ the $L^2_{\mathcal{H}}$-limit*

$$\lim_{n\in N} \int_X \xi(x)\nu_n(dx) = \mathbf{M}^{(r)}(\xi) \tag{1.2}$$

exists;

$c_{\alpha,B}$) *for any measurable strong-sense l-homogeneous B-valued random field $\xi(x)$, $x \in X$, with $E\|\xi(x)\|^\alpha < \infty$ the L^α_B-limit (1.2) exists and the equality (1.2) holds;*

$d_{\alpha,B}$) *let $\alpha > 1$; for any measurable right dynamical system $\{T_x, x \in X\}$ in the phase space $(\Lambda, \mathcal{F}, m)$ and for any $f \in L^\alpha_B\ (\Lambda, \mathcal{F}, M)$ the following L^α_B-limit exists:*

$$(L^\alpha_B)\text{–}\lim_{n\in N} \int_X f(\omega T_x)\nu_n(dx) = \mathbf{M}^{(r)}(f); \tag{1.3}$$

$d_{1,B}$) property $d_{\alpha,B}$) is valid in the case $\alpha = 1$ if $m(\Lambda) < \infty$.

If X is a group, Properties a)–d) are equivalent to any of the following two properties of the net $\{\nu_n\}$:

e) the net $\{\nu_n\}$ r–averages all measurable positively definite functions on X;

f) the net $\{\nu_n\}$ r–averages the coefficients of all measurable unitary l–representations of X.

Properties a)–d) (a)–f) if X is a group) of a net $\{\nu_n\}$ remain equivalent if we consider only continuous measurable objects.

Proof. *a*) $\Rightarrow$ $d_{\alpha,B}$) for any $\alpha > 1$ and any B. Evidently, *a*) $\Rightarrow$ $d_{2,\mathbf{C}}$). Let $0 < \gamma < 2\min(1, \alpha - 1)$; let $\Lambda \in \mathcal{F}$ with $m(\Lambda) < \infty$ and $\varphi_n(\omega) :=$ $:= \int_X \mathcal{X}_\Lambda(\omega T_x)\nu_n(dx) - \mathbf{M}(\mathcal{X}_\Lambda)$ (by Theorem 1.4.13, $\mathbf{M}(\mathcal{X}_\Lambda) = \prod_{I^\alpha}^{L^\alpha} \mathcal{X}_\Lambda$ does not depend on α and $\|\mathbf{M}(\mathcal{X}_\Lambda)\|_{L^\alpha} \le \|\mathcal{X}_\Lambda\|_{L^\alpha}$, $\alpha > 1$). Using the Hölder Inequality we obtain:

$$\int_\Omega |\varphi_n|^\alpha dm = \int_\Omega |\varphi_n|^\gamma |\varphi_n|^{\alpha-\gamma} dm \le$$
$$\le \left(\int_\Omega |\varphi_n|^2 dm\right)^{\gamma/2} \left(\int_\Omega |\varphi_n|^\delta dm\right)^{(2-\gamma)/2},$$

where $\delta = 2(\alpha - \gamma)(2 - \gamma)^{-1} > 1$. Now,

$$\int_\Omega |\varphi_n|^\delta dm \le \int_\Omega \left[\int_X |\mathcal{X}_\Lambda(\omega T_x) - \mathbf{M}(\mathcal{X}_\Lambda)|^\delta \nu_n(dx)\right] m(d\omega) =$$
$$= \int_X \left[\int_\Omega |\mathcal{X}_\Lambda(\omega T_x) - \mathbf{M}(\mathcal{X}_\Lambda)|^\delta m(d\omega)\right] \nu_n(dx) =$$
$$= \int_\Omega |\mathcal{X}_\Lambda(\omega) - \mathbf{M}(\mathcal{X}_\Lambda)|^\delta m(d\omega) \le 2^\delta m(\Lambda) < \infty;$$

the change of the order of integration performed above is valid in view of property *b*) of measurable dynamical systems (see App. (9.F)). Thus, $\lim_{n\in N} \int_\Lambda |\varphi_n|^\alpha dm = 0$, and for $f = \mathcal{X}_\Lambda$ with $m(\Lambda) < \infty$ (1.3) is proved. This implies that (1.3) is true for all B–valued integrable simple functions, i.e. for functions of the form: $f = \sum_{i=1}^n b_i \mathcal{X}_{\Lambda_i}$, where

$n = \overline{1,\infty}$, $b_i \in B$, $m(\Lambda_i) < \infty$, $i = \overline{1,n}$. For any function $f \in L_B^\alpha$ and any $\varepsilon > 0$ there exists a function f_ε such that $\|f - f_\varepsilon\|_{L_B^\alpha} < \varepsilon$ (see Dunford and Schwartz [1], Ch. 3, § 3). Hence $\|\mathbf{M}^{(r)}(f - f_\varepsilon)\|_{L_B^\alpha} < \varepsilon$, and for any $n > n_\varepsilon$,

$$\Big\| \int_X f(\omega T_x)\nu_n(dx) - \mathbf{M}^{(r)}(f)\Big\|_{L_B^\alpha} \le \Big\| \int_X (f - f_\varepsilon)(\omega T_x)\nu_n(dx)\Big\| +$$
$$+ \Big\| \int_X f_\varepsilon(\omega T_x)\nu_n(dx) - \mathbf{M}^{(r)}(f_\varepsilon)\Big\|_{L_B^\alpha} + \|\mathbf{M}^{(r)}(f - f_\varepsilon)\|_{L_B^\alpha} \le 3\varepsilon$$

Consequently, $a) \Rightarrow d_{\alpha,B})$ for $1 < \alpha < \infty$.

$d_{\alpha,B}) \Rightarrow d_{1,B})$ by the inequality

$$\int_\Omega \Big| \int_X f(\omega T_x)\nu_n(dx) - \mathbf{M}^{(r)}(f)\Big| dm \le$$
$$\le m(\Omega)^{(\alpha-1)/\alpha} \left(\int_\Omega \Big| \int_X f(\omega T_x)\nu_n(dx) - \mathbf{M}^{(r)}(f)\Big|^\alpha m(d\omega) \right)^{1/\alpha}.$$

$d_{\alpha,B}) \Rightarrow c_{\alpha,B})$: this is obvious.

$c_{2,\mathbf{C}}) \Rightarrow b_{\mathcal{H}})$. Let $\mathcal{H}$ be a Hilbert space and let $\xi(\cdot)$ be a measurable wide–sense l–homogeneous $\mathcal{H}$–valued random field on X. Let us consider the $\mathbf{C}$–valued Gaussian random field $\eta(x)$ on X with $\mathbf{E}\eta(x) \equiv \equiv 0$, $\mathbf{E}\eta(x)\overline{\eta(y)} = \mathbf{E}\langle \xi(x), \xi(y)\rangle_{\mathcal{H}}$ and $\mathbf{E}\eta(x)\eta(y) = 0$ (see, e.g., Doob [1], Ch. 2, Theorem 3.1). Since all finite–dimensional distributions of $\eta(\cdot)$ are uniquely defined by these conditions, this field is strong–sense l–homogeneous. It is clear that the correspondence $\xi(x) \Leftrightarrow \eta(x)$, $x \in X$, can be extended to an isometric isomorphism between the spaces H_ξ and H_η. Clearly, the field $\eta(\cdot)$ is measurable, $\int_X \xi d\nu_n \Leftrightarrow \Leftrightarrow \int_X \eta d\nu_n$ and $\mathbf{M}^{(r)}(\xi) \Leftrightarrow \mathbf{M}^{(r)}(\eta)$. Thus the implication $c_B) \Rightarrow b_{\mathcal{H}})$ is proved.

$b_{\mathcal{H}}) \Rightarrow a)$. Let U be an isometric representation of X in H and let $h \in H$. Let us consider a Gaussian l–homogeneous field $\eta(\cdot)$ on X with $\mathbf{E}\eta(x) \equiv 0$, $\mathbf{E}\eta(x)\overline{\eta(y)} = \langle U_x h, U_y h\rangle_H$ (note that $R(x,y) := := \langle U_x h, U_y h\rangle_H$ is a positive definite function on $X \times X$). Denote by H_h the subspace of H spanned by the elements $U_x h$, $x \in X$. Clearly, the correspondence $U_x h \Leftrightarrow \eta(x)$ can be extended to an isometric isomorphism between H_h and H_η and we have: $\int_X U_x h \nu_n(dx) \Leftrightarrow$ $\Leftrightarrow \int_X \eta(x)\nu_n(dx)$ and $\mathbf{M}(h) \Leftrightarrow \mathbf{M}^{(r)}(\eta)$. This implies that $b_{\mathcal{H}}) \Rightarrow a)$ for any $\mathcal{H}$.

Now suppose that X is a group.

$a) \Rightarrow f)$ by Lemma 1.1.

$f) \Rightarrow e)$: this is evident.

$e) \Rightarrow a)$ by Lemma 1.1.

□

Definition 1.1. A net $\{\nu_n, n \in N\}$ with the equivalent properties a)–*d)* (a)–*f)* if X is a group) stated in Lemma 1.2, with respect to *measurable* objects, will be called a *mean right–averaging net*; if $\{\nu_n, n \in N\}$ has these properties only for *continuous measurable* objects, it is called a *C– mean r–averaging net. Mean l–averaging nets* of measures are defined similarly. Let Q be a class of objects for which the corresponding relation (1.1), (1.2) or (1.3) holds; then we say that $\{\nu_n\}$ *r–averages the class* Q. Finally, if μ is a measure on $\mathfrak{B}$ and if $\{A_n, n \in N\}$ is a net of measurable sets such that the net of measures $\nu_{A_n}(B) := \mu(A_n \cap B)/\mu(A_n)$ r–averages some class Q or is mean (or (C–)mean) r–averaging, we shall say that the net $\{A_n, n \in N\}$ has the corresponding property (with the weight measure μ).

Note 1.1. Evidently, any mean r–averaging net $\{\nu_n\}$ is also C–mean r–averaging.

Note 1.2. On any open r–regular locally compact semigroup (in particular, on any locally compact group) the classes of mean r–averaging nets of measures and of C–mean r–averaging nets coincide. This follows from Note 1.1 and the continuity of all measurable objects considered in Lemma 1.2, on such a semigroup (see App., § 9).

Note 1.3. If X is σ–compact, then any completely measurable right dynamical system and any completely measurable l–homogeneous random field are measurable (see App., (9.J)). Therefore, if $\{\nu_n\}$ is an arbitrary mean r–averaging net of measures, then the properties *b)–d)* of Lemma 1.2 are valid for such dynamical systems and random fields.

Note 1.4. If $\{\nu_n\}$ is a mean r–averaging net on a separable semigroup, the properties a)–*d)* (see Lemma 1.2) are valid for all continuous objects: all such objects are measurable (see App., § 9).

If $x \mapsto U_x$ is a right representation of a group X, then $x \mapsto \widetilde{U}_x := := U_{x^{-1}}$ is a left representation. This implies the following statement.

Proposition 1.3. *Let X be a group. A net $\{\nu_n,\ n \in N\}$ is a mean l– (resp. r–) averaging net iff the net $\{\bar{\nu}_n,\ n \in N\}$, where $\bar{\nu}_n(A) = = \nu(A^{-1})$, is mean r– (resp. l–) averaging.*

Proposition 1.4. *On a non–compact locally compact group any mean r–averaging net of measures has the right version of property (A_0) (§ 3.6).*

Proof. Let $\{\nu_n\}$ be a mean r–averaging net of measures and let μ be the left Haar measure on X. Choose an arbitrary compact set $M \subset X$ and a compact set $M_1 \supset M$ such that $\mu(M_1) > 0$ (such an M_1 exists). We shall consider the r–dynamical system $T : yT_x = x^{-1}y,\ x \in X$, in the space $(X, \mathfrak{B}, \mu)$; this dynamical system is completely measurable. Therefore the associated unitary representation U: $(U_x f)(y) = = f(x^{-1}y)$, $f \in L^2\ (X, \mathfrak{B}, \mu)$ is measurable, and by Lemma 1.2 $\{\nu_n\}$ r–averages the coefficient of this representation:

$$\varphi(x) = \mu(xM_1 \cap M_1 \cdot M_1) = \int_X \mathcal{X}_{M_1}(x^{-1}y)\mathcal{X}_{M_1 \cdot M_1}(y)\mu(dy).$$

It follows from App., (9.G) that $\varphi(\cdot) \in C(X)$; it is evident that in fact $\varphi(\cdot) \in C_0(X)$ and, consequently, $\lim_{n\in N} \sup_{x\in X} \int \varphi(xy)\nu_n(dy) = 0$ (see § 3.6). It remains to note that

$$\int \varphi(xy)\nu_n(dy) \geq \nu_n(x^{-1}M)\mu(M_1), \qquad x \in X.$$

□

We shall see in §§ 3 and 4 that the condition (A_0) is also sufficient in order that a net $\{\nu_n\}$ would average some classes of objects considered in Lemma 1.2; on SQM–groups it even implies that the net $\{\nu_n\}$ is a "universally" mean averaging (see § 5).

Theorem 3.5.1 implies the following statement.

Theorem 1.5. *Let X be a topological semigroup. Let $q \in \mathcal{P}(\mathfrak{B})$ and let the l–complete subsemigroup Y_q generated by the carrier $c(q)$ of q coincide with X. Then the sequence $\lambda_n^{(q)} = \frac{1}{n}\sum_{k=1}^{n} q^k$, $n = \overline{1,\infty}$, is C–mean l–averaging.*

Example 1.1. Let X be a separable topological semigroup and let F be a countable dense subset of X. If $q \in \mathcal{P}(X)$ and $c(q) = F$, then $\{\lambda_n^{(q)}\}$ is a C–mean l– and r–averaging sequence.

Example 1.2. Let X be a connected topological group. If $q \in \mathcal{P}(X)$ and $c(q)$ contains an open neighborhood V of the identity e, then $\{\lambda_n^q\}$ is a C–mean l– and r–averaging sequence. Indeed, the group X is generated by V (see, e.g., Pontryagin [1], § 22) and therefore by $c(q)$.

In Notes 1.5–1.9 we present some results related to averaging by "pure" convolution powers $\{q^n\}$.

Note 1.5. Suppose G is a topological group, $T = \{T_g,\ g \in G\}$ is a measurable dynamical system in the space $(\Lambda,\ \mathcal{F},\ m)$ and $\{q_n,\ n = \overline{1,\infty}\} \subset \mathcal{P}(\mathfrak{B})$. For any $q \in \mathcal{P}(\mathfrak{B})$ we define in $L^\alpha\ (\Lambda,\ \mathcal{F},\ m)$, $1 \leq \alpha < \infty$, a continuous linear operator P_q: $P_q(f)(\omega) := \int_G f(T_g\omega)q(dg)$ and the symmetric measure $\bar{q}$: $\bar{q}(A) := q(A^{-1})$, $A \in \mathfrak{B}$. It is easy to verify that $P_{\bar{q}} = P_q^*$ and $P_{q_1}^* \cdot \ldots \cdot P_{q_n}^* P_{q_n} \cdot \ldots \cdot P_{q_1} f = \int_G f(T_g\omega)\nu_n(dg)$ where $\nu_n := q_1 * \ldots * q_n * \bar{q}_n * \ldots * \bar{q}_1$. By Rota's theorem (see Rota [1], Doob [2]) if $1 < \alpha < \infty$ or if $\alpha = 1$ and $m(\Lambda) < \infty$, then $(L^\alpha)\ \lim_{n\to\infty} \int_G f(T_g\omega)\nu_n(dg)$ exists.

Note 1.6. Let G be a topological group, let $q \in \mathcal{P}(\mathfrak{B})$, $q = \bar{q}$, and $Y_{q^2} = G$. Theorem 1.5 and Note 1.5 imply that the sequence $\{q^{2n},\ n = \overline{1,\infty}\}$ is C–mean l–averaging; since the operator P_q is continuous and $P_q[\mathrm{M}^{(l)}(f)] = \mathrm{M}^{(l)}(f)$, we deduce that the sequence $\{q^n\}$ is also

C–mean l–averaging. By symmetry it is also C–mean r–averaging.

Note 1.7. Let G, T and q be as in Note 1.5 and let, moreover, G be locally compact, $q = \bar{q}$, $Y_q = G$. For $M \subset X$ we put $\mathfrak{I}_M(T) = \{\Lambda : m(T_g\Lambda \Delta \Lambda) = 0,\ g \in M\}$. An insignificant modification of our argument used in Note 1.6 allows us to prove the following result, due to Oseledec [1]: the sequence $\{q^n\}$ is T–averaging iff $\mathfrak{I}_{c(q)\cdot c(q)}(T) = \mathfrak{I}_{c(q)}(T)$.

Note 1.8. **(Lin)** Let G be a locally compact commutative group and let $q \in \mathcal{P}(\mathfrak{B})$ satisfy the following conditions: $Y_q = G$ and $c(q)$ is not contained in a coset of any subgroup $\Gamma \neq G$. Then $\{q^n\}$ is a mean averaging sequence. This assertion is contained in a more general result due to Lin [2].

Note 1.9. **(Derrienic–Lin)** Let G be a σ–compact locally compact group. Let $q \in \mathcal{P}(\mathfrak{B})$. If *1)* $Y_q = G$; *2)* $c(q)$ is not contained in some F–coset of a normal subgroup $\Gamma \neq G$; and *3)* for some n the measure q^n is not singular with respect to the Haar measure, then $\{q^n\}$ is mean averaging. This statement is a special case of Theorem 2.8 in Derrienic, Lin [1].

Theorems 3.3.4 and 3.5.2 imply the following two theorems.

Theorem 1.6. *On any non–discrete locally compact group there exist mean r–averaging (with the weight $\mu^{(l)}$*)), the left Haar measure nets of Baire compact sets.*

Theorem 1.7. *On any non–discrete separable locally compact group there exist mean r–averaging (with the weight measure $\mu^{(l)}$*)) increasing sequences of compact sets.*

Of course, the symmetric "left" version of Theorems 1.6 and 1.7 are also true.

---

*) We denote by $\mu^{(l)}$ the left Haar measure.

1.2. Mean averaging nets on homogeneous spaces. Let $Y = G/K$ be a locally compact left homogeneous space, where K is a compact stationary subgroup, let μ be the invariant measure on Y, and let $\mathfrak{B}(Y)$ and $\mathfrak{B}(G)$ be the σ–algebras of Borel set in Y and G, resp.

We denote by y_0 the K–invariant point in Y; if $\nu \in \mathcal{P}(\mathfrak{B}(Y))$, then $\widetilde{\nu}$ means some preimage of ν in $\mathcal{P}(\mathfrak{B}(G))$. Let U: $g \to U_g$ be a measurable unitary l–representation of class 1 (with respect to K) of G in H; $I_K := \{h : h \in H, U_K h = h, k \in K\}$. If $h \in I_K$ and $g_1 y_0 = g_2 y_0$ (or, in other words, $g_1^{-1} g_2 \in K$), then $U_{g_1} h = U_{g_2} h$, and we may put $U_y h := U_g h$ where $y = g y_0$; it is clear that $\int_Y U_y h \nu(dy) = \int_G U_g h \widetilde{\nu}(dg)$. Let $\{T_g,\ g \in G\}$ be a measurable right dynamical system of class 1 in $(\Lambda, \mathcal{F}, m)$; let $\mathfrak{J}_K$ be the σ–algebra of all sets in $\mathcal{F}$ which are invariant with respect to the dynamical system $\{T_k,\ k \in K\}$ and let m_K be the restriction of m to $\mathfrak{J}_K$. If $f \in L^\alpha\ (\Lambda, \mathfrak{J}_K, m_K)$, $\alpha \geq 1$, then $f(\omega T_{g_1}) = f(\omega T_{g_2})$ m.–a.e. as soon as $g_1 y_0 = g_2 y_0$; therefore we may put $f(\omega T_y) := f(\omega T_g)$ where $y = g y_0$. Clearly, $\int_Y f(\omega T_y) \nu(dy) = \int_G f(\omega T_g) \widetilde{\nu}(dg)$. We recall also that we can associate with any measurable homogeneous random field $\xi(y)$, $y \in Y$, a measurable l–homogeneous random field ξ on G: $\widetilde{\xi}(g) = \xi(y)$ if $g \in y$, i.e. if $y = g y_0$ we have $\int_Y \xi(y) \nu(dy) = \int_G \widetilde{\xi}(g) \widetilde{\nu}(dg)$. Note that if $Y = G$ (i.e. $K = \{e\}$), then $I_K = H$, $\mathfrak{J}_K = \mathcal{F}$, $m_K = m$.

The following statement is a simple analogue of Lemma 1.2.

Lemma 1.8. *Let B run through the class of all Banach spaces and let $\mathcal{H}$ and H run through the class of all Hilbert spaces; $1 \leq \alpha < \infty$. The following properties of a net $\{\nu_n,\ n \in N\} \in \mathcal{P}(\mathfrak{B}(Y))$ are equivalent:*

a) for any measurable unitary left class–1 representation $U : g \mapsto U_g$ of G in a Hilbert space H and for any $h \in I_K$ we have

$$\lim_{n \in N} \int_Y U_y h \nu_n(dy) = \mathbf{M}(h);$$

$b_{\mathcal{H}}$) for any measurable wide–sense homogeneous $\mathcal{H}$–valued random field $\xi(y)$, $y \in Y$, we have $(L^2_{\mathcal{H}})\ \lim_{n \in N} \int_Y \xi(y) \nu_n(dy) = \mathbf{M}^r(\xi)$;

$c_{\alpha,B}$) for any measurable strong sense homogeneous B–valued ran-

dom field $\xi(y)$, $y \in Y$, *with* $\mathbf{E} \mid \xi(y_0) \mid^\alpha < \infty$, *we have*

$$(L^\alpha_B) \lim_{n \in N} \int_G \xi(y)\nu_n(dy) = \mathbf{M}^{(r)}(\xi);$$

$d_{\alpha,B}$) $(\alpha > 1)$: *for any measurable right dynamical system* $\{T_g, g \in G\}$ *of class 1 in the phase space* $(\Omega, \mathcal{F}, m)$ *and for any* $f \in L^\alpha_B\ (\Omega, \mathfrak{I}_K, m_K)$ *we have* $(L^\alpha_B) \lim_{n \in N} \int_Y f(\omega T_y)\nu_n(dy) = \mathbf{M}^{(r)}(f)$;

$d_{1,B}$) *the property* $(d_{\alpha,B})$ *is valid in the case* $\alpha = 1$ *if* $m(\Omega) < \infty$;

e) the net $\{\nu_n\}$ *averages all spherical positive definite functions on* Y.

Definition 1.3. A net of measures $\{\nu_n\} \subset \mathcal{P}(\mathfrak{B}(Y))$ is called a *mean averaging net (m.a.n.)* on Y if it has the properties *a)–e)* stated in Lemma 4.1.8; a net of sets $\{A_n\}$ is called a *m.a.n.* on Y if $A_n \subset \mathfrak{B}(Y)$, $0 < \mu(A_n) < \infty$ and the net of measures $\nu_{A_n}(A) := \mu(A_n \cap A)/\mu(A_n)$ is a m.a.n.

Note 1.8. According to Note 1.2, the notion of "C–mean averaging net" on Y coincides with the notion of m.a.n..

The problem on the existence and construction of m.a.n.'s on Y is positively solved by theorems 1.5–1.7, Examples 1.1, 1.2, and the following statement.

Proposition 1.9. *If a net of preimages* $\{\widetilde{\nu}_n,\ n \in N\}$ *is a mean l–averaging net on* G, *then the net of their projections* $\{\nu_n,\ n \in N\}$ *is a m.a.n. on* Y.

This is a simple consequence of the statement App. (2.K).

Note that on many homogeneous spaces the class of m.a.n.'s is considerably larger than the class of nets found in Proposition 1.9; for example, as we shall see in § 5 such is the situation on the Euclidean spaces E_n.

Statements which are connected with finding m.a.n.'s on semigroups or homogeneous spaces are usually called *mean ergodic theorems*.

§ 2. Mean Ergodic Theorems on Amenable Semigroups

Clearly, any universally $A_B^{(r)} \cap F_B$–r–averaging net is a mean r–averaging net, and hence any universal ergodic theorem of § 3.3 implies a mean ergodic theorem. As an example we state the following evident corollary of Theorem 3.3.1.

Theorem 2.1. *Any r–ergodic (Følner) net of sets is mean r–averaging.*

Example 2.1. Let $G = \mathbb{R}^2$; $A_n := \{(x,y) : |x| < a, |y| \leq \leq b_n\}$, $n = 1,2,\ldots$, $b_n \uparrow \infty$. Clearly this sequence is not ergodic. Let $\xi(x)$, $x \in \mathbb{R}$, be an ergodic stationary random process; $\eta(x,y) := \xi(x)$, $(x,y) \in \mathbb{R}^2$. Then $\frac{1}{|A_n|}\int_{A_n}\eta(x,y)dx\,dy = \frac{1}{2a}\int_{-a}^{a}\xi(x)dx$, and $\{A_n\}$ does not average the field η.

On the other hand, Example 3.3.2 shows that the ergodicity condition can be weakened (see also Examples 8.1–8.3 below).

§ 3. Mean Ergodic Theorems for Weakly Converging Unitary Representations and for Strongly Quasimixing Dynamical Systems

Let $Y = G/K$ be a noncompact locally compact space; let $\mathfrak{B}(Y)$ (resp. $\mathfrak{B}(G)$) be the σ–algebra of Borel sets in Y (resp. in G).

In § 1 it was proved that the (right) condition

$(A_\circ)$ $\lim_{n\in N}\sup_{g\in G}\nu_n(gM) = 0$ for any compact set $M \subset Y$

is necessary in order that a net $\{\nu_n, n \in N\}$ $(\subset \mathcal{P}(\mathfrak{B}))$ would be mean r–averaging. In §§ 3 and 4 we shall characterize the mean objects that are averaged by all nets with the property $(A_\circ)$.

Let $\mathfrak{B}$ be $\mathfrak{B}(G)$ or $\mathfrak{B}(Y)$, let μ be the invariant measure on $\mathfrak{B}$. Consider the following classes of nets of measures and sets on G and Y:

C_1 is the class of all nets $\{\nu_n, n \in N\} \subset \mathcal{P}(\mathfrak{B})$ with the property $(A_\circ)$;

C_2 is the class of all sequences $\{\nu_n, n = \overline{1,\infty}\} \subset \mathcal{P}(\mathfrak{B})$ with the property $(A_\circ)$;

C_3 is the class of all nets $\{A_n, n \in N\} \subset \mathfrak{B}$ with the property

$(A'_\circ)$ $0 < \mu(A_n) < \infty$ and $\lim_{n\to\infty} \mu(A_n) = \infty$;

C_4 is the class of all sequences $\{A_n, n = \overline{1,\infty}\}$ with the property $(A'_\circ)$.

C_5 is the class of all sequences $\{A_n, n = \overline{1,\infty}\}$ such that the set $\bigcup_{n=1}^{\infty} A_n$ is a quasi-lattice*) and such that

$(A''_\circ)$ $0 < |A_n| < \infty$, $n = \overline{1,\infty}$, and $\lim_{n\to\infty} |A_n| = \infty$;

C_6 is the class of all sequences $\{A_n\}$ in C_5 for which the set $\bigcup_{n=1}^{\infty} A_n$ is a lattice*).

Let us recall that to any set $A \in \mathfrak{B}(Y)$ with $0 < \mu(A) < \infty$ there corresponds a measure $\lambda_A(B) := \mu(A \cap B)/\mu(A)$, $B \in \mathfrak{B}(Y)$, "uniformly distribution on A" and we have: $\frac{1}{\mu(A)} \int_A f\, d\mu = \int f\, d\lambda_A$. If $0 < |A| < \infty$, we set $\lambda_A(B) := |A \cap B|/|A|$, $B \in \mathfrak{B}(Y)$; then $|A|^{-1} \sum_{y\in A} f(y) = \int f\, d\lambda_A$.

It follows from Examples 3.7.1′ and 3.7.2′ that if a net $\{A_n, n \in N\}$ belongs to one of the classes C_3–C_6, then the net $\{\lambda_{A_n}, n \in N\}$ belongs to the class C_1 and, if $N = \mathbb{N}$, then it belongs to the class C_2. In the sequel the words "the net of measures $\{\nu_n, n \in N\}$ belongs to the class C_i" where $i = \overline{3,6}$ mean that $\nu_n = \lambda_{A_n}$, $n \in N$, and the net $\{A_n, n \in N\}$ belongs to C_i. We use here the notations $U_y h$ and $f(\omega T_y)$ introduced in § 1.

Lemma 3.1. *Let $U : g \mapsto U_g$ be a measurable K-class-1 unitary representation of G in a Hilbert space H; at any $h \in I_K$ the following conditions are equivalent:*

a) *the following weak limit exists* $(w) \lim\limits_{y\to\infty} U_y h (= M(h)$ *by Corollary* 4.18);

a′) *for any $h_1 \in I_K$, $\lim\limits_{y\to\infty} \langle U_y h, h_1\rangle$ exists;*

b_i) *any net $\{\nu_n, n \in N\} \in \mathcal{P}(\mathfrak{B})$ of class C_i averages h, i.e.*

---

*) See Definition 3.7.1.

$$\lim_{n\in N}\int U_y h\nu_n(dy) = M(h) \qquad (i=\overline{1,6}).$$

Proof. $a) \Rightarrow a')$; this is evident.

$a') \Rightarrow a)$. If $a')$ holds and $h_1 \in I_K$, $g \in G$, then $\lim_{y\to\infty}\langle U_y h, h_1\rangle = = \lim_{y\to\infty}\langle U_{g^{-1}y}h, h_1\rangle = \lim_{y\to\infty}\langle U_y h, U_g h\rangle$. This implies that $\lim_{y\to\infty}\langle U_y h, h'\rangle$ exists for any $h' \in \mathcal{L}_K$, the linear span of the set $\{U_g h_1,\ g_1 \in G,\ h_1 \in I_K\}$. Since $\mathcal{L}_K$ is dense in H and $\|U_y h\| = \|h\|$, $y \in Y$, the limit $(w)\lim_{y\to\infty} U_y h$ exists (see, e.g., Ahiezer, Glazman [1], § 27, Th. 2).

$a) \Rightarrow b_1)$ by Theorem 3.6.1 and Lemma 4.1.1.

$b_1) \Rightarrow b_2)$, $b_3) \Rightarrow b_4)$ and $b_5) \Rightarrow b_6)$: this is evident.

$b_2) \Rightarrow b_5)$ by Example 3.6.2$'$.

$b_4) \Rightarrow b_6)$. Let $\{A_n,\ n=\overline{1,\infty}\} \in C_6$ and let $\bigcup_{n=1}^{\infty} A_n$ be a $(1, V)$–quasi–lattice. Since U is a representation of class *1* and since it is uniformly continuous (see App., Subsect. 9.1), there exists a sequence of spherically symmetric neighborhoods $V_i \subset V$ of the K–fixed point y_0 such that $\|U_{y_1}h - U_{y_2}h\| < i^{-1}$ as soon as $y_2 \in y_1 V_i$. Let us consider an increasing sequence of positive integers $\{k_i,\ i=\overline{1,\infty}\}$ such that $k_i\mu(V_i) \to \infty$, and let us set $B_n := \bigcup_{y\in A_n} yV_{i_n}$ if $k_{i_n} \le |A_n| < k_{i_n+1}$. It is clear that $i_n \to \infty$ and $\mu(B_n) = |A_n|\mu(V_{i_n}) \ge k_{i_n}\mu(V_{i_n})$; thus $\{B_n, n=\overline{1,\infty}\} \in C_4$. Our assertion follows from the following relations:

$$\begin{aligned}
&\Big\|\frac{1}{|A_n|}\sum_{y\in A_n} U_y h - \frac{1}{\mu(B_n)}\int_{B_n} U_z h\mu(dz)\Big\| = \\
=&\Big\|\frac{1}{\mu(B_n)}\sum_{y\in A_n}\int_{yV_{i_n}} U_y h\mu(dz) - \frac{1}{\mu(B_n)}\sum_{y\in A_n}\int_{yV_{i_n}} U_z h\mu(dz)\Big\| \le \\
\le&\frac{1}{\mu(B_n)}\sum_{y\in A_n}\int_{yV_{i_n}} \|U_y h - U_z h\|\mu(dz) \le i_n^{-1}.
\end{aligned}$$

$b_6) \Rightarrow a')$. Let us suppose that $a')$ is not fulfilled. Then there exists an element $h_0 \in I_K$ such that the function $y \mapsto \varphi(y) := \langle U_y h, h_0\rangle$

has no limit as $y \in \infty$. Since φ is bounded, there exist numbers a, b, $a \neq b$, and nets $\{y_\alpha,\ \alpha \in A\} \subset Y$, $\{z_\beta,\ \beta \in B\} \subset Y$ such that $\lim_{\alpha\in A} y_\alpha = \lim_{\beta\in B} z_\beta = \infty$ and $\lim_{\alpha\in A}\varphi(y_\alpha) = a$, $\lim_{\beta\in B}\varphi(z_\beta) = b$. Let $\varepsilon < |b-a|/3$ and let $\alpha_\varepsilon \in A$, $\beta_\varepsilon \in B$ be such that $|\varphi(y_\alpha) - a| < \varepsilon$ and $|\varphi(z_\beta) - b| < \varepsilon$ if $\alpha > \alpha_\varepsilon$ and $\beta > \beta_\varepsilon$. The function φ is spherically symmetric (i.e. $\varphi(ky) = \varphi(y)$, $k \in K$, $y \in Y$), uniformly continuous and $|\varphi(y_\alpha) - \varphi(z_\beta)| > \varepsilon$, $\alpha > \alpha_\varepsilon$, $\beta > \beta_\varepsilon$. Therefore there exists a spherically symmetric neghborhood V of the K–fixed point y_0 such that $y_\alpha \notin z_\beta V$ and $z_\beta \notin y_\alpha V$ if $\alpha > \alpha_\varepsilon$, $\beta > \beta_\varepsilon$. Since the sets $\{y_\alpha,\ \alpha > \alpha_\varepsilon\}$ and $\{z_\beta,\ \beta > \beta_\varepsilon\}$ are unbounded, we can choose $(1,V)$–quasi–lattices $\{y_{\alpha_k},\ k = \overline{1,\infty}\} \subset \{y_\alpha,\ \alpha > \alpha_\varepsilon\}$ and $\{z_{\beta_k},\ k = \overline{1,\infty}\} \subset \{z_\beta,\ \beta > \beta_\varepsilon\}$ (see Example 3.6.2$'$). Let $\{n_m,\ m = 1, 2, \ldots\}$ be an increasing sequence of positive integers such that $\lim_{m\to\infty} n_m/n_{m-1} = \infty$. Let us set $\widetilde{y}_i := y_{\alpha_i}$ if $n_{m-1} \leq i < n_m$ and m is odd; $\widetilde{y}_i := z_{\beta_i}$ if $n_{m-1} \leq i < n_m$ and m is even ($m = \overline{1,\infty}$). It is clear that $\{\widetilde{y}_i\}$ is a $(1,V)$–quasi–lattice in Y. Suppose that $\lim_{n\to\infty} \frac{1}{n}\sum_{i=1}^{n}\varphi(\widetilde{y}_i) =: \alpha$ exists. Then

$$\lim_{r\to\infty} \frac{1}{n_{2r+1}} \sum_{i=1}^{n_{2r+1}} \varphi(\widetilde{y}_i) = \alpha \lim_{r\to\infty} n_{2r}/n_{2r+1} +$$

$$+ \lim_{r\to\infty} \frac{n_{2r+1} - n_{2r}}{n_{2r+1}} \frac{1}{n_{2r+1} - n_{2r}} \sum_{i=n_{2r}}^{n_{2r+1}} \varphi(\widetilde{y}_i) = a,$$

and similarly $\lim_{r\to\infty} \frac{1}{n_{2r}} \sum_{i=1}^{n_{2r}} \varphi(\widetilde{y}_i) = b$. Thus $\lim_{n\to\infty} \frac{1}{n}\sum_{i=1}^{n}\varphi(\widetilde{y}_i)$ does not exist. Therefore $\lim_{n\to\infty}\sum_{i=1}^{n} U_{\widetilde{y}_i} h$ does not exist either; this contradicts b_6), since the sequence $A_n = \{\widetilde{y}_1, \ldots, \widetilde{y}_n\}$, $n = \overline{1,\infty}$, belongs to the class C_6.

□

Theorem 3.2. *Let U be a measurable unitary representation of class 1 of G. The following conditions are equivalent:*

a) *for any $h \in I_K$, $(w) \lim\limits_{g\to\infty} U_g h (= \mathrm{M}(h))$ exists;*

b) *the weak limit $(w) \lim U_g$ $(= \mathrm{M})$ exists, where M is the averaging operator with respect to U;*

c_i) *any net $\{\nu_n,\ n \in N\} \subset \mathcal{P}(Y)$ of class C_i averages U over Y, i.e. for any $h \in I_K$, $\lim\limits_{n\in N} \int_Y U_y h \nu_n(dy) = M(h)$, $i = \overline{1,6}$;*

d_i) any net $\{\lambda_n,\ n \in N\} \subset \mathcal{P}(G)$ of class C_i averages U over G, i.e. for any $h \in H$, $\lim \int_G U_g h \lambda_n(dg) = \mathbf{M}(h)$, $i = 1, 6$.

Proof. It is clear that $b) \Rightarrow a$. The relations $a) \Leftrightarrow c_i)$, $i = \overline{1,6}$, follow from Lemma 3.1; the relations $b) \Leftrightarrow d_i)$, $i = \overline{1,6}$, are also implied by Lemma. Now, let U have property a). Since $\langle U_g h, h' \rangle = \langle h, U_{g^{-1}} h' \rangle$, $h, h' \in H$, the limit $\lim_{f \to \infty} \langle U_f h', h \rangle$ exists for any $h' \in H$ and $h \in I_K$. In order to prove that $a) \Rightarrow b)$ we have only to repeat the argument used in the proof of the relation $a') \Rightarrow a$ in Lemma 3.1.

□

Let $T = \{T_g,\ g \in G\}$ be a measurable r–dynamical system of class *1* (with respect to K) with probability phase space $(\Omega,\ \mathcal{F},\ P)$, let $\mathfrak{I}_K$ be the σ–algebra of all sets in $\mathcal{F}$ that are invariant with respect to the dynamical system $\{T_k,\ k \in K\}$, let P_K be the restriction of P to I_K, and let B be a Banach space. We recall that $I_K = \mathcal{F}$, $P_K = P$ if $Y = G$.

Theorem 3.3. *(i) If T is strongly quasimixing, then for any net $\{\nu_n,\ n \in N\} \subset \mathcal{P}(\mathfrak{B})$ with the property $(A_\circ)$ and any function $f \in L^\alpha_B$ $(\Lambda,\ I_K,\ P_K)$, $\alpha \geq 1$, the following limit exists:*
(L^α_B) $\lim_{n \in N} \int_Y f(\omega T_y) \nu_n(dy) = \mathbf{M}(f)$, and for any net $\{\nu_n,\ n \in N\} \subset \mathcal{P}(G)$ with the property $A_\circ$ and any $f \in L^\alpha_B$ $(\Omega, \mathcal{F}, P)$ the following limit exists: (L^α_B) $\lim\limits_{n \in N} \int_G f(\omega T_g) \nu_n(dg) = \mathbf{M}(f)$.

(ii) Suppose that for any function $f \in L^2$ $(\Omega, \mathfrak{I}_K, P_K)$ and all nets $\{\nu_n,\ n \in N\} \subset \mathcal{P}(Y)$, belonging to one of the classes C_i, $i = \overline{1,6}$, we have: (L^2) $\lim\limits_{n \in N} \int f(\omega T_y) \nu_n(dy) = \mathbf{M}(f)$. Then T is strongly quasimixing.

Proof. *(i)* If $B = \mathbb{C}$, $\alpha = 2$ this assertion is a simple consequence of Theorem 3.2. Now, if $\alpha > 1$ and B is an arbitrary Banach space, the proof is carried out as the proof of the relation $d_{2,\mathbb{C}}) \Rightarrow d_{\alpha,B})$ in Lemma 1.2. Finally, in the case $\alpha = 1$ the proof is similar to the proof of the relation $d_{\alpha,B}) \Rightarrow d_{1,B})$ in Lemma 1.2.

(ii) This is a simple consequence of Theorem 3.2.

□

§ 4. Mean Ergodic Theorems for Strongly Quasimixing Random Fields

Theorems 3.2 and 3.3 together with App. (8.G) imply the following two theorems. Let Y be a non–compact locally compact homogeneous space.

Theorem 4.1. *Let $\xi(\cdot)$ be a wide–sense homogeneous random field on Y.*

(i) If $\xi(\cdot)$ is strongly quasimixing in the wide sense, then for any net $\{\nu_n,\ n \in N\} \subset \mathcal{P}(\mathfrak{B}(Y))$ with the property $(A_\circ)$ (see § 4.3) the limit $(L^2) \lim\limits_{n\in N} \int \xi d\nu_n = \mathrm{M}^{(r)}(\xi)$ exists.

(ii) If for any net of some class C_i, $i = \overline{1,6}$, $(L^2) \lim\limits_{n\in N} \int \xi d\nu_n = \mathrm{M}^{(r)}(\xi)$, then the field $\xi(\cdot)$ is strongly quasimixing in the wide sense.

Let B be a Banach space.

Theorem 4.2. *If $\xi(\cdot)$ is a strongly quasimixing strict–sense homogeneous measurable B–valued random field on Y with $\mathrm{E}\|\xi(y)\|^\alpha < \infty$ for $1 \le \alpha < \infty$, then any net $\{\nu_n,\ n \in N\} \subset \mathcal{P}(\mathfrak{B})$ with the property $(A_\circ)$ averages $\xi(\cdot)$, i.e. there exists $(L^\alpha_B) \lim\limits_{n\in N} \int \xi(x)\nu_n(dx) = \mathrm{M}(\xi)$. In particular, all nets and sequences of class C_i, $i = \overline{1,6}$, average $\xi(\cdot)$.*

§ 5. Mean Ergodic Theorem on *SQM*–Spaces

Let $Y = G/K$ be a non–compact, locally compact separable homogeneous space. Theorem 2.3.2 and the results of § § 3 and 4 imply the following statement.

Theorem 5.1. *If Y is a SQM–space, then all nets of the classes C_1–C_6 (see § 3) on Y are mean averaging nets (m.a.n.'s).*

Conversely: if all nets of some class C_i, $i = \overline{1,6}$, on Y are m.a.n.'s, then Y is an SQM–space.

Let us recall that the homogeneous Euclidean space E_m and Lobachevsky space L_m, $m \geq 2$, as well as all non–compact irreducible Riemannian symmetric spaces are SQM–spaces; all non–compact connected quasisimple groups with finite center are SQM–groups (note that all these groups are not amenable and therefore do not possess any Følner nets, occurring in Theorem 2.1, but we do not need such nets here: on such a group any net $\{A_n, n \in N\}$ with $\mu(A_n) \to \infty$ is a m.a.n.).

Note 5.1. Homogeneous random fields on E_m are usually called "isotropic and homogeneous" random fields on $\mathbb{R}^m$. As we have seen, such fields are averaged by any net $\{A_n\} \subset \mathfrak{B}(\mathbb{R})$ with $\mu(A_n)$. But this fails if we remove the "isotropy" condition, as Example 2.1 shows. On E_m even sequences of "strips of zero width" are mean averaging: let $\overrightarrow{a} \in \mathbb{R}^m$ and $B_n \in \mathfrak{B}(\mathbb{R})$ with Lebesgue measures $l(B_n) \to \infty$; the sequence of measures $\nu_n(C) = [l(B_n)]^{-1} l(C \cap B_n \overrightarrow{a})$, $C \in \mathbb{R}^2$, is mean averaging.

§ 6. Mean Averaging Nets on Group Extensions and Products

Let G be a σ–compact locally compact topological group, $\mu_G^{(l)}$ and $\mu_G^{(r)}$ its left and right Haar measures. In this subsection we shall apply the results of § 3.2 to construct mean averaging nets. Theorems 6.1–6.6 below are obvious consequences of Theorems 3.2.3–3.2.9, since factor representations, restrictions and trivial extensions of unitary representations are unitary representations; we state these results here mainly for the sake of quotation in the sequel. We denote by $\mathrm{MAN}(G, \mu)$ the class of all mean r–averaging nets of sets (with the weight measure μ). Consider in G a normal subgroup Γ; $\Gamma := G/F$.

Theorem 6.1. *Let $\{F_n\} \in \mathrm{MAN}(G, \mu_F^{(l)})$, $\{\Gamma_n\} \in \mathrm{MAN}(\Gamma, \mu_\Gamma^{(l)})$ and let $\widetilde{F}_n$ be a section in G such that the set $G_n := \widetilde{F}_n \Gamma_n \in \mathfrak{B}(G)$, $n \in N$. Then $\{G_n\} \in \mathrm{MAN}(G, \mu_G^{(l)})$.*

Theorem 6.2. *Let Γ be a central normal subgroup of G and $\{F_n\} \in \in \mathrm{MAN}(G, \mu_F^{(r)})$, $\{\Gamma_n\} \in \mathrm{MAN}(\Gamma, \mu_G)$. Let $\widetilde{F}_n$ be a section in G over F_n such that $G_n := \widetilde{F}_n\Gamma$ is measurable, $n \in N$. Then $\{\Gamma_n\} \in \in \mathrm{MAN}(\Gamma, \mu_G^{(r)})$.*

Theorem 6.3. *Let Γ be compact.*

(i) If $\{F_n\} \in \mathrm{MAN}(F|\mu_F^{(l)})$ and $G_n := \pi^{-1}(F_n)$, then $\{G_n\} \in \mathrm{MAN}(G|\mu_G^{(l)})$.

(ii) If $\{F_n\} \in \mathrm{MAN}(F|\mu_F^{(r)})$, then $\{G_n\} \in \mathrm{MAN}(G, \mu_G^{(r)})$.

Lemma 6.4. *Let $\{F_n\} \subset \mathfrak{B}(F)$, $\{G_n\} \in \mathfrak{B}(\Gamma)$ and $G_n = \widetilde{F}_n\Gamma_n \in \in \mathfrak{B}(G)$, $n \in N$.*

(i) If $\{G_n\} \in \mathrm{MAN}(G, \mu_G^{(l)})$, then $\{F_n\} \in \mathrm{MAN}(F, \mu_F^{(l)})$.

(ii) If $\{G_n\} \in \mathrm{MAN}(G, \mu_G^{(r)})$, then $\{F_n\} \in \mathrm{MAN}(F, \mu_G^{(r)})$.

Theorem 6.5. *Let G be a left almost semidirect product of subgroups: $G = G^{(1)} \odot G^{(2)} \odot \cdots \odot G^{(m)}$ (see App., (1.9)).*

(i) If $\{G_n^{(i)}\} \in \mathrm{MAN}(G^{(i)}, \mu_G^{(l)})$, $i = \overline{1,m}$, and $G_n := G_n^{(1)} \ldots G_n^{(m)}$, then $\{G_n\} \in \mathrm{MAN}(G, \mu_G^{(l)})$.

(ii) If $\{G_n^{(i)}\} \in \mathrm{MAN}(G^{(i)}, \mu_G^{(r)}$, $i = \overline{1,m}$, and $G_n' = G_n^{(m)} \cdots G_n^{(1)}$, then $\{G_n'\} \in \mathrm{MAN}(G, \mu_G^{(r)})$.

Theorem 6.6. *Let G be decomposed into an almost direct product of subgroups $G^{(1)}, \ldots, G^{(m)}$ (see App., (1.9)).*

(i) If $\{G_n^{(i)}\} \in \mathrm{MAN}(G^{(i)}, \mu_{G^{(i)}}^{(l)})$, then $\{G_n\} = \{G_n^{(1)} \ldots G_n^{(m)}\} \in \in \mathrm{MAN}(G, \mu_G^{(l)})$.

(ii) If $\{G_n^{(i)}\} \in \mathrm{MAN}(G^{(i)}, \mu_{G_i}^{(r)})$, then $\{G_n\} \in \mathrm{MAN}(G, \mu_G^{(r)})$.

§ 7. Construction of Mean Averaging Sequences of Sets on Non–Amenable Groups

In this section G denotes a connected locally compact topological group, $\mu_G^{(l)}$ and $\mu_G^{(r)}$ are its left and right Haar measures (we write μ_G if G is unimodular); by MAS(μ) we denote the family of all mean right averaging sequences of compact sets on G (with the weight measure μ). Now our aim is to give a *constructive* proof of the fact that $\mathrm{MAS}(\mu_G^{(l)}) \neq \neq \emptyset$ and $\mathrm{MAS}(\mu^{(r)}) \neq \emptyset$. In the amenable case this was done in § 3.3. Now we consider non–amenable groups. Any theorem below is a step towards our final aim, and any proof gives a construction of MAS–sequences on some class of groups.

Below, in Chapter 6, we construct sequences of sets which are not only mean but also pointwise averaging; but the construction given here is considerably simpler.

7.1. Mean averaging sequences on semisimple Lie groups. First we shall state a simple consequence of Theorems 2.4.6 and 5.1.

Theorem 7.1. *A non–compact connected Lie group G is almost simple and has finite center iff any sequence of sets $\{G_n\}(\in \mathfrak{B}(G))$ with $\mu(G_n) \to \infty$ is mean averaging.*

Theorem 7.2. *Let G be a semisimple connected Lie group with finite center Z: then* $\mathrm{MAS}(G, \mu_G) \neq \emptyset$.

Proof. By App. (1. B), $G = G^{(0)} \cdot G^{(1)} \cdot \ldots \cdot G^{(m)}$ where $G^{(0)}$ is a compact subgroup. $G^{(i)}$ are almost simple subgroups of G with centers $Z^{(i)} \subset Z$, $i = \overline{1,m}$, $G^{(i)}$ and $G^{(j)}$ commute and thus $G^{(i)} \cap G^{(j)} \subset Z$ if $i \neq j$, $i, j = \overline{0,m}$. So G is an almost direct product of $G^{(0)}, \ldots, G^{(m)}$. If we take $\{G_n^{(i)}\} \subset \mathfrak{B}(G^{(i)})$ with $\mu_{G^{(i)}}(G_n^{(i)}) \to \infty$ and $G_n := G^{(0)} \cdot G_n^{(1)} \cdot \ldots \cdot G_n^{(m)}$, then $\{G_n\} \in \mathrm{MAS}(G, \mu_G)$.

Theorem 7.3. *If G is a semisimple connected Lie group, then*

$\mathrm{MAS}(G, \mu_G) \neq \emptyset$.

Proof. If the center Z is finite, the statement coincides with Theorem 7.2. So let us consider the case when Z is infinite. Let $Z^{(\infty)}$ be the subgroup of Z generated by the elements of infinite order. Note that the group $F = G/Z^{(\infty)}$ has finite center (see, e.g., Helgason [1], Corollaries 5.3 and 6.2, Ch. 2), and by Theorem 7.2 we can choose a sequence $\{F_n\} \in \mathrm{MAS}(G, \mu_F)$. Since $Z^{(\infty)}$ is commutative we can also consider a sequence $\{Z_n^{(\infty)}\} \in \mathrm{MAS}(Z^{(\infty)}, \mu_{Z^{(\infty)}})$. Let $\widetilde{F}_n$ be compact cross–sections in G over F_n (see App. (1.J)); then $G_n := \widetilde{F}_n Z_n^{(\infty)}$ is compact. Theorem 6.1 implies that $\{G_n\} \in \mathrm{MAS}(\mu_G)$.

□

Let G be a semisimple connected Lie group, G^* its universal covering group, π the natural homomorphism $G^* \mapsto G$; Φ the fundamental group of G (i.e. $\Phi \in Z(G^*)$ and $G^*/\Phi \cong G$). Consider the Iwasawa decompositions $G = KNA$ and $G^* = KNA$ (here $K \cong K^*/\Phi$, N and A are simply connected).

Lemma 7.4. *Let K_0^* be a compact subset of K^* such that $\mu(K_0^*) > 0$ and $(K_0^*)^{-1}K_0^* \cap \Phi = \{e\}$ (if G is simply connected we can set $K_0^* = K^*$); $K_0 := \pi(K_0^*)$. Then the restriction π_0 of π to K_0^*NA is a one–to–one homeomorphism $K_0^*NA \mapsto K_0NA$, and thus the sets K_0 and K_0NA are simply connected.*

Proof. Suppose that π_0 is not one–to–one; then there exist $k_1, k_2 \in K_0^*$, $k_1 \neq k_2$, with $k_1^{-1}k_2 \in \Phi$; since $k_1^{-1}k_2 \in \Phi \cap (K_0^*)^{-1}K_0^*$, we obtain a contradiction: $k_1 = k_2$. Since K_0 is compact, the induced mapping $K_0^* \overset{\pi_0}{\mapsto} K_0$ is a homeomorphism and the statement is proved.

□

Lemma 7.5. *Let G be a semisimple connected Lie group and K_0 the set defined in Lemma 7.4; denote by $Z = Z(G)$ the center of G. There exist mean averaging sequences of the form $K_0N_nA_nZ_n^{(\infty)}$, where $\{Z_n^{(\infty)}\} \subset \mathrm{MAS}(\mu_{Z^{(\infty)}})$ if Z is infinite and $Z_n^{(\infty)} \equiv \{e\}$ if Z is finite.*

Proof. This statement follows from the proofs of Theorems 7.2 and 7.3.

□

Corollary 7.6. *If a semisimple connected Lie group G has finite center Z, then there exists an r–averaging sequence of simply connected compact sets $\{G_n\}$.*

7.2. Mean averaging sequences on general connected groups. Let G be a simply connected Lie group. By App. (1.D) we have the right semidirect factorization $G = R \odot S$, where R is the "amenable radical" of G and S is a semisimple Lie group of non–compact type; both R and S are simply connected. Theorem 6.5 implies the following statement.

Theorem 7.7. *Let G be simply connected. If $\{R_n^{(l)}\} \in \mathrm{MAS}(\mu_R^{(l)})$, $\{R_n^{(r)}\} \in \mathrm{MAS}(\mu_R^{(r)})$ and $\{S_n\} \in \mathrm{MAS}(\mu_S)$, then $\{S_n R_n^{(l)}\} \in \mathrm{MAS}(\mu_G^{(l)})$ and $\{R_n^{(r)} S_n\} \in \mathrm{MAS}(\mu_G^{(r)})$.*

Now suppose G is a connected Lie group and consider the universal covering group G^* of G. Let π be the natural homomorphism $G^* \mapsto \mapsto G$; then $R := \pi(R^*)$ is the maximal amenable subgroup of G, and $S = \pi(S^*)$ is a simple connected group of non–compact type. We have $G = R \bullet S = S \bullet R$, but now R and S can intersect. Let $D := R \cap S$; denote by Φ the fundamental group of G, i.e. the (discrete) kernel of π: $\Phi \in Z(G^*)$ (the center of G^*) and $G^*/\Phi \cong G$ of G.

Lemma 7.8. *Let $S = KNA$ and $S^* = K^*NA$ be the Iwasawa decompositions of S, S^*. Choose K_0^* and K_0 as in Lemma 7.4. Let N_0 and A_0 be compact subsets of N and A, $S_0 := K_0 N_0 A_0$, and let $\widetilde{S} = K_0^* N_0 A_0$ be the cross–section in S over S_0; consider a compact set $R_0^* \subset R^*$ and set $R_0 = \pi(R_0^*)$; then $R_0^* \widetilde{S}_0$ (resp. $\widetilde{S}_0^*$) is a cross–section in G^* over $R_0 S_0$ (resp. over $S_0 R_0$).*

Proof. We shall consider, e.g., the sets $R_0 S_0$ and $R_0^* \widetilde{S}_0$. Let us suppose that there exist $r_1, r_2 \in R_0^*$ and $s_1, s_2 \in \widetilde{S}_0$ such that $\pi(r_1 s_1) = \pi(r_2 s_2)$.

Then $\pi(r_2)^{-1}\pi(r_1) = \pi(s_2)\pi(s_1)^{-1} \in D$, and since $D \subset K$ this implies $s_2 s_1^{-1} \in \pi^{-1}(D) \cap S^* \subset K^*$. Consider the decomposition $s_i = k_i^* n_i a_i$, $k_i^* \in K^*$, $n_i \in N$, $a_i \in A$, $i = 1, r$; $n_1 = n_2$, $a_1 = a_2$, $s_2 s_1^{-1} \in K^*$. We obtain $\pi(k_1^*) = \pi(k_2^*)$, i.e. $(k_2^*)^{-1} k_1^* \in \Phi \cap K_\circ^{-1} K^* = \{e\}$, and consequently $k_1^* = k_2^*$; thus $s_1 = s_2$, $r_1 = r_2$ and $r_1 s_1 = r_2 s_2$.

□

Theorem 7.9. *If G is a connected Lie group, then* $\mathrm{MAS}(\mu_G^{(l)}) \neq \emptyset$ *and* $\mathrm{MAS}(\mu_G^{(r)}) \neq \emptyset$.

Proof. We again consider the decomposition $G = RS$.

Let $\{R_n^*\} \in \mathrm{MAS}(\mu_{R^*}^{(r)})$ (see Theorem 2.1), $\{S_n\} \in \mathrm{MAS}(\mu_S)$. We can take the sets $S_n = K_0 N_n A_n$, where K_0 is as in Lemma 7.4 and $\{\Phi_n\} \in \mathrm{MAS}(\mu_\Phi)$. By Lemma 7.4, the set $\widetilde{S}_n = K_0^* N_n A_n$ is a compact cross–section in S^* over S and, by Theorem 6.2, $\{R_n^* \widetilde{S}_n \Phi_n Z_n^{(\infty)}\} \in \mathrm{MAS}(\mu_{R^*}^{(r)})$, while Lemma 7.8 states that $G_n^* := R_n^* \Phi_n Z_n^{(\infty)}$ is a cross–section in G^* over $R_n S_n Z_n^{(\infty)}$ where $R_n = \pi(R_n^*)$, $n = 1, 2, \ldots$. Lemma 6.4 implies that $\{R_n S_n Z_n^{(\infty)}\} \in \mathrm{MAS}(\mu_G^{(r)})$. It can be proved similarly that $\{S_n R_n Z_n^{(\infty)}\} \in \mathrm{MAS}(\mu_G^{(l)})$.

□

Theorem 7.10. *If G is a σ–compact locally compact connected group, then* $\mathrm{MAS}(\mu_G^{(l)}) \neq \emptyset$ *and* $\mathrm{MAS}(\mu_G^{(r)}) \neq \emptyset$.

Proof. It is well–known (see, e.g., Kaplansky [1], Ch. 2, §§ 10, 11) that there exists a compact normal subgroup Γ such that $F = G/\Gamma$ is a connected Lie group. If $\{F_n\} \in \mathrm{MAS}(\mu_F^{(l)})$ (resp. $\{F_n\} \in \mathrm{MAS}(\mu_F^{(r)})$), then by Theorem 6.3, $\{\pi^{-1}(F_n)\} \in (\mu_G^{(l)})$ (resp. $\{\pi^{-1}(F_n)\} \in \mathrm{MAS}(\mu_G^{(r)})$).

□

§ 8. Spectral Criterion

Let G be a topological group with a spectral system $(\Lambda, \mathcal{D}, U(\cdot))$ (see App., § 5).

We denote by λ_1 the element of Λ corresponding to the trivial representation $\mathbf{I}$.

Let $\nu \in \mathcal{P}(\mathfrak{B})$. For any $\lambda \in \Lambda$ we define an operator $\widetilde{\nu}(\lambda)$ in the space $H(\lambda)$ by $\widetilde{\nu}(\lambda)h(\lambda) = \int_G U_g(\lambda)h(\lambda)\nu(dg)$ (recall that the $H(\lambda)$ are one–dimensional if the group G is commutative, and in this case the $\widetilde{\nu}(\lambda)$ are complex numbers). It is clear that the $\widetilde{\nu}(\lambda)$ are linear operators and $\|\widetilde{\nu}(\lambda)\| \le 1$. The function $\lambda \to \widetilde{\nu}(\lambda)$ is the *Fourier transform* of ν.

With any unitary representation U in a Hilbert space H we associate the ν–average

$$\widehat{U} = \int U_g \nu(dg).$$

The following lemma shows the connection between the operator $\widehat{U}$ and the Fourier transform $\widetilde{\nu}(\cdot)$.

Lemma 8.1. *Let U be a unitary representation of G in H with spectral measure Φ. Then**)

$$\widehat{U}h = \int_\Lambda \oplus \widetilde{\nu}(\lambda)h(\lambda)\Phi(d\lambda), \quad h \in H.$$

Proof. By Fubini's theorem we get for any $h' \in H$,

*) Here and in the sequel we identify H with the space $\int_\Lambda \oplus H(\lambda)\Phi(d\lambda)$.

$$
\begin{aligned}
\langle \widehat{U}h, h' \rangle_H &= \int_G \langle U_g h, h' \rangle_H \nu_n(dg) = \\
&= \int_G \int_\Lambda \langle U_g(\lambda)h(\lambda), h'(\lambda) \rangle_{H(\lambda)} \Phi(d\lambda) \nu_n(dg) = \\
&= \int_\Lambda \int_G \langle U_g(\lambda)h(\lambda), h'(\lambda) \rangle_{H(\lambda)} \nu_n(dg) \Phi(d\lambda) = \\
&= \int_\Lambda \langle \widehat{U}(\lambda)h(\lambda), h'(\lambda) \rangle \Phi(d\lambda) = \\
&= \langle \int_\Lambda \oplus \widehat{U}(\lambda)h(\lambda)\Phi(d\lambda), h' \rangle_H .
\end{aligned}
$$

□

On many groups all the irreducible representations $U(\lambda)$, $\lambda \in \Lambda$, and the elementary positive definite functions are known (see, e.g., Vilenkin [1]). On such groups the following criterion can be useful; it is based on von Neumann's "spectral" method used by him to prove the classical mean ergodic theorem.

In this section we suppose that N is a direction containing a cofinal sequence $n_1, n_2, \ldots$; under this assumption the Lebesgue dominated convergence theorem holds for nets of functions $\{f_n,\ n \in N\}$.

Theorem 8.2. *The following properties of a net $\{\nu_n,\ n \in N\}$ $(\subset \mathcal{P}(\mathfrak{B}))$ are equivalent:*

a) $\lim_{n \in N} \widetilde{\nu}_n(\lambda)h(\lambda) = 0$ *for any* $h(\lambda) \in H(\lambda)$, $\lambda \in \Lambda \backslash \{\lambda_1\}$. (8.1)

b) $\{\nu_n\}$ r-averages the class **EP**, *i.e. for any elementary positive definite function $\varphi \not\equiv 1$,*

$$\lim_{n \in N} \int_G \varphi(yx)\nu_n(dx) = \mathbf{M}(\varphi)$$

uniformly with respect to $y \in G$.

c) $\{\nu_n\}$ is a mean r-averaging net.

Proof. The equivalence of a) and b) is implied by Lemma 1.1 and, obviously, c) implies a). So it remains to show the implication a)⇒d).

Consider a unitary representation U in a Hilbert space H with spectral measure F. We have $H = \int_{\Lambda} \oplus H(\lambda)\Phi(d\lambda)$. Let I be the subspace of all U–invariant elements in H; then $\mathbf{M}(h) = \prod_{H(\lambda_1)}^{H} h$. Put $\widehat{U}_n = \int U_g \nu(dg)$. We have $h - \mathbf{M}(h) = \int_{\Lambda\setminus\{\lambda_1\}} \oplus h(\lambda)\Phi(d\lambda)$. Therefore, by Lemma 8.1, $\widehat{U}_n h - \mathbf{M}(h) = \int_{\Lambda\setminus\{\lambda_1\}} \oplus \widehat{U}_n(\lambda)h(\lambda)\Phi(d\lambda)$ and $\|\widehat{U}_n - \mathbf{M}(h)\|^2 = \int_{\Lambda\setminus\{\lambda_1\}} \|\widehat{U}(\lambda)h(\lambda)\|^2_{H(\lambda)}\Phi(d\lambda)$. Since $\|\widehat{U}_n(\lambda)h(\lambda)\|_{H(\lambda)} \leq \|h(\lambda)\|_{H(\lambda)}$ and $\int_{\Lambda} \|h(\lambda)\|^2\Phi(d\lambda) = \|h\|^2 < \infty$, application of the Lebesgue dominated convergence theorem completes the proof.

□

We can easily generalize this theorem to homogeneous spaces, by considering the class *SEP* of spherical *EP*–functions instead of the class *EP* and the class of left irreducible unitary representations of class *1* instead of the class *IUR*.

Corollary 8.3. **(Blum–Eisenberg)** *Let G be a commutative group. The following properties of a separable net* $\{\nu_n,\ n \in N\} \in \mathcal{P}(\mathfrak{B}(G))$ *are equivalent:*

a) $\lim\limits_{n\in N} \int_G \varphi(x)\nu_n(dx) = 0$ *for any character* $\varphi \neq \mathbf{1}$*;*

b) $\{\nu_n,\ n \in N\}$ *averages the class* $AP(G)$*;*

c) $\{\nu_n,\ n \in N\}$ *is a mean averaging sequence.*

Proof. *c)* $\Rightarrow$ *a)*; this is evident.

a) $\Rightarrow$ *b)* by the theorem on approximation of almost periodic functions (see, e.g., Levitan [1], Ch. 6, § 3).

b) $\Rightarrow$ *a)* since all characters $\varphi \in AP(G)$.

a) $\Rightarrow$ *c)* is contained in Theorem 8.1.

□

Let $G = \mathbb{R}^m$ and let $\langle\cdot,\cdot\rangle$ be the inner product in $\mathbb{R}^m$. Consider in $\mathbb{R}^m$ the class Π of sets $E_{\sigma,a,d,\alpha}$ ($\alpha, a, d \in \mathbb{R}$, $0 \leq a < d$, $0 < \alpha < d$, $\sigma \in \mathbb{R}^m$, $\|\sigma\| = 1$) which are unions of periodic systems of parallel "strips": $E_{\sigma,a,d,\alpha} = \bigcup_{k=-\infty}^{\infty}\{x_i : a + kd < \langle\sigma, x\rangle < a + kd + \alpha\}$; if $G = \mathbb{R}$, we can always set $\sigma = 1$, and the sets $E_{1,a,d,\alpha}$ are unions of periodic systems of intervals.

Corollary 8.4. *On $\mathbb{R}^m$ the following properties of a separable net of measures $\{\nu_n,\ n\in N\}\subset\mathcal{P}(\mathfrak{B})$ are equivalent:*

a) $\{\nu_n\}$ *is mean averaging;*

b) $\{\nu_n\}$ *averages the class AP;*

c) $\lim\limits_{n\in N}\int_{\mathbb{R}^m} e^{i\langle\lambda,x\rangle}\nu_n(dx)=0,\ \lambda\in\mathbb{R}^m,\ \lambda\neq 0;$

d) $\lim\limits_{n\in N}\nu_n(E_{\sigma,a,d,\alpha})=\frac{\alpha}{d},\ E_{\sigma,a,d,\alpha}\in\Pi;$

e) $\lim\limits_{n\in N}(\nu_n(E)-\nu_n(E+x))=0,\ E\in\Pi,\ x\in\mathbb{R}^m.$

Proof. The equivalence of the conditions *a)* and *c)* is implied by Theorem 8.2. Let us show that *b)* $\Rightarrow$ *d)*. Obviously, for any $\varepsilon>0$, σ, a, d, α we can find continuous periodic functions ψ_ε and φ_ε such that $\psi_\varepsilon(x)\le\mathcal{X}_{E_{\sigma,a,d,\alpha}}(x)\le\varphi_\varepsilon(x)$ and $M(\varphi_\varepsilon-\psi_\varepsilon)<\varepsilon$. It is easy to check that $M(\psi_\varepsilon)\le\alpha/d\le M(\varphi_\varepsilon)$. If $\{\nu_n\}$ is an AP–averaging net, then

$$(\alpha/d)-\varepsilon\le\varliminf_{n\to\infty}\nu_n(E_{a,d,\alpha})\le\varlimsup_{n\to\infty}\nu_n(E_{a,d,\alpha})\le(\alpha/d)+\varepsilon.$$

Thus *b)* $\Rightarrow$ *d)*. The implication *d)* $\Rightarrow$ *e)* is evident. Let $\{\nu_n\}$ possess the property *e)*. Consider the function $\varphi(x)=\cos\langle\lambda,x\rangle$. It is obvious that the set $\{x : k/N\le\varphi(x)\le(k+1)/N\}\in\Pi$. Therefore φ can be uniformly approximated by linear combinations of functions $\mathcal{X}_E$, $E\in$ $\in\Pi$. This implies the existence of the limit $\lim\limits_{n\to\infty}\int\cos\langle\lambda,x\rangle\nu_n(dx)=$ $=\mathcal{X}_{\{0\}}(\lambda)$. Similarly it can be proved that $\lim\limits_{n\in N}\int\sin\langle\lambda,x\rangle\nu_n(dx)=0$, $\lambda\in\mathbb{R}^m$. Thus *e)* $\Rightarrow$ *c)*.

□

Note that condition *e)* in Corollary 8.2 is considerably weaker than the weak ergodicity condition (compare with Theorem 2.1).

Corollary 8.3 allows us to find non–ergodic mean averaging nets in $\mathbb{R}^m$. We shall give three simple examples.

Example 8.1. (Averaging by spheres) Let $S_R^{(m)}$ and B_R^m be the sphere and the ball of radius R in $\mathbb{R}^m$, resp. $(m\le 2)$; $\alpha_m:=\sigma(S_1^{(m)})$, the surface area of $S_1^{(m)}$. Let us consider the measure $\nu_R(A)=\frac{\sigma(S_R^m\cap A)}{\alpha_m R^{m-1}}$, $A\in S_R^m$, and calculate the mean $I_R(\lambda)=$

$= \int_{\mathbb{R}^m} e^{i\langle\lambda,x\rangle}\nu_R(dx) = \frac{1}{\alpha_m R^{m-1}}\int_{S_R^{(m)}} e^{i\langle\lambda,x\rangle}\sigma(dx)$ when $\lambda \in \mathbb{R}^m$, $\lambda \neq 0$. Let us fix some $\lambda \neq 0$; without loss of generality we may assume that $\|\lambda\| = 1$. Let y^{m-1} denote the projection of $y \in \mathbb{R}^m$ onto the $(m-1)$–dimensional subspace orthogonal to λ; $r = \|y^{(m-1)}\|$. Then $\langle\lambda, y\rangle = \pm\sqrt{1-r^2}$; we set $S_1^{(1)} = \{1\} \in \mathbb{R}^1$ and $\alpha_1 = 1$. If $y = R^{-1}x$ we have: $I_R(\lambda) = \frac{1}{\lambda_m}\int_{S_1^{(m)}} e^{iR\langle\lambda,y\rangle}\sigma(dy) = \frac{2}{\lambda_m}\int_{B_1^{(m-1)}} e^{\pm iR(\sqrt{1-r^2}}(1 - -r^2)^{\frac{1}{2}}dy^{(m-1)} = 2\alpha_m^{-1}\int_{S_1^{(m-1)}} d\sigma \int_0^1 r^{m-2}e^{\pm iR\sqrt{1-r^2}}(1-r^2)dr = = 2\alpha_m^{-1}\int_0^1(1-t^2)^{\frac{m-3}{2}}e^{\pm iRt}dt$. By the Riemann–Lebesgue convergence theorem (see, e.g., Franklin [1], § 291), $\lim_{R\to\infty} I_R(\lambda) = 0$, and Corollary 4.8.3 implies that $\{\nu_R, R > 0\}$ is a mean averaging net of measures.

Example 8.2. **(Averaging by annuli in $\mathbb{R}^2$ and spherical layers in $\mathbb{R}^m$, $m \geq 3$)** Let $A_R = B^{(m)}\backslash B_R^{(m)}$, $m \geq 2$ $(a > 0)$; $\widetilde{A}_R = B_1^{(m)}\backslash B_{1-a(R)}^{(m)} = R^{-1}\cdot A_R$, where $a(R) = \frac{d(R)}{R}$; $\beta_R = l(\widetilde{A}_R)$, the volume of $\widetilde{A}_R$. We have $I_R(\lambda) := \frac{1}{l(A_R)}\int_{A_R} e^{i\langle\lambda,x\rangle}dx = = \beta_R^{-1}\int_{\widetilde{A}_R} e^{iR\langle\lambda,y\rangle}dy = \beta_R^{-1}\int_{1-a(R)}^1 \rho^{m-1}d\rho\int_{S_1^{(m)}} e^{iR\rho\langle\lambda,z\rangle}dz$. According to Example 8.1, $\lim_{R\to\infty}\int_{S_1^{(m)}} e^{iR\rho\langle\lambda,z\rangle}dz = 0$; of course, this convergence is uniform with respect to all $\rho \geq \varepsilon > 0$, and therefore $\{A_R, R : a(R) \leq 1-\varepsilon\}$ is a mean averaging subnet. If $a(R) > 1-\varepsilon$, then $\beta(R) > \gamma$, $I_R(\lambda) = \frac{1}{\gamma}\int_0^1 \rho^{m-1}d\rho|\int_{S_1^{(m)}} e^{iR\rho\langle\lambda,z\rangle}dz|$, and $\{A_R, a(R) > 1-\varepsilon\}$ is a mean averaging subnet by the Lebesgue convergence theorem. Thus $\{a_R, R > 0\}$ is mean averaging.

Example 8.3. Let $G = R \odot R_{++}^{\times}$ be the group of affine transformations of the real line $\mathbb{R}$ (G is isomorphic to the group $\{g_{a,\sigma} = \left(\begin{smallmatrix}\sigma & a\\ 0 & 1\end{smallmatrix}\right)$, $a \in \mathbb{R}$, $\sigma \in \mathbb{R}_{++}\}$). The "dual" space $\Lambda = \mathbb{R} \cup \{+\} \cup \{-\}$ and $U(\lambda)$, $\lambda \in \mathbb{R}$, are one–dimensional representations: $U_{g_{a,\sigma}}(\lambda) = \sigma^{i\lambda}$; the left irreducible unitary representations $U(+)$ and $U(-)$ act in $H(+) := := L^2([0,+\infty))$ and $H(-) := L^2((-\infty, 0])$, respectively, as follows: for any $t \to f(t)$, $(U_{g_{a,\sigma}}(\pm)f)(t) = e^{ita}f(t\sigma)\sigma^{\frac{1}{2}}$.

We consider the right averaging by the "rectangles" $A_n = = \{g_{a,\sigma} : -a_n \leq a \leq a_n, \sigma_n^{-1} \leq \sigma_n \leq \sigma_n\}$, $n \in N$. The right

Haar measure $\mu^{(r)}(dad\sigma) = \sigma^{-1}dad\sigma$, and $\mu^{(r)}(A_n) = 4\sigma_n \ln a_n$. The measure $\nu_n(\beta) = \frac{\mu(A_n \cap B)}{\mu(A_n)}$ has Fourier transform

$$\widetilde{\nu}_n(\lambda) = (4a_n \ln \sigma_n)^{-1} \int_{-a_n}^{a_n} \int_{\sigma_n^{-1}}^{\sigma_n} \sigma^{i\lambda}\sigma^{-1} dad\sigma = \\ = (\ln \sigma_n)^{-1} \sin(\lambda \ln \sigma_n) \cdot \lambda^{-1}$$

if $\lambda \in \mathbb{R}\backslash\{1\}$, and for $f(t) := \mathcal{X}_{(t_1,t_2)}(t) \in H(\pm)$, $\widetilde{\nu}_n(\pm)f(t) = = (4a_n \ln \sigma_n)^{-1} \int_{-a_n}^{a_n} \int_{\sigma_n^{-1}}^{\sigma_n} e^{ita}\mathcal{X}_{(t_1,t_2)}(t\sigma)\sigma^{-\frac{1}{2}} dad\sigma$. Elementary calculations show that $\widetilde{\gamma}_n(\lambda) \to 0$ and $|\widetilde{\nu}_n(\pm)\mathcal{X}_{t_1,t_2}| \to 0$, if $\sigma_n \to \infty$. Since the $\mathcal{X}_{t_1,t_2}$ span the space $H(\pm)$ and $\|\widetilde{\nu}(\pm)\| \le 1$ we have $\widetilde{\nu}(\pm)f \to \to 0$ for any $f \in H(\pm)$. Thus the net $\{A_n\}$ is mean r–averaging if $\sigma_n \to \infty$. Under this condition $\{A_n\}$ is also l–averaging; to verify this we have to consider the representations $U(\pm) : (fU_{g_{a,\sigma}}(\pm))(t) = = e^{-it\sigma^{-1}a} f(t\sigma^{-1})\sigma^{-\frac{1}{2}}$ acting in the same spaces (compare with Example 6.1).

§ 9. Mean Ergodic Theorem for Homogeneous Generalized Random Fields

Let $I(\varphi)$, $\varphi \in K(R^m)$, be a wide–sense homogeneous generalized random field; let $F(d\lambda)$ and $\Phi(d\lambda)$ be, resp., the mean square and the random spectral measures of $I(\cdot)$, and let $I(\cdot) \in \mathfrak{S}_p$, i.e.

$$\int_{R^m} \frac{F(d\lambda)}{(1+\|\lambda\|^2)^p} < \infty \tag{9.1}$$

for some integer $p \ge 0$ (see App., Subsect. 8.7). We put $\widetilde{\psi}(\lambda) := := \int_{R^m} e^{i\langle\lambda,x\rangle}\psi(x)dx$, $\psi \in L^1(\mathbb{R}^m)$. Let us consider the linear space $M_F := \{\psi : \psi \in L^1(\mathbb{R}^m),\ \widetilde{\psi} \in L^2(F)\}$. We introduce in M_F the norm $\|\psi\| = (\int_{R^m} |\psi(\lambda)|^2 F(d\lambda))^{\frac{1}{2}}$, $\psi \in M_F$. The spectral representation $I(\varphi) = \int_{R^m} \widetilde{\varphi}(\lambda)\Phi(d\lambda)$ allows us to extend the field I to a mean–square continuous linear functional on M_F; moreover, $E|I(\psi)|^2 = = \|\psi\|^2$, $\psi \in M_F$. Any measurable wide–sense homogeneous random field $\xi(x)$, $x \in \mathbb{R}^m$, can be considered as a homogeneous generalized random field of class $\mathfrak{S}_0$, defined as $I_\xi(\varphi) = \int_{R^m} \xi(x)\varphi(x)dx$, $\varphi \in \mathbb{R}^m$.

The extension of $I(\varphi)$ to M_F is given by

$$I_\xi(\psi) = \int \xi(x)\psi(x)dx, \qquad \psi \in M_F \tag{9.2}$$

(it is easy to check that $\|\psi\|^2 = \iint R(x-y)\psi(x)\overline{\psi(y)}dxdy$ and therefore (see, e.g., Gnedenko [1]) the class of functions $\psi \in L^1(\mathbb{R}^m)$ for which the stochastic integral is defined coincides with M_F).

We shall state a simple sufficient condition under which $\psi \in M_F$ for all $I \in \mathfrak{S}_p$. For this we shall need notion of variation of a function on $\mathbb{R}^m$, $m \geq 1$. One of the possible ways to introduce this notion is the following. Denote $x^{(i)} = (x_1, \ldots, x_{i-1}, x_{i+1}, \ldots, x_m) \in \mathbb{R}^{m-1}$, $i = \overline{1,m}$; $(\text{var } f)(x^{(i)})$ is the usual variation of the function $f(x)$ with respect to the variable x_i when the vector $x^{(i)}$ is fixed. The *variation* of $f(x)$, $x \in \mathbb{R}^m$, is the number $\text{var } f := \max_{1\leq i \leq m} \int_{\mathbb{R}^{m-1}} (\text{var } f)(x^{(i)})dx^i$; if $\text{var } f < \infty$ we say that f is a *function of bounded variation*. It is clear that if $\int \left|\frac{\partial f}{\partial x_i}(x)\right| dx < \infty$, $i = \overline{1,m}$, then $\text{var } f < \infty$.

For any natural $p \geq 0$ we denote by $D^{p,1}$ the set of all functions ψ which possess integrable partial derivatives of any order $r \leq p$*) and, moreover, all partial derivatives of order p are of finite variation; $D^{-1,1} := L^1$. Evidently, if ψ has integrable partial derivatives of order $r \leq p+1$, then $\psi \in D^{p,1}$.

Lemma 9.1. *If* (9.1) *is fulfilled, then* $D^{p-1,1} \subset M_F$ $(p \geq 0)$.

Proof. Let $\psi \in D^{p-1,1}$; then $\sup_{\lambda \in \mathbb{R}^m} |\widetilde{\psi}(\lambda)| \leq \int_{\mathbb{R}^m} |\psi(x)|dx < \infty$. Consequently, for $p = 0$ our assertion is true. Let $p \geq 1$. Put $\psi^{(p)}(x) = \frac{\partial^p \psi(x)}{\partial x_1^{r_1} \ldots \partial x_m^{r_m}}$, where $r_1 + \ldots + r_m = p$. Using the formula

$$\widetilde{\psi^{(p-1)}}(\lambda) = (-1)^{p-1}\lambda_1^{r_1} \ldots \lambda_m^{r_m} \widetilde{\psi}(\lambda), \quad (r_1 + \ldots + r_m = p-1),$$

*) We assume here that $\psi^{(0)}$, the derivative of order 0, coincides with ψ.

we obtain

$$\begin{aligned}
&\sup_{\lambda\in\mathbb{R}^m} |\lambda_1^{r_1}\dots\lambda_m^{r_k}\dots\lambda_m^{r_m}\widetilde{\psi}(\lambda)| = \\
&= \sup_{\lambda\in\mathbb{R}^m} |\lambda_k|\cdot\Big|\int e^{i\langle\lambda,x\rangle}\psi^{(p-1)}(x)dx\Big| \le \\
&\le \sup_{\lambda\in\mathbb{R}^m} |\lambda_k|\int\Big|\int_{-\infty}^{\infty} e^{i\langle\lambda_k,x_k\rangle}\psi^{(p-1)}(x)dx_k\Big|dx^{(k)} = \\
&= \sup_{\lambda\in\mathbb{R}^m} \int\Big|\int_{-\infty}^{\infty} e^{i\langle\lambda_k,x_k\rangle}dx_k\psi^{(p-1)}\Big|dx^{(k)} \le \\
&\le \int_{R^{m-1}}\Big\{\int_{-\infty}^{\infty}|dx_k\psi^{(p-1)}(x)|\Big\}dx^{(k)} \le \\
&\le \mathrm{var}\,\frac{\partial^{p-1}\psi(x)}{\partial x_1^{r_1}\dots\partial x_M^{r_m}} < \infty.
\end{aligned}$$

Consequently,

$$\sup_{\lambda\in\mathbb{R}^m} |\lambda|^{2p}|\widetilde{\psi}(\lambda)|^2 = \sup_{\lambda\in\mathbb{R}^m}\sum_{p_i=p}\lambda_1^{2p_1}\dots\lambda_m^{2p_m}|\widetilde{\psi}(\lambda)|^2 < \infty,$$

and $\widetilde{\psi}\in L^2(F)$, i.e. $\psi\in M_F$.

□

Lemma 9.2. *Let I be a homogeneous generalized random field with spectral measure F and let $\{\psi_n,\, n\in N\}$ be a net of probability densities in M_F. Suppose that*

a) $\{\psi_n,\, n\in N\}$ is mean averaging;

b) the family $\{|\widetilde{\psi}_n(\lambda)|^2,\, n\in N\}$ is uniformly integrable with respect to F.

Then

$$\text{l.i.m.}_{n\in N}\, I(\psi_n) = \Phi(\{0\})^{*)}. \tag{9.3}$$

Proof. Let us fix a real number $\varepsilon > 0$ and let $r > 0$ be such that

$$\sup_{n\in N} \int_{|\lambda|>r} |\widetilde{\psi}_n(\lambda)|^2 F(d\lambda) < \varepsilon. \tag{9.4}$$

Denote by F_r and Φ_r the restrictions of F and Φ to $B_r^m := \{\lambda : \|\lambda\| \leq \leq r\}$; $\xi(x) := \int_{R^m} e^{i(\lambda,x)}\Phi_r(d\lambda)$ is a wide–sense homogeneous continuous random field on $\mathbb{R}^m$. Therefore $\lim_{n\in N} \int_{S_r^{(m)}} |\widetilde{\psi}_n(\lambda) - \mathcal{X}(\lambda)|^2 dF = = \lim_{n\in N} \mathbf{E}|\int_{R^m} \xi(x)\psi_n(x)dx - \Phi(\{0\})|^2 = 0$. Together with (9.4) this implies $\lim_{n\in N} \mathbf{E}|I(\psi_n) - \Phi(0)|^2 = \int_{R^m} |\widetilde{\psi}_n - \xi_{\{0\}}(\lambda)|^2 dF = 0$.

□

Arguments similar to those used in the proof of Lemma 9.1 show that if $\psi_n \in D^{p-1,1}$ and $\sup\limits_{n\in N} \operatorname{var} \psi_n^{(p-1)} < \infty$, then $\sup\limits_{n\in N} |\widetilde{\psi}_n(\lambda)| \in \in C_1(1 + \|\lambda\|^2)^{\frac{-p}{2}}$. Thus we arrive at the following result.

Theorem 9.3. *Let $\{\psi_n,\ n \in N\}$ be a net of densities in M_F with the following properties:*

a) $\{\psi_n,\ n \in N\}$ is mean averaging;

b) $\sup\limits_{n\in N} \operatorname{var} \psi_n^{(p-1)} < \infty$.

Then for any $I \in \mathfrak{S}_p$ relation (9.3) is true.

If $p = 0$, the assertion of Theorem 9.3 is contained in Theorem 2.1.

Corollary 9.4. *Let $B_R^{(m)}$ be balls of radii R in $\mathbb{R}^m$; $\psi_R(x) = = \frac{1}{l_m(B_r^{(m)})} \mathcal{X}_{B_R^{(m)}}(x)$, $R > 0$. Then for any $I \in \mathfrak{S}_1$ (9.3) holds.*

---

*) l.i.m. means “limit in the mean square”.

Proof. According to Theorem 2.2 $\{\psi_R,\ R>0\}$ is a mean averaging net; $\sup_{R>0} \operatorname{var} \psi_R \le 2\sup_{0<R<\infty}[l_{m-1}(B_R^{(m-1)})/l_m(B_R^{(m)})] = 2l_{m-1}(B_1^{(m-1)})/l_m(B_1^{(m)}) < \infty$. It remains to use Theorem 9.3.

□

If $\xi(\cdot)$ is a continuous homogeneous random field on $\mathbb{R}^m$, $\varphi \in K$ and ψ is a probability density, then $\int_{\mathbb{R}^m} \xi(x)(\varphi * \psi)(x)dx = \int_{\mathbb{R}^m} \left(\int_{\mathbb{R}^m} \xi(x)\psi(x-y)dx\right)\varphi(y)dy$, i.e. if we introduce the random field $\eta(y) = \int_{\mathbb{R}^m} \xi(x)\psi(x-y)dx$, then $I_\xi(\varphi * \psi) = I_\eta(\varphi)$. Therefore, in general, if ψ is fixed we may treat the generalized field $I_1(\varphi) = I(\varphi * \psi)$ as the integral of I with the weight function ψ and denote $I(\varphi * \psi) = \int I(\varphi)\psi dx$. Another approach to the notion of a generalized field –"integral" can be given in the case $m = 1$, $\psi_T = \frac{1}{2T}\mathcal{X}_{[-T,T]}$. Let $L(\cdot)$ be the primitive of $I(\cdot)$, i.e. $L' = I$. It is natural to set $\frac{1}{2T}\int_{x-T}^{x+T} I(\varphi)dt := \frac{1}{2T}L(\varphi(x-T)) - L(\varphi(x+T)) = \frac{1}{2T}I\left(\int_{x-T}^{x+T}\varphi(t)dt\right) = I(\varphi * \psi_T) =: \int I(\varphi)\psi_T(t)dt$.

□

Corollary 9.5. *Let $\{\psi_n,\ n \in N\}$ be a mean averaging net. Then for any homogeneous generalized field $I(\cdot)$ and any $\varphi \in K$ there exist*

$$\text{l.i.m.}_{n\in N} \int_{R^m} I(\varphi)\psi_n(x)dx = \Phi(\{0\}) \int_{-\infty}^{\infty} \varphi(t)dt$$

(the equality holds with probability 1); in other words, the net of generalized fields $\{\int_{\mathbb{R}^m} I(\varphi)\psi_n(x)dx,\ n \in N\}$ converges in $L^2\ (\Lambda, \mathcal{F}, P)$ to the random constant $\Phi(\{0\})$.

Proof. Let $\psi_n^* = \varphi * \psi_n$. Since $\widetilde{\psi}_n^* = \widetilde{\varphi}\cdot\widetilde{\psi}_n$, the net $\{\psi_n^*,\ n \in N\}$ is, by Corollary 8.4, mean averaging. Moreover,

$$\begin{aligned}\operatorname{var}\ (\psi_n^{*(p-1)}) &\le \sup_{p_1+\ldots+p_m=p} \iint |\varphi^{(p)}(x-y)||\psi_n(y)|dydx = \\ &= \sup_{p_1+\ldots+p_m=p} \int_{\mathbb{R}^m} |\varphi^{(p)}(x)|dx < \infty.\end{aligned}$$

Our assertion follows from Theorem 9.3.

□

§ 10. Mean Ergodic Theorem on the Free Group F_2

In the previous two sections we used the spectral decomposition of the representations and homogeneous random fields into irreducible components for proofs of mean ergodic theorems. In this section we consider another version of the spectral approach: spectral decompositions of the averages (as operators in the Hilbert space) are used. This method effectively used by Y. Guivarc'h and it goes back to an idea of V.I. Arnold and A.L. Krylov [1]; it seems to be effective when the study of irreducible representations of the group in consideration is complicated.

Consider the free group F_2 with generators a and b. Let $g \mapsto U_g$ be a unitary representation of F_2 in a Hilbert space H. Let R_n be the set of words of length n in F_2 (see Example 3.5.1).

$$\widehat{U}_n := \frac{1}{|R_n|} \sum_{g \in R_n} U_g$$

where $|R_n|$ is the cardinality of R_n. We have

$$\widehat{U}_0 = I, \quad \widehat{U}_1 = \frac{1}{4}(U_a + U_b + U_{a^{-1}} + U_{b^{-1}}).$$

Lemma 10.1. **(Arnold–Krylov)** *The following recurrence relation holds:*

$$\widehat{U}_{n+1} = \frac{4}{3}\widehat{U}_1\widehat{U}_n - \frac{1}{3}\widehat{U}_{n-1} \tag{10.1}$$

Proof. Let $R^x_{n-1} = \{g : g \in R_{n-1}, g = x_1^{k_1} \dots x_r^{k_r}, x_1 \neq x\}$ where $x \in \{a, b, a^{-1}, b^{-1}\}$. Note the following simple relations: $R_{n+1} \subset aR_n \cup \cup bR_n \cup a^{-1}R_n \cup b^{-1}R_n$, $xR_n \cap R_{n+1} = xR_n \setminus R_{n-1} = xR_n \setminus (xR_n \cap \cap R_{n-1}) = xR_n \setminus R^x_n$ where $x \in \{a, b, a^{-1}, b^{-1}\}$. Therefore $R_{n+1} = = (aR_n \setminus R^a_{n-1}) \cup (bR_n \setminus R^b_{n-1}) \cup (a^{-1}R_n \setminus R^{a^{-1}}_{n-1}) \cup (b^{-1}R_n \setminus R^{b^{-1}}_{n-1})$ where the sets in the parenthesis are disjoint. We have:

$$\sum_{g\in R_{n+1}} U_g = \sum_{g\in aR_n} U_g - \sum_{g\in R^a_{n-1}} U_g + \sum_{g\in bR_n} U_g - \sum_{g\in R^b_{n-1}} U_g +$$
$$+ \sum_{g\in a^{-1}R_n} U_g - \sum_{g\in R^{a^{-1}}_{n-1}} U_g + \sum_{g\in b^{-1}R_n} U_g - \sum_{g\in R^{b^{-1}}_{n-1}} U_g =$$
$$= \sum_{g\in R_1} U_g \cdot \sum_{g\in R_n} U_g - 3 \sum_{g\in R_{n-1}} U_g,$$

since any g in R_{n-1} belongs to three of the sets R^a_{n-1}, R^b_{n-1}, $R^{a^{-1}}_{n-1}$, $R^{b^{-1}}_{n-1}$. This implies the following equality

$$|R_{n+1}|\widehat{U}_{n+1} = 4|R_n|\widehat{U}_1\widehat{U}_n - 3|R_{n-1}|\widehat{U}_{n-1} \tag{10.2}$$

Let U be the identity representation. Then $\widehat{U}_n = I$ and equality (10.2) implies:

$$|R_{n+1}| = 4|R_n| - 3|R_{n-1}|.$$

Since $|R_0| = 1$, $|R_1| = 4$, the solution of this difference equation is $|R_n| = 4 \cdot 3^{n-1}$. Hence (10.2) implies (10.1).

□

Denote by I the subspace of U–invariant vectors, and let $H_1 := \{h : U_a h = U_{a^{-1}} h = U_b h = U_{b^{-1}} h = -h\}$, $I := \{h : U_g h = h,\ g \in F_2\}$, $H_0 := H \ominus H_1$.
Arguments used in §3.6 show that $I = \{h : \widehat{U}_1 h = h\}$ and $H_1 = \{h : \widehat{U}_1 h = -h\}$.

Lemma 10.2. **(Guivarc'h)** *For any $h \in H_0$, $\lim_{n\to\infty} \widehat{U}_n h = \Pi_I^{H_0} h$.*

Proof. Let $\widehat{U}_1^0$ be the restriction of $\widehat{U}_1$ to H_0. It is evident that $\widehat{U}_1^0$ is an Hermitian contraction. Therefore for any polynomial f the operator function $f(\widehat{U}_1^0)$ has the spectral representation $f(\widehat{U}_1^0) = \int_{-1}^{1} f(\lambda)E(d\lambda)$ where $E(d\lambda)$ is the spectral operator measure of $\widehat{U}_1^0$.

Of course, $E(\{-1\}) = 0$, $E(\{1\}) = \Pi_I^{H_0}$. It is easy to check that $\widehat{U}_n^0 = P_n(U_1^0)$ where P are polynomials defined as follows:

$$P_0(\lambda) = 1, \quad P_1(\lambda) = \lambda \quad \text{and}$$
$$P_{n+1}(\lambda) = \frac{4}{3}\lambda P_n(\lambda) - \frac{1}{3}P_{n-1}(\lambda).$$

Therefore $\widehat{U}_n^0 = \int\limits_{-1}^{1} P_n(\lambda)E(d\lambda)$ and $\widehat{U}_n^0 - \Pi_I^{H_0} = \int\limits_{1}^{1-0} P_n(\lambda)E(d\lambda)$. Hence

$$\|(\widehat{U}_n^0 - \Pi_I^{H_0})h\|^2 = \int\limits_{-1}^{1-0} [P_n(\lambda)]^2 \mu_h(d\lambda), \quad h \in H_0, \tag{10.3}$$

where $\mu_h(d\lambda) = (E(d\lambda)h, h)$. Now the problem is reduced to the study of the limit behaviour of the polynomials P_n. Consider the one–dimensional representation $U : U_a h = U_b h = e^{i\alpha}h$, $U_{a^{-1}}h = U_{b^{-1}}h = = e^{-i\alpha}h$, $h \in \mathbb{C}$. Here $\widehat{U}_1 1 = \cos\alpha$. Put $\lambda := \cos\alpha$; λ ranges over the whole interval $[-1, 1]$ if α ranges over $[0, \pi]$. We have

$$P_n(\lambda) = \widehat{U}_n 1 = P_n(\widehat{U}_1) = \frac{1}{|R_n|}\sum_{k=0}^{n} \alpha_{k,n} e^{ik\alpha} e^{-i(n-k)\alpha}$$

where $\alpha_{k,n} \geq 0$ and $\sum_{k=1}^{n} \alpha_{k,n} = |R_n|$. Thus $|P_n(\lambda)| \leq 1$, $-1 \leq \lambda \leq 1$. For any λ, $-1 \leq \lambda \leq 1$, we consider the difference equation

$$3P_{n+1}(\lambda) - 4\lambda P_n(\lambda) + P_{n-1}(\lambda) = 0.$$

The characteristic equation is

$$3x^2 - 4\lambda x + 1 = 0.$$

Hence

$$P_n(\lambda) = c_1 x_1^n + c_2 x_2^n \quad \text{if} \quad \lambda \neq \pm\frac{\sqrt{3}}{2},$$

and

$$P_n(\lambda) = (c_1' + nc_2')x_1^n \quad \text{if} \quad \lambda = \pm\frac{\sqrt{3}}{2}, \tag{10.4}$$

where $x_{1,2}(\lambda) = (2\lambda \pm \sqrt{4\lambda^2 - 3})/3$ and c_1, c_2, c_1', c_2' are constants (depending on λ). If $|\lambda| < 1$, then $\alpha(\lambda) := \max(|x_1|, |x_2|) < 1$, and (10.4) implies $|P_n(\lambda)| \leq c(\lambda)n[\alpha(\lambda)]^n$. Thus $P_n(\lambda) \to 0$, $|\lambda| < 1$, and, since $\sup_n |P_n(\lambda)| \leq 1$, $\mu_h([-1,1)) = \|h\|^2$ we deduce the statement of the theorem from (10.3), using the Lebesgue majorated convergence theorem.

□

Consider the sets

$$S_n := \{x_1 \dots x_k, x_i \in \{a, b, a^{-1}, b^{-1}\}, 0 \leq k \leq n\} = \bigcup_{j=0}^{n} R_j,$$

$$A_n = R_n \cup aR_n, \quad B_n = R_n \cup bR_n,$$
$$C_n = S_n \cup aS_n, \quad D_n = S_n \cup bS_n.$$

Theorem 10.3. *The sequences $\{A_n\}$, $\{B_n\}$, $\{C_n\}$, and $\{D_n\}$ are mean averaging.*

Proof. It is clear that the sets R_n and aR_n are disjoint and $|R_n| = |aR_n|$. If $h_0 \in H_0$, then, by Theorem 10.1, $\lim_{n\to\infty} \frac{1}{|R_n|} \sum_{g\in R_n} U_g = \Pi_I^{H_0}$ and therefore

$$\lim_{n\to\infty} \frac{1}{|R_n|} \sum_{g\in aR_n} U_g h_0 =$$
$$= \lim_{n\to\infty} \frac{1}{|R_n|} U_a \sum_{f\in R_n} U_f h_0 = U_a \Pi_I^{H_0} h_0 = \Pi_I^{H_0} h_0.$$

Hence

$$\lim_{n\to\infty} \frac{1}{|A_n|} \sum_{g\in A_n} U_g h_0 =$$
$$= \lim_{n\to\infty} \frac{1}{2|R_n|} \left(\sum_{g\in R_n} U_g h_0 + \sum_{g\in aR_n} U_g h_0 \right) = \Pi_I^{H_0} h_0.$$

If $h_1 \in H_1$

$$\sum_{g \in A_n} U_g h_1 = |R_n|(-1)^n h_1 + |R_n|(-1)^{n+1} h_1 = 0.$$

Therefore for any $h \in H$ we have

$$\lim_{n\to\infty} \frac{1}{|A_n|} U_g h = \Pi_I^{H_0} \Pi_{H_0}^H h = \Pi_I^H h.$$

It is evident that $|S_n| = \sum_{j=0}^n |R_j|$. Therefore

$$\lim_{n\to\infty} \frac{1}{|S_n|} \sum_{g\in S_n} U_g h =$$

$$= \lim_{n\to\infty} \frac{1}{|S_n|} \sum_{j=0}^{n} |R_j| \frac{1}{|R_j|} \sum_{g\in R_j} U_g h = \lim_{n\to\infty} \frac{1}{|R_n|} \sum_{g\in R_n} U_g h$$

and

$$\lim_{n\to\infty} \frac{1}{|S_n|} \sum_{g\in aS_n} U_g h = \lim_{n\to\infty} \frac{1}{|R_n|} \sum_{g\in aR_n} U_g h, \quad h \in H.$$

Hence

$$\lim_{n\to\infty} \frac{1}{|C_n|} \sum_{g\in C_n} U_g h = \lim_{n\to\infty} \frac{1}{|A_n|} \sum_{g\in A_n} U_g h = \Pi_I^H h, \quad h \in H.$$

Thus we have proved that the sequences $\{A_n\}$ and $\{C_n\}$ are mean averaging, and the same arguments show that $\{B_n\}$ and $\{D_n\}$ also have this property.

□

BIBLIOGRAPHICAL NOTES

1. The first generalizations of the von Neumann ergodic theorem for various kinds of representations of $\mathbb{Z}$ were given in Yosida [2] (1938), Krotkov, Galperin [1] (1953), Lorch [1] (1939), Riesz [1] (1938), [2], [3],

[4] (1941), [5] (1942) (for more extended bibliography see, e.g., Jacobs [1], Dunford, Schwartz [1], Krengel [2]).

2. In Wiener [1] (1939) and Dunford [1], [2] (1939) the von Neumann theorem was extended to the groups $\mathbb{Z}^m$ and $\mathbb{R}^m$ $(m > 1)$ (averaging with concentric balls and cubes was considered). Pitt [1] (1942) proved the mean ergodic theorem in $\mathbb{R}^m$ for an extensive class of "regular" nets of sets.

3. Calderón [1] (1953) has extended the mean and the pointwise ergodic theorems to a wide class of unimodular amenable groups.

4. Theorem 2.1 is contained in Tempelman [2] (1961), [3], [4] (1962); this theorem was repeatedly rediscovered in various forms (see, e.g., Greenleaf [2] (1973), Matthes, Kerstan, Mecke [1] (1978), Renaud [1] (1971), Chatard [1] (1970), Fleischmann [1] (1978)).

5. Theorem 1.5 and the results stated in Notes 1.5 and 1.6 are close to Theorems 3 and 4 and Corollary 1 in Tempelman [5] (1967) (see also Tempelman [14] (1975)); the result stated in Note 1.7 was proved in Oseledec [1] (1965) and the ones stated in Notes 1.8 and 1.9 are special cases of results contained in Lin [2] (1977) and in Derriennic, Lin [1] (1989).

6. Theorems 1.6 and 1.7 are special cases of theorems 6 and 7 in Tempelman [19] (1981) (see also Tempelman [14] (1975)).

7. The study of sequences of measures which average all weakly convergent (at infinity) representations of $\mathbb{Z}$ was begun in Blum, Hanson [1] (1960) and continued in Akcoglu, Sucheston [1] (1972), [2] (1975), Jones, Kuftinec [1] (1971), Fong, Sucheston [1] (1974) and in Hanson, Pledger [1](1969). Sato [5] (1974), [4] (1981) and Fong [1] (1976) obtained results for $\mathbb{R}$.

Iwanik [1] (1979) has proved the equivalence of assertions (a), (b_1) and of the assertion "(b_1) is valid for all nets $\{\nu_n\} \in \widetilde{\mathcal{P}}(\mathfrak{B})$" in Lemma 3.1, as well as the equivalence of similar statements in Theorem 3.2

for mixing dynamical systems in the case when G is a group. The remaining results of §3 appeared in Tempelman [24] (1986); in the proofs ideas of H. Fong and A. Iwanik were essentially used.

8. In Tempelman [23] (1983) SQM-spaces were characterized as spaces on which nets of sets with property (A_0) are mean averaging nets (see also Savichev and Tempelman [2] (1984)). §§3–5 are contained in the book Tempelman [24] (1986) and §6 is an extension of Ch.4, §6, of that book. The results of §7 appear to be new.

9. The spectral method of proving the mean ergodic theorem goes back to von Neumann and Riesz, and is often used in the text–books (see, e.g., Doob [1]). The statement of Corollary 8.3 was proved in Blum, Eisenberg [2] (1974). Theorem 9.1 is a version of Theorem 9.1 of Chapter 4 in Tempelman [24] (1986).

10. The results of §9 are contained in Tempelman [4] (1962) (see also Tempelman [14] (1975), [24] (1986)). Urbanik [1] (1956), [2] (1958) has proved an ergodic theorem similar to Corollary 9.5, for stationary generalized processes in the sense of Mikusinski.

The results of §10 are based on the method due to Arnold, Krylov [1]; the statement of Lemma 10.1 is contained in this work.

Lemma 10.2 was proved in Guivarc'h [1]. For finite–dimensional unitary representations Theorem 10.3 was proved in Greenleaf [2]. Another class of sets in free groups, averaging all finite–dimensional unitary representations was found in Gorbis, Tempelman [1].

Close to the material of this chapter are: Blum, Eisenberg, Hahn [1], Vershik, Kaimanovich [1], C. Ionescu Tulcea [1], Yoshimoto [1], Takahashi [1], Tempelman [5, 7–9, 11–13, 15] and Yadrenko [1].

CHAPTER 5
MAXIMAL AND DOMINATED ERGODIC THEOREMS

§ 1. Preliminaries

1.1. Maximal and dominant nets. Let $(X, \mathfrak{B})$ be a measurable semigroup, N a linearly ordered set and $\{\nu_n, n \in \mathfrak{B}\}$ a net of probability measures on $\mathfrak{B}$.

In this chapter we discuss inequalities of two kinds which can be naturally associated with any completely measurable linear representation in a Lebesgue space $L^\alpha(\Omega, \mathcal{F}, m)$ and, in particular, with any completely measurable dynamical system. We shall consider only "left objects": left dynamical systems and right representations; all the statements can be reformulated for "right objects" in a quite obvious way. So let S be a completely measurable linear representation of X in $L^\alpha(\Omega, \mathcal{F}, m)$. With a function $f \in L^\alpha(\Omega, \mathcal{F}, m)$ we associate the *averages* $\widehat{f}_n(\omega) = \int (fS_x)(\omega)\nu_n(dx)$ and the *dominant function* $f^*(\omega) = \sup\limits_{n\in N} |\widehat{f}_n(\omega)|$.

Definition 1.1. We say that $\{\nu_n, n \in N\}$ is a *maximal net* with index α and constant C'_α with respect to S (notation: $\{\nu_n\} \in \mathrm{MAX}_\alpha(S)$) if for any $f \in L^\alpha(\Omega, \mathcal{F}, m)$ the function $f^*(\omega)$ is measurable*) and for any $\varepsilon > 0$,

$$m\{\omega : f^*(\omega) > \varepsilon\} \le \frac{C'_\alpha}{\varepsilon^\alpha} \int |f(\omega)|^\alpha m(d\omega).$$

We say that $\{\nu_n, n \in N\}$ is a *dominant net* with index α and constant C'_α (notation: $\{\nu_n\} \in \mathrm{DOM}_\alpha(S)$) if for any $f \in L^\alpha(\Omega, \mathcal{F}, m)$ the

*) Clearly $f^*(\omega)$ is measurable if N is countable (see also Subsect. 1.5 below).

function $f^*(\omega)$ is measurable and

$$\int_A [f^*(\omega)]^\alpha m(d\omega) \leq C''_\alpha \int_A |f(\omega)|^\alpha m(d\omega), \tag{1.2}$$

i.e.

$$\|f^*\|_{L^\alpha} \leq (C''_\alpha)^{\frac{1}{\alpha}} \|f\|_{L^\alpha}. \tag{1.2'}$$

If S is associated with a completely measurable left dynamical system in the space $(\Omega, \mathcal{F}, m)$, i.e. if $(fS_x)(\omega) = f(T_x\omega)$, then $\widehat{f}_n(\omega) = \int f(T_x\omega)\nu_n(dx)$, and we write simply $\{\nu_n\} \in MAX_\alpha(T)$ or $\{\nu_n\} \in \mathrm{DOM}_\alpha(T)$. In the case when all ν_n are uniformly distributed on sets $A_n \in \mathfrak{B}$ with respect to some measure μ on $\mathfrak{B}$, i.e. when $\nu_n(B) = \frac{\mu(A_n \cap B)}{\mu(A_n)}$, our averages can be written as

$$\widehat{f}_n(\omega) = \frac{1}{\mu(A_n)} \int_{A_n} fS_x(\omega)\mu(dx).$$

If $\{\nu_n\} \in \mathrm{MAX}_\alpha(S)$, ($\{\nu_n\} \in \mathrm{DOM}_\alpha(S)$), we write $\{A_n\} \in \mathrm{MAX}_\alpha(S|\mu)$ (resp., $\{A_n\} \in \mathrm{DOM}_\alpha(S|\mu)$).

Propositions stating conditions under which nets $\{\nu_n\}$ belong to $\mathrm{MAX}_\alpha(S)$ or $\mathrm{DOM}_\alpha(S)$ are called *maximal*, respectively, *dominated, ergodic theorems*. It always suffices to prove such a theorem for non-negative functions only, since $f^*(\omega) \leq |f|^*(\omega)$.

The following simple statement presents an important property of MAX– and DOM–nets.

Proposition 1.1. *If* $\{\nu_n\} \in \mathrm{MAX}_\alpha(S)$ *or* $\mathrm{DOM}_\alpha(S)$, *then for any* $f \in L^\alpha(\Omega, \mathcal{F}, m)$,

$$f^*(\omega) = \sup_{n \in N} \left| \int fS_x(\omega)\nu_n(dx) \right| < \infty \quad m\text{–a.e.} \tag{1.3}$$

Proof. This is evident if $\{\nu_n\} \in \mathrm{DOM}_\alpha(S)$. If $\{\nu_n\} \in \mathrm{MAX}_\alpha(S)$, then $0 \leq m(\{\omega : f^*(\omega) = +\infty\}) = \lim_{k\to\infty} m(\{\omega : f^*(\omega) > k\}) \leq \lim_{k\to\infty} \frac{C'_\alpha}{k^\alpha} \int_\Omega |f(\omega)|^\alpha m(d\omega) = 0$.

□

Under some mild additional conditions, the maximality of $\{\nu_n\}$ is necessary for (1.3) (see Burkholder [1]). In its turn, property (1.3) is, clearly, necessary for the validity of the corresponding "pointwise ergodic theorem": $\lim_{n\in N} \int_G fS_x(\omega)\nu_n(dx)$ exists m–a.e. (see Ch. 6). One can therefore expect that a maximal or a dominated ergodic theorem can be a natural preparatory step in proving pointwise ergodic theorems. The Banach convergence principle shows that this is a crucial step indeed.

We conclude this subsection by a trivial example of DOM–sequences, which we will need in the sequel.

Example 1.1. Let $\nu \in \mathcal{P}(\mathfrak{B})$ and $\nu_n \equiv \nu$, $n \in N$. Then $\{\nu_n\} \in$ $\in \mathrm{DOM}_\alpha(S)$ for any $\alpha \geq 1$ and any strongly measurable bounded representation in $L^\alpha(\Omega, \mathcal{F}, m)$ $(\alpha \geq 1)$. Indeed, by the Hölder inequality and Fubini's theorem

$$\int_\Lambda \Big| \int_X S_x f(\omega) d\nu \Big|^\alpha m(d\omega) \leq \int_\Omega \left(\int_X |S_x f(\omega)|^\alpha d\nu \right) m(d\omega) \leq$$
$$\leq \int_X \left(\int_\Omega |S_x f(\omega)|^\alpha m(d\omega) \right) \nu(dx) \leq \sup_{x\in X} \|S_x\|^\alpha \int_\Omega |f(\omega)|^\alpha m(d\omega).$$

1.2. Relation between the classes $\mathbf{MAX}_\alpha$ and $\mathbf{DOM}_\alpha$.

Proposition 1.2. *(i)* $\mathrm{DOM}_\alpha(S) \subset \mathrm{MAX}_\alpha(S)$, $\alpha \geq 1$.

(ii) If $1 \leq \alpha < \beta < \infty$, *then* $\mathrm{MAX}_\alpha(S) \subset \mathrm{MAX}_\beta(S)$ *and* $\mathrm{DOM}_\alpha(S) \subset \mathrm{DOM}_\beta(S)$.

Proof. *(i)* Apply the Chebyshev inequality.

(ii) By the Hölder inequality, for any $f \in L^\beta_+(\Omega, \mathcal{F}, m)$,

$$\int |fS_x(\omega)|\nu_n(dx) \leq \Big\{ \int |fS_x(\omega)|^{\frac{\beta}{\alpha}} \nu_n(dx)\Big\}^{\frac{\alpha}{\beta}},$$

while for $\{\nu_n\} \in \mathrm{MAX}_\alpha(S)$ we have

$$m\{\omega : f^*(\omega) > \varepsilon\} \leq m\{(f^{\frac{\beta}{\alpha}})^* > \varepsilon^{\frac{\beta}{\alpha}}\} \leq$$
$$\leq \frac{C'_\alpha}{\left(\varepsilon^{\frac{\beta}{\alpha}}\right)^\alpha} \int \left(f^{\frac{\beta}{\alpha}}\right)^\alpha dm \leq \frac{C'_\alpha}{\varepsilon^\beta} \int f^\beta dm,$$

and thus $\{\nu_n\} \in \mathrm{MAX}_\beta(S)$; the second inclusion can be proved similarly.

□

The more refined Proposition 1.5 below shows that the inclusions $\mathrm{MAX}_1(S) \subset \mathrm{DOM}_\alpha(S)$ for $\alpha > 1$ hold. The method of proof used here is due to Wiener [1].

Lemma 1.4. *If $\{\nu_n\} \in \mathrm{MAX}_1(S)$ with constant C'_1, then $m(\{\omega : f^*(\omega) \geq \varepsilon\} \leq \frac{2C'_1}{\varepsilon} \int_{\{\omega: f(\omega) > \frac{\varepsilon}{2}\}} |f(\omega)| m(d\omega)$ for any $f \in L^1(\Omega)$.*

Proof. Consider the function

$$g(\omega) = \begin{cases} f(\omega) & \text{if} \quad f(\omega) > \varepsilon/2, \\ 0 & \text{if} \quad f(\omega) \leq \varepsilon/2. \end{cases}$$

It is evident that $f(\omega) \leq g(\omega) + \varepsilon/2$, and therefore $f^*(\omega) \leq g^*(\omega) + \varepsilon/2$. Since $\{\nu_n\} \in \mathrm{MAX}_1(S)$, we have

$$m(\{\omega : f^*(\omega) \geq \varepsilon\}) \leq m(\{g^*(\omega) \geq \varepsilon/2\}) \leq$$
$$\leq 2\frac{C'_1}{\varepsilon} \int_{\{\omega: f(\omega) > \frac{\varepsilon}{2}\}} |f(\omega)| m(d\omega).$$

□

Proposition 1.5. $\mathrm{MAX}_1(S) \subseteq \mathrm{DOM}_\alpha(S)$, $\alpha > 1$. *More precisely, if f is a non-negative measurable function and $\nu_n \in \mathrm{MAX}_1(S)$ with constant C'_1, then for $1 < \alpha < \infty$,*

$$\int_\Omega [f^*(\omega)]^\alpha m(d\omega) \leq \frac{2^\alpha \alpha C'_1}{\alpha - 1} \int_\Omega [f(\omega)]^\alpha m(d\omega);$$

for $\alpha = 1$,

$$\int_\Omega f^*(\omega) m(d\omega) \le 2[m(\Omega) + C_1' \int_\Omega f(\omega) \ln^+ f(\omega) m(d\omega)]^{*)};$$

and for $0 < \alpha < 1$,

$$\int_\Omega [f^*(\omega)]^\alpha m(d\omega) \le 2^\alpha [m(\Omega) + \frac{\alpha}{1-\alpha} \int_\Omega f(\omega) m(d\omega)].$$

Proof. By Fubini's theorem, for every real number α,

$$\int_\Omega [f^*(\omega)]^\alpha m(d\omega) = \alpha \int_\Omega \left(\int_0^{f^*(\omega)} s^{\alpha-1} ds \right) m(d\omega) =$$

$$= \alpha \iint_{\{(s,\omega): f^*(\omega) \ge s\}} s^{\alpha-1} ds\, m(d\omega) = \alpha \int_0^\infty s^{\alpha-1} m(\Lambda_s) ds$$

(recall that $\Lambda_s = \{\omega : f^*(\omega) \ge s\}$). On making the substitution $s = 2t$ we obtain

$$\int_\Omega [f^*(\omega)]^\alpha m(d\omega) = 2^\alpha \alpha \int_0^\infty t^{\alpha-1} m(\Omega_{2t}) dt. \tag{1.4}$$

Let us estimate this integral. Using Lemma 1.4 and Fubini's theorem, we obtain for every $\delta \ge 0$

$$\int_\delta^\infty t^{\alpha-1} m(\Lambda_{2t}) dt \le C_1' \int_\delta^\infty t^{\alpha-2} \left\{ \int_{\{\omega: f(\omega) > t\}} f(\omega) m(d\omega) \right\} dt =$$

$$= C_1' \int_{\{\omega: f(\omega) > \delta\}} f(\omega) \left(\int_\delta^{f(\omega)} t^{\alpha-2} dt \right) m(d\omega).$$

Thus,

---

*) $\ln^+ f(\omega) = \max\{0, \ln f(\omega)\}$.

$$\int_{\delta}^{\infty} t^{\alpha-1} m(\Lambda_{2t})dt \leq C_1' \int_{\{\omega: f(\omega)\geq \delta\}} f(\omega)K(\omega,\delta)m(d\omega), \qquad (1.5)$$

where $K(\omega,\delta) = \int_{\delta}^{f(\omega)} t^{\alpha-2}dt$. For $\alpha > 1$ we can put $\delta = 0$; obviously, $K(\omega, 0) = [f(\omega)]^{\alpha-1}/(\alpha-1)$. This, in connection with (1.4) and (1.5) (for $\delta = 0$), immediately gives us the first inequality to be proved. Suppose now that $0 < \alpha \leq 1$ and $m(\Omega) < \infty$; we put $\delta = 1$. On the set $\{\omega : f(\omega) \geq 1\}$ we have

$$K(\omega,1) = \begin{cases} \ln f(\omega) & \text{for} \quad \alpha = 1, \\ \frac{1}{1-\alpha}\{1 - [f(\omega)]^{\alpha-1}\} \leq \frac{1}{1-\alpha} & \text{for} \quad \alpha < 1; \end{cases}$$

and (1.5) gives

$$\int_{1}^{\infty} t^{\alpha-1} m(\Lambda_{2t})dt \leq \begin{cases} C_1' \int_{\Omega} f(\omega)\ln^{+} f(\omega)m(d\omega) & \text{for} \quad \alpha = 1, \\ \frac{C_1'}{1-\alpha}\int_{\Omega} f(\omega)m(d\omega) & \text{for} \quad \alpha < 1. \end{cases}$$

Also, for $\alpha > 0$,

$$\int_{0}^{1} t^{\alpha-1} m(\Lambda_{2t})dt \leq m(\Omega)\int_{0}^{1} t^{\alpha-1}dt = \frac{m(\Omega)}{\alpha}.$$

It remains to use (1.4) again.

□

1.3. Extension of classes of maximal nets. Here we give some methods to construct new maximal nets starting from given ones. The first method, which is rather simple, is often useful and we shall quote it as the "majorization principle".

Proposition 1.6. *Let S be a completely measurable positive representation, λ_n, $\nu_n \in \mathcal{P}$, $\{\nu_n, n \in N\} \in \mathrm{MAX}_{\alpha}(G)$ with constant C_{α}' ($\alpha \geq 1$), and let there exist a constant C such that $\lambda_n(A) \leq C\nu_n(A)$, $n \in N$, $A \in \mathfrak{B}$. Then $\{\lambda_n\} \in \mathrm{MAX}_{\alpha}(S)$ with constant $C_{\alpha}'C^{\alpha}$.*

Proof. Since $\int (fS_x)(\omega)\lambda_n(dx) \leq C \int (fS_x)(\omega)\nu_n(dx)$, for any measurable nonnegative function f we only have to use the maximal property of $\{\nu_n\}$ with $\varepsilon' = \varepsilon/C$.

□

Definition 1.2. Let $\mathcal{A} = \{A_n, n \in N\}$ and $\mathcal{A}' = \{A'_n, n \in N\}$ be nets of measurable sets with $\mu(A_n) > 0$, $\mu(A'_n) < +\infty$, $n \in N$. $\mathcal{A}$ is said to be *regular with respect to* $\mathcal{A}'$ (denoted by $\mathcal{A} \prec \mathcal{A}'$) if

(i) $A_n \subseteq A'_n$, $n \in N$;

(ii) $\sup\limits_{n \in N} [\mu(A'_n)/\mu(A_n)] =: \gamma(\mathcal{A}|\mathcal{A}') < \infty$;

$\gamma(\mathcal{A}|\mathcal{A}')$ is called the *relative index of* $\mathcal{A}$ *with respect to* $\mathcal{A}'$.

We have $\nu_{A_n}(A) \leq \gamma(\mathcal{A}|\mathcal{A}')\nu_{A'_n}(A)$ for the uniformly distributed measures $\nu_{A_n}(A) = \frac{\mu(A_n \cap A)}{\mu(A_n)}$ and $\nu_{A'_n}(A) = \frac{\mu(A'_n \cap A)}{\mu(A'_n)}$. Therefore we obtain the following "set version" of Proposition 1.6.

Proposition 1.6′. *Let S be a completely measurable uniformly bounded positive representation and $\mathcal{A}' \in \mathrm{MAX}_\alpha(S, \mu)$ with constant C'_α. If $\mathcal{A} \prec \mathcal{A}'$, then $\mathcal{A} \in \mathrm{MAX}_\alpha(S, \mu)$ with constant $C'_\alpha[\gamma(\mathcal{A}|\mathcal{A}')]^\alpha$.*

Proposition 1.7. *Let S be a completely measurable uniformly bounded positive representation of X in $L^\alpha(\Lambda, \mathcal{F}, m)$; suppose that $\mathcal{A} = \{A_n,\ n = 1, 2, \ldots\} \in \mathrm{MAX}_\alpha(S|\mu)$ and $\sum_{n=1}^\infty [\frac{\mu(B_n)}{\mu(A_n)}]^\alpha < \infty$. Then $\{A_n \cup B_n\} \in \mathrm{MAX}_\alpha(S|\mu)$.*

Proof. Let us fix some $f \in L^\alpha(\Lambda, \mathcal{F}, m)$. By the Chebyshev and Hölder inequalities,

$$m(\{\omega : \sup_{n\in N} \frac{1}{\mu(A_n \cup B_n)} | \int_{B_n} S_x f(\omega)\mu(dx)| > \frac{\varepsilon}{2}\}) \leq$$

$$\leq \sum_{k=1}^{\infty} m(\{\omega : \frac{1}{\mu(A_k \cup B_k)} \int_{B_k} |S_x f(\omega)|\mu(dx) > \frac{\varepsilon}{2}\}) \leq$$

$$\leq (\frac{2}{\varepsilon})^\alpha \sum_{k=1}^{\infty} \frac{[\mu(B_k)]^{\alpha-1}}{[\mu(A_k \cup B_k)]^\alpha} \int_\Lambda \int_{B_k} |S_x f(\omega)|^\alpha \mu(dx) m(d\omega) \leq$$

$$\leq (\frac{2}{\varepsilon})^\alpha \sum_{k=1}^{\infty} \left[\frac{\mu(B_k)}{\mu(A_k \cup B_k)}\right]^\alpha \max_{x\in B_k} \|S_x\|^\alpha_{L_\alpha} \int\limits_\Lambda |f(\omega)|^\alpha m(d\omega).$$

Therefore

$$m(\{\omega : | \sup_{n\in N} \frac{1}{\mu(A_n \cup B_n)} \int_{A_n \cup B_n} S_x f(\omega)\mu(dx)| \geq \varepsilon\}) \leq$$

$$\leq m(\{\omega : | \sup_{n\in N} \frac{1}{\mu(A_n \cup B_n)} \int_{B_n} S_x f(\omega)\mu(dx)| \geq \frac{\varepsilon}{2}\}) +$$

$$+ m(\{\omega : \sup_{n\in N} | \frac{1}{\mu(A_n)} \int_{A_n} S_x f(\omega)\mu(dx)| \geq \frac{\varepsilon}{2}\}) \leq$$

$$\leq (\frac{2}{\varepsilon})^\alpha \sum_{k=1}^{\infty} \left[\frac{\mu(B_k)}{\mu(A_k)}\right]^\alpha \max_{x\in B_k} \|S_x\| \int_\Omega |f|^\alpha dm +$$

$$+ \frac{C_1'}{\varepsilon^\alpha} \int_\Omega |f|^\alpha dm \leq \frac{C}{\varepsilon^\alpha} \int_\Omega |f(\omega)|^\alpha (d\omega),$$

where $C = 2^\alpha \left[\sum_{k=1}^{\infty} \max_{x\in B_k} \|S_x\| \left[\frac{\mu(B_k)}{\mu(A_k)}\right]^\alpha + C_1'\right]$.

□

Proposition 1.8. *Let* $\alpha \geq 1$. *If* $\mathcal{A}^{(k)} = \{A_n^{(k)}, n \in N\} \in \in \mathrm{MAX}_\alpha(S|\mu)$ *with* $C_\alpha'^{(k)}$, $k \in \overline{1,r}$, *then* $\{\bigcup_{k=1}^r A_n^{(k)}\} \in \mathrm{MAX}_\alpha(S|\mu)$ *with* $C_\alpha' = r^\alpha \sum_{k=1}^r C_\alpha'^{(k)}$.

Proof. Let $f \in L^\alpha_+$, $\varepsilon > 0$. Denote $A_n := \bigcup_{k=1}^r A_n^{(k)}$, $n \in N$. We have

$$m(\{\omega : \frac{1}{\mu(A_n)} \int_{A_n} S_x f(\omega)\mu(dx) \geq \varepsilon\}) \leq$$
$$\leq m(\{\omega : \sum_{k=1}^r \frac{1}{\mu(A_n^{(k)})} \int_{A_n^{(k)}} S_x f(\omega)\mu(dx) \geq \varepsilon\}) \leq$$
$$\leq \sum_{k=1}^r m(\{\omega : \frac{1}{\mu(A_n^{(k)})} \int_{A_n^{(k)}} S_x f(\omega)\mu(dx) \geq \frac{\varepsilon}{r}\}) \leq$$
$$\leq \frac{r^\alpha}{\varepsilon^\alpha} \sum_{k=1}^r C_\alpha'^{(k)} \int_\Omega |f(\omega)|^\alpha m(d\omega).$$

□

Proposition 1.9. *Suppose that $S : x \mapsto S_x$ is a completely measurable bounded representation of X; F_n, K_1, K_2 are compact sets in X; $G_n = K_1F_nK_2$, ν_{G_n}, ν_{F_n}, ν_{K_1}, and ν_{K_2} are Borel probability measures on G_n, F_n, K_1, K_2, and ν_{G_n} is the image of $\nu_{K_1} \times \nu_{F_n} \times \nu_{K_2}$ under the mapping $(k_1, f, k_2) \mapsto k_1fk_2$ from $K_1 \times F_n \times K_2$ into G_n. If $\{\nu_{F_n}\} \in \mathrm{MAX}_\alpha(S)$ and either $K_2 = \{e\}$ or $\alpha > 1$, then $\{\nu_{G_n}\} \in \mathrm{MAX}_\alpha(S)$.*

Proof. Let $\varphi \in L^\alpha(\Lambda, \mathcal{F}, m)$. We denote:

$$S_{K_i}\varphi(\omega) = \int_{K_i} S_{k_i}\varphi(\omega)\nu_{K_i}(dk_i), \quad S_{F_n}\varphi(\omega) = \int_{F_n} S_f\varphi\nu_{F_n}(df).$$

By Fubini's theorem we obtain

$$S_{G_n}\varphi(\omega) := \int_{G_n} S_g\varphi(\omega)\nu_{G_n}(dg) = \int_{G_n} S_{k_1fk_2}\varphi(\omega)\nu_{G_n}(dg) =$$
$$= \int_{K_1 \times F_n \times K_2} S_{k_1}S_fS_{k_2}\varphi(\omega)\nu_{K_1}(dk_1)\,\nu_{F_n}(df)\,\nu_{K_2}(dk_2) =$$
$$= S_{K_1}S_{F_n}S_{K_2}\varphi(\omega).$$

According to Example 1.1,

$$\int_\Omega |S_{K_2}\varphi(\omega)|^\alpha m(d\omega) < \sup_{x\in K_2} \|S_x\| \int |\varphi|^\alpha m(d\omega);$$

therefore

$$m\{\omega : \sup_{n\in N} |S_{F_n} S_{K_2}\varphi(\omega)| > \varepsilon\} \leq$$

$$\leq \sup_{x\in K_2} \|S_x\| \frac{C(\{F_n\})}{\varepsilon^\alpha} \int |f(\omega)|^\alpha m(d\omega), \quad \alpha \geq 2.$$

Thus our statement is proved when $K_2 = \{e\}$. Let $K_2 \neq \{e\}$ and $\alpha > 1$. By Proposition 1.5, $\int \sup_{n\in N} |S_{F_n} S_{K_2}\varphi|^\alpha dm \leq C''_\alpha \int |\varphi|^\alpha dm$. Let P be the linear modulus of the operator S_{K_1} (cf. Dunford and Schwartz [1], Lemma 8.6.4). P is a positive linear operator and has the following properties: $\|P\| \leq \|S_{K_1}\| \leq \sup_{x\in K_1} \|S_x\|$ and $|S_{K_1}\gamma(\omega)| \leq P(\gamma(\omega))$, $\omega \in \Omega$, $\gamma \in L^\alpha_+(\Lambda, \mathcal{F}, m)$. Therefore

$$\sup_{n\in N} |S_{K_1} S_{F_n} S_{K_2}\varphi(\omega)| \leq \sup_{n\in N} |P S_{F_n} S_{K_2}\varphi(\omega)| \leq$$
$$\leq P(\sup_{n\in N} |S_{F_n} S_{K_2}\varphi(\omega)|)$$

and

$$\int_\Omega \sup_{n\in N} |S_{K_1} S_{F_n} S_{K_2}\varphi(\omega)|^\alpha m(d\omega) \leq$$

$$\leq \|P\|^\alpha \int_\Omega |\sup_{n\in N} S_{F_n} S_{K_2}\varphi|^\alpha dm \leq \frac{\|P\|^\alpha C''_\alpha}{\varepsilon^\alpha} \int_\Omega |\varphi|^\alpha dm$$

Thus $\{\nu_{G_n}\} \in \mathrm{DOM}_\alpha(S)$; $\{\nu_{G_n}\} \in \mathrm{MAX}_\alpha(S)$, by Proposition 1.2.

□

1.4. Hardy–Littlewood type inequalities and the Transfer Principle. Let $(X, \mathfrak{B})$ be a measurable semigroup and μ a σ–finite measure on $\mathfrak{B}$. Consider the right representation in $L^\alpha(X, \mathfrak{B}, \mu)$ by the left translations: $(\varphi L_x)(y) = \varphi(xy)$, $\varphi \in L^\alpha(X, \mathfrak{B}, \mu)$ $(\alpha \geq 1)$. Again we assume that N is linearly ordered. Let $\{\nu_n, n \in N\}$ be a net in $\mathcal{P}(\mathfrak{B})$. The statement "$\{\nu_n\} \in MAX_\alpha(L)$" is equivalent to:

1) the function $\varphi^*(y) := \sup\limits_{n\in N} \int \varphi(xy)\nu_n(dx)$ is measurable, $\varphi \in \in L^\alpha(X)$,

2) the following "Hardy–Littlewood type maximal inequality" holds:

$$\mu\{y : |\varphi^*(y)| \geq \varepsilon\} \leq \frac{C'_\alpha}{\varepsilon^\alpha} \int_X |\varphi(y)|^\alpha \mu(dy), \quad \varphi \in L^\alpha(X, \mathfrak{B}, \mu). \quad (1.6)$$

The statement "$\nu_n \in \mathrm{DOM}_\alpha(L)$" is equivalent to the union of *1)* and

2') the "Hardy–Littlewood type dominated inequality"

$$\int |\varphi^*(y)|^\alpha \mu(dy) < C''_\alpha \int_X |\varphi(y)|^\alpha \mu(dy), \quad \varphi \in L^\alpha(X, \mathfrak{B}, \mu), \quad (1.7)$$

holds, i.e.

$$\|\varphi\|_{L_\alpha} \leq (C''_\alpha)^{\frac{1}{\alpha}} \|\varphi\|_{L^\alpha}, \quad \varphi \in L^\alpha(X, \mathfrak{B}, \mu).$$

Note 1.1. Since $|\varphi^*(x)| \leq |\varphi|^*(x)$, it suffices to establish (1.6) and (1.7) for nonnegative functions.

Note 1.2. Inequalities (1.6) and (1.7) are trivial if φ is measurable but does not belong to $L^\alpha(\Omega, \mathcal{F}, m)$. But inequality (1.6) admits a more general "local", version, which makes sense for any "locally integrable" function φ. Let again A_n be sets such that $\nu_n(A_n) = 1$. We fix some subset $\widetilde{N}$ of N, denote $\widetilde{A} := \bigcup_{n\in\widetilde{N}} A_n$ and consider sets K and M with the properties: K, $M \subset \mathfrak{B}$, $\widetilde{A} \cdot M \subset K$. Since $\mathcal{X}_{\widetilde{A}}(x)\mathcal{X}_M(y) \leq \leq \mathcal{X}_K(xy)$ we get

$$\begin{aligned} M \cap \{y : \sup_{n\in\widetilde{N}} \int \varphi(xy)\nu_n(dx) \geq \varepsilon\} \subseteq \\ \subseteq \{y : \sup_{n\in\widetilde{N}} \int \mathcal{X}_{\widetilde{A}}(x)\mathcal{X}_M(y)\varphi(xy)\nu_n(dx) \geq \varepsilon\} \subseteq \\ \subseteq \{y : \sup_{n\in\widetilde{N}} \int \mathcal{X}_{K_n}(xy)\varphi(xy)\nu_n(dx) \geq \varepsilon\}. \end{aligned}$$

If (1.6) holds, we apply it to the function $\mathcal{X}_K(x)\varphi(x)$ and to the subnet $\{A_n, n \in \widetilde{N}\}$. We have:

$$\mu(M \cap \{y : \sup_{n \in \widetilde{N}} \int \varphi(xy)\nu_n(dx) \geq \varepsilon\}) \leq \frac{C'_\alpha}{\varepsilon^\alpha} \int_K |\varphi(y)|^\alpha \mu(dy). \quad (1.8)$$

This inequality makes sense if $\mathcal{X}_K(\cdot)\varphi(\cdot) \in L^\alpha(X, \mathfrak{B}, \mu)$.
Of course, if $\varphi \in L^\alpha(X, \mathfrak{B}, \mu)$ we can put $\widetilde{A} = K = X$, $\widetilde{N} = N$ and return to (1.6).

The following "local" version of the dominated inequality (1.7) can be deduced similarly:

$$\int_M |\varphi^*(y)|^\alpha \mu(dy) \leq C''_\alpha \int_K |\varphi(y)|^\alpha \mu(dy). \quad (1.9)$$

A remarkable idea, due to N. Wiener (see Wiener [1], Calderón [1]), allows one to transfer inequalities (1.6) and (1.7), respectively, to maximal and dominated theorems for all dynamical systems with time X; A. P. Calderón has extended this "transfer principle" to more general, so called "semilocal", operators (see Calderón [2]).

Let $\{\nu_n, n \in N\} \in \mathcal{P}(\mathfrak{B})$; let T be a dynamical system in $(\Omega, \mathcal{F}, m)$. Denote

$\widehat{f}_n(\omega) := \int f(T_x\omega)\nu_n(dx)$,

$f^*(\omega) = \sup_{n \in N} |\widehat{f}_n(\omega)|$,

$f^*_n(\omega) = \sup_{k \leq n} |\widehat{f}_k(\omega)|$, $f \in L^\alpha(\Omega, \mathcal{F}, m)$, $\alpha \geq 1$.

The functions $\widehat{f}_n(\omega)$ are $\mathcal{F}$–measurable.

The proof of the following fundamental statement is based on the above mentioned methods of A. P. Calderón and N. Wiener.

Theorem 1.10. **(Transfer Principle)** *Let* $\{\nu_n, n \in N\} \subset \mathcal{P}(\mathfrak{B})$ *and let* A_n *be sets with* $\nu_n(A_n) = 1$*; denote* $\widetilde{A}_n := \bigcup_{k \leq n} A_k$*. Suppose that the condition*

(Meas) *for any dynamical system T in $(\Omega, \mathcal{F}, m)$ and any function $f \in L^\alpha(\Omega, \mathcal{F}, m)$, the functions $f_n^*(\omega)$, $n \in N$, and $f^*(\omega)$ are $\mathcal{F}$-measurable**)
and the following "transfer condition" holds:
(T) *there exists a net $\{M_n\} \subset \mathfrak{B}$ such that*

$$\varlimsup_{n \in N} \frac{\mu^*(\widetilde{A}_n \cdot M_n)}{\mu(M_n)} =: C_{\mathbf{T}} < \infty.$$

We then have:
(i) if $\{\nu_n\} \in \mathrm{MAX}_\alpha(L)$ with C'_α, then $\{\nu_n\} \in \mathrm{MAX}_\alpha(T)$ with $C_{\mathbf{T}} C'_\alpha$ for any T;
(ii) if $\{\nu_n\} \in \mathrm{DOM}_\alpha(L)$ with C''_α, then $\{\nu_n\} \in \mathrm{DOM}_\alpha(T)$ with $C_{\mathbf{T}} C''_\alpha$ for any T.

Proof. *(i)* Consider a dynamical system T in $(\Omega, \mathcal{F}, m)$ and a function $f \in L_+^\alpha(\Omega)$. Denote
$f^*(\omega) := \sup_{n \in N} \int f(T_x\omega)\nu_n(dx)$;
$f_n^*(\omega) := \sup_{l \leq n} \int f(T_x\omega)\nu_l(dx)$;
$\Lambda_n^{(\varepsilon)} := \{\omega : f_n^*(\omega) \geq \varepsilon\} \subset \Omega$, $\Lambda_n^{(\varepsilon)}(y) := \{\omega : f_n^*(T_y\omega) \geq \varepsilon\} \subset \Omega$,
and
$M_n^{(\varepsilon)}(\omega) := \{y : \sup_{k \leq n} \int f(T_{xy}\omega)\nu_k(dx) \geq \varepsilon\} = \{y : f_n^*(T_y\omega) \geq \geq \varepsilon\} \subset X$.
It is obvious that $\Lambda_n^{(\varepsilon)}(y) = T_y^{-1}(\Lambda_n^{(\varepsilon)})$ and therefore $m(\Lambda_n^{(\varepsilon)}(y)) = = m(\Lambda_n^{(\varepsilon)})$, $y \in X$. Applying Fubini's theorem to the function $(y, \omega) \mapsto \mathcal{X}_{M_n \times \Lambda_n^{(\varepsilon)}}(y, T_y\omega)$ we obtain

$$\begin{aligned} m(\Lambda_n^{(\varepsilon)}) &= \frac{1}{\mu(M_n)} \int_{M_n} m(\Lambda_n^{(\varepsilon)}(y))\mu(dy) = \\ &= \frac{1}{\mu(M_n)} \int_\Omega \mu(M_n^{(\varepsilon)}(\omega) \cap M_n) m(d\omega). \end{aligned}$$

Consider an arbitrary set $K_n \in \mathfrak{B}$ such that $\mu(K_n) < \infty$ and $\widetilde{A}_n M_n \subset \subset K_n$. Inequality (1.8) with $K = K_n$, $M = M_n$, $\widetilde{N} = \{k : k \leq n\} \subset$

---

*) Clearly, this condition is fulfilled if the direction N is countable.

$\subset N$, $\varphi_\omega(x) = f(T_x\omega)$ $(\omega \in \Omega)$, $\widetilde{A} = \widetilde{A}_n = \bigcup_{l\leq n} A_l$, and Fubini's theorem imply:

$$m(\Lambda_n^{(\varepsilon)}) \leq \frac{1}{\mu(M_n} \frac{C'_\alpha}{\varepsilon^\alpha} \int_\Omega \left\{ \int_{K_n} [f(T_x\omega)]^\alpha \mu(dx) \right\} m(d\omega) =$$
$$= \frac{C'_\alpha}{\varepsilon^\alpha} \frac{\mu(K_n)}{\mu(M_n)} \int_\Omega |f(\omega)|^\alpha m(d\omega).$$

Since $\mu^*(\widetilde{A}_n \cdot M_n) = \inf\{\mu(K_n) : K_n \in \mathfrak{B},\ \widetilde{A}_n \cdot M_n \subset K_n\}$, we have $m(\Lambda_n^{(\varepsilon)}) \leq \frac{C'_\alpha}{\varepsilon^\alpha} \frac{\mu^*(\widetilde{A}_n \cdot M_n)}{\mu(M_n)} \int_\Omega |f(\omega)|^\alpha m(d\omega)$.

We now take the limits of both sides of this inequality when n runs over all cofinal subdirections of N. Since N is linearly ordered, we obtain:

$$m(\{\omega : f^*(\omega) \geq \varepsilon\}) \leq \frac{C'_\alpha C_{\mathbf{T}}}{\varepsilon^\alpha} \int_\Omega |f(\omega)|^\alpha m(d\omega).$$

(ii)
$$\int_\Omega |f_n^*(\omega)|^\alpha m(d\omega) = \int_\Omega |f_n^*(T_y\omega)|^\alpha m(d\omega) =$$
$$= \frac{1}{\mu(M_n)} \int_{M_n} \left[\int_\Omega |f_n^*(T_y\omega)|^\alpha m(d\omega) \right] \mu(dy) =$$
$$= \frac{1}{\mu(M_n)} \int_\Omega \left[\int_{M_n} |f_n^*(T_y\omega)|^\alpha \mu(dy) \right] m(d\omega) =$$
$$= \frac{1}{\mu(M_n)} \int \left[\int_{M_n} |\varphi_{\omega,n}^*(y)|^\alpha \mu(dy) \right] m(d\omega),$$

$\varphi_\omega(y) = f(T_y\omega)$. We apply formula (1.9) to $\varphi_\omega(y)$ to get:

$$\int_\Omega |f_n^*(\omega)|^\alpha m(d\omega) \leq$$
$$\leq \frac{C''_\alpha}{\mu(M_n)} \int_\Omega \left| \int_{K_n} |\varphi_\omega(y)|^\alpha \mu(dy) \right| m(d\omega) =$$
$$= \frac{C''_\alpha}{\mu(M_n)} \int_{K_n} \left[\int_\Omega |f(T_y\omega)|^\alpha (d\omega) \right] \mu(dy) =$$
$$= \frac{C''_\alpha \mu(K_n)}{\mu(M_n)} \int_\Omega |f(\omega)|^\alpha m(d\omega).$$

We conclude the proof as in part *(i)*.

□

1.5. Hardy–Littlewood type inequalities and local ergodic theorems. Let $(X, \mathfrak{B})$ be a measurable group and μ a σ–finite measure in $\mathfrak{B}$.

Proposition 1.11. *Consider a cofinally separable net of measurable nonnull sets $\mathcal{A} = \{A_n,\ n \in N\}$. Denote*

$$\widehat{\varphi}_n(x) := \frac{1}{\mu(A_n)} \int_{A_n} \varphi(xy)\mu(dy), \quad \widetilde{A} := \bigcup_{n \in N} A_n.$$

Let $M, K \in \mathfrak{B}$, $0 < \mu(M) < \infty$ and $\widetilde{A} \cdot M \subset K$. Suppose also that

a) *for any $\varphi \in L^\alpha(K)$ the relation* (1.9) *holds;*

b) *there exists a measure $\nu \in \mathcal{P}(\mathfrak{B})$ such that for any φ belonging to some dense subset $D \subset L^\alpha(K)$,*

$$\lim_{n \in N} \widehat{\varphi}_n(x) = \int \varphi(xy)\nu(dy) \quad \mu\text{–a.e. on} \quad M. \tag{1.10}$$

Then (1.10) *holds for all $\varphi \in L^\alpha(K)$.*

This is a simple consequence of the Banach Convergence Principle (App., (12.A)) and of App., (13.D). Sometimes the following more general statement is useful.

Proposition 1.12. *Consider a uniformly cofinally separable family of nets of measurable nonnull sets $\mathcal{A}^{(\gamma)} = \{A_n^{(\gamma)},\ n \in N\}$, $\gamma \in \Gamma$. Denote $\varphi_n^*(x) := \sup_{\gamma \in \Gamma} |\widehat{\varphi}_n^{(\gamma)}(x)|$; $\varphi^{**}(x) := \sup_{n \in N} |\varphi_n^*(x)|$.*

(i) Suppose that

a) *φ^{**} is measurable and*

$$\int_M |\varphi^{**}(x)|^\alpha \mu(dx) \le C'_\alpha \int_K |\varphi(x)|^\alpha \mu(dx), \quad \varphi \in L^\alpha(K);$$

b) *there are a family of measures $\{\nu^{(\gamma)},\ \gamma \in \Gamma\}$ and a dense subset D in $L^\alpha(K)$ such that for all $\varphi \in D$,*

$$\lim_{n\in N} \varphi_n^*(x) = \sup_{\gamma\in\Gamma} \Big| \int \varphi(xy)\nu^{(\gamma)}(dy) \Big| \quad \text{a.e. on} \quad M; \tag{1.11}$$

c) $\int_M |\sup_\gamma \int \varphi(xy)\nu^\gamma(dy)|^\alpha \mu(dx) \le C''_\alpha \int_K |\varphi(x)|^\alpha \mu(dx)$,

$\varphi \in L^\alpha(K)$.
Then (1.11) *holds for all* $\varphi \in L^\alpha(K)$.

(ii) *Suppose that conditions* a') *and* c) *are fulfilled as well as*
b') *for all* $\varphi \in D$,

$$\lim_{n\in N} \sup_{\gamma\in\Gamma} \Big| \frac{1}{\mu(A_n)} \int_{A_n^{(\gamma)}} \varphi(xy)\mu(dy) - \int \varphi(xy)\nu^{(\gamma)}(dy) \Big| = 0 \tag{1.12}$$

a.e. on M.
Then (1.12) *holds for all* $\varphi \in L^\alpha(K)$.

Proof. (i) Let $N_0 = \{n_1, n_2, \ldots\}$ be a uniformly c-separability set*) in N. Consider the operators T_k: $T_k\varphi(x) = \varphi_{n_k}^*(x)$ from the space $L^\alpha(K)$ into the space $L^\alpha(M)$. Since $|T_k(\varphi-\psi)| = |\varphi_{A_k}^* - \psi_{n_k}^*| \le (\varphi_{n_k} - \psi_{n_k})^*$, property a) implies that $\sup_{1\le k<\infty} |T_k\varphi(x)| < \infty$ a.e. on M. It is easy to check that our conditions imply all the conditions of the Banach Convergence Principle (App., (12.A)). Therefore the limit in (1.11) exists a.e. on M for all $\varphi \in L^\alpha(K)$, i.e. T_k converges strongly to a continuous operator T: $L^\alpha(K) \mapsto L^\alpha(M)$. But condition c) implies that the operator T_1: $(T_1\varphi)(x) = \sup_{\gamma\in\Gamma} \big| \int \varphi(xy)\nu_\gamma(dy) \big|$ from $L^\alpha(K)$ into $L^\alpha(M)$ is continuous. Of course, condition b) implies that $T = T_1$ on D, and since D is dense in $L^\alpha(K)$ we have $T = T_1$ on $L^\alpha(K)$. It remains to use App., (13.F).

(ii) Consider the operators $\widetilde{T}_k$,

$$\widetilde{T}_k\varphi(x) := \sup_{\gamma\in\Gamma} \Big| \frac{1}{\mu(A_n^{(\gamma)})} \int_{A_n^{(\gamma)}} \varphi(xy)\mu(dy) - \int \varphi(xy)\nu^{(\gamma)}(dy) \Big|,$$

*) See Appendix, §13.

instead of T_k. Our assertions imply all the conditions of App., (12.A). Since $\widetilde{T}_k\varphi(x) \to 0$ m–a.e. on M if $\varphi \in D$ and D is dense in $L^\alpha(K)$, our statement is proved for the subdirection N_0. We have to apply App., (13.F) once more.

□

Note 1.3. If G is a locally compact group, K is compact and $D = C(K)$, the space of all continuous functions on K, (1.10) means that for μ–almost all $x \in M$ the measures $\nu_{A_n}^{(x)}$ converge weakly to the measure $\nu^{(x)}$ (here $\nu^{\infty}(B) := \nu(x^{-1}B)$ and $\nu^*_{A_n}(B) := \frac{\mu(A_n \cap x^{-1}B)}{\mu(A_n)}$).

Note 1.4. The classical form of Proposition 1.10, "differentiation of integrals", deals with the case when $\gamma = \delta_e$ (the measure concentrated at the identity e). Then (1.10) turns into

$$\lim_{n \in N} \widehat{\varphi}_n(x) = \varphi(x) \quad \mu\text{–a.e. on} \quad M. \tag{1.13}$$

If G is a locally compact topological group, K and M are compact, μ is the left Haar measure, and any A_n is contained in some open neighbourhood of e, then (1.10) is fulfilled with $D = C(k)$, the set of all continuous functions on K. Indeed, $y \to \varphi(xy)$ is uniformly continuous with respect to $x \in M$ and thus (1.13) is fulfilled for all $x \in M$. We do not go into details; an extended discussion of this problem can be found in Edwards and Hewitt [1], Hewitt and Ross [1], Guzman [1].

The following simple statement is a "local transfer principle"; it allows one to transfer statements similar to Propositions 1.10 and 1.11 to "local ergodic theorems" for arbitrary dynamical systems.

Proposition 1.13. *Let $T = \{T_x,\ x \in X\}$ be a completely measurable right dynamical system in $(\Omega, \mathcal{F}, m)$ and $f \in L^\alpha\ (\Omega, \mathcal{F}, m)$. Let $\{A_n,\ n \in N\}$ and $\{A_n^{(\gamma)},\ n \in N\}$, $\gamma \in \Gamma$, be the same as in Propositions* 1.10 *and* 1.11. *Denote $\widehat{f}_n(\omega) = \frac{1}{\mu(A)} \int_{A_n} f(\omega T_x)\mu(dx)$; $\widehat{f}_n^{(\gamma)}(\omega) = \frac{1}{\mu(A_n^{(\gamma)})} \int_{A_n} f(\omega T_x)\mu(dx)$; $\widehat{f}_n^*(\omega) = \sup_{\gamma \in \Gamma} |\widehat{f}_n^{(\gamma)}(\omega)|$.*

(i) If (1.10) holds for all $\varphi \in L^\alpha(M)$, then

$$\lim_{n\in N} \widehat{f}_n(\omega) = \int f(\omega T_x)\nu(dx) \quad m\text{—a.e.} \tag{1.14}$$

(ii) If (1.11) holds for all $\varphi \in L^\alpha(M)$, then

$$\lim_{n\in N} f_n^*(\omega) = \sup_{\gamma\in\Gamma} \Big| \int f(\omega T_x)\nu^{(\gamma)}(dx)\Big| \quad m\text{—a.e.} \tag{1.15}$$

(iii) If (1.12) holds for all $\varphi \in L^\alpha(M)$, then

$$\lim_{n\in N} \sup_{\gamma\in\Gamma} \Big| \frac{1}{\mu(A_n^{(\gamma)})} \int_{A_n^{(\gamma)}} f(\omega T_x)\mu(dy) - \int f(\omega T_x)\nu^{(\gamma)}(dx)\Big| \tag{1.16}$$
$$m\text{—a.e.}$$

Proof. Consider the functions $\varphi_\omega(x) = f(T_x\omega)$, $\omega \in \Omega$. We have $\varphi_\omega(\cdot) \in L^\alpha(M)$. Thus (1.10) implies

$$\lim_{n\in N} \widehat{f}_n(\omega T_y) = \lim_{n\in N} (\widehat{\varphi}_\omega)_n(y) =$$
$$= \int \varphi_\omega(x)\nu(dx) = \int f(\omega T_x)\nu(dx) \quad m\text{—a.e.} \quad \text{on } \Omega\times M.$$

Therefore there exists a point y_0 such that

$$\lim_{n\in N} \widehat{f}_n(\omega T_{y_0}) = f(\omega T_x)\nu(dx) \quad m\text{—a.e.}$$

Since T_{y_0} is measure preserving, we have (1.14). Relations (1.15) and (1.16) can be proved similarly.

□

§ 2. Regular Nets of Sets

2.1. Definitions and discussion. In this section, X is a non–compact, σ–compact, locally compact topological semigroup; $\mathfrak{B}$ is the

σ–algebra of Borel sets in X; μ is a right relatively semiinvariant measure on $\mathfrak{B}$. If $\{A_n, n \in N\} \in \mathfrak{B}$, we denote $\widetilde{A}_n = \cup_{k \leq n} A_k$. Recall that $A^{-1}B = \{z : z = x^{-1}y,\ x \in A,\ y \in B\}$.

Definition 2.1. We say that a net of sets $\mathcal{A} = \{A_n, n \in N\}$ is *(left) regular**) if all $A_n \in \mathfrak{B}$, $0 < \mu(A_n) < \infty$ and

(R) there exists a constant $C_R < \infty$ such that

$$\mu^*(\widetilde{A}_n^{-1}(A_n y)) \leq C_R \mu(A_n) \Delta_\mu^{(r)}(y), \quad y \in X.$$

The greatest lower bound of these C_R is denoted by $\gamma_R(\mathcal{A})$ and called the *(left) regularity index of* $\mathcal{A}$; we set $\gamma_R(\mathcal{A}) = +\infty$ if $\mathcal{A}$ is not regular.

Definition 2.2. We say that a net $\mathcal{A} = \{A_n, n \in N\}$ is *(left) transfer regular* if it is (left) regular and the following "(left) transfer condition" is fulfilled:

(**T**) there exists a net of Borel sets $\{M_n,\ n \in N\}$ such that $\underline{\lim}_{n \in N} \frac{\mu^*(\widetilde{A}_n M_n)}{\mu(M_n)} =: C_{\mathbf{T}} < \infty$.

We set $\gamma_{\mathbf{T}R}(\mathcal{A}) := C_{\mathbf{T}} \gamma_R(\mathcal{A})$ and $\gamma_{\mathbf{T}R}(\mathcal{A}) := \infty$ if $\mathcal{A}$ is not transfer–regular.

Definition 2.3. We say that a net $\{A_n,\ n \in N\}$ is *(left) piecewise (transfer) regular* if $\{A_n\} \prec \{A'_n\}$ and $A'_n = \cup_{k=1}^r A_n^{(k)}$, where $\{A_n^{(k)},\ n \in N\}$ are (transfer) regular nets.

Note 2.1. Condition (R) is fulfilled if the following two conditions are both satisfied:

(M) monotonicity: $A_{n_1} \subset A_{n_2}$ if $n_1 < n_2$;

and

(R) there exists a constant C'_R such that

$$\mu(\widetilde{A}_n)^{-1}(A_n y)) \leq C'_R \mu(A_n) \Delta_\mu^{(r)}(y), \quad y \in X, \quad n \in N.$$

---

*) In what follows we usually omit the word "left".

Condition (M) can be weakened: it is obvious that $\mathcal{A}$ is regular iff the conditions

(WM) (weak monotonicity): $\mu^*(\widetilde{A}_{n-1}^{-1}(A_n y)) \leq C_1' \mu(A_n)\Delta_\mu^{(r)}(y)$, $n \in N$, $y \in X$;

and

$$(R')\ \mu^*(A_n^{-1}(A_n y)) \leq C_1'' \mu(A_n)\Delta_\mu^{(r)}(y), \quad n \in N, \quad y \in X,$$

are fulfilled (then $C_R \leq C_1' + C_1''$).

If X is a group, the conditions R, R' and (WM) can be rewritten in the following form:

(R) $\mu(\widetilde{A}_n^{-1} A_n) \leq C_1 \mu(A_n)$, $n \in N$;

(R') $\mu(A_n^{-1} A_n) \leq C\mu(A_n)$, $n \in N$;

(WM) $\mu(\widetilde{A}_{n-1}^{-1} A_n) \leq C\mu(A_n)$, $n \in N$.

Condition (R') deals only with algebraic–geometric properties of the sets A_n while (WM) also includes some restrictions on the rate of "non–monotonicity" of the net $\mathcal{A}$: the sets A_n are allowed to "run away" but their "speed" must be relatively small (as compared to the "speed" of increase of $\mu(A_n)$). However, if (R') is not fulfilled, (WM) can be of a quite different nature (see Example 2.12 below).

We shall illustrate this by several simple examples.

Example 2.1. Any net of intervals (a_n, b_n) satisfies condition (R') with $C = 2$. However, a sequence of disjoint intervals $A_n = (a_n, b_n)$, $n = 1, 2, \ldots$ of length $b_n - a_n \sim \ln n$ or $b_n - a_n \sim n^\gamma$, $\gamma > 0$, is not regular.
Indeed, if $b_n - a_n \sim \ln n$, then $l(\widetilde{A}_n) = l(-\widetilde{A}_n) \sim n \ln n$ and for large n we have $\frac{(l(-\widetilde{A}_n + A_n)}{l(A_n)} \geq Cn$; the case $b_n - a_n \sim n^\gamma$ is similar.

The case when $b_n - a_n$ grows exponentially is more delicate, as the following example shows.

Example 2.2. Let $a_n = 2^{n+1}$, $b_n - a_n = 2^n$. Then $(-b_k, -a_k) = (-2^{k+1} - 2^k, -2^{k+1})$ and the intervals $(-b_k, -a_k) + (a_n, b_n) = (2^{n+1} - 2^{k+1} - 2^k, 2^{n+1} + 2^n - 2^{k+1})$, $1 \leq k \leq n$, overlap. Therefore $-\widetilde{A}_n + A_n = (-2^n, 2^{n+1} + 2^n - 4)$ and $\frac{l(-\widetilde{A}_n + A_n)}{l(A_n)} \leq 2$, and the sequence $\{A_n\}$

is regular. Now let $a_n = 4^n$, $b_n - a_n = 2^n$; the intervals $(-b_k, -a_k) + (a_n, b_n) = (4^n - 4^k - 2^k, 4^n - 4^k + 2^n)$ do not overlap if $k > \frac{n}{2}$. Therefore $l(-\widetilde{A}_n + A_n) \geq 2^n \cdot (\frac{n}{2} - 1)$, and this sequence is not regular.

Example 2.3. Let again $A_n = (a_n, b_n)$ where $a_n = 2^{2^n}$, $b_n - a_n = \sqrt{a_n} = 2^{2^{n-1}}$; $(-b_k, -a_k) + (a_n, b_n) = (2^{2^n} - 2^{2^k} - 2^{2^{k-1}}, 2^{2^n} + 2^{2^{n-1}} - 2^{2^k})$. Therefore $(-\widetilde{A}_{n-1} + A_n) \subset (2^{2^n} - 2^{2^{n-1}} - 2^{2^{n-2}}, 2^{2^n} + 2^{2^{n-1}})$ and $l(-\widetilde{A}_{n-1} + A_n) \leq 3 \cdot 2^{2^{n-1}} \leq 3 \cdot l(A_n)$. Thus condition (WM) is fulfilled with $C_1' = 2$; $\{A_n\}$ is regular and $\gamma_R \leq 4$.

Example 2.4. Let $A_n = (n, n + \sqrt{n})$. The intervals $(-k - \sqrt{k}, -k) + (n, n + \sqrt{n}) = (n - k - \sqrt{k}, n + \sqrt{n} - k)$ overlap, therefore $-\widetilde{A}_n + A_n = (-\sqrt{n}, n + \sqrt{n} - 1)$, and $l(-\widetilde{A}_n + A_n) \geq n = \sqrt{l}(A_n)$. Thus this sequence is not regular.

We shall see later, in § 5 and in Ch. 6, that all regular sequences of intervals are maximal and pointwise averaging. It is also known (see, e.g., Bellow, Jones and Rosenblatt [1] and the literature quoted there) that all non–regular intervals considered in Examples 2.2–2.4 are *not* pointwise averaging.

It is interesting to compare Condition (R') with the ergodicity (Følner) condition

$$(E)\ \lim_{n\to\infty} \frac{\mu(A_n \cap xA_n)}{\mu(A_n)} = 0, \quad x \in X$$

(see App., § 3). Condition (E) is sufficient in the mean ergodic theorem (see § 4.2) and, as we see in § 6.3, plays an important role in the pointwise ergodic theorem. These conditions reflect various aspects of "regularity" and "simplicity" of the form of the sets A_n. (R) is a condition of "compactness" in the ordinary sense of the word: it requires that the sets A_n have no distant isolated parts and no "long" but relatively "thin" "*shoots*". Condition (E) allows such "*shoots*" and isolated parts, but requires that they grow, like the whole set A_n, "in all directions" as n increases. We give some examples.

Example 2.5. In $\mathbb{Z}$ the sequence $A_n = \{x : x = 2k,\ k = 1, 2, \ldots, n\}$, $n = 1, 2, \ldots$, satisfies condition (R'), but not condition (E).

Example 2.6. In $\mathbb{R}^2$ the sequences of rectangles $A_n = \{x = (x_1, x_2);\ 0 \le x_1 \le n,\ 0 \le x_2 \le 1\}$ satisfies condition (R'), but not (E).

Example 2.7. In $\mathbb{R}^2$ the sequence of "crosses"

$$A_n = \{(-n, n) \times (-\ln n, \ln n)\} \cup (-\ln n, \ln n) \times (-n, n)\}$$

satisfies condition (E) but not condition (R'). The sequence $B_n = = \{(-n, n) \times (-1, 1)\} \cup \{(-1, 1) \times (-n, n)\}$ satisfies neither of these conditions. Nevertheless, both sequences are piecewise transfer regular.

Note 2.1. If X is a locally compact group and $\mathcal{A}$ is a sequence of compact neighborhoods of the identity, then $\gamma_R(\mathcal{A}) \ge 1$, and equality holds iff $\mathcal{A}$ is an increasing sequence of open subgroups in X. In this case $\gamma_{TR}(\mathcal{A}) = 1$, and (E) is equivalent to the condition $\bigcup_1^\infty A_n = X$.

The following simple statements sometimes help to prove regularity or transfer regularity of nets of sets.

Lemma 2.1. *If $\mathcal{A} \prec \mathcal{A}'$ and $\mathcal{A}'$ is regular or transfer regular, then $\mathcal{A}$ also has the corresponding property, and $\gamma_R(\mathcal{A}) \le \gamma(\mathcal{A}|\mathcal{A}')\gamma_R(\mathcal{A}')$.*

Proof. We have

$$\gamma_R(\mathcal{A}) = \inf_n \frac{\mu(\widetilde{A}_n^{-1}(A_n y))}{\mu(A_n)\Delta^{(r)}(y)} \le \inf_n \frac{\mu((\widetilde{A}'_n)^{-1}(A'_n y))}{\mu(A'_n)\Delta^{(r)}(y)} \cdot \gamma(\mathcal{A}|\mathcal{A}'),$$

and

$$\frac{\mu^*(\widetilde{A}_n M_n)}{\mu(M_n)} \le \frac{\mu^*((\widetilde{A}_n)M_n)}{\mu(M_n)}.$$

□

Lemma 2.2. *If $X = \prod_{i=1}^k X^{(i)}$ is a direct product of locally compact semigroups and the nets of sets $\mathcal{A}^{(i)} = \{A_n^{(i)},\ n \in N\}$ $(\subset X^{(i)})$,*

$1 \leq i \leq k$, have the properties (E), (R) or (**T**) *with respect to measures $\mu^{(i)}$, then the net $\mathcal{A} = \prod_{i=1}^{n} A_n^{(i)}$ in X has the same properties, with respect to $\mu = \prod_{i=1}^{k} \mu^{(i)}$, and $\gamma_R(\mathcal{A}) = \prod_{i=1}^{k} \gamma(\mathcal{A}^{(i)})$, $\gamma_{TR}(\mathcal{A}) = \prod_{i=1}^{k} \gamma_{TR}(\mathcal{A}^{(i)})$.*

This is a simple consequence of Fubini's theorem.

2.2. Two notes on transfer regularity on groups. We shall make two remarks concerning transfer regularity. The first one shows that *transfer regularity is possible only with respect to the left Haar measure*; the second one shows that *on amenable groups regularity (with respect to the left Haar measure) implies transfer regularity.*

Note 2.1. Let X be a non–discrete group, μ a relatively invariant measure on X and $\{A_n,\ n \in N\}$ a net in $\mathfrak{B}$. Suppose that the set $\widetilde{A}_\infty := \bigcup_{n\in N} A_n$ is unbounded (this is the case when $\{A_n\}$ is ergodic). Conditions (R) and $(\mathbf{T})$ are compatible only if $\mu = \mu^{(l)}$, the left Haar measure on X. Indeed, suppose that $\mu \neq \mu^{(l)}$, i.e. $\Delta_\mu^{(l)}(x) \not\equiv 1$. The homomorphism $x \to \Delta_\mu^{(l)}(x)$ from X into the multiplicative group $\mathbb{R}_{++}^\times$ is open (see Pontryagin [1], Th. 3.20.12), and therefore the set $\Delta_\mu^{(l)}(\widetilde{A}_\infty)$ is unbounded. Hence at least one of the following cases takes place:

(i) There exists a net $\{a_n\}$, $a_n \in \widetilde{A}_n$, such that $\Delta_\mu^{(l)}(a_n) \to \infty$; in this case $\lim_{n\in N} \frac{\mu^*(\widetilde{A}_n M_n)}{\mu(M_n)} \geq \lim_{n\in N} \frac{\mu(a_n M_n)}{\mu(M_n)} = \lim_{n\in N} \Delta_\mu^{(l)}(a_n) = +\infty$, and $(\mathbf{T})$ does not hold.

(ii) There exists a net $\{b_n\}$, $b_n \in \widetilde{A}_n$, such that $\lim_{n\in N} \Delta_\mu^{(l)}(b_n) = 0$; then $\sup_{n\in N} \frac{\mu(\widetilde{A}_n^{-1} A_n)}{\mu(A_n)} \geq \lim_{n\in N} \frac{\mu(b_n^{-1} A_n)}{\mu(A_n)} = \lim_{n\to\infty} \Delta_\mu^{(l)}(b_n^{-1}) = +\infty$, and condition (R) is not satisfied.

Note 2.2. Let X be an amenable locally compact group, $\mu = \mu^{(l)}$. If a net $\mathcal{A} = \{A_n,\ n \in N\} \subset \mathfrak{B}$ and the sets $\{\widetilde{A}_n\}$ are bounded, then condition $(\mathbf{T})$ is fulfilled and $C_{\mathbf{T}} = 1$. Let $x_n \in A_n$, $\widetilde{A}_n \subset K_n$ and K_n compact. Since $e \in K_n x_n^{-1}$, $n \in N$, by the Emerson–Greenleaf theorem (see App. (3.I)) there exist compact sets U_n such that

$$\lim_{n\in N} \mu^{(l)}(K_n x_n^{-1} U_n \Delta U_n)/\mu^{(l)}(U_n) = 0.$$

Let us set $M_n = x_n^{-1} U_n$. Then

$$\begin{aligned}
&|\mu^{(l)}(K_n M_n)/\mu^{(l)}(M_n) - 1| = \\
=&|\mu^{(l)}(K_n M_n) - \mu^{(l)}(x_n M_n)|/\mu^{(l)}(M_n) = \\
=&[\mu^{(l)}(M_n)]^{-1} \Big| \int (\chi_{K_n M_n} - \chi_{x_n M_n}) d\mu^{(l)} \Big| \leq \\
\leq&[\mu^{(l)}(M_n)]^{-1} \int |\chi_{K_n M_n} - \chi_{x_n M_n}| d\mu^{(l)} = \\
=&[\mu^{(l)}(U_n)]^{-1} \mu^{(l)}(K_n x_n^{-1} U_n \Delta U_n);
\end{aligned}$$

therefore

$$\underline{\lim}_{n\in N} [\mu^{(l)}(M_n)]^{-1} \mu^{(l)}(\tilde{A}_n M_n) \leq \lim_{n\in N} [\mu^{(l)}(M_n)]^{-1} \mu^{(l)}(K_n M_n) = 1.$$

□

2.3. Regular sets in $\mathbb{R}^m$ and $\mathbb{Z}^m$: Examples. In the following examples we take for μ Lebesgue measure l in $\mathbb{R}^m$ and the counting measure in $\mathbb{Z}^m$.

Example 2.8. Lemma 2.2 shows that any net of "parallelepipeds" $A_n = \{x = (x_1, \dots, x_m),\ a_i^{(n)} \leq x_i \leq b_i^{(n)})\}$ in $\mathbb{R}^m$ and $\mathbb{Z}^m$ is transfer regular with $\gamma_{TR}(\mathcal{A}) = 2^m$; if $b_i^{(n)} - a_i^{(n)} \to \infty$ such a net is also ergodic.

Let us consider less trivial examples

Example 2.9. Let $X = \mathbb{R}^m$ and A a Borel set with $0 < l(A) < \infty$. We fix a point $x_0 \in \mathbb{R}^m$ and let A' be the "star–shaped hull" of A with respect to x_0; $A' := \{x : x = \lambda(y - x_0) + x_0,\ 0 \leq \lambda \leq 1,\ y \in A\}$. Then any set of homotetic sets $A_n = t_n(A - x_0) + x_0$ is transfer regular and $\gamma_{TR} = \gamma_R \leq \frac{l(-A'+A)}{l(A)}$. Indeed, $\widetilde{A}_n \subset A'_n := t_n(A' - x_0) + x_0$ and $-A'_n + A_n = t_n(-A' + A)$. Therefore $\frac{l(-A'_n+A_n)}{l(A_n)} = \frac{l(-A'+A)}{l(A)}$. Note that $\{A_n\}$ is also ergodic, since

$$\frac{l(A_n \Delta(A_n + x))}{l(A_n)} = \frac{l(A\Delta(A + t_n^{-1}x))}{l(A)} \to 0 \quad \text{if } t_n \to \infty.$$

Example 2.10. Any net of convex bodies $\mathcal{A} = \{A_n,\, n \in N\}$ is regular. It is easy to check that $\gamma_R = 2^m$ if the A_n are symmetric; and we always have $\gamma_{TR} = \gamma_R \leq \binom{2m}{m}$ (see, e.g., Leichtweiß[1], Theorem 20.3). Let $r(A)$ be the least upper bound of the radii of the balls contained in A; if $r(A_n) \to \infty$, then $\mathcal{A}$ is also ergodic.

Example 2.11. Let $\{A'_n,\, n \in N\}$ be an increasing net of convex sets in $\mathbb{R}^n$ with $\lim_{n\in N} r(A'_n) = \infty$ (see the previous example); then the sequence of sets $A_n = \mathbb{Z}^m \cap A'_n$ is regular and ergodic in $\mathbb{Z}^m$. Such are, in particular, the sequences of "parallelepipeds" $B_n = \{z = (z_1, \ldots, z_m) : |z_i| \leq a_n^{(i)},\ i = 1, \ldots, m\}$ and "balls" $C_n = \{z : \sum_1^m z_i^2 \leq r_n^2\}$ with $a_n^{(i)} \to \infty$ and $r_n \to \infty$; the sequences $B_n \cap \mathbb{Z}_+^n$ and $C_n \cap \mathbb{Z}_+^n$ have the same properties in $\mathbb{Z}_+^m$.

Example 2.12. Let $m \geq 2$. Consider spherical layers $A_n = L_{r_n}^{R_n} = \{x : x \in \mathbb{R}^m,\ r_n \leq |x| \leq R_n\}$, $n \in N$. It is easy to check that $-\widetilde{A}_n + A_n \subset B_{2R_n}$, and the net $\{A_n\}$ is regular if $2^m R_n^m \leq C(R_n^m - r_n^m)$, i.e. if $r_n \leq \left(\frac{C-2^m}{C}\right)^{\frac{1}{m}} R_n$. Thus, a net $\{L_{r_n}^{R_n}\}$ is regular if $\sup\limits_{n\in N} \frac{r_n}{R_n} = \alpha < 1$, then $\gamma_R(\mathcal{A}) \leq 2^m(1-\alpha^m)^{-1}$. Now suppose that $N = \mathbb{N}$, $r_n > R_{n-1}$ and consider condition (WM). Since $-\widetilde{A}_{n-1} + A_n \subset L_{r_n - R_{n-1}}^{R_n + R_{n-1}}$, (WM) is satisfied if $(R_n + R_{n-1})^m - (r_n - R_{n-1})^m \leq C(R_n^m - r_n^m)$, $n \in N$. In its turn, this inequality is true if $R_n + 2R_{n-1} - r_n \leq C'R_n - C'r_n$ with $C' = (2^m)^{-1}C$, and thus (WM) is fulfilled if $\sup_{n\in N} \frac{R_{n-1}}{R_n - r_n} < \infty$. If $\frac{r_n}{R_n} \to 1$ and $r_n^{-1}R_{n-1} = O\left(r_n^{-1}R_n - 1\right)$, the net $\{L_{r_n}^{R_n}\}$ is not regular but satisfies condition (WM).

2.4. Regular nets on general groups: Examples.

Example 2.13. Any commutative compactly generated group G is isomorphic to a group $G' = K \times \mathbb{R}^m \times \mathbb{Z}^k$ where K is a compact group (see Pontryagin [1], § 39). Lemma 2.2. shows that if $\{A_n\}$ and $\{C_n\}$ are regular nets in $\mathbb{R}^m$ and $\mathbb{Z}^k$, then $\{G'_n\} = \{K \times B_n \times C_n\}$ is a regular net in G'. $\{G'_n\}$ is ergodic if $\{B_n\}$ and $\{C_n\}$ are ergodic. Any locally compact group G contains an open subgroup of compact origin FG'', and any regular $\{G''_n\}$ in G'' is regular in G.

Example 2.14. Let G be a locally compact group and μ its Haar measure. Let U be a compact symmetric neighbourhood of "polynomial growth", i.e. there exist positive constants C_1, C_2 and α such that $C_1 k^\alpha \le \mu(U^k) \le C_2 k^\alpha$, $k = 1, 2, \ldots$ (here $U^n = U^{n-1}$, $n = 1, 2, \ldots$). Then the sequence $\{U^n\}$ is transfer regular. Indeed, it is increasing and $\mu(U^n \cdot U^n) = \mu(U^{2n}) \le 2^\alpha C_1^{-1} C_2 \mu(U^n)$. Such sets U exist in nilpotent discrete groups; the existence of such sequences in connected groups has been studied, e.g., in Jenkins [1].

2.5. The Covering Lemma. Let $A_1, \ldots, A_L$ $(L < \infty)$ be integrable nonnull sets in X with $0 < \mu(A_i) < \infty$, $i \in \overline{1, L}$; let M be an arbitrary set in X and $n(x)$ a mapping (not necessarily single–valued) from X into the set $[1, \ldots, L]$. We call a family $\Sigma = \{A_{n(x)}x,\ x \in M\}$ a *quasi–covering* of the set M (by translates of the sets $A_1, \ldots, A_L$)*).

Lemma 2.3. *Consider a finite net of integrable nonnull sets $\mathcal{A} = \{A_1, \ldots, A_L\}$. Let $M \subset X$ and $\mu^*(\bigcup_{l=1}^L A_l \cdot M) < \infty$. If Σ is an arbitrary quasi–covering of M by translates of the sets A_i, then there exists a finite family*

$$\Sigma' : A_{k_1} x_1, \ldots, A_{k_m} x_m$$

*) If X contains the identity e and $e \in \cap_{i=1}^L A_i$, then any element $x (\in M)$ belongs to the set $A_{n(x)}x$, and Σ is a covering in the usual sense.

of disjoint μ-mod 0 *sets in* Σ *such that*

$$\mu^*(M) \le \sum_{i=1}^{m} \mu(\widetilde{A}_{k_i}^{-1}(A_{k_i}x_i)) \le \gamma_R(\mathcal{A}) \sum_{i=1}^{m} \mu(A_{k_i})\Delta_\mu^{(r)}(x_i). \qquad (2.1)$$

Proof. We put $k_1 = \max_{x \in M} n(x)$ and take an $x_i \in n^{-1}(k_1)$; then $A_{k_1}x_1 \in \Sigma$; we take $A_{k_1}x_1$ as the first set in Σ'. The remaining sets are constructed inductively. Suppose that we have already found disjoint mod 0 sets $A_{k_1}x_1, \ldots, A_{k_r}x_r$. We put

$$\Gamma_r = M \cap \bigcap_{i=1}^{r} \{x : \mu(\widetilde{A}_{k_i}x \cap A_{k_i}x_i) = 0\}.$$

If $\Gamma_r = \emptyset$, we have completed our construction of Σ' and we can put $m = r$. Indeed, if $\Gamma_l = \emptyset$ then $M \subset \bigcup_{i=1}^{l} \{x : \mu(\widetilde{A}_{k_i}x \cap A_{k_i}x_i) > 0\}$, whence

$$\mu^*(M) \le \sum_{i=1}^{l} \mu^*(\{x : \mu(A_{k_i}x \cap A_{k_i}x_i) > 0\}) \le$$

$$\le \sum_{i=1}^{l} \mu[\widetilde{A}_{k_i}^{-1}(A_{k_i}x_i)] \le \gamma_R(\mathcal{A}) \sum_{i=1}^{l} \mu(A_{k_i})\Delta_\mu(x_i).$$

If $\Gamma_r \neq \emptyset$, we put $k_{r+1} = \max_{x \in \Gamma_r} n(x)$ and take an $x_{r+1} \in n^{-1}(k_{r+1}) \cap \cap \Gamma_r$. Let $1 \le i < j \le r$. It is clear that $\Gamma_1, \ldots, \Gamma_r$ is a decreasing sequence; therefore the sequence of numbers $k_s = \max_{x \in \Gamma_s} n(x)$, $s = 1, \ldots, r$, is nonincreasing, and we have $\widetilde{A}_{k_j} \subseteq \widetilde{A}_{k_i}$. As $x_j \in \in \Gamma_j \subset \Gamma_i$, we have $\mu(A_{k_j}x_j \cap A_{k_i}x_i) \le \mu(\widetilde{A}_{k_i}x_j \cap A_{k_i}x_i) = 0$; hence the sets $A_{k_i}x_1, \ldots, A_{k_r}x_r$ are disjoint mod 0. The quantity $C := \inf_{x \in X} \Delta_\mu(x) > 0$, since M is bounded. If condition (2.1) is still not fulfilled for $l = r$, then

$$r\mu(A_{k_1}) \le \sum_{i=1}^{r} \mu(A_{k_i}) \le C^{-1} \sum_{i=1}^{r} \mu(A_{k_i}x_i)\Delta_\mu(x_i) \le \gamma_R^{-1} C^{-1} \mu^*(M),$$

whence $r \le \gamma_R^{-1} C^{-1} \mu^*(M)/\mu(A_{k_1})$. Consequently our process for the construction of Σ' stops at a certain step $k \le \gamma_R^{-1} C^{-1} \mu^*(M)/\mu(A_{k_1}) + 1 < \infty$. As we have seen, (2.1) is then fulfilled.

□

§ 3. Hardy–Littlewood Type Inequalities for Regular Nets

In this section we shall apply the results of § 2 to prove Hardy–Littlewood type inequalities for nets $\{A_n,\ n \in N\}$ in general semigroups. We use the traditional approach based on the Covering lemma established in Subsect. 2.7. (see, e.g. Wiener [1], Calderón [1], Edwards and Hewitt [1] (Theorm 2.3)).

We suppose that X is a σ–compact, locally compact topological semigroup.

Lemma 3.1. *Let $\mathcal{A} = \{A_1, \ldots, A_L\}$ be a finite sequence of integrable nonnull sets, M a measurable set, φ a nonnegative measurable function on X, and*

$$M_\varepsilon = \{y : \sup_{n \le L} \frac{1}{\mu(A_n)} \int_{A_n} \varphi(xy)\mu(dx) \ge \varepsilon\}.$$

Then for all $\varepsilon > 0$

$$\mu(M \cap M_\varepsilon) \le \frac{\gamma_R(\mathcal{A})}{\varepsilon} \int_K \varphi(x)\mu(dx), \tag{3.1}$$

where K is an arbitrary measurable set containing $\widetilde{A}_L M$.

Proof. We can assume that φ is integrable on K. At first we suppose that M is bounded. For each element $y \in M_\varepsilon$ there is at least one $n(y) \in \{1, \ldots, L\}$ such that

$$\frac{1}{\mu(A_{n(y)})} \int_{A_{n(y)}} \varphi(xy)\mu(dx) \ge \varepsilon. \tag{3.2}$$

The family of sets $\Sigma = \{A_{n(y)}y,\ y \in M \cap M_\varepsilon\}$ is a quasi–covering of the set $M \cap M_\varepsilon$. We apply Theorem 3.1 to obtain a finite selection of disjoint mod 0 sets from Σ:

$$\Sigma' : A_{k_1}y_1, \ldots, A_{k_m}y_m;$$

moreover,

$$\mu(M \cap M_\varepsilon) \le \gamma_R \sum_{i=1}^{m} \mu(A_{k_i})\Delta_\mu^{(r)}(y_i). \tag{3.3}$$

By (3.2), (3.3) and App. (1.E), we have

$$\mu(M \cap M_\varepsilon) \le \frac{\gamma_R}{\varepsilon} \sum_{i=1}^{m} \int_{A_{k_i}} \varphi(xy_i)\mu(dx)\Delta_\mu^{(r)}(y_i) \le$$

$$\le \frac{\gamma_R}{\varepsilon} \sum_{i=1}^{m} \int_{A_{k_i}y_i} \varphi(x)\mu(dx) \le \frac{\gamma_R}{\varepsilon} \int_K \varphi(x)\mu(dx).$$

Now let M be unbounded and let $M^{(r)} \uparrow M$, where the $M^{(r)}$ are bounded sets.
We have proved that $\mu(M^{(r)} \cap M_\varepsilon) \le \frac{\gamma_R(\mathcal{A})}{\varepsilon} \int_K \varphi(x)\mu(dx)$. It remains to let $r \uparrow \infty$.

□

Theorem 3.2. *Let N be a linearly ordered set, $\{A_n,\ n \in N\}$ a separable net of integrable nonnull sets, M a measurable set, φ a locally integrable function, and*

$$M_\varepsilon = \{y : \sup_{n \in N} \Big|\frac{1}{\mu(A_n)} \int_{A_n} \varphi(xy)\mu(dx)\Big| \ge \varepsilon\}.$$

Then for all $\varepsilon > 0$ and $\alpha \ge 1$ the following Hardy–Littlewood type inequality holds:

$$m(M \cap M_\varepsilon) \le \frac{\gamma_R(\mathcal{A})}{\varepsilon^\alpha} \int |\varphi(x)|^\alpha \mu(dx). \tag{3.4}$$

Proof. For finite N and $\alpha = 1$ the assertion coincides with Lemma 3.1. If N is infinite and countable there exists a sequence $\{N_k\}$ of finite subdirections such that $N_k^{(k)} \uparrow N$ as $k \to \infty$. We certainly have $\gamma_R(\mathcal{A}^{(k)}) \le \gamma_R(\mathcal{A})$, where $\mathcal{A}^{(k)} = \{A_n, n \in N^{(k)}\}$. Therefore
$m(M \cap \{y : \sup\limits_{n \in N^{(k)}} |\frac{1}{\mu(A_n)} \int\limits_{A_n} \varphi(xy)\mu(dx)| \ge \varepsilon\}) \le \frac{\gamma_R(\mathcal{A})}{\varepsilon} \int\limits_K |\varphi(x)|\mu(dx)$.
If we now let $k \uparrow \infty$, we obtain (3.4) with $\alpha = 1$. If $\{A_n, n \in N\}$ is separable we take a separability set $N_0 \subset N$, use App. 13.D and obtain (3.4) with $\alpha = 1$ again. It remains to use Proposition 1.2.

□

§ 4. Hardy–Littlewood Type Inequalities for some Non–regular and Singular Weights

In this section we set forth Hardy–Littlewood type inequalities for averages with singular weight measures concentrated on concentric spheres in $\mathbb{R}^m$, $m \ge 3$, or homothetic convex curves in $\mathbb{R}^2$; we consider here averages on spherical layers, which form a non–regular net if their widths are small in comparison with the outer radii.

We shall state the main results.

Theorem 4.1. **(E. M. Stein for m ≥ 3, J. Bourgain for m = 2)** *Let σ_r be the rationally invariant measure on the sphere $S_r = S_r^{(m)}$ of radius r and center at 0. For any $\mathfrak{B}$–measurable nonnegative function we denote:*

$$\varphi_r(x) = \int \varphi(x+y)\sigma_r(dy), \quad \varphi^*(x) = \sup_{r>0} \varphi_r(x).$$

If $m \ge 2$, then $\varphi^*(x)$ is measurable, and for any $\alpha > \frac{m}{m-1}$,

$$\int_{\mathbb{R}^m} [\varphi^*(x)]^\alpha dx \le C_\alpha \int_{\mathbb{R}^m} |\varphi(x)|^\alpha dx. \tag{4.1}$$

A simple consequence of this result can be stated as follows.

Theorem 4.2. *Consider spherical layers in* $\mathbb{R}^m$ *(annuli in* $\mathbb{R}^2$*):* $L_{r,\delta(r)} = \{x : r - \delta(r) \le |x| \le r\}$, $r > 0$, $0 < \delta(r) \le r$. *For any* $\mathfrak{B}$*-measurable nonnegative* φ *define* $\varphi_{r,\delta(r)} := \frac{1}{|L_{r,\delta(r)}|} \int_{L_{r,\delta(r)}} \varphi(x+y)dy$ *(*$|A|$ *is the Lebesgue measure of* $A \in \mathfrak{B}$*) and* $\varphi_L^*(x) = \sup_{r>0} \varphi_{r,\delta(r)}$. *Then for any* $\alpha > \frac{m}{m-1}$

$$\int |\varphi_L^*(x)|^\alpha dx \le C_\alpha \int |\varphi(x)|^\alpha dx. \tag{4.2}$$

In the two–dimensional case Bourgain [2] proved the following extension of Theorem 4.1.

Theorem 4.3. (Bourgain) *Let* Γ *be the boundary of a compact convex centrally symmetric body in* $\mathbb{R}^2$. *For any measurable nonnegative* φ *denote*

$$\varphi_r(x) = \int \varphi(x+ry)\sigma(dy).$$

(here $\sigma(dy)$ *is the arc length measure on* Γ*) and*

$$\varphi_\Gamma^*(x) = \sup_{r>0} \varphi_r(x).$$

If Γ *is smooth up to order 5 and has everywhere positive Gaussian curvature, then for any* $\alpha > 2$,

$$\int |\varphi_\Gamma^*|^\alpha dx \le C(\Gamma, \alpha) \int |\varphi|^\alpha dx. \tag{4.3}$$

We omit the proof of Theorem 4.3, the interested reader can find it in Bourgain [2] (in this work only bounded functions are considered; the general case can be proved by using Lemma 4.4 (see below) and the Levi monotone convergence theorem). We shall give the proofs of Theorems 4.1 and 4.2 in the simplest case $m \ge 4$, $\alpha = 2$ only; we follow, with minor modifications, Stein and Wainger [1]. This work also contains the complete proof of Theorem 4.1 in the case $m \ge 3$.

Lemma 4.4. *(i) If $\varphi(x)$ and $\varphi_n(x)$ are nonnegative functions in $L^\alpha(\mathbb{R}^m, \mathfrak{B}, l)$ and $\varphi_n(x) \uparrow \varphi(x)$, $x \in X$, then $\varphi_n^*(x) \uparrow \varphi^*(x)$. Therefore, if the φ_n^* are $\mathfrak{B}^*$–measurable and $\|\varphi_n^*\|_{L^\alpha} \leq C_\alpha^{\frac{1}{\alpha}} \|\varphi_n\|_{L^\alpha}$, then φ^* is measurable and (4.1) is fulfilled.*

(ii) Let $\Phi^{(i)} \subset L_+^\alpha(\mathbb{R}^m)$, $i = 1, 2, 3$, $\alpha \geq 1$; suppose that $\varphi^{(1)} - \varphi^{(3)} \in \Phi^{(2)}$ if $\varphi^{(1)} \in \Phi^{(1)}$, $\varphi^{(3)} \in \Phi^{(3)}$ and $\varphi^{(1)}(x) \geq \varphi^{(3)}(x)$, $x \in X$. We also assume that for all $\varphi (\in \Phi^{(1)} \cup \Phi^{(2)})$ the conclusion of Theorem 4.1 holds. If $\varphi \in \Phi^{(3)}$, $\varphi_n \in \Phi^{(1)}$ and $\varphi_n \downarrow \varphi$, then φ satisfies the conclusion of Theorem 4.1.

Proof. *(i)* It is easy to check that $\psi_1^* \geq \psi_2^*$ if $\psi_1 > \psi_2$. Therefore the sequence φ_n^* is increasing and $\varphi_n^*(x) \leq \varphi^*(x)$. Hence

$$\lim_{n\to\infty} \varphi_n^*(x) \leq \varphi^*(x). \tag{4.4}$$

By the Levi monotone convergence theorem, $(\varphi_n)_r \uparrow \varphi_r$ as $n \to \infty$, $r > 0$. Since $\varphi_n^* \geq (\varphi_n)_r$, $r > 0$, we have $\underline{\lim}_{n\to\infty} (\varphi_n^*)_r(x) \geq \varphi_r(x)$, and thus

$$\underline{\lim}_{n\to\infty} (\varphi_n^*)_r(x) \geq \varphi^*(x), \quad x \in X. \tag{4.5}$$

Inequalities (4.4) and (4.5) imply statement *(i)*.

(ii) At first, note that by the Levi theorem, $\|\varphi_n\|_{L^\alpha} \downarrow \|\varphi\|_{L^\alpha}$. Since $\varphi_n - \varphi \in \Phi^{(2)}$, we have $\|(\varphi_n - \varphi)^*\|_{L^\alpha} \leq C_2 \|\varphi_n - \varphi\|_{L^\alpha}$. Employing the Levi theorem again we get $\lim_{n\to\infty} \|\varphi_n - \varphi\|_{L^\alpha} = 0$ and hence $\lim_{n\to\infty} \|(\varphi_n - \varphi)^*\|_{L^\alpha} = 0$. From this relation and the inequality $\varphi_n^*(x) - \varphi^*(x) \leq (\varphi_n - \varphi)^*(x)$ we deduce

a) there exists a subsequence $\{n_k\}$ such that $\lim_{k\to\infty} \varphi_{n_k}^*(x) = \varphi^*(x)$ a.e. and thus $\varphi^*(x)$ is measurable;

b) $\|\varphi_n^*\|_{L^\alpha} \downarrow \|\varphi^*\|_{L^\alpha}$, and (4.1) holds.

□

Lemma 4.5. *For any $\varphi \in L^1(\mathbb{R}^m)$ the Fourier transform*

$$\widetilde{\varphi}_r(\lambda) := \int_{\mathbb{R}^m} e^{i((\lambda, x)} \varphi_r(x) dx = \gamma(r|\lambda|) \widetilde{\varphi}(\lambda)$$

where $\gamma(t) = \frac{C_m}{t^{\frac{m-2}{2}}} J_{\frac{m-2}{2}}(t)$ *and* $J_n(t)$ *is the Bessel function of order* n.

The function γ_m *has the following properties:*

1) $|\gamma_m(t)| \leq C_1(1 + t^{\frac{m-1}{2}})^{-1}$, $t > 0$.

2) $|\gamma'_m(t)| \leq C_2(1 + t^{\frac{m-1}{2}})^{-1}$, $t > 0$.

Proof. Using Fubini's theorem we get:

$$\begin{aligned}\widetilde{\varphi}_r(\lambda) = \int e^{i(\lambda,x)}\Big(\int \varphi(x+y)\sigma_r(dy)\Big)dx = \\ = \int\Big(\int e^{i(\lambda,x)}\varphi(x+y)dx\Big)\sigma_r(dy) = \\ = \int\Big(\int e^{i(\lambda,(z-y))}\varphi(z)dz\Big)\sigma_r(dy) = \\ = \widetilde{\varphi}(\lambda)\widetilde{\sigma}_r(-\lambda)\end{aligned}$$

where $\widetilde{\sigma}_r(\lambda) = \int e^{i(\lambda,x)}\sigma_r(dx) = C_3\gamma(r|\lambda|)$ (see Example 4.8.1). Properties *1)* and *2)* of γ are implied by well–known properties of the Bessel function.

□

Lemma 4.6. **(E. M. Stein)** *Let* φ *be a smooth function,* $\varphi \in \in L^2(\mathbb{R}^m)$, $m \geq 4$, *and*

$$g_\varphi(x) := \Big[\int_0^\infty \Big|\frac{d}{dr}\varphi_r(x)\Big|^2 r\, dr\Big]^{\frac{1}{2}}.$$

Then

$$\int |g_\varphi(x)|^2 dx \leq C \int |\varphi(x)|^2 dx. \tag{4.6}$$

Proof. By the Plancherel equality and Lemma 4.5,

$$\int_{\mathbb{R}^m} |g_\varphi(x)|^2 dx = \int_0^\infty r \int_{\mathbb{R}^m} \left|\frac{d}{dr}\varphi_r(x)\right|^2 dx\, dr =$$
$$= \int_0^\infty r \int_{\mathbb{R}^m} \left|\widetilde{\frac{d}{dr}\varphi_r(\lambda)}\right| d\lambda\, dr =$$
$$= \int_{\mathbb{R}^m} |\widetilde{\varphi}(\lambda)|^2 \int_0^\infty r\left|\frac{d}{dr}\gamma(r|\lambda|)\right|^2 dr\, d\lambda.$$

Now,

$$\int_0^\infty r\left|\frac{d}{dr}\gamma(rt)\right|^2 dr = \int_0^{t^{-1}} r\left|\frac{d}{dr}\gamma(rt)\right|^2 dr +$$
$$+ \int_{t^{-1}}^\infty r\left|\frac{d}{dr}\gamma(rt)\right|^2 dr = \mathfrak{J}_1 + \mathfrak{J}_2.$$

By Lemma 4.5, $\left|\frac{d}{dr}\gamma(rt)\right| \le C_1 t$, whence $\mathfrak{J}_1 \le C_1|t|^2 \int_0^{t^{-1}} r\, dr = \frac{1}{2}C_1$ and $\mathfrak{J}_2 \le t^2 \int_{t-1}^\infty r\left[C_2 \frac{1}{(tr)^{m-1}}\right]^2 dr = C_2 t^{3-m} \int_{t-1}^\infty r^{2-m} dr = \frac{C_2}{m-3}$. Thus $\int |g_\varphi|^2 dx \le C \int |\widetilde{\varphi}(\lambda)|^2 dx\lambda = C \int |\varphi(x)|^2 dx$.

□

Lemma 4.7. *Theorem* 4.1 *is true if* φ *is smooth.*

Proof. It is easy to verify that the function $r \to \varphi_r(x)$ is continuous for any $x \in \mathbb{R}^m$, and thus $\varphi_s^*(x)$ is $\mathfrak{B}$–measurable. It is evident that

$$r^m \varphi_r(x) = \int_0^r \frac{d}{ds}(\varphi_s(x) \cdot s^m) ds =$$
$$= m \int_0^r \varphi_s(x) s^{m-1} ds + \int_0^r s^m \frac{d}{ds}\varphi_s(x) ds;$$

hence

$$\varphi_r(x) = r^{-m} m \int_0^r \varphi_s(x) s^{m-1} ds + r^{-m} \int_0^r s^m \frac{d}{ds}\varphi_s(x) ds.$$

Denote

$$\psi_1(x) := \sup_{r>0}\{r^{-m}\int_0^r \varphi_s(x)s^{m-1}ds\},$$
$$\psi_2(x) := \sup_{r>0}\{r^{-m}\int_0^r s^m \frac{d}{ds}\varphi_s(x)ds\}.$$

As usual, $B_r := \{x : \|x\| \leq r\}$. It is easy to check that

$$\varphi_{B_r}(x) := \frac{1}{|B_r|}\int_{B_r} \varphi(x+y)dy = Cr^{-m}\int_0^r \varphi_s(x)s^{m-1}ds,$$

and thus $\psi_1(x) \leq C_1\varphi_B^*(x)$ where $\varphi_B^*(x) = \sup_{r>0}\frac{1}{|B_r|}\int_{B_r}\varphi(x+y)dy$. Since the net $\{B_r,\ r>0\}$ is regular, by Theorem 3.2,

$$\|\psi_1\|_{L^2} \leq C_1\|\varphi_B^*\|_{L^2} \leq C_2\|\varphi\|_{L^2};$$
$$\psi_2 \leq \Big(\int_0^\infty s\Big|\frac{d}{ds}\varphi_s(x)\Big|^2 ds\Big)^{\frac{1}{2}} \cdot \sup_r \Big[r^{-m}\Big(\int_0^r s^{2m-1}ds\Big)^{\frac{1}{2}}\Big] \leq C_3 g_\varphi(x).$$

Finally, by Lemma 4.6,

$$\|\varphi^*\|_{L^2} \leq |\psi_1\|_{L^2} + \|\psi_2\|_{L^2} \leq C_4\|\varphi\|_{L^2}.$$

□

Proof of Theorem 4.1. (case $\alpha = 2$, $m \geq 4$) Denote by M the class of all functions $\varphi \in L_+^2$ for which the conclusion of Theorem 4.1 is true; if $\mathcal{A} \subset \mathfrak{B}$, $\mathcal{X}_{\mathcal{A}} := \{\mathcal{X}_A(\cdot),\ A \in \mathcal{A}\} \cap L_+$. G and F are the classes of all open and closed subsets of $\mathbb{R}^m$, respectively. First we shall show that $\mathcal{X}_G \subset M$. Since any function $\varphi(\in \mathcal{X}_G)$ can be approximated from below by an increasing sequence of smooth functions, Lemmas 4.7 and 4.4 *(i)* imply $\mathcal{X}_G \subset M$.

We can set $\Phi^{(1)} = \mathcal{X}_G$, $\Phi^{(2)} = \mathcal{X}_G$, $\Phi^{(3)} = F$ in Lemma 4.4 *(ii)* and, since $F \subset G_\delta$, we get: $\mathcal{X}_F \subset M$. Lemma 4.1 *(i)* shows that $\mathcal{X}_{F_\sigma} \subset$ $\subset M$. We once more use Lemma 4.1 *(ii)*, with $\Phi^{(1)} = \mathcal{X}_G$, $\Phi^{(2)} =$ $= \mathcal{X}_{F_\sigma}$, $\Phi^{(3)} = \mathcal{X}_{G_\delta}$, to get $\mathcal{X}_{G_\delta} \in M$. By Lemma 4.4 *(i)*, $\mathcal{X}_{G_{\delta\sigma}} \subset M$. Note that $F_\sigma \subset G_{\delta\sigma}$; if $A \in F_{\sigma\delta}$, then $\mathbb{R}^m\backslash A \in G_{\delta\sigma}$ and, therefore, if $A \subset B \in F_\sigma$ then $B\backslash A = B \cup (\mathbb{R}^m\backslash A) \subset G_{\delta\sigma}$. Lemma 4.4 *(ii)* with

$\Phi^{(1)} = \mathcal{X}_{F_\sigma}$, $\Phi^{(2)} = \mathcal{X}_{G_{\delta\sigma}}$ implies $\mathcal{X}_{F_{\sigma\delta}} \in M$. This procedure allows one to show that all classes $\mathcal{X}_{F\ldots}$ and $\mathcal{X}_{G\ldots}$ are contained in M and thus, by transfinite induction, $\mathcal{X}_{\mathfrak{B}} \subset M$, and (4.1) holds for every $\varphi \in \mathcal{X}_B$ with some constant C_2. Now consider nonnegative simple functions $\varphi(x) = \sum_{i=1}^m a_i \mathcal{X}_{A_i}(x)$ where all $A_i \in \mathfrak{B}$ are disjoint. We have proved that $\varphi \in M$ if $m = 1$. Suppose $\varphi \in M$ if $m \le k$. If $m = k+1$, then $\varphi = \varphi^{(k)} + a_{k+1}\mathcal{X}_{A_{k+1}}$ where $\varphi^{(k)} \in M$; reasoning as in the case $m = 1$ ($\varphi^{(k)}$ is fixed) we prove that $\varphi \in M$. Therefore Theorem 4.1 holds for all simple functions. Since any nonnegative $\mathfrak{B}$–measurable function can be approximated from below by an increasing sequence of simple functions, it remains to apply Lemma 4.4 *(i)* once more.

□

Proof of Theorem 4.2. According to Proposition 1.11, the net $\{L_{r,\delta(r)},\ r > 0\}$ is separable, and Proposition 1.13 implies that for any $\varphi \in L^2(\mathbb{R}^m)$ the dominant function φ_L^* is $\mathfrak{B}$–measurable.

Now, for any $x \in \mathbb{R}^m$

$$\begin{aligned}\varphi_L^*(x) &= \sup_{r>0} \varphi_{r,\delta(r)}(x) = \\ &= \sup_{r>0} \frac{1}{|L_{r,\delta(r)}|} \int_{r-\delta(r)}^{r} t^{m-1}\varphi_t(x)dt \le \sup_{t>0} \varphi_t(x) = \varphi^*(x),\end{aligned}$$

and (4.2) is implied by (4.1).

□

As a simple application of the general ergodic theorems stated in Subsect. 1.7 we state the following:

Theorem 4.4. *Let* $\alpha > \frac{m}{m-1}$, $0 < R \le \infty$ *and* $\varphi \in L^\alpha(\mathfrak{B}_{2R})$

$(B_\infty := \mathbb{R}^m)$. Denote

$$\widehat{\varphi}_r(x) = \int \varphi(x+y)\nu_r(dy),$$

$$\widehat{\varphi}_{r,d}(x) := \frac{1}{l(L_{r,d})}\int_{L_{r,d}} \varphi(x+y)dy =$$

$$= \frac{1}{l(L_{r,d})}\int_{r-d}^{r} t^{m-1}\widehat{\varphi}_t(x)dt,$$

$$\varphi^*(x) := \sup_{0<r\leq R} \widehat{\varphi}_r(x),$$

$$\varphi^*_d(x) := \sup_{0<r\leq R} \varphi_{r,d}(x.)$$

We have

(*i*) $\lim_{d\to 0} \widehat{\varphi}_{r,d}(x) = \widehat{\varphi}_r(x)$ *a.e. on* B_R *for all* $0 < r \leq R$.

(*ii*) $\lim_{d\to 0} \varphi^*_d(x) = \varphi^*(x)$ *a.e. on* B_R.

(*iii*) $\lim_{d\to 0} \sup_{0<r\leq R} |\widehat{\varphi}_{r,d}(x) - \widehat{\varphi}_r(x)| = 0$ *a.e on* B_R, $R > 0$.

Proof. We put $N = \{d : 0 < d \leq R\}$ with the downward order, $\Gamma = \{r : 0 \leq r \leq R\}$, $N^{(r)} = (0,\infty]$. It is easy to check that any net $\{L_{r,d},\ d \in (0,r]\}$ is cofinally separable and the family $\{L_{r,d},\ d \in [0,r]\}$, $0 < r \leq R$, is uniformly cofinally separable; any dense subset $N_0 \subset \subset (0, R]$ is a uniformly c–separability set.

(*i*) Theorem 4.1 implies condition a) of Proposition 1.10 with $K = = B_{2R}$, $M = B_R$. Now set $D := C(B_{2R})$, the space of all continuous functions on B_{2R}. If $\varphi \in C(B_{2R})$, the function $X \to \varphi_r(x)$ is uniformly continuous with respect to r, $0 < r \leq R$, whence $\lim_{d\to 0} \widehat{\varphi}_{r,d}(x) = = \widehat{\varphi}_r(x)$ uniformly with respect to r, $0 < r \leq R$. It remains to apply Proposition 1.10.

(*ii*) and (*iii*). We apply Proposition 1.15 with r and d instead of γ and n. The proof of assertion (*i*) also implies condition b') of Proposition 1.11; moreover, $\lim_{d\to 0} \varphi^*_d(x) = \varphi^*(x)$, $\varphi \in C(B_{2R})$, so condition *b)* of Proposition 1.11 is also fulfilled. Condition *c)* is involved by Theorem 4.1; the validity of condition a) can be proved by a reasoning as in the proof of Theorem 4.2.

□

§ 5. Maximal and Dominated Ergodic Theorems: The Transfer Approach

We have finished the preparatory part of this chapter and are now able to state the main results. First, combining Theorem 3.2 with the Transfer Principle (Theorem 1.10), we arrive at the following statement.

Theorem 5.1. *Let X be a measurable semigroup with a σ-finite measure; let N be a linearly ordered set and $\mathcal{A} = \{A_n,\ n \in N\}$ a separable net of integrable nonnull sets in X. Consider an arbitrary completely measurable dynamical system T in $(\Omega, \mathcal{F}, m)$, and for any nonnegative $\mathcal{F}$-measurable function f denote*

$$f^*(\omega) = \sup_{n \in N} \frac{1}{\mu(A_n)} \int_{A_n} f(T_x\omega)\mu(dx).$$

Then for all $\varepsilon > 0$ and $\alpha \geq 1$,

$$m\{\omega : f^*(\omega) > \varepsilon\} \leq \frac{\gamma_{TR}(\mathcal{A})}{\varepsilon^\alpha} \int_\Omega [f(\omega)]^\alpha m(d\omega). \tag{5.1}$$

Thus, if $\mathcal{A}$ is transfer-regular, $\mathcal{A} \in MAX_\alpha(T)$ for any $\alpha \geq 1$, and $\mathcal{A} \in DOM_\alpha(T)$ for any $\alpha > 1$ (with constant $\frac{2^\alpha \alpha}{\alpha - 1}\gamma_{TR}(\mathcal{A})$).

As a simple consequence of this theorem and Propositions 1.5′ and 1.7 we have the following:

Corollary 5.2. *Any piecewise transfer regular separable net $\mathcal{A} \in$ $\in \mathrm{MAX}_\alpha(S)$ for any $\alpha \geq 1$.*

Recall that if X is an amenable locally compact group and $\mu = \mu^{(l)}$ (the left Haar measure), then $\gamma_{TR}(\mathcal{A}) = \gamma_R(\mathcal{A})$ (see Note 2.2).

Now we shall consider the non-regular nets studied in § 4. In Theorems 4.1 and 4.2 we can take $\widetilde{A}_2 = B_r = \{|x| \leq r\}$ and $M_r = B_{r^2}$. Therefore the "transfer condition" (**T**) is fulfilled with $C_{\mathbf{T}} = 1$, and the Transfer Principle and Theorem 4.2 as well as Propositions 1.11 and 1.12 allow us to state the following proposition.

Theorem 5.3. *In* $\mathbb{R}^m$ *the net of spherical layers* $\{L_{r,\delta(r)}, r > 0\} \in \in \mathrm{DOM}_\alpha(T)$ *for any completely measurable system* T *and any* $\alpha > \frac{m}{m-1}$.

Note that if $\inf_{r>0} \frac{\delta(r)}{r} > 0$, the net $\{L_{r,\delta(r)}\}$ is regular and Theorem 5.3 is contained in Theorem 5.1.

The following general local ergodic theorem is a simple consequence of Theorem 4.4 and the "local transfer principle" (Proposition 1.14). Let T_x, $x \in X$, be a completely measurable dynamical system in $(\Omega, \mathcal{F}, m)$. Denote

$$\widehat{f}_{r,d}(\omega) = \frac{1}{l(L_{r,d})} \int_{L_{r,d}} f(T_x\omega)dx,$$

$$\widehat{f}_r(\omega) = \int f(T_x\omega)\sigma_r(dx),$$

$$\widehat{f}^*_d(x) = \sup_{0<r\leq R} \widehat{f}_{r,d}(x),$$

$$f^*(x) = \sup_{0<r\leq R} \widehat{f}_r(x).$$

Theorem 5.4. *If* $\alpha > \frac{m}{m-1}$ *and* $f \in L^\alpha(\Omega, \mathcal{F}, m)$, *then*

(i) $\lim_{d\to 0} \widehat{f}_{r,d}(\omega) = \widehat{f}_r(\omega)$ *m–a.e.;*

(ii) $\lim_{d\to 0} f^*_d(\omega) = f^*(\omega)$ *m–a.e.;*

(iii) $\lim_{d\to 0} \sup_{0<r\leq R} |\widehat{f}_{r,d}(\omega) - \widehat{f}_r(\omega)| = 0$ *m–a.e.,* $R > 0$.

Theorem 5.3 implies that the functions $f^*_d(\omega)$ are measurable. Assertion *(ii)* of Theorem 5.4 shows that the function $f^*(\omega)$ is measurable too. Now the Transfer Principle and Theorem 4.2 allow us to state the following proposition.

Theorem 5.5. **(R. Jones)** *Let* $X = \mathbb{R}^m$, $S_r = \{x : |x| = r\}$ *and let* ν_r *be the rotation invariant probability measure on* S_r. *The net* $\{\sigma_r, r > 0\} \in \mathrm{DOM}_\alpha(T)$ *for any completely measurable dynamical system* T *and any* $\alpha > \frac{m}{m-1}$.

Let us apply the Transfer Principle to Theorem 4.3. In this case $\widetilde{A}_r = \bigcup_{0<\gamma\leq r} \gamma\Gamma$; $\{\widetilde{A}_r\}$ is a net of convex bodies. Condition (**T**) is satisfied with $M_r = \bigcup_{\gamma<k_r} A_r$ if $k_r/r \to \infty$. The measurability of the dominant function $f^*(\omega)$ can be proved as in Theorem 5.5. Therefore we obtain

Theorem 5.6. *If Γ is a smooth boundary of a compact centrally symmetric convex body in $\mathbb{R}^2$, $A_r = r\Gamma$ and ν_r is the arc length measure on A_r, then $\{\nu_r,\ r>0\} \in \mathrm{DOM}_\alpha(T)$ for any $\alpha > 2$.*

§ 6. Maximal Ergodic Theorems: The Dual Space Approach

In this section we set forth another approach to proving the maximal ergodic theorem for regular sets. This approach was developed by A. Shulman [1], who used ideas due to A. Cordoba and R. Duncan.

We start with the following basic lemma, proved by A. Shulman in the case of positive operators associated with averages of a dynamical system; it is closely related to a statement by Duncan [1] (see Note 2.1 below).

Lemma 6.1. *Let $\{T_k\}$ be a sequence of bounded linear operators in the space $L^\alpha(\Omega,\mathcal{F},m)$, $\alpha \geq 1$. For $f \in L^\alpha(\Omega,\mathcal{F},m)$, denote $f^*(\omega) = \sup_{1\leq k<\infty} |T_k f(\omega)|$. For every $\alpha > 1$ the following properties of $\{T_k\}$ are equivalent:*

(i) there exists a constant C_α such that for any $f \in L^\alpha(\Omega,\mathcal{F},m)$ and for any $\varepsilon > 0$,

$$m\{\omega : f^*(\omega) \geq \varepsilon\} \leq \frac{C_\alpha}{\varepsilon^\alpha} \int |f|^\alpha dm; \tag{6.1}$$

(ii) there exists a constant C_β such that for every collection of disjoint m–mod 0 integrable sets $\{\Omega_i\}_1^N$

$$\|\sum_{n=1}^N T_n^* \mathcal{X}_{\Omega_n}(\omega)\|_{L^\beta} \leq C_\beta [m(\bigcup_{n=1}^N \Omega_n)]^{\frac{1}{\beta}}, \quad N = 1, 2, \ldots, \tag{6.2}$$

where β is the number conjugate to α $(\alpha^{-1}+\beta^{-1}=1)$ and T_n^* is the adjoint operator in $L^\beta(\Omega,\mathcal{F},m)$.
Moreover, (ii)⇒(i) if $\alpha=1$, $\beta=\infty$.

Proof. a) *(ii)⇒(i)* with $C_\alpha=2^\alpha C_\beta^\alpha$ ($C_\alpha=C_\beta^\alpha$ if all T_k are positive). Let $f\in L^\alpha$, $\varepsilon>0$, $N\in\mathbb{N}$. Denote $\Lambda_n^\varepsilon=\{\omega:|T_nf(\omega)|\geq\varepsilon\}\cap\Omega_0$, $\Omega_n^\varepsilon:=\Lambda_n^\varepsilon\setminus\bigcup_{k=1}^{n-1}\Lambda_k^\varepsilon$; $\Omega^\varepsilon:=\bigcup_{k=1}^N\Lambda_k^\varepsilon=\bigcup_{k=1}^N\Omega_k^\varepsilon$. It is clear that $\Omega^\varepsilon=\{\omega\sup_{n\leq N}|T_nf(\omega)|>\varepsilon\}$ and the sets Ω_n^ε, $n=1,2,\ldots$ are disjoint. Denote $\Gamma_n^+=\{\omega:T_nf(\omega)\geq 0\}$, $\Gamma_n^-=\{\omega:T_nf(\omega)<0\}$, $\Omega_n^{\varepsilon+}=\Omega_n^\varepsilon\cap\Gamma_n^+$, $\Omega_n^{\varepsilon-}=\Omega_n^\varepsilon\cap\Gamma_n^-$. By the Chebyshev inequality,

$$
\begin{aligned}
m(\Omega^\varepsilon)=&\sum_{n=1}^N m(\Omega_n^\varepsilon)\leq\frac{1}{\varepsilon}\sum_{n=1}^N\int_{\Omega_n^\varepsilon}|T_nf(\omega)|m(d\omega)=\\
=&\frac{1}{\varepsilon}\sum_{n=1}^N\int_{\Omega_n^{\varepsilon+}}T_nf\,dm-\frac{1}{\varepsilon}\sum_{n=1}^N\int_{\Omega_n^{\varepsilon-}}T_nf\,dm\leq\\
\leq&\frac{1}{\varepsilon}\int\varphi_\varepsilon^+f\,dm-\frac{1}{\varepsilon}\int\varphi_\varepsilon^-f\,dm
\end{aligned}
$$

where $\varphi_\varepsilon^\pm m=\sum_{n=1}^N T_n^*\mathcal{X}_{B_n^{\varepsilon\pm}}$.
By applying the Hölder inequality we get

$$
m(\Omega^\varepsilon)\leq\frac{1}{\varepsilon}(\|\varphi_{\varepsilon+}\|_{L^\beta}+\|\varphi_{\varepsilon-}\|_{L^\beta})\|f\|_{L^\alpha}. \tag{6.4}
$$

Condition *(ii)* implies: $\|\varphi_{\varepsilon\pm}\|_{L^\beta}\leq C_\beta(m(\bigcup_{n=1}^\infty\Omega_n^{\varepsilon\pm}))^{\frac{1}{\beta}}$, and therefore

$$
m(\Omega^\varepsilon)\leq\frac{2C_\beta}{\varepsilon}m(\Omega^\varepsilon)^{\frac{1}{\beta}}\|f\|_{L^\alpha} \tag{6.5}
$$

(if all T_k are positive, $\varphi_{\varepsilon-}=0$ and the factor 2 in (6.5) can be dropped). Since $m(\Omega^\varepsilon)<\infty$, we deduce from (6.4) and (6.5) the following inequality:

$$
[m(\Omega^\varepsilon)]^{1-\frac{1}{\beta}}\leq\frac{2}{\varepsilon}C_\beta\|f\|_{L^\alpha},
$$

which is equivalent to

$$m(\Omega^\varepsilon) \leq \frac{2^\alpha}{\varepsilon^\alpha} C_\beta^\alpha \int |f|^\alpha dm.$$

Since N is an arbitrary integer, this implies *(i)* with $C_\alpha = 2^\alpha C_\beta^\alpha$.

b) If $\alpha > 1$, then *(i)*$\Rightarrow$*(ii)* with $C_\beta = 2C_\alpha^{\frac{1}{\alpha}}(\frac{\alpha}{\alpha-1})^{\frac{1}{\alpha}}$. Let *(i)* be fulfilled and let $\{\Omega_n\}_1^N$ be a collection of disjoint m–mod 0 integrable sets in $\mathcal{F}$. We suppose that $m(\bigcup_{n=1}^N \Omega_n) < \infty$ (otherwise the statement is trivial). Consider the linear operator U: $L^\alpha(\Omega, \mathcal{F}, m) \mapsto F(\Omega, \mathcal{F}, m)$ (the space of all measurable functions $\varphi(\omega)$, $\omega \in \Omega$) defined as follows: $(Uf)(\omega) := := \sum_{n=1}^N \mathcal{X}_{\Omega_n} T_n f$. Since $|(Uf)(\omega)| \leq f^*(\omega)$, according to Proposition 1.5, Property *(i)* implies $\|Uf\|_{L^\alpha} \leq 2C_\alpha^{\frac{1}{\alpha}} \|f\|_{L^\alpha}$. Thus U is a bounded linear operator in $L^\alpha(\Omega, \mathcal{F}, m)$ with $\|U\| \leq 2C_\alpha^{\frac{1}{\alpha}}(\frac{\alpha}{\alpha-1})^{\frac{1}{\alpha}} = C_\beta$. The adjoint operator U^* acts in $L^\beta(\Omega, \mathcal{F}, m)$ and $\|U^*\| = \|U\| \leq C_\beta$. Thus

$$\|U^* h\|_{L^\beta} \leq C_\beta \|h\|_{L^\beta}, \quad h \in L^\beta. \tag{6.3}$$

For any $f \in L^\alpha$ and $g \in L^\beta$,

$$\int U^* hf\, dm = \int hUf\, dm = \int h \sum_{n=1}^N \mathcal{X}_{\Omega_n} T_n f\, dm =$$

$$= \int f \sum_{n=1}^N T_n(h\mathcal{X}_{\Omega_n}) dm;$$

hence $U^* h = \sum_{n=1}^N T_n^*(h\mathcal{X}_{\Omega_n})$.
We have $U^*(\bigcup_{k=1}^N \mathcal{X}_{\Omega_k}) = \sum_{n=1}^N T_n^*(\bigcup_{k=1}^N \mathcal{X}_{\Omega_k}\mathcal{X}_{\Omega_n}) = \sum_{n=1}^N T_n^* \mathcal{X}_n$. So it remains to use (6.3) with $h = \sum_{k=1}^N \mathcal{X}_{\Omega_k}$.

□

Note 6.1. Lemma 6.1 is close to the following statement, due to R. Duncan (see Duncan [1]). Under the conditions of Lemma 6.1 the following properties of $\{T_n\}$ are equivalent for any $1 \leq \alpha < \infty$:

(i') $\int f^* dm \leq C'_\alpha \|f\|_{L^\alpha}$, $f \in L^\alpha(\Omega, \mathcal{F}, m)$;

(ii') there exists a constant C'_β such that for every sequence $\{\Omega_n\}$ of disjoint m–mod 0 sets:

$$\sum_{k=1}^{N} \|T_k^* \mathcal{X}_{\Omega_k}\|_{L^\beta} \le C'_\beta, \quad N = 1, 2, \dots$$

A comparison of Lemma 6.1 with Duncan's theorem shows that if $m(\Omega) < \infty$, then for any $\alpha > 1$ conditions *(i′)* and *(ii′)* are equivalent. So in this case both statements coincide. As we shall see below, Lemma 6.1 is more suitable in applications when the measure m is infinite.

Let G be a locally compact group and μ some relatively invariant Borel measure on G; $\Delta(\cdot)$ denotes its left modular function.

Consider a completely measurable bounded linear right representation S of G in $L^\alpha(\Omega, \mathcal{F}, m)$ and the operators T_n : $(T_n f)(\omega) = \frac{1}{\mu(A_n)} \int_{A_n} (f S_g)(\omega) \mu(dg)$. $S^* : g \to S_g^*$ is a left representation in $L^\beta(\Omega, \mathcal{F}, m)$ and it is easy to check that $(T_n^* h)(\omega) = \frac{1}{\mu(A_n)} \int_{A_n} (S_g^* h)(\omega) \mu(dg)$ (here $\alpha^{-1} + \beta^{-1} = 1$). If S is associated with a left dynamical system $\{T_g,\, g \in G\}$, then S^* is associated with the right system $\{T_{g^{-1}},\, g \in G\}$. Thus, in this case condition (6.2) assumes the following form:

$$\int_\Omega \Big|\sum_{n=1}^{N} \frac{1}{\mu(A_n)} \int_{A_n} S_g^* \mathcal{X}_{\Omega_n}(\omega) \mu(dg)\Big|^\beta m(dw) \le C_\beta^\beta m\Big(\bigcup_{n=1}^{N} \Omega_n\Big). \quad (6.2')$$

In his work [1], A. Shulman proved Lemma 6.1 in the case of a dynamical system and deduced from it a version of Theorem 5.1 under somewhat weaker conditions. We shall give a complete discussion of his results only for $\alpha = 2$ (Theorem 6.2 below), when the statement and proof are rather simple, and then give the statement for all $\alpha > 1$ with a hint of the proof.

Let $\psi(A) := \int_A \Delta(g) \mu(dg)$, $A \in \mathfrak{B}$.

Theorem 6.2. **(A. Shulman)** *Let $\{A_n,\, n = 1, 2, \dots\}$ be a sequence of nonnull integrable Borel sets. Suppose that*

$$\sup_{1\le n_1<\infty}\frac{1}{\mu(A_{n_1})}\int_G\Big(\sup_{1\le n_2<n_1}\frac{1}{\mu(A_{n_2})}\psi(A_{n_1}g_1^{-1}\cap A_{n_2})\Big)\mu(dg_1)\colon=$$
$$:=\gamma<\infty.$$

Then $\{A_n\}\in \mathrm{MAX}_2(T)$ *with constant* $C_2'=2\gamma+1$ *for any completely measurable dynamical system* T.

Proof. Let $\{T_g,\, g\in G\}$ be a dynamical system in $(\Omega,\mathcal{F},m)$ and S the associated representation. We shall use only the following properties of S:

a) S is unitary in $L^2(\Omega,\mathcal{F},m)$;

b) $\|S_g\|_{L^\infty}=1$, $g\in G$.

By the Riesz convexity theorem, a) and b) imply that S is isometric in all L^α, $\alpha\ge 2$. We shall prove that condition *(ii)* of Lemma 6.1 is fulfilled with $C_2=(C_2')^{\frac12}$. Consider an arbitrary sequence of disjoint m–mod 0 sets $\{\Omega_n\}\subset\mathcal{F}$ with $\sum_{n=1}^N m(\Omega_n)<\infty$. We have

$$\int_\Omega\Big|\sum_{n=1}^N\frac{1}{\mu(A_n)}\int_{A_n}S_g^*\mathcal{X}_{\Omega_n}\mu(dg)\Big|^2 m(d\omega)=\mathcal{I}_1+2\mathcal{I}_2,$$

where

$$\mathcal{I}_1:=\sum_{n=1}^N\int_\Omega\Big|\frac{1}{\mu(A_n)}\int_{A_n}S_x^*\mathcal{X}_{\Omega_n}\mu(dx)\Big|^2 m(d\omega)$$
$$\mathcal{I}_2:=\sum_{n_1=1}^N\sum_{n_2=1}^{n_1-1}\int_\Omega\Big[\frac{1}{\mu(A_{n_1})\mu(A_{n_2})}\cdot$$
$$\cdot\int_{A_{n_1}}\int_{A_{n_2}}S_{g_1}^*\mathcal{X}_{\Omega_{n_1}}S_{g_2}^*\mathcal{X}_{\Omega_{n_2}}\mu(dg_1)\mu(dg_2)\Big]m(d\omega)$$

Using the Cauchy inequality, Fubini's theorem and Property a) of S, we obtain

$$\begin{aligned}
\mathcal{I}_1 &\le \sum_{n=1}^{N} \int_{\Omega} \frac{1}{\mu(A_n)} \int_{A_n} |S_g^* \mathcal{X}_{\Omega_n}|^2 \mu(dg) m(d\omega) = \\
&= \sum_{n=1}^{N} \int_{A_n} \int_{\Omega} |S_g^* \mathcal{X}_{\Omega_n}|^2 m(d\omega) \mu(dg) \le \\
&\le \sum_{n=1}^{N} \frac{1}{\mu(A_n)} \int_{A_n} \left(\int_{\Omega} \mathcal{X}_{\Omega_n} m(d\omega) \right) \mu(dg) = \sum_{n=1}^{N} m(\Omega_n);
\end{aligned}$$

$$\begin{aligned}
\mathcal{I}_2 = &\sum_{n_1=1}^{N} \sum_{n_2=1}^{n_1-1} \frac{1}{\mu(A_{n_1}) \mu(A_{n_2})} \cdot \\
&\cdot \int_{A_{n_1}} \int_{A_{n_2}} \left(\int_{\Omega} S_{g_1}^* \mathcal{X}_{\Omega_{n_1}} \cdot S_{g_2}^* \mathcal{X}_{\Omega_{n_2}} dm \right) \mu(dg_1) \mu(dg_2).
\end{aligned}$$

Now by Property *a)* again,

$$\int_A \varphi(g_2^{-1} g_1) \mu(dg_1) = \Delta(g_2) \int_{g_2^{-1} A} \varphi(g_1) \mu(dg_1)$$

(see App., Subsect. 1.6) and $\mathcal{X}_{g_2^{-1} A_{n_1}}(g_1) = \mathcal{X}_{A_{n_1}}(g_1 g_2) = \mathcal{X}_{A_{n_1} g_1}(g_2)$:

$$\begin{aligned}
\mathcal{I}_2 = &\sum_{n_1=1}^{N} \sum_{n_2=1}^{n_1-1} \frac{1}{\mu(A_{n_1}) \mu(A_{n_2})} \cdot \\
&\cdot \int_{A_{n_2}} \left[\int_{A_{n_1}} \cdot \left(\int_{\Omega} S_{g_2^{-1} g_1}^* \mathcal{X}_{\Omega_{n_1}} \cdot \mathcal{X}_{\Omega_{n_2}} dm \right) \mu(dg_1) \right] \mu(dg_2) = \\
= &\sum_{n_1=1}^{N} \sum_{n_2=1}^{n_1-1} \frac{1}{\mu(A_{n_1}) \mu(A_{n_2})} \int_G \left[\int_G \mathcal{X}_{g_2^{-1} A_{n_1}}(g_1) \mathcal{X}_{A_{n_2}}(g_2) \Delta(g_2) \cdot \right. \\
&\left. \cdot \left(\int_{\Omega} S_{g_1}^* \mathcal{X}_{\Omega_{n_1}} \cdot \mathcal{X}_{\Omega_{n_2}} dm \right) \mu(dg_1) \right] \mu(dg_2) =
\end{aligned}$$

$$= \sum_{n_1=1}^{N} \int_G \sum_{n_2=1}^{n_1-1} \frac{1}{\mu(A_{n_2})} \left(\int_\Omega S_{g_1}^* \mathcal{X}_{\Omega_{n_1}} \cdot \mathcal{X}_{\Omega_{n_2}} dm \right) \cdot$$

$$\cdot \frac{\psi(A_{n_1} g_1^{-1} \cap A_{n_2})}{\mu(A_{n_2})} \mu(dg_1) \leq$$

$$\leq \sum_{n_1=1}^{N} \frac{1}{\mu(A_{n_1})} \int_G \left(\int_\Omega |S_{g_1^{-1}} \mathcal{X}_{\Omega_{n_1}}| \sum_{n_2=1}^{n_1-1} \mathcal{X}_{\Omega_{n_2}} dm \right) \cdot$$

$$\cdot \max_{1 \leq n_2 < n_1} \frac{\psi(A_{n_1} g_1 \cap A_{n_2})}{\mu(A_{n_2})} \mu(dg_1).$$

Since Ω_n are disjoint m–mod 0, $\sum_{n_2=1}^{n_1-1} \mathcal{X}_{\Omega_{n_2}} \leq 1$, and by Property *b)* we get

$$\mathcal{I}_2 \leq \gamma \sum_{n_1=1}^{N} m(\Omega_{n_1}).$$

It remains to apply Lemma 6.1.

□

Corollary 6.3. **(A. Shulman)** *If* $\mu = \mu^{(l)}$, *the left Haar measure, and*

$$\gamma := \sup_{1 \leq n_1 < \infty} \frac{1}{\mu(A_{n_1})} \int_G \max_{1 \leq n_2 < n_1} \frac{\mu(A_{n_1} g_1^{-1} \cap A_{n_2})}{\mu(A_{n_2})} \mu(dg_1) < \infty,$$

then $\{A_n\} \in \mathrm{MAX}_2(T)$ *for every dynamical system* T.

Proof. $\mu^{(l)} = \psi$.

□

Corollary 6.4. **(A. Shulman)** *Let* $\mu = \mu^{(l)}$, $\widetilde{A}_n := \bigcup_{l=1}^{n} A_l$. *If*

$$\widetilde{\gamma} := \sup_{1 \leq n < \infty} \frac{\mu(\widetilde{A}_{n-1} \cdot A_n)}{\mu(A_n)} < \infty, \tag{6.4}$$

then the conclusion of Theorem 6.2 holds with $C_2' = 2\widetilde{\gamma} + 1$.

Proof. It is easy to verify that $A \cap g^{-1}A_l = \emptyset$ for all $g \notin A_l^{-1}A_n$; therefore $\mu(A_n g^{-1} \cap A_l) \leq \mathcal{X}_{\widetilde{A}_{n-1}^{-1}\cdot A_n}(g)\mu(A_l)$, for all $g \in G$ and $1 \leq \leq l < n$. Thus

$$\gamma = \sup_{1 \leq n < \infty} \frac{1}{\mu(A_n)} \int_G \max_{1 \leq l < n} \frac{\mu(A_n g^{-1} \cap A_l)}{\mu(A_l)} \mu(dg) \leq$$

$$\leq \sup_{1 \leq n < \infty} \frac{\mu(\widetilde{A}_{n-1} \cdot A_n)}{\mu(A_n)} = \widetilde{\gamma}.$$

□

Note that condition (6.4) coincides with condition (WM) (see § 2) and is weaker than the regularity condition (R) (see Example 2.12).

Now we shall state the general theorem.

Denote $\mathcal{N}_{\leq}^{(k)} := \{(n_1, \ldots, n_k) \in \mathbb{N}^k : n_k \leq n_{k-1} \leq \ldots \leq n_1\}$; $\mathcal{N}_{=}^{(k)} = \{(n_1, \ldots, n_k) \in \mathbb{N}^*;\ n_1 = n_2 = \ldots = n_k\}$; $\mathcal{N}_1 = \mathcal{N}_{\leq}^{(k)} \setminus \mathcal{N}_{=}^{(k)}$; $\mathcal{N}_i(n_1, \ldots, n_{i-1}) := \{n \in \mathbb{N} : (n_1, \ldots, n_{i-1}, n, n_{i+1}, \ldots, n_k) \in \mathcal{N}_1$ for some $n_j \in \mathbb{N},\ i < j \leq k\}$; $\psi_i(A) = \int_A [\Delta(g)]^{i-1}\mu(dg)$, $i = \overline{1,k}$; $A_{n_1,\ldots,n_k}(g_1, \ldots, g_{k-1}) := A_{n_1} g_1^{-1} g_2^{-1} \ldots g_{k-1}^{-1} \cap \ldots \cap A_{n_{k-1}} g_{k-1}^{-1} \cap A_{n_k}$.

Theorem 6.5. **(A. Shulman)** *Suppose that for any* $k \in \mathbb{N}$,

$$\gamma_k := \sup_{n_1 \in \mathbb{N}} \frac{\int \psi_1(dg_1)}{\mu(A_{n_1})} \Big\{ \cdots \Big\{ \sup_{n_{k-1} \in \mathcal{N}_{k-1}(n_1, \ldots, n_{k-2})} \frac{\int \psi_{k-1}(dg_{k-1})}{\mu(A_{n_{k-1}})} \cdot$$

$$\cdot \sup_{n_k \in \mathcal{N}_k(n_1, \ldots, n_{k-1})} \frac{\psi_k(A_{n_1,\ldots,n_k}(g_1, \ldots, g_{k-1}))}{\mu(A_{n_k})} \Big\} \cdots \Big\};$$

then $\{A_n\} \in \mathrm{MAX}_\alpha(T)$ *for every completely measurable dynamical system* T *and any* $\alpha > 1$.

Proof. At first the theorem is proved for all α with $\beta = k \in \mathbb{N}$, $\alpha = \frac{k}{k-1}$. According to Lemma 6.1, we have to show that under the conditions of the theorem,

$$\int \Big|\sum_{n=1}^{N} \frac{1}{\mu(A_n)} \int_{A_n} S_g^* \mathcal{X}_{\Omega_n}(\omega)\Big|^k m(d\omega) \leq C_k^k m(\bigcup_{n=1}^{\infty} \Omega_n); \tag{6.5}$$

this is done in the same way as in the proof of Theorem 6.2; however, the calculations are tedious and we omit them. It turns out that (6.5) holds with $C_k \leq (k!)^{\frac{1}{k}} \gamma_k^{\frac{1}{k}}$. For any $\alpha > 1$ we find a k such that $\alpha > \frac{k}{k-1}$, and then we use Proposition 1.2.

□

Corollary 6.6. **(A. Shulman)** *Let $\mu = \mu^{(l)}$. Suppose that for any $k \in N$,*

$$\gamma_k = \sup_{n_1 \in \mathbf{N}} \int \mu(dg_1)\{\cdots\{\sup_{1<n_{k-1}\leq n_{k-2}} \int \mu(dg_{k-1}) \cdot \\ \cdot \sup_{1\leq n_k \leq n_{k-1}} \frac{\mu(A_{n_1} g_1^{-1} \cdots g_{k-1}^{-1} \cap \ldots \cap A_{n_{k-1}} g_{k-1}^{-1} \cap A_{n_k}}{\mu(A_{n_1}) \ldots \mu(A_{n_{k-1}}) \mu(A_{n_k})} \} \cdots \}. \tag{6.6}$$

Then $\{A_n\} \in \mathrm{MAX}_\alpha(T)$ with $C_\alpha = \gamma_R(\{A_n\})$ for any T and any $\alpha > 1$. If $\gamma_\infty := \underline{\lim}_{k\to\infty} \gamma_k^{\frac{1}{\gamma}} < \infty$, then $\{A_n\} \in \mathrm{MAX}_1(T)$ and $C_\alpha = C_\beta = \gamma_\infty$.

Corollary 6.7. *) *If $\mu = \mu^{(l)}$ and the sequence $\{A_n\}$ is regular, then $\{A_n\} \in MAX_\alpha(T)$ for any T and any $\alpha > 1$.*

Proof. **(A. Shulman)** It is easy to verify that

$$A_{n_1,\ldots,n_k}(g_1,\ldots,g_{k-1}) \neq \emptyset \quad \text{only if} \quad g_i \in A_{n_{i+1}}^{-1} \cdot A_n, \quad i = \overline{1, k-1}.$$

Therefore $\mu(A_{n_1,\ldots,n_k}(g_1,\ldots,g_{k-1}) \leq \prod_{i=1}^{k-1} \mathcal{X}_{A^{-1}_{n_{i=1}} \cdot A_{n_i}}(g_i)\mu(A_{n_k}) \leq$ $\leq \prod_{i=1}^{k-1} \mathcal{X}_{\widetilde{A}^{-1}_{n_i} \cdot A_{n_i}}(g_i)\mu(A_{n_k})$ if $n_{i+1} \leq n_i$. Now we estimate γ_k in

*) Compare with Corollary 5.2.

Corollary 6.6,

$$\gamma_k \le \sup_{n_1 \in N} \frac{\mu(\widetilde{A}_{n_1} \cdot A_{n_1})}{\mu(A_{n_1})} \cdot \dots \cdot \sup_{1 \le n_{k-2} \le n_{k-3}} \frac{\mu(\widetilde{A}^{-1}_{n_{k-2}} \cdot A_{n_{k-2}})}{\mu(A_{n_{k-2}})} \cdot$$

$$\cdot \sup_{1 \le n_{k-1} \le n_{k-2}} \frac{\mu(\widetilde{A}_{n_{k-1}} \cdot A_{n_{k-1}})}{\mu(A_{n_{k-1}})} \le \Big(\sup_{n \in \mathbb{N}} \frac{\mu(\widetilde{A}^{-1}_n \cdot A_n)}{\mu(A_n)} \Big)^{k-1} < \infty.$$

□

BIBLIOGRAPHICAL NOTES

1. Inequalities of the maximal type were used in the first proofs of the pointwise ergodic theorem in G.D. Birkhoff [1] (1931), Chintschin [1] (1932), Kolmogorov [1] (1938). Yosida and Kakutani [1] (1939) have distinguished and proved the maximal ergodic theorem as an independent statement and showed its key role in the proof of the pointwise ergodic theorem. This was brilliantly manifested also in Wiener [1] (1939), where the deap connection between the maximal and dominated ergodic theorems and Hardy–Littlewood type inequalities was shown. In this work N. Wiener used the covering method in order to prove Hardy–Littlewood type inequalities for balls and cubes in $\mathbb{R}^m$ and $\mathbb{Z}^m$, and "transfered" these inequalities to general maximal and dominated ergodic theorems in $\mathbb{R}^m$ and $\mathbb{Z}^m$; the latter were used to prove mean and pointwise ergodic theorems in $\mathbb{R}^m$ and $\mathbb{Z}^m$. Pitt [1] (1942) extended Wiener's results to more general "regular" sequences of sets.

2. In Calderón [1] (1953) the maximal and dominated ergodic theorems were extended to a wide class of groups, and averaging with nets of sets under conditions close to regularity was considered. Further generalizations are given in Cotlar [1] (1956).

3. In Calderón [2] (1968) the general Transfer Principle was stated and proved.

4. The "Local Transfer Principle" (Proposition 1.12) is based on an idea due to Wiener [1] (1939).

5. The main results of §2 and §3 were announced in Tempelman [5] (1968); the proofs were published in Tempelman [7] (1972). Proofs of these or similar results were published also in Bewley [1] (1971) and Chatard [1] (1970).

6. The notion of a regular net of sets (§2) generalizes the notions of "regularity" and "ergodicity" of nets of sets introduced by Calderón [1] (1953), Cotlar [1] (1956), Pitt [1] (1947).

In Tempelman [5] (1967) increasing regular sequences of sets in semi-groups were considered, and in Tempelman [7] (1972) the results of §2 and §3 were proved under a condition which reduces to

$$\mu(\widetilde{A}_n^{-1} \cdot \widetilde{A}_n) \leq C\mu(A_n), \quad n \in \mathbb{N}, \tag{1}$$

if X is a group. The weaker condition

$$\mu(\widetilde{A}_n^{-1} \cdot A_n) \leq C\mu(A_n), \quad n \in \mathbb{N}, \tag{2}$$

considered in §§2,3,5 was introduced in Shulman [1] (1988), were the maximal ergodic theorems are proved under condition (2) by a quite different method (see §6 and Note 10 below). A. Shulman noticed also that the sequence $\{A_n\}$ in Example 2.3 satisfies condition (2) (oral communication). The possibility of pointwise averaging with "moving" sequences, considered in Examples 2.2–2.4, was studied by other methods in Bellow, Jones, Rosenblatt [1] (1990); M. Akcoglu and A. del Junco (see Akcoglu, del Junco [1] (1975) have shown that the non-regular sequence of sets considered in Example 2.4 is not pointwise averaging.

7. The Covering Lemma (Lemma 2.3) generalizes the classical Vitali covering theorem (see, e.g., Natanson [1], Saks [1]). It is close to the covering theorems due to Alfsen [1] (1965), and Edwards, Hewitt [1] (1965) (see also Hewitt, Ross [1]).

8. The Hardy–Littlewood type maximal inequalities for regular nets (§ 3) generalize results due to Edwards, Hewitt [1] (1965).

9. The dominated inequality for the net of spheres (Theorem 4.1) is due to Stein [1] (1976) (see also Stein, Wainger [1] (1978)); Theorem 4.3 is due to Bourgain [1].

Theorem 5.1 is contained in Tempelman [5] (1968) (a proof is in Tempelman [7] (1972)). Theorem 5.5 is contained in R. Jones [1] (to appear).

10. The dual space approach to the maximal ergodic theorem for regular nets of sets (see § 6) was developed in Shulman [1] (1988) where ideas of Cordoba [1] (1975) and Duncan [1] (1977) are used.

11. In Jones, Olsen [1], Olsen [1] – [3] and Sucheston [1] other approaches to ergodic theorems for multi–parameter dynamical systems and linear representations were developed.

CHAPTER 6
POINTWISE ERGODIC THEOREMS

§ 1. Pointwise Averaging Nets: Definition and Preliminary Discussion

We denote by $\widetilde{F}_B = \widetilde{F}_B(\Omega, \mathcal{F}, m)$ the space of all measurable B-valued functions with the seminorm

$$|f|_{P_B} = \inf_{\alpha \geq 0} \arctan[\alpha + m(\{\omega : |f(\omega)| > \alpha\})];$$

convergence in $\widetilde{F}_B$ is the same as convergence in m.

Lemma 1.1. *Let S be a linear manifold in $L_B^\alpha\ (\Omega, \mathcal{F}, m)$ $(\alpha \geq 1)$, and $\{U_n,\ n \in \mathbb{N}\}$ a sequence of linear operators from L_B^α into $\widetilde{F}_B$ such that*

1) if $f \in S$, then the limit $\lim_{n \in \mathbb{N}} (U_n f)(\omega)$ exists for almost all $\omega \in \Omega$;

2) if $f \in [S]_{L_B^\alpha}$, then $\sup_{n \in \mathbb{N}} |(U_n f)(\omega)| < \infty$ for almost all $\omega \in \Omega$.

Then for every function $f \in [S]_{L_B^\alpha}$ the limit $\lim_{n \in \mathbb{N}} (U_n f)(\omega)$ exists for almost all $\omega \in \Omega$.

This is a version of the Convergence Principle (see App. (12.A)).

Let $(X, \mathfrak{B})$ be a measurable semigroup and $T = \{T_x,\ x \in X\}$ a dynamical system in $(\Omega, \mathcal{F}, m)$. For any $f \in L^\alpha(\Omega, \mathcal{F}, m)$ $(\alpha \geq 1)$ we define the average

$$\widehat{f}_n(\omega) = \int f(T_x \omega) \nu_n(dx).$$

According to App. (9, G), $\widehat{f}_n \in L_B^\alpha$.

We denote also by L_B the set of all integrable simple functions; D_B^α the subspace of $L_B^\alpha(\Omega, \mathcal{F}, m)$ generated by the functions of the form

$f_g(\omega) = g(\omega) - g(T_x\omega)$, $g \in L_B^\alpha$, $x \in X$; Q_B the set of all functions f_g with $g \in L_B$; and D_B the linear hull of Q_B. It is obvious that D_B is dense in D_B^α. Recall that $L_B^\alpha = D_B^\alpha \oplus I_B^\alpha$ if $\alpha > 1$ or if $\alpha = 1$ and $m(\Omega) < \infty$.

For $\Lambda \in \mathcal{F}$, $\omega \in \Omega$, we consider the sets $D_{\omega,\Lambda} = \{x : T_x\omega \in \Lambda\}$ (the "time" spent in Λ by the trajectory of ω). By Fubini's theorem, $D_{\omega,\Lambda} \in \mathfrak{B}$ for almost all $\omega \in \Omega$.

Lemma 1.2. *Let $1 \le \alpha < \infty$ and let B be a Banach space. Consider the following properties of a sequence $\{\nu_n\} \in \mathfrak{B}(B)$:*

1) for any $f \in L_B^\alpha$,

$$\lim_{n\in\mathbb{N}} \widehat{f}_n(\omega) = \mathbf{M}(f) \quad \text{for almost all} \quad \omega \in \Omega; \tag{1.1}$$

2) for any $f \in L_+^\alpha(\Omega, \mathcal{F}, m)$ and almost all $\omega \in \Omega$,

$$\sup_{n\in\mathbb{N}} \widehat{f}_n(\omega) < \infty; \tag{1.2}$$

3) for any $x \in X$, any $\Lambda \in \mathcal{F}$ with $m(\Lambda) < \infty$ and almost all $\omega \in \Omega$ we have $\lim_{n\in\mathbb{N}}(\nu_n(D_{\omega,\Lambda}) - \nu_n(x^{-1}D_{\omega,\Lambda})) < 0$;

4) relation (1.1) holds for all f belonging to some linear manifold S which is dense in L_B^α.

The following implications take place.

(i) Property 1) implies properties 2) and 3).

(ii) Properties 2) and 4) imply 1).

(iii) If $\alpha > 1$ or $\alpha = 1$ and $m(\Omega) < \infty$, then 2) and 3) imply 1).

(iv) If 2) and 3) are fulfilled for $\alpha = 1$ and some $\alpha > 1$, then 1) holds with $\alpha = 1$.

Proof. *(i)* Suppose that *1)* holds. Then Property *2)* follows from the fact that $f(\cdot) = b\varphi(\cdot) \in L_B^\alpha$ if $\varphi \in L_+^\alpha$, $b \in B$. Let us prove property *3)*. We have $\mathcal{X}_{D_{\omega,\Lambda}}(y) - \mathcal{X}_{x^{-1}D_{\omega,\Lambda}}(y) = \mathcal{X}_\Lambda(T_y\omega) - \mathcal{X}_\Lambda(T_{xy}\omega) =: \varphi_{x,\Lambda}(T_y\omega)$ where $\varphi_{x,\Lambda}(\omega) = \mathcal{X}_\Lambda(\omega) - \mathcal{X}_\Lambda(T_x\omega) \in D_{\mathbb{R}}^\alpha \subset L_B^\alpha(\Omega, \mathcal{F}, m)$; consequently, $\mathbf{M}(\varphi_{X,\Lambda}|T) = 0$. It remains to apply (1.1) to the function $f(\omega) = b\varphi_{x,\Lambda}$ where $b \in B$ (in the proof of Theorem 1.6.5 it was shown that $\mathbf{M}(b\varphi_{x,\Lambda}) = b\mathbf{M}(\varphi_{x,\Lambda})$ and thus $\mathbf{M}(b\varphi_{x,\Lambda}|T) = 0$).

(ii) Suppose conditions *2)* and *4)* are fulfilled. We define in L_B^α operators U_n $(U_n f)(\omega) = \widehat{f}_n(\omega)$. These operators are linear and $\|U_n\| \le 1$ (see App., (9.G)). Since the topology of L_B^α is stronger than the topology of $\widetilde{F}_B$, all U_n are also continuous linear operators from L_B^α into $\widetilde{F}_B$. It is clear that they satisfy the conditions of Lemma 4.1. Therefore, for any $f \in L_B^\alpha$ the limit $\lim_{n\to\infty} \widehat{f}_n(\omega) =: \widehat{f}(\omega)$ exists m–a.e.. If $m(\Omega) = \infty$ we consider along with m the finite measure

$$m_1(\Lambda) := (m(\Lambda_1) + 1) \sum_{k=1}^{\infty} \frac{m(\Lambda \cap \Lambda_k)}{2^{k-1}(m(\Lambda_k) + 1)}$$

where Λ, $\Lambda_k \in \mathcal{F}$, $m(\Lambda_k) < \infty$ and $\bigcup_{k=1}^{\infty} \Lambda_k = \Omega$. This measure is equivaled to $m(\cdot)$. If $m(\Omega) < \infty$ we can take $\Lambda_1 = \Omega$, $\Lambda_k = \emptyset$ for $k > 1$; then $m_1 = m$. It is easy to verify that all U_n are continuous linear operators from $L_B^\alpha(\Omega, \mathcal{F}, m)$ into $\widetilde{F}_B(\Omega, \mathcal{F}, m_1)$. Since $m_1(\Omega) < < \infty$, $\lim_{n\to\infty} U_n f = \widehat{f}(\omega)$ holds also in $\widetilde{F}_B(\Omega, \mathcal{F}, m_1)$, and therefore U: $Uf = \widehat{f}$ is a continuous linear operator from L_B^α into $\widetilde{F}_B(\Omega, \mathcal{F}, m_1)$ (see, e.g. Dunford and Schwartz [1], Th. 2.1.17). But the averaging operator $\mathbf{M}$ can also be considered as a continuous linear operator from $L_B^\alpha(\Omega, \mathcal{F}, m)$ into $\widetilde{F}_B(\Omega, \mathcal{F}, m_1)$. Since $\mathbf{M}(f) = U(f)$ for $f \in S$ and $[S]_{L_B^\alpha} = L_B^\alpha$, we have $\widehat{f} = U(f) = \mathbf{M}(f)$, $f \in L_B^\alpha$; Statement *(ii)* is proved.

(iii) If condition *3)* is fulfilled, then (1.1) holds for all functions $f = = b\varphi_{x,\Lambda}$, $x \in X$, $b \in B$, $\Lambda \in \mathcal{F}$, $m(\Lambda) < \infty$ (the function $\omega \mapsto \varphi_{x,\Lambda}(\omega)$ was defined in the proof of Statement *(i)*). Therefore (1.1) holds for all $f \in Q_B$ and thus for all $f \in Q_B + I_B^\alpha$. If $\alpha > 1$ or if $\alpha = 1$ and $m(\Omega) < \infty$, this set is dense in L_B^α, and Statement *(iii)* follows from *(ii)*.

(iv) Statement *(iii)* shows that under the stated conditions Property *4)* holds with $S = L_B^{\alpha_0} \cap L_B^1 \subset L_B^1$; it remains to apply Statement *(ii)*.

□

Definition 1.1. A net of Borel probability measures $\{\lambda_n,\ n \in N\}$ is called *pointwise left–averaging* if for any completely measurable left dynamical system $T = \{T_x,\ x \in X\}$, any $\alpha \ge 1$, B and $f \in L_B^\alpha$ property (1.1) is fulfilled. A net of Borel sets $\{A_n,\ n \in N\}$ is called *pointwise left-averaging* with respect to a measure μ if $\{\lambda_n(A) := := \frac{\mu(A_n \cap A)}{\mu(A_n)}\}$ is a pointwise left-averaging net of measures. Statements connected with finding or constructing pointwise left- or right-averaging nets are called *pointwise ergodic theorems*. Such theorems constitute the main content of this chapter.

The following statement represents the probabilistic aspect of pointwise ergodic theorems.

Proposition 1.3. *If $\{\lambda_n,\ n \in N\}$ is a pointwise left-averaging net of measures on X, then for any strict-sense homogeneous with respect to the right shifts completely measurable B-valued random field $\xi(\cdot)$ with $E\|\xi(x)\|_B < \infty$ the following limit exists a.s.:*

$$\lim_{n\to\infty} \int \xi(x)\lambda_n(dx) = \mathbf{E}[\xi(x)|\mathfrak{I}_\xi].$$

Proof. It clearly suffices to prove this statement for the coordinate image $\widetilde{\xi}(\cdot)$ of $\xi(\cdot)$ (see App., § 8). If $\widetilde{T}$ is the translation dynamical system in $(\Omega^\xi, \mathcal{F}^\xi, P^\xi)$, then $\widetilde{\xi}(x) = f(T_x\omega)$ where $f(\omega)\widetilde{\xi}(e)$, and our statement follows immediately from Definition 1.1.

□

§ 2. Pointwise Averaging in Dense Subsets of L^2: Spectral Approach

Lemma 1.2 asserts that each of the properties *3)* and *4)* together with *2)* imply the pointwise ergodic theorem (property 1). Property *2)* is satisfied by any maximal net and was explicitly discussed in the previous chapter. Of course, any ergodic net $\{\nu_n,\ n \in N\}$ satisfies condition *3)*. In this section we present a spectral method for proving property *4)* for non-ergodic sequences; it can be considered as a "pointwise" version of the spectral criterion for the validity of the mean ergodic theorem (see Ch. 4, § 9). This approach goes back to Reich [1] and Blum and Reich [1].

2.1. The main theorem. Consider a separable topological group G with spectral system $(\Lambda, \mathcal{D}, U(\cdot))$ (see App., § 5). We denote by λ_1 the point with $U(\lambda_1) = \mathbf{I}$, the unit representation. Let ν be a Borel probability measure on G. We use here the notations introduced in § 8 of Ch. 4; in particular,

$$\widetilde{\nu}(\lambda) := \int U_g(\lambda)\nu(dy), \quad \lambda \in \Lambda.$$

It is clear that $\widetilde{\nu}(\lambda)$ is a linear operator in $H(\lambda)$ and $\|\widetilde{\nu}(\lambda)\| \leq 1$; the function $\lambda \to \widetilde{\nu}(\lambda)$ is the *Fourier transform* of the measure ν.

Definition 2.1. Let U be a unitary left representation in a Hilbert space H; Φ its spectral measure. We say that a sequence $\{\nu_n, n \in N\}$ $(\subset \mathcal{P}(\mathfrak{B}))$ *satisfies the spectral condition* with respect to U if there exists a generating set H_0 in $H = \int_\Lambda \oplus H(\lambda)\Phi(d\lambda)$ such that

$$\sum_{n=1}^{\infty} \|\widetilde{\nu}_n(\lambda)h(\lambda)\|^2_{H(\lambda)} < \infty \tag{2.1}$$

for any $h(\cdot) \in H_0$ for Φ–almost all $\lambda \in \Lambda \setminus \{\lambda_1\}$.

Consider the elementary positive definite functions $\varphi_{\lambda,h}(g) = \langle U_g(\lambda)h(\lambda), h(\lambda)\rangle_{H(\lambda)}$, $\lambda \in \Lambda$. Relation (2.1) can be rewritten in the following way:

$$\sum_{n=1}^{\infty} \int \varphi_\lambda(g_1^{-1}g_2)\nu_n(g_1)\nu_n(g_2) < \infty, \tag{2.1$'$}$$

or

$$\sum_{n=1}^{\infty} \int \varphi_\lambda(g)(\bar{\nu}_n * \nu_n)(dg) < \infty \tag{2.1$''$}$$

where $\bar{\nu}_n(B) = \nu_n(B^{-1})$.

If G is commutative, then $H(\lambda) = \mathbb{C}$, and the spectral condition (with $H_0 = H$) reduces to:

$$\sum_{n=1}^{\infty} |\widetilde{\nu}_n(\lambda)|^2 < \infty \tag{2.1$'''$}$$

for Φ–almost all $\lambda \in \Lambda\backslash\{\lambda_1\}$.

Let us consider some examples.

Example 2.1. Let $G = \mathbb{R}$ and $A_n = (a_n, b_n)$. $\nu_n(B) = \frac{|A_n \cap B|}{|A_n|}$. Then $\widetilde{\nu}_n(\lambda) = \frac{1}{b_n - a_n}\int_{a_n}^{b_n} e^{i\lambda x}dx = (b_n - a_n)^{-1}\frac{e^{i\lambda b_n} - e^{i\lambda b_n}}{i\lambda}$. Thus $|\nu_n(\lambda)| \leq$ $\leq C(\lambda)(b_n - a_n)^{-1}$, and the spectral condition is fulfilled if $\sum_{n=1}^{\infty}(b_n - a_n)^{-1} < \infty$.

Example 2.2. Let σ be a rotation invariant measure on the m–dimensional sphere $S_r = \{x : x \in \mathbb{R}^m, \|x\| = r\}$. Then $\widehat{\sigma}_r(\lambda) =$ $= \int e^{i\langle\lambda,x\rangle}\sigma_r(dx) = C_m(\|\lambda\|r)^{\frac{-m-2}{2}}\mathcal{J}_{\frac{m-2}{2}}(\|\lambda\|r)$, where $\mathcal{J}_k(\lambda)$ is the Bessel function of order k. Therefore $|\widehat{\sigma}_r(\lambda)| \leq C_m\|\lambda\|^{-\frac{m-1}{2}}r^{-\frac{m-1}{2}}$. The same is true if σ_r is the surface area measure on $\Gamma_r = r\Gamma$ where Γ is the ($C^{[m+7/2]}$–smooth) boundary of a compact convex body (see Herz [1]). The spectral condition is fulfilled for a sequence $\{S_{r_n}\}$ with $\sum_{n=1}^{\infty} r_n^{-m+1} < \infty$. For spherical layers $L_{r,d(r)} = \{r - d(r) \leq \|x\| \leq r\}$ the spectral condition is fulfilled if $\sum_{n=1}^{\infty} r_n^{-(m+1)} < \infty$.

Example 2.3. Let $G = R \odot R_{++}^{\times}$, the group of affine transformations of $\mathbb{R}$. Using the results of Example 4.8.3 we deduce that a sequence $A_n := \{(a, \sigma) : |a| \leq a_n, \sigma_n^{-1} \leq \sigma < \sigma_n\}$ satisfies the spectral condition if $\sum_{n=1}^{\infty}(\log \sigma_n)^2 < \infty$

Let U be a unitary left representation in $H \subset L^2(\Omega, \mathcal{F}, m)$.

Theorem 2.1. *If $\{\nu_n\}$ satisfies the spectral condition with respect to U, then*

$$\lim_{n\to\infty}\int (U_g f)(\omega)\nu_n(dg) = \mathbf{M}(f) \quad m\text{–a.e.} \tag{2.2}$$

for any f belonging to a dense subset Γ in H.

Proof. According to Theorem 1.5.12, $H = D \oplus I$ where I is the subspace of all invariant functions and $M(f) = 0$ if $f \in D$. It is clear that $I \cong H(\{\lambda_1\})$ and $D \cong \int_{\Lambda\backslash\{\lambda_1\}} \oplus H(\lambda)\Phi(d\lambda) = \int_{\Lambda} \oplus H(\lambda)\Phi_0(d\lambda)$ where

$\Phi_o(A) = \Phi(A\backslash\{\lambda_1\})$. The set $D_0 := \Pi^H_D H_0$ generates the subspace D. Of course, if the set $I \oplus D_0$ generates H and $\widetilde{D}$ is an "averageable" dense subset of D, then $I + \widetilde{D}$ is such a set in H. So let us construct the set $\widetilde{D}$. Consider some $h \in D_0$. Denote:

$$\widehat{U}_n h := \int_G U_g h \nu_n(dg),$$

$$\Lambda_k :=\{\lambda : \|h(\lambda)\|^{-2} \sum_{n=1}^{\infty} \|\widetilde{\nu}_n(\lambda)h(\lambda)\|^2 < k\},$$

$$D_k = \int_{\Lambda_k} \oplus H(\lambda)\Phi_0(d\lambda),$$

$$h_k := \int_{\Lambda_k} \oplus h(\lambda)\Phi_o(d\lambda) = \Pi^H_{D_k} h.$$

$M^k_{n,\varepsilon} = \{\omega : \Omega : (\widehat{U}_n h_k)(\omega) > \varepsilon\}$, $k \in \mathbb{N}$. By Lemma 4.8.1, $h\widehat{U}_n = \int_\Lambda \oplus\widetilde{\nu}_n(\lambda)h(\lambda)\Phi_o(d\lambda)$. Let Λ_o be the subset on which (2.1) does not hold. In view of the spectral condition, $m(\Lambda_0) = 0$ and $\Lambda_k \uparrow \Lambda\backslash\Lambda_o$. Since $h - h_k = \int_{\Lambda\backslash\Lambda_k} \oplus h(\lambda)\Phi_o(d\lambda)$, this implies $h_k \to h$ and thus the set $\{h_k,\ h \in D_0\}$ generates D. For any $k \in \mathbb{N}$ and $h \in D_0$,

$$\begin{aligned} m(M^k_{n,\alpha}) \le& \frac{1}{\varepsilon^2} \int_\Omega |\widehat{\nu}_n h_k(\omega)|^2 m(d\omega) = \\ =& \frac{1}{\varepsilon^2} \|\widehat{U}_n h_k\|^2_D = \frac{1}{\varepsilon^2} \int_{\Lambda_k} \|\widetilde{U}_n(\lambda)h(\lambda)\|^2_{H(\lambda)}\Phi_o(d\lambda). \end{aligned} \tag{2.3}$$

Hence, by Levi's theorem,

$$\begin{aligned} \sum_{n=1}^{\infty} m(M^k_{n,\varepsilon}) \le& \varepsilon^{-2} \int_{\Lambda_k} \sum_{n=1}^{\infty} \|\widetilde{\nu}_n(\lambda)h(\lambda)\|^2_{H(\lambda)}\Phi_o(d\lambda) = \\ =& \varepsilon^{-2} \int_{\Lambda_k} \|h(\lambda)\|^2 \frac{1}{\|h(\lambda)\|^2} \sum_{n=1}^{\infty} \|\widetilde{\nu}_n(\lambda)h(\lambda)\|^2 \Phi_o(d\lambda) \le \\ \le& \varepsilon^{-2} k\|h\|^2_D < \infty. \end{aligned} \tag{2.4}$$

Now it follows from the Borel–Cantelli lemma that for almost all $\omega \in \Omega$,

$$\lim_{n\to\infty} \int U_g h_k \nu_n(dg) = 0.$$

So we have (2.2) for the linear hull $\widetilde{D}$ of the set $\{h_k,\ h \in D_0,\ k \in \mathbb{N}\}$; of course, $\widetilde{D}$ is dense in D.

If $H_0 = H$ then $D_0 = D$ and (2.3) and (2.4) are valid for all $h \in D$; therefore in this case we can set $\widetilde{D} = \bigcup_{k=1}^{\infty} D_k$.

□

Corollary 2.2. *If* (2.1) (*or* (2.1‴) *in the commutative case*) *is fulfilled for all* $\lambda \in \Lambda\backslash\{\lambda_1\}$ *then the conclusion of Theorem* 2.1 *holds for any unitary representation* U.

2.2. Spectral condition in cyclic spaces. If H contains a cyclic vector h_0 we can set $H_0 = \{U_g h_0,\ g \in G\}$ in Definition 2.1. If we also assume $\|h_0(\lambda)\|_{H(\lambda)} = 1$, $\lambda \in \Lambda$, then the spectral condition can be rephrased as follows:

$$\sum_{n=1}^{\infty} \Big\| \int U_{gf}(\lambda) h_0(\lambda) \nu_n(dg) \Big\|_{H(\lambda)}^2 < \infty \quad \Phi\text{–a.e.} \tag{2.5}$$

for every $f \in G$. If μ is relatively invariant Borel measure on G and $\nu_n(B) = \frac{\mu(A_n \cap B)}{\mu(A_n)}$, relation (2.5), in turn, can be rewritten in the following form:

$$\sum_{n=1}^{\infty} \Big\| \frac{1}{\mu(A_n)} \int_{A_n f} U_g(\lambda) h(\lambda) \mu_n(dg) \Big\|_{H(\lambda)}^2 < \infty \quad \Phi\text{–a.e.} \tag{2.5'}$$

for any $f \in G$. We consider the measures $\nu_n^{(f)}(A) = \nu_n(f^{-1}A)$. The spectral condition can be stated in the following way: for any $f \in G$,

$$\sum_{n=1}^{\infty} \iint \varphi_{\lambda,h_0}(g_1^{-1} g_2) \nu_n^{(f)}(dg_1) \nu_n^{(f)}(dg_2) < \infty, \tag{2.6}$$

or

$$\sum_{n=1}^{\infty} \int \varphi_{\lambda,h_0}(g)(\bar{\nu}_n^{(f)} * \nu_n^{(f)})(dg) < \infty$$

for Φ–almost all $\lambda \neq \lambda_I$.

2.3. Spectral condition in the case of an L^α–spectrum (uniform weights). Let μ be a relatively invariant measure on G and let $\nu_{A_n}(B) = \frac{\mu(A_n \cap B)}{\mu(A_n)}$, $B \in \mathfrak{B}$, where A_n are integrable nonnull sets, $n = 1, 2, \ldots$. Thus we are concerned with means

$$f_n(\omega) = \frac{1}{\mu(A_n)} \int_{A_n} (U_g f)(\omega)\mu(dg).$$

For any unitary representation U the space $H = \sum_{i=1}^{\infty} \oplus H_i$, where each H_i is invariant and has a cyclic vector ($1 \leq n \leq \infty$). So we may assume that H has such a vector h. Then the spectral condition acquires the form (2.5′), or

$$\sum_{n=1}^{\infty} \frac{1}{[\mu(A_n)]^2} \int_{A_n f} \varphi_\lambda(g_1^{-1} g_2)\mu(dg_1)\mu(dg_2) < \infty \quad \Phi\text{–a.e.} \qquad (2.5'')$$

for any $f \in G$, where $\varphi_\lambda(\cdot)$ is the elementary positive definite function corresponding to $U(\lambda)$.

Lemma 2.3. *Let $\mu = \mu^{(l)}$, the left–Haar measure and $\{A_n\}$ and $\{B_n\}$ sequences of nonnull integrable sets. Denote*

$$\widehat{\varphi}_\lambda^{(n)} := \frac{1}{[\mu(A_n)]^2} \int_{B_n} \int_{B_n} \varphi_\lambda(g_1^{-1} g_2)\mu(dg_1)\mu(dg_2).$$

If $\varphi_\lambda(\cdot) \in L^\alpha(G, \mathfrak{B}, \mu)$ for some $\alpha \geq 1$, then

$$\widehat{\varphi}_\lambda^{(n)} \leq \frac{[\mu(B_n)]^{2-\frac{1}{\alpha}}}{[\mu(A_n)]^2} \|\varphi_\lambda\|_{L^\alpha}.$$

Proof. By using the Hölder inequality we get

$$\begin{aligned}\widehat{P}_{\lambda}^{(n)} \leq & \frac{[\mu(B_n)]^{2-\frac{2}{\alpha}}}{[\mu(A_n)]^2}\left[\int_{B_n}\int_{B_n}|\varphi(g_1^{-1}g_2)|^{\alpha}\mu(dg_1)\mu(dg_2)\right]^{\frac{1}{\alpha}} \leq \\ \leq & \frac{[\mu(B_n)]^{2-\frac{2}{\alpha}}}{[\mu(A_n)]^2}\left[\int\limits_{B_n}\mu(dg_2)\int_{G}|\varphi(g_1^{-1}g_2)|^{\alpha}\mu(dg_1)\right]^{\frac{1}{\alpha}} = \\ = & \frac{[\mu(B_n)]^{2-\frac{1}{\alpha}}}{[\mu(A_n)]^2}\|\varphi_{\lambda}\|_{L^{\alpha}}.\end{aligned}$$

□

Now we can state some simple consequences of Theorem 2.1. At first we use Lemma 2.3 with $B_n = A_n$.

Corollary 2.4. *Suppose that H contains a generating set H_0 such that for any $h \in H_0$, $\varphi_{h,\lambda} \in L^{\alpha}(G, \mathfrak{B}, \mu)$ with some $\alpha = \alpha(h, \lambda) \geq 1$. Denote $\nabla = \{\alpha(h, \lambda),\ \lambda \in \Lambda,\ h \in H_0\}$. If $\sum_{n=1}^{\infty}[\mu(A_n)]^{-\frac{1}{\alpha}}$ for $\alpha \in \nabla$, then $\{A_n\}$ satisfies the spectral condition and thus the conclusion of Theorem 2.1 is true for any U.*

Now we set $B_n = A_n f$, $f \in G$; then $\mu(B_n) = \Delta_{\mu}^{(r)}(f)\mu(A_n)$.

Corollary 2.5. *Assume that H posseses a cyclic element h such that $\varphi_{h,\lambda} \in L^{\alpha}(G, \mathfrak{B}, \mu)$ for some $\alpha = \alpha(\lambda) \geq 1$ $(\lambda \in \Lambda)$. Denote $\nabla =$ $= \{\alpha(\lambda) : \lambda \in \Lambda\}$. If $\sum_{n=1}^{\infty}[\mu(A_n)]^{-\frac{1}{\alpha}} < \infty$, $\alpha \in \nabla$, then $\{A_n\}$ satisfies the spectral condition for any U and the conclusion of Theorem 2.1 is true for any U.*

Corollary 2.6. *Denote $\Lambda^{(d)} := \{\lambda : U(\lambda)$ belongs to the discrete series$\}$. Suppose that $\Phi_U(\Lambda^{(d)}) = \Phi_U(\Lambda\backslash\{\lambda_1\})$. If $\sum_{n=1}^{\infty}[\mu(A_n)]^{-\frac{1}{2}} <$ $< \infty$, then $\{A_n\}$ satisfies the spectral condition for U.*

Proof. For any $\lambda \in \Lambda^{(d)}$ and any $h(\lambda) \in H(\lambda)$ we have $\varphi_{h,\lambda} \in$ $\in L^2(G, \mathfrak{B}, \mu)$. It remains to apply Corollary 2.5 with $H_0 = H$

and $\nabla = \{2\}$.

$\square$

2.4. Spectral condition for sets in almost simple Lie groups. Suppose that G is an almost simple group with finite center and consider a nontrivial unitary representation U of G in H. Let K be a maximal compact subgroup of G. Then $H = \sum_{i=1}^{\infty} \oplus H_i$ where H_i are K–invariant finite–dimensional subspaces of H (see, e.g., Pontryagin [1]). Let $H^{(k)} = \sum_{i=1}^{k} \oplus H_i$. Of course, $H^{(k)}(\lambda) := H^{(k)} \cap H(\lambda)$ are K–invariant subspaces in $H(\lambda)$. For any $h \in H$ we have $h_k := \Pi_{H^{(k)}}^{H} h \to h$ and $h_k(\lambda) = \Pi_{H^{(k)}(\lambda)}^{H(\lambda)} h(\lambda) \in H^{(k)}(\lambda)$. Thus, by Wallach's theorem (see Borel, Wallach [1] and Howe, Moore [1]), $\varphi_{\lambda,h_k} \in L^{\alpha}(G, \mathfrak{B}, \mu)$, for Φ_U–almost all $\lambda \in \Lambda$, where $\alpha = \alpha(\lambda, h_k)$. The set $H_0 = \{h_k,\ h \in H,\ k \in \mathbb{N}\}$ is dense in H and therefore Corollary 2.4 implies the following statement.

Corollary 2.7. *If* $\sum_{n=1}^{\infty} [\mu(A_n)]^{-\frac{1}{\alpha}} < \infty$ *for any* $\alpha \geq 1$, *then the conclusion of Theorem* 2.1 *holds.*

Example 2.4. Let $G = SO(m, 1)$ and $B_n = \{g : gx_0 \in \widetilde{B}_n\}$, where the $\widetilde{B}_n$ are balls of radii n with center x_0 in the Lobachevsky space L_m. Then $\mu(B_n) \asymp e^n$, and $\sum_{n=1}^{\infty} [\mu(B_n)]^{-\frac{1}{\alpha}} < \infty$ for any $\alpha > 1$.

§ 3. Pointwise Averaging with Regular Ergodic Nets of Sets

3.1. Main results. Let N be a linearly ordered set. Lemma 1.2, Proposition 5.1.1 and Corollary 5.5.2 imply the following statement.

Theorem 3.1. *Any piecewise left–regular weakly left–ergodic cofinally separable**) *net of sets* $\{A_n,\ n \in N\}$ *is pointwise left–averaging.*

---

*) Recall that $\{A_n,\ n \in N\}$ is cofinally separable if, e.g., N is countable.

The following statement follows from Theorem 5.2 and Notes 1.1–1.3 (it was proved in Tempelman [1] in the unimodular case and in Emerson [2] for general amenable groups).

Corollary 3.2. *Let X be an amenable locally compact group and let $\mu = \mu^{(l)}$ be its left Haar measure. Suppose a net $\{A_n,\ n \in N\}$ of integrable nonnull sets in X has the following properties:*

(i) $\lim\limits_{n\in N} \frac{\mu(A_n \Delta x A_n)}{\mu(A_n)} = 0,\ x \in X$ *(left ergodicity);*

(ii) $\{A_n\}$ *is increasing;*

(iii) $\sup\limits_{1\le n<\infty} \frac{\mu^*(A_n^{-1} A_n)}{\mu(A_n)} < \infty.$

Then $\{A_n,\ n \in N\}$ is a pointwise left-averaging net.

Recall that condition (i) is sufficient for $\{A_n\}$ to be a mean averaging net.

Ornstein and Weiss [1] have given another proof of this theorem for discrete groups.

Note 3.1. It seems rather doubtful that nets possessing both properties *(i)* and *(iii)* exist in non-unimodular groups (such nets are unknown even in the group $\mathbb{R} \odot \mathbb{R}^{\times}_{++}$). Nevertheless, Emerson and Greenleaf [1] have given a construction of L^{1+}–pointwise averaging sequences of sets that is suitable for any connected amenable locally compact group; in § 8 such sequences will be constructed in any connected locally compact group.

We shall state now two simple consequences related to the case $X = \mathbb{R}^m$ and the nets considered in Examples 5.2.9 and 5.2.10.

Corollary 3.3. *Any net of bounded convex sets $\{A_n,\ n \in N\}$ in $\mathbb{R}^m$ with $r(A_n) \to \infty$ is a mean averaging net; under the additional condition of monotonicity it is also pointwise averaging.*

Corollary 3.4. *A net of sets $A_n = t_n(A - x_0) + x_0$ in $\mathbb{R}^m$, which are homothetic with respect to a bounded non-zero set A, with $t_n \uparrow \infty$ is mean and pointwise averaging.*

As a special case of these results we obtain the law of large numbers for independent identically distributed B–valued random variables $\xi_1, \xi_2, \ldots$ (see Grenander [2], Beck [1–3], Mourier [1,2]).

The following interesting consequence of Corollary 3.2 is implied by Note 5.2.1.

Corollary 3.5. *If $X = \cup_{i=1}^{\infty} A_n$ where $\{A_n\}$ is an increasing sequence of compact subgroups, then $\{A_n\}$ is a mean and pointwise left– and right–averaging sequence in X.*

Note 3.2. There is another way to prove this corollary, based on the martingale convergence theorem. Let $\{A_n\}$ be as in Corollary 3.5 and $m(\Omega) = 1$. Let $\{T_x : x \in X\}$ be a dynamical system. Denote by $\mathfrak{I}_n$ the σ–algebra of sets invariant under the dynamical system $\{T_x : x \in \in A_n\}$. Obviously $\mathfrak{I}_n$ is a decreasing sequence. According to Theorem 1.5.7 we have $\frac{1}{\mu(A_n)} \int_{A_n} f(T_x\omega)\mu(dx) = \mathbf{M}(f|T) = \mathbf{E}(f|\mathfrak{I}_n)$. Hence in the present case the maximal, mean and pointwise ergodic theorems also follow from well–known theorems on the convergence and dominants of conditional mathematical expectations (martingales) (see Doob [1], Chatterji [1] and Scalora [1]). This interesting relation between ergodic theorems and theorems on the convergence of martingales, in the case when X is the group of proper dyadic fractions with the usual addition mod 1 and A_n is the cyclic subgroup of the elements $1/2^n$, was observed by Pitckel' and Stepin [1].

We conclude this subsection by an example of a non–monotone sequence of intervals which is, nevertheless, pointwise averaging.

Example 3.1. Let $X = \mathbb{Z}$; the sequence of intervals $[2^{2^k}, 2^{2^k} + 2^{2^{k-1}}]$ is regular (see Example 5.2.3) and ergodic, therefore it is pointwise averaging.

This was shown by another method in Below, Jones, Rosenblatt [1] where various conditions are found ensuring that "moving" intervals in $\mathbb{Z}$ are pointwise averaging.

3.2. Counterexamples. In this subsection we shall justify the restrictions imposed on the sequence $\{A_n\}$ in Theorem 3.1 and Corollary

3.2. Clearly, our aim will be fulfilled if we show that in the absence of any one of the conditions, Corollary 3.2 is false.

Example 3.2. We show that Corollary 3.2 is not true without the assumption *(i)*. Let $X = Z_+$, let μ be the counting measure, let $N = \{1, 2, \ldots\}$, and let $A_n = \{1, 2, \ldots, 2^n\}$ if n is even and $A_n = \{1, 2, \ldots\} \cup \{2^n + 2i,\ i = 1, \ldots, 2^{n-1}\}$ if n is odd. It is clear that $\{A_n\}$ satisfies the conditions *(ii)* and *(iii)* and that $A_n \uparrow Z_+$; however, $\{A_n\}$ is not even weak ergodic: if D is the set of all even numbers and n and x are odd,then $D - x$ is the set of odd numbers, and

$$\frac{\mu(A_n \cap D)}{\mu(A_n)} = \frac{2}{3}, \quad \frac{\mu(A_n \cap (D - x)}{\mu(A_n)} = \frac{1}{3}.$$

Let ξ_0 be a random variable taking the values -1 or $+1$ each with probability $\frac{1}{2}$, and let $\xi_i = (-1)^i \xi_0$, $i = 1, 2, \ldots$. Clearly $\xi_0, \xi_1, \ldots$ is a stationary random sequence and $\widehat{\xi}_n = 0$ for n even and $\widehat{\xi}_n = \xi_0/3$ for n odd. Clearly $\lim_{n\to\infty} \widehat{\xi}_n$ exists neither in the sense of convergence almost everywhere nor in the sense of L^α–convergence.

Example 3.3. Let $X = Z_+$, let μ be the counting measure, let $\xi_0, \xi_1, \ldots$ be a sequence of independent random variables on the probability space $(\Omega, \mathfrak{S}, m)$ with $m(\{\omega : \xi_i = -1\}) = m(\{\omega : \xi_i = +1\}) = \frac{1}{2}$ for all i, let $N = \{2, 3, \ldots\}$, and let $A_n = [a_n, b_n]$ be a sequence of disjoint intervals of Z_+ of "length" $\mu(A_n) = b_n - a_n + 1 = [\ln n]$. This sequence satisfies condition *(i)* and *(iii)*; *(ii)* is not satisfied; furthermore, this sequence is not regular (see § 5.2). We have

$$\widehat{\xi}_n = \frac{1}{[\ln n]} \sum_{k=a_n}^{b_n} \widehat{\xi}_k, \quad \mathbf{E}\widehat{\xi}_n = 0, \quad \mathbf{E}(\widehat{\xi}_n)^2 = \frac{1}{[\ln n]}.$$

Thus $\lim_{n\to\infty} \mathbf{E}(\widehat{\xi}_n)^2 = 0$; that is, $\lim_{n\to\infty} \widehat{\xi}_n = 0$ in L^2, in complete agreement with Theorem 5.2.1. Furthermore,

$$m(\{\omega : \widehat{\xi}_n = 1\}) = m(\{\omega : \xi_i = 1, \quad i \in [a_n, b_n]\}) = \frac{1}{2^{[\ln n]}},$$

and therefore $\sum_{n=1}^{\infty} m(\{\omega : \widehat{\xi}_n = 1\}) = \infty$. Since the random variables $\widehat{\xi}_n$ are independent, by the Borel–Cantelli Lemma we have $\widehat{\xi}_n = 1$ infinitely often with probability *1*. At the same time $|\widehat{\xi}_n|$ converges in probability to 0. From this it follows with probability *1* that the limit $\lim_{n\to\infty} \widehat{\xi}_n$ does not exist.

There are other examples which show the essence of the monotonicity condition *(ii)* in Corollary 3.2. For example, Akcoglu and del Junco [1] have shown that the sequence of intervals $\{[k, k+\sqrt{k}]\}$ is not pointwise averaging; Bellow and Losert [1] have shown the same for $\{[4^k, 4^k + 2^k]\}$. In § 5.2 we have seen that these sequences are not regular (they do not satisfy the weak monotonicity condition (EM)).

In the beginning of Subsect. 3.1 we have assumed that N is linearly ordered. Therefore the following modification of Example 3.3 is of interest. We put $B_n = [0, [\ln \ln n]] \cup A_n$ $(n \geq 3)$ and $N = \{B_n\}$, and (partially) order N by set inclusion. Thus N is a directed set and the net $\{A,\ A \in N\}$ satisfies the conditions *(i)*–*(iii)*; however, with probability *1*,

$$\lim_{A\in N} \frac{1}{\mu(A)} \sum_{i\in A} \xi_i$$

does not exist. Thus the individual ergodic theorem (Theorem 6.1) does not hold when N is a directed but not linearly ordered set; the mean ergodic theorem (Theorem 5.2.1) is true for this case.

Note 3.2 shows a connection between condition *(ii)* and the monotonicity condition on the sequence of σ–algebras in the theorem about the convergence of conditional expectations.

Example 3.4. Let $X = Z_+^2$, let μ be the counting measure, let $N = \{2, 3, \ldots\}$, let $S_k = [a_k, b_k] \times [0, h_k]$, where $h_k = [e^{k^2}]$ and a_k and b_k are as in the previous example, and let $A_n = \bigcup_2^n S_k$. Condition *(i)* is satisfied because the sequence $\{S_n\}$ satisfies the condition and

$$\lim_{n\to\infty} \frac{\mu(A_n)}{\mu(S_n)} = \lim_{n\to\infty} \frac{\sum_{k=1}^{n} e^{k^2} \ln k}{e^{n^2} \ln n} = 1;$$

condition *(ii)* is also satisfied; condition *(iii)* is not satisfied. We take the sequence $\xi_0, \xi_1, \ldots$ as in the previous example and put $\xi(i,j) = \xi_i$ for all $(i,j) \in Z^2_+$. Along with $\widehat{\xi}_n$ we consider the random variables

$$\eta = \frac{1}{\mu(A_n)} \sum_{(i,j)\in A_{n-1}} \xi(i,j)$$

and

$$\zeta_n = \frac{1}{\mu(A_n)} \sum_{(i,j)\in S_n} \xi(i,j);$$

it is clear that $\widehat{\xi}_n = \eta_n + \zeta_n$. With probability *1* we have

$$|\eta_n| \leq \frac{\mu(A_{n-1})}{\mu(A_n)} \quad \text{and} \quad \zeta_n \sim \frac{1}{[\ln n]} \sum_{k=a_n}^{b_n} \xi_k.$$

Consequently $\eta_n \to 0$ with probability *1* and $\lim_{n\to\infty} \zeta_n$ does not exist (see Example 3.3). So with probability *1* the limit $\lim_{n\to\infty} \widehat{\xi}_n$ also does not exist. Clearly this is also the case if $A_n = \bigcup_{k=2}^n S_k \cup B_n$, where $B_n = [0,n] \times [0,n]$; in this case $A_n \uparrow Z^2_+$ (another counterexample to condition *(iii)* is given in Greenleaf [3]).

§ 4. Ergodic Theorems with Regular Ergodic Weights

So far we have considered averages of realizations obtained using the simplest uniform "weights" $\nu_n : \nu_n(B) = \frac{\mu(A_n \cap B)}{\mu(A_n)}$. However, the ergodic theorems proved above automatically yield generalizations concerning averages with "weights" of a more general form. Let $\{T_x, x \in X\}$ be a dynamical system. We consider the auxiliary dynamical system $\{T'_y, y \in Y\}$, where $Y = X \times \mathbb{R}$ and $T'_y = T_x$ for $y = (x,z)$ with $x \in X$ and $z \in \mathbb{R}$. Let $\{\Theta_n(x), n \in N\}$ be an arbitrary cofinally separable net of nonnegative integrable functions on X, $\bar{\mu} = \mu \times l$, where l is the Lebesgue measure on R, and $A'_n = \{y = (x,z) : 0 < z \leq \leq m_n \Theta_n(x)\}$, where m_n is some sequence of positive numbers. For every $f \in L^p_B$ we have

$$\frac{1}{\bar{\mu}(A'_n)} \int_{A'_n} f(T'_{(x,z)}\omega)\mu(dx)l(dz) = \int_X f(T_x\omega)\Theta^*_n(x)\mu(dx)$$

where $\Theta^*_n(x) = \Theta_n(x) \cdot \left[\int_X \Theta_n(x)\mu(dx) \right]^{-1}$.
On applying Theorem 3.1 to the dynamical system $\{T'_y,\ y \in Y\}$ we get the following result.

Theorem 4.1. *If the function f and (for some choice of m_n) the net $\{A'_n,\ n \in N\}$ satisfy the conditions of Theorem 3.1, then the following limit exists a.e.:*

$$\lim_{n\in N} \int_X f(T_x\omega)\Theta^*_n(x)\mu(dx) = \mathbf{M}(f|T).$$

Note that if X is a unimodular locally compact group, then for $m_n \to \infty$ condition (E) on the sequence $\{A'_n\}$ is equivalent to the condition

$$\lim_{n\to\infty} \int |\Theta^*_n(y) - \Theta^*_n(xy)|\mu(dy) = 0 \quad \text{for all} \quad x \in X.$$

We generalize Corollaries 3.3 and 3.4 in the same manner.

Corollary 4.2. *Let $\{\Theta_n,\ n \in N\}$ be a net of nonnegative functions on $\mathbb{R}^m$ $(m \geq 1)$ satisfying the following conditions:*

1) For each n the support A_n of Θ_n is a convex body.

2) For each n the function Θ_n on A_n is convex upwards.

3) $\lim_{n\in N} r(A_n) = \infty$.

4) There exists a sequence of positive numbers $\{k_n,\ n \in N\}$ such that $k_{n_1}\Theta_{n_1}(x) \leq k_{n_2}\Theta_{n_2}(x)$ for all $x \in X$ and $n_1 < n_2$.

If $f \in L^\alpha_B$ $(\alpha \leq 1)$, then the limit

$$\lim_{n\in N} \int_{\mathbb{R}^m} f(T_x\omega)\Theta^*_n(x)dx = \mathbf{M}(f|T) \tag{4.1}$$

exists for almost all $\omega \in \Omega$ and is finite; if conditions 1)–3) are satisfied and $f \in L^\alpha_B$ for $\alpha > 1$ or for $\alpha = 1$ and $m(\Omega) < \infty$, then the limit (4.1) exists in the sense of L^α_B-convergence.

Corollary 4.3. *Let $\Theta(x)$, $x \in \mathbb{R}^m$, be a bounded nonnegative measurable function with bounded support A for which $\int_{\mathbb{R}^m} \Theta(x)dx = 1$; let $\{t_n, n \in N\}$ be a sequence of positive numbers with $\lim_{n\in N} t_n = \infty$. If $f \in L^\alpha_B$ $(\alpha \geq 1)$, then the limit*

$$\lim_{n\in N} \frac{1}{t_n} \int_{\mathbb{R}^m} f(T_x\omega)\Theta(\frac{x}{t_n})dx = \mathbf{M}(f|T)$$

exists for almost all $\omega \in \Omega$. If $f \in L^\alpha_B$ $(\alpha > 1$ or $\alpha = 1$ and $m(\Omega) < < \infty)$, then this limit exists in the sense of convergence in L^α_B.

Proof. We put $\Theta_n(x) = \Theta(x/t_n)$ and $m_n = t_n$. Since in this case the sets A'_n are homothetic with respect to the set $A' = \{(x,z) : 0 < z \leq \leq \Theta(x)\}$, it remains to apply Corollary 3.4.

□

§ 5. Pointwise Averaging with Spheres and Spherical Layers

Lemma 1.2 allows one to combine the results of § 5.4 and § 2 and get pointwise ergodic theorems for spheres and other singular and non-regular sets.

At first we present a version of a result due to R. Jones [1].

Theorem 5.1. *Let σ_r be the rotation invariant probability measure on spheres S_r in $\mathbb{R}^m$, $m \geq 3$, and $\sum_{n=1}^\infty r_n^{-(m-1)} < \infty$. Then for any dynamical system T in $(\Omega, \mathcal{F}, m)$ and any $f \in L^\alpha(\Omega, \mathcal{F}, m)$, $\alpha > \frac{m}{m-1}$, the following limit exists m–a.e.:*

$$\lim_{n\to\infty} \int f(T_x\omega)\sigma_{r_n}(dx) = \mathbf{M}(f|T). \tag{5.1}$$

Proof. In view of Lemma 2.2, Theorem 5.3 and Example 2.2 imply our statement for $\alpha = 2$. Since $L^2 \cap L^\alpha$ is dense in L^α, it remains to use Theorem 5.3 and Lemma 2.2 again.

□

Using another approach, R. Jones [1] has shown that (5.1) is true for the whole net $\{S_r, r \in \mathbb{R}\}$.

The same arguments as used in proof of Theorem 5.1 (with Theorem 5.5.4 instead of Theorem 5.5.3) prove the following:

Theorem 5.2. *(i) Let $L_{r,d(r)} = \{x : x \in \mathbb{R}^m,\ r - d(r) \leq \|x\| \leq r\}$, $m \geq 3$; suppose that $\sum_{n=1}^{\infty} r_n^{-m-1} < \infty$. Then for any completely measurable dynamical system T and any $f \in L^\alpha\ (\Omega, \mathcal{F}, m)$, $\alpha > \frac{m}{m-1}$, the following limit exists m–a.e.:*

$$\lim_{n\to\infty} \frac{1}{l(L_{r_n,d(r_n)})} \int_{L_{r_n,d(r_n)}} f(T_x\omega)dx = \mathbf{M}(f|T).$$

Corollary 5.3. *Consider in $\mathbb{R}^m$ a net of spherical layers $\{L_{r,d(r)}, r > 0\}$. Suppose that $d_0 := \inf\{d(r),\ r > 0\} > 0$ and there is some $\beta > 0$ such that on the subdirection $N_\beta := \{r : d(r) < \beta r\}$ the Hölder condition is fulfilled for some γ, $0 < \gamma \leq 1$:*

$$|d(r_1) - d(r_2)| \leq C|r_1 - r_2|^\gamma, \quad r_1, r_2 \in N_\alpha.$$

Then for any completely measurable dynamical system T and any $f \in L^\alpha\ (\Omega, \mathcal{F}, m)$, $\alpha > \frac{m}{m-1}$,

$$\lim_{r\to\infty} \frac{1}{l(L_{r,d(r)})} \int_{L_{r,d(r)}} f(T_x\omega)dx = \mathbf{M}(f|T) \quad m\text{–a.e.} \tag{5.4}$$

Proof. The net $\{L_{r,d(r)}, r \in N\backslash N_\alpha\}$ is regular and ergodic, therefore, by Theorem 3.1 the equality (5.4) holds when r runs over $N\backslash N_\beta$. So we consider the subnet N_β. We denote $N = N_\beta$. According to Theorem 5.2 any subset $\{L_{\delta_n,d(\delta n)},\ n = 1, 2, \ldots\}$ is pointwise averaging in L^α, $\alpha > \frac{m}{m-1}$. So, by App., (13.G), it remains to show that for some $\delta = \delta_\varepsilon$ the sequence $\{L_{\delta_n,d(\delta n)}\}$ is an ε–subnet. We shall use the following estimates for $l(L_{r,d(r)})$: $\gamma_m d(r)(r - d(r))^{m-1} \leq l(L_{r,d(r)}) \leq$

$\leq \gamma_m d(r) r^{m-1}$. Let $r_l > 0$ and $0 < \delta < 1$. If $(n-1)\delta < r < n\delta$, we have

$$l(L_{r,d(r)} \Delta L_{n\delta, d(n\delta)}) \leq l(L_{n\delta,r} \cup L_{\rho_{n,\delta,r}, d_{n,\delta,r}})$$

where $\rho_{n,\delta,r} := \max\{\delta nk - d(\delta n),\ r - d(r)\} \leq r + \delta$; $d_{n,\delta,r} := |\delta n - -r + d(r) - d(\delta n)| \leq C\delta^\gamma + \delta < (C+1)\delta^\gamma$. Thus $l(L_{r,d(r)} \Delta L_{n\delta,d(n\delta)}) \leq \leq \gamma_m(\delta(r+\delta)^{m-1} + (C+1)\delta^\gamma (r+\delta)^{m-1}) \leq (C+2)\gamma_m \delta^\gamma (r+1)^{m-1}$. Now, $l(L_{r,d(r)}) \geq \gamma_m (1-\beta)^{m-1} d_0 r^{m-1}$. So we get

$$\rho'(L_{r,d(r)}, L_{n\delta,d(n\delta)}) \leq \frac{(C+2)\delta^\gamma (r+1)^{m-1}}{(1-\beta)^{m-1} r^m d_0}$$

where ρ' is the semimetrics defined in App., §13.
Therefore we can choose a δ_ε such that $\{L_{n\delta_\varepsilon, d(n\delta_\varepsilon)},\ n > \delta_\varepsilon^{-1}\}$ is an ε–subnet of the net $\{L_{r,d(r)},\ r \geq 1\}$.

□

§ 6. Pointwise Averaging with Convolution Weights

Consider a separable topological semigroup X. Let $q \in \mathcal{P}(\mathfrak{B})$, $\lambda_n^{(q)} = \frac{1}{n}\sum_{k=1}^n q^k$; Y_q denotes the r–full subsemigroup generated by the carrier $c(q)$ of q.

Theorem 6.1. *Let $x \to S_x$ be a strongly contracting right representation of X in $L^1(\Omega, \mathcal{F}, m)$. Then:*

(i) $\{\lambda_n^{(q)}\} \in \mathrm{MAX}_\alpha(S)$.

(ii) If $Y_q = X$, then for any $f \in L^\alpha(\Omega, \mathcal{F}, m)$, $\alpha \geq 1$ the following limit exists m–a.e.:

$$\lim_{n\to\infty} \int (fS_x)(\omega) \lambda_n^{(q)}(dx) = \mathbf{M}(f). \tag{6.1}$$

Proof. Consider the operator S_q:

$$(fS_q)(\omega) = \int (fS_x)(\omega) q(dx).$$

It is easy to check that $(fS_q^k) = \int (fS_x)(\omega) q^k(dx)$ and thus $\frac{1}{n}\sum_{k=1}^{n}(fS_q^k)(\omega) = \int (fS_x)(\omega)\lambda_n^{(l)}(dx)$; S_q is a strong contraction, and therefore statement *(i)* and the convergence in (6.1) are implied by the Dunford–Schwartz ergodic theorem (see Introduction, §2). By Theorem 4.1.5 this limit coincides with $\mathbf{M}(f)$.

□

Note 6.1. Being a simple consequence of the classical Dunford–Schwartz theorem, Theorem 6.1 not only includes this theorem as a special case ($X = \mathbb{Z}_+$, $q(\{1\}) = 1$, $q(\{k\}) = 0$ if $k \neq 1$) but also leads to its generalizations to other semigroups and groups in the same "classical" form - with uniform weights. We illustrate this possibility with the case $X = \mathbb{R}^m$ and balls $B_r^+ = B_r \cap \mathbb{R}^m$. Let q be the standard Gaussian measure with distribution density $\varphi(x) = (2\pi)^{-\frac{m}{2}}\exp\{-\frac{1}{2}|x|^2\}$. Then q^k has the density $\varphi_k(x) = (2\pi k)^{-\frac{m}{2}}\exp\{-\frac{1}{2k}|x|^2\}$. It is easy to check that $\psi_{n^2}(x) \geq C_m r^{-m}\mathcal{X}_{B_r}(x)$ for $n-1 \leq r \leq n$. Indeed, if $|x| \leq n$, then

$$n^m\psi_{n^2}(x) \geq n^2\psi_{n^2}(n) =$$

$$= \frac{1}{n^2}\sum_{k=1}^{n^2}\left(\frac{n^2}{k}\right)^{\frac{m}{2}}\exp\left(-\frac{n^2}{2k}\right) \overset{n\to\infty}{\longrightarrow} \int_0^1 t^{-\frac{m}{2}}\exp\left(-\frac{t}{2}\right)dt.$$

Therefore the majorization principle (Proposition 5.1.6) implies that $\{B_r,\ r > 0\} \in \mathrm{MAX}_\alpha(S)$ with respect to the Lebesgue measure l ($l \geq 1$). Certainly, the same is true for any other net $\{A_r\} \prec \{B_r\}$ (e.g. the net of cubes $P_r = \{x : -r \leq x_i \leq r\}$ is regular with respect to the net $\{B_{\frac{r\sqrt{m}}{2}},\ r > 0\}$). Now we suppose that the net $\{A_r\}$ is also ergodic. Consider any simple function $f(\omega)$ and put $h = f - fS_y$. Then

$$\left|\frac{1}{l(A_r)}\int\limits_{A_r}(hS_x)(\omega)dx\right| =$$
$$=\frac{1}{l(A_r)}\left|\int_{A_r}(fS_x)(\omega)dx - \int_{A_r}(fS_{x+y})(\omega)dx\right| \leq$$
$$\leq\frac{1}{l(A_r)}\int_{A_r\Delta(A_r+y)}|(fS_x)(\omega)|dx \leq$$
$$\leq\frac{l(A_r\Delta(A_r+y))}{l(A_r)}\sup_{\omega\in\Omega}|f(\omega)|.$$

Thus

$$\lim_{r\to\infty}\frac{1}{l(B_r)}\int_{B_r}(hS_x)(\omega)dx = 0 \quad m\text{—a.e.} \tag{6.2}$$

So we have established (6.2) for a generating set of functions in D^α ($\alpha > 1$). Since $\{A_r\} \in \mathrm{MAX}_\alpha(S)$, Lemma 1.1 implies that (6.2) is true for any $f \in D^\alpha$, and hence for all $f \in L^\alpha$, $\alpha > 1$,

$$\lim_{r\to\infty}\int\limits_{A_r}(fS_x)(\omega)dx = \mathbf{M}(f) \quad m\text{—a.e.} \tag{6.3}$$

Since $L^2 \cap L^1$ is dense in L^1 one more application of Lemma 1.1 shows that (6.3) is true for $\alpha = 1$ too.

In the rest of this section we assume that X is a left–regular separable locally compact topological semigroup.

Corollary 6.2. *Let B be a Banach space. If $Y_q = X$ for any completely measurable dynamical system $\{T_x,\ x \in X\}$ in $(\Omega, \mathcal{F}, m)$ and any $f \in L^\alpha_B$, $\alpha \geq 1$, the following limit exists m–a.e.:*

$$\lim_{n\to\infty}\int f(T_x\omega)\lambda_n^{(q)}(dx) = \mathbf{M}(f|T).$$

Proof. The validity of this statement with $B = \mathbb{C}$ follows from Theorem 6.1. Assertion *(i)* of Lemma 1.2 shows that $\{\lambda_n^{(q)}\}$ satisfies condition *3)* of this lemma; condition *2)* for any $\alpha \geq 1$ is implied by Theorem 6.1. So it remains to apply statements *(iii)* and *(iv)* of Lemma 1.2. □

Note 6.2. If we suppose that Y_q is an l–full subsemigroup, we obtain the "right" versions of Theorem 6.1 and Corollary 6.2 for right objects: left representations and right dynamical systems. Examples 4.1.1 and 4.1.2 show two ways of constructing measures q such that Y_q is both r– and l–full.

Finally, in Notes 6.2–6.4 we discuss the possibility of pointwise averaging by "pure" convolutions of measures q. These Notes are "pointwise" version of Notes 4.1.5–4.1.7.

Note 6.3. We use the notations introduced in Note 4.1.5; in addition we put
$L^{1+}(\Omega, \mathcal{F}, m) := \bigcup_{1<\alpha<\infty} L^{\alpha}(\Omega, \mathcal{F}, m)$ if $m(\Omega) = \infty$
and
$L^{1+}(\Omega, \mathcal{F}, m) := \{f : \int |f| \log^{+} |f| dm < \infty\}$ if $m(\Omega) < \infty$.
As in Note 4.1.5, Rota's theorem implies that for any dynamical system $\{T_g,\ g \in G\}$ the limit $\lim_{n\to\infty} \int f(T_g\omega)\nu_n(dg)$ exist m–a.e.

Note 6.4. Under the condition of Note 4.1.6, for any $\{T_g,\ g \in G\}$ and any $f \in L^{1+}$ we have m–a.e.: $\lim_{n\to\infty} \int f(T_g\omega)q^n(dx) = \mathbf{M}(f)$; this is a simple consequence of Note 6.1 and Corollary 6.2.

Note 6.5. **(Oseledec [1])** Under the conditions of Note 4.1.7,

$$\lim_{n\to\infty} \int f(T_g\omega)q^n(dg) = \mathbf{M}(f) \quad m\text{–a.e.}$$

for any $f \in L^{1+}$ iff $\mathfrak{J}_{c(q)\cdot c(q)}(T) = \mathfrak{J}_{c(q)}(T)$.

§ 7. General Ratio Ergodic Theorems

7.1. Krengel's counterexample. Let X be a separable locally compact topological semigroup and μ the left Haar measure on G. Consider a completely m–measurable dynamical system $\{T_x,\ x \in X\}$ in a σ–finite measure space $(\Omega, \mathcal{F}, m)$. One might hope to generalize the Hopf ergodic theorem (see Introduction, Theorem 1.3. A) in the following way. Let $\{A_n\}$ be some "good" sequence of sets and let $f, g \in L^1(\Omega, \mathcal{F}, m)$, $g(x) > 0$ a.e.; then

$$\lim_{n\to\infty} \frac{\int_{A_n} f(T_x\omega)\mu(dx)}{\int_{A_n} g(T_x\omega)\mu(dx)}$$

exists and is finite m–a.e. But the following counterexample, due to Krengel [1], shows that such an "unweighted" version of the Hopf ergodic theorem is false, even in the case $X = \mathbb{Z}^2$.

Example 7.1. Let $X = \mathbb{Z}^2$, $\Omega = \mathbb{Z}$ and let m be the counting measure on $\mathbb{Z}$. Define $T_{(i,j)}\omega = \omega + i + j$, $(i,j) \in \mathbb{Z}^2$. Consider the functions $g(\omega) = |\omega|^{-3}$, $f(\omega) = \omega^{-2}$ and the sets $A_n := \{(i,j) : 0 \le i, j \le$ $\le n\}$. It is easy to check that $\lim_{n\to\infty} \sum_{x\in A_n} f(T_x\omega) = \infty$ (although the dynamical system is "dissipative"!) but $\lim\limits_{n\to\infty} \sum_{x\in A_n} g(T_x\omega) < \infty$.

Moreover, the notion of conservativity is closely connected with the one–dimensional time, and difficulties arise if we try to generalize this notion to the multiparameter case.

Nevertheless, we shall see that conservativity can be naturally defined and the ratio ergodic theorems are valid if we use appropriate convolution weights.

7.4. General convergence theorem. Let X be a separable locally compact semigroup. Let q be a Borel probability measure on X and $\lambda_n := \sum_{k=1}^n q^k$.

Consider some positive contractive right representation of X in $L^1(\Omega, \mathcal{F}, m)$. Clearly, the operator $S_q := \int S_g q(dg)$ is a positive contraction and $\sum_{k=1}^n S_q^k = \int S_g \lambda_n(dg)$. Therefore the Chacon–Ornstein theorem (see Introduction, Theorem 2.2. A) implies the following statement.

Theorem 7.1. *Let $g \in L^1_+(\Omega, \mathcal{F}, m)$. Suppose that S is a measurable*

positive contractive right representation of X *in* Ω.
Let $\lim_{n\to\infty}\int(gS_x)(\omega)\lambda_n(dx) > 0$ *m–a.e.. If* $f \in L^1(\Omega, \mathcal{F}, m)$ *then the limit*

$$\lim_{n\to\infty}\frac{\int fS_x\lambda_n(dx)}{\int gS_x\lambda_n(dx)} := h_{f,g}(\omega) \tag{7.1}$$

exists m–a.e. and is finite.

If $\alpha > 1$ *and* $\frac{f}{g} \in L^\alpha(\Omega, \mathcal{F}, m_g)$, *then* (7.1) *holds also in the sense of convergence in* $L^\alpha(\Omega, \mathcal{F}, m_g)$ (*here* $m_g(d\omega) = g(\omega)m(d\omega)$).

If $\{T_x,\ x \in X\}$ is a completely measurable dynamical system, then (7.1) can be written in the following form:

$$\lim_{n\to\infty}\frac{\int f(T_x\omega)\lambda_n(dx)}{\int g(T_x\omega)\lambda_n(dx)} =: h_{f,g}(\omega) \tag{7.1'}$$

Note that for $X = \mathbb{Z}_+$ we can put here $q(\{1\}) = 1$, then (7.1) and (7.1′) reduce to the Chacon–Ornstein and Hopf theorems, respectively.

7.5. Identification of the limit. Conservative representations and conservative dynamical systems.

Definition 7.1. A set Λ is said to be *S–absorbing* if it is absorbing for any S_x, $x \in X$. We denote by $\mathcal{A}$ the class of all S–absorbing sets.

Let Y_q denote the r–full semigroup generated by the carrier $c(q)$ of q.

Lemma 7.2. *Suppose* $Y_q = X$. *A set* Λ *is S–absorbing iff it is* S_q*–absorbing.*

Proof. Let Λ be S–absorbing. Then $f(\omega)\mathcal{X}_\Lambda = 0$ m–a.e. implies $(fS_x)(\omega)\mathcal{X}_\Lambda(\omega) = 0$ m–a.e.; we have

$$(fS_q)(\omega)\mathcal{X}_\Lambda(\omega) \equiv \int (fS_x)(\omega)\mathcal{X}_\Lambda(\omega)q(dx) = 0 \quad m\text{–a.e.} \tag{7.2}$$

So any S–absorbing set is S_q–absorbing.

Now suppose Λ is S_q–absorbing and $f \in L^1_+$. (7.2) implies
$(fS_x)(\omega)\mathcal{X}_\Lambda(\omega) = 0$ q–a.e. for m–almost all $\omega \in \Omega$.
Since the representation S is continuous, $(fS_x)(\omega)\mathcal{X}_\Lambda(\omega) = 0$ for all $x \in c(q)$. Thus Λ is S_x–absorbing for all $x \in c(q)$ and, as is easy to verify, also for all $x \in X$.

□

Definition 7.2. We say that a continuous positive contractive representation $S : x \to S_x$ is *q–conservative* if $\lim_{n\to\infty} \int (fS_x)(\omega)\lambda_n(dx) = \infty$ m–a.e. for any $f \in L^1_+(\Omega, \mathcal{F}, m)$ as soon as $\lim_{n\to\infty} \int (fS_x)(\omega)\lambda_n(dx) > 0$ m-a.e.; S is *q–dissipative* if $\lim_{n\to\infty} \int (fS_x)(\omega)\lambda_n(dx) < \infty$ for every $f \in L^1_+$.

This is clearly equivalent to the conservativity (dissipativity) of the operator S_q. A theorem of Hopf (see Introduction, Theorem 2.2. B) immediately implies the following statement.

Theorem 7.2. *Consider some $q \in \mathcal{P}(\mathfrak{B})$ with $Y_q = X$. For any continuous positive contractive representation S there is a unique decomposition $\Omega = C \cup D$ with the following properties:*

a) $C \cap D = \emptyset$;

b) $C \in \mathcal{A}$;

c) the restriction of S to $L^1(C)$ is q–conservative;

d) $\lim_{n\to\infty} \int (fS_x)(\omega)\lambda_n(dx) < \infty$ for m–almost all $\omega \in D$ ($f \in L^1_+(\Omega)$).

Theorem 7.2.5 reduces the problem of identification of the limit $h_{f,g}$ to the "conservative" part $C \subset \Omega$. The answer is given by the following simple consequence of Theorem 2.2. C (see Introduction).

Theorem 7.3. *Let $Y_q = X$; then for any q–conservative continuous positive contractive representation S the equality $h_{f,g} = E_g(\frac{f}{g}|\mathcal{A})$ holds; thus for any q–conservative dynamical system T we have $h_{f,g} = \mathbf{E}_g(\frac{f}{g}|\mathfrak{J})$ where $\mathfrak{J}$ is the σ–algebra of T–invariant sets and $\mathbf{E}_g(\cdot|\mathfrak{J})$ is the conditional expectation with respect to the measure $m_g(d\omega) = g(\omega)(d\omega)$.*

7.6. Dynamical systems: conservativity and recurrence; conservatizing weights. We shall discuss here the connection between q–conservativity of a dynamical system T and the recurrence properties of the random walk (q, T) associated with T and q.

Any Borel probability measure q defines a random walk on X: at the time moment $t = n$ the wandering "particle" is at the point $\eta_n = \xi_1 \cdot \ldots \cdot \xi_n$, where the random "steps" ξ_i are independent random elements in X with distribution q. A dynamic system T induces the corresponding random walk (q, T) on the orbit $O(\omega|T)$ of any point $\omega \in \Omega$: at $t = 0$ the "particle" is at the point ω and at $t = n$ it is a $T_{\eta_n}\omega$.

Definition 7.3. We say that the random walk (q, T) is *recurrent* if for m–almost all $\omega\,(\in \Omega)$ and for any $\Lambda\,(\in \mathcal{F})$ the wandering "particle" on $O\,(\omega|T)$ either misses Λ with probability *1* or visits Λ infinitely often with probability *1*.

Example 7.1. If $X = \mathbb{Z}_+$ and $q(\{1\}) = 1$, then the walk (q, T) is non–random; it is recurrent iff the endomorphism T is conservative.

We shall prove now that the same is true in the general case.

Proposition 7.4. *The dynamical system T is q–conservative iff the random walk (q, T) is recurrent.*

Proof. Suppose T is q–conservative; then for any $\Lambda \in \mathcal{F}$ either $\sum_{n=1}^{\infty} \mathbf{E}\mathcal{X}_\Lambda(T_{\eta_n}\omega) = \infty$ (and the "particle" visits Λ infinitely often) or $\sum_{n=1}^{\infty} \mathbf{E}\mathcal{X}_\Lambda(T_{\eta_n}\omega) = 0$ (and the "particle" misses Λ). So (q, T) is recurrent.

Converse, if (q, T) is recurrent, then for any $f = \mathcal{X}_\Lambda$, $\Lambda \in \mathcal{F}$, m–a.e. either

$$\sum_{n=1}^{\infty} \int f(T_x\omega) q^n(dx) = \infty$$

or

$$\sum_{n=1}^{\infty} \int f(T_x\omega) q^n(dx) = 0; \tag{7.3}$$

It is evident that (7.3) is true for any integrable positive simple function f, and since any $f \in L^1_+$ can be approximated from below by simple functions, (7.3) holds for all $f \in L^1_+$. Thus recurrence implies conservativity.

□

Example 7.2. Let $X = \mathbb{Z}$, and let q have the following properties: Y_q, the group generated by $c(q)$, coincides with $\mathbb{Z}$, $\sum_{x=-\infty}^{\infty} |x| q(x) < \infty$ and $\sum_{x=-\infty}^{\infty} x q(x) = 0$. Then the random walk $\eta_n = \xi_1 + \ldots + \xi_n$ in $\mathbb{Z}$ is recurrent and the "particle" visits any point $x \in \mathbb{Z}$ infinitely often (the simplest case is $q(\{-1\}) = q(\{1\}) = \frac{1}{2}$; then $\lambda_n(\{x\}) = 2^x \sum_{m=0}^{[\frac{x+n}{2}]} \binom{2m-x}{m} 4^{-m}$ if $|x| \leq n$ and $\lambda_n(\{x\}) = 0$ if $|x| > n$). It is obvious that for any dynamical system T, the random walk $T^{\eta_n}\omega$ in $O(\omega|T)$ is recurrent, and thus any dynamic system T is q–conservative. Such "conservatizing weights" exist also in $\mathbb{R}$, $\mathbb{R}^2$ and $\mathbb{Z}^2$.

§ 8. Construction of Pointwise Averaging Sets on General Connected Groups

In this section we shall consider pointwise averaging of strongly isometric representations. We recall that a strongly isometric representation of a group G is in fact a *family* of consistent representation S^α in $L^\alpha(\Omega, \mathcal{F}, m)$ with $\|S_g^{(\alpha)}\| = 1$, $1 \leq \alpha < \infty$. Certainly, if $\{T_g, g \in G\}$ is a right dynamical system in $(\Omega, \mathcal{F}, m)$, then the associated left representations $S^{(\alpha)}$: $(S_g^{(\alpha)}\varphi)(\omega) = \varphi(\omega T_g)$ from a strongly isometric left representation. Like Greenleaf and Emerson [1], we shall have to consider generalized strongly isometric representations, i.e. strongly contractive representations of G (as a *semi*group) of the form: $S_g^{(\alpha)} = \widetilde{S}_g^{(\alpha)} \Pi^{(\alpha)}$ where $\widetilde{S}_g^{(\alpha)}$ is a strong isometric representation of G and the $\Pi^{(\alpha)}$ are projections in L^α, $\alpha \geq 1$, commuting with all $\widetilde{S}_g^{(\alpha)}$. The spaces $(\Omega, \mathcal{F}, m)$ will be supposed to be σ–finite measure spaces;

G will be a connected non–compact locally compact topological group; μ will be a Borel measure on G (it will usually be either the left Haar measure $\mu_G^{(l)}$ or the right Haar measure $\mu_G^{(r)}$, or the Haar measure μ_G in the unimodular case).

Our aim is to construct on G sequences of compact sets $\{G_n\}$ with $\mu(G_n) > 0$, $n = 1, 2, \ldots$, and with the following properties: for any σ–finite measure space $(\Omega, \mathcal{F}, m)$ and any generalized strongly isometric measurable representation S,

(M) $(L^\alpha)\text{–}\lim_{n\to\infty} \frac{1}{\mu(G_n)} \int_{G_n} S_g\varphi\mu(dg) = \mathbf{M}^{(\alpha)}(\varphi|S)$ if $\varphi \in$ $\in L^\alpha(\Omega, \mathcal{F}, m)$, $1 < \alpha < \infty$, or $\alpha = 1$ and $m(\Omega) < \infty$;

(P_α) m–a.e. there exists $\lim\limits_{n\to\infty} \frac{1}{\mu(G_n)} \int_{G_n} S_g\varphi(\omega)\mu(dg) = \mathbf{M}^{(\alpha)}(\varphi|S)$, $\varphi \in L^\alpha(\Omega, \mathcal{F}, m)$;

(D) $\sup\limits_{1\le n<\infty} \frac{1}{\mu(G_n)} \left|\int_{G_n} S_g\varphi(\omega)\mu(dg)\right| \in L^\alpha(\Omega, \mathcal{F}, m)$, $1 < \alpha < \infty$.

Certainly, property (M) implies that $\{G_n\}$ is a mean r–averaging sequence in the sense of § 4.1; if this property holds we shall write $\{G_n\} \in M^{(r)}(\mu)$. (P_α) means that the pointwise ergodic theorem holds; and (D) is a dominated ergodic theorem. If the G_n are compact and $\{G_n\}$ has properties (M), (D) and (P_α) for all $1 \le \alpha < \infty$ (all $1 <$ $< \alpha < \infty$) we write $\{G_n\} \in \mathrm{MDP}_1^{(r)}(\mu)$ (resp. $\{G_n\} \in \mathrm{MDP}_{1+0}^{(r)}(\mu)$). The mark "$r$" means, of course, that we consider "right" objects here: right dynamical systems, left representations and right means. The necessary changes for the "left" objects will be evident, so we do not discuss them.

Like in § 4.7 we shall move to our final aim by several intermediate steps, each of which gives a construction of MDP–sequences on some wide class of groups and deserves to be singled out. This will be done in Subsections 8.2 and 8.3. The following subsection contains some auxiliary statements.

8.1. Pointwise averaging on group extensions. Auxiliary statements. First we shall state "pointwise" analogues of Theorems 3.2.5 and 3.2.7.

Theorem 8.1. *Let Γ be a compact normal subgroup of G and $F :=$*

$:= G/\Gamma$. *If* $\{F_n\} \in \mathrm{MDP}_1^{(r)}(\mu_F^{(l)})$ (*resp.* $\{F_n\} \in \mathrm{MDP}_1^{(r)}(\mu_F^{(r)})$), *then* $\{G_n\} := \{\pi^{-1}(F_n)\} \in \mathrm{MDP}_1^{(r)}(\mu_G^{(l)})$ (*resp.* $\{G_n\} := \{\pi^{-1}(F_n)\} \in$ $\in \mathrm{MDP}_1^{(r)}(\mu_G^{(r)})$). *The case with the classes* $\mathrm{MDP}_{1+0}^{(r)}$ *is similar.*

Proof. We shall use the well–known formula

$$\int \psi(g)\mu)G^{(l)}(dg) = \int_F \int_\Gamma \psi(f\gamma)\mu_\Gamma(d\gamma)\mu^{(l)}(df)$$

(note that the function $g \to \int_\Gamma \psi(g\gamma)\mu_\Gamma(d\gamma)$ is constant on the left Γ–cosets f). So we get

$$\begin{aligned} S_{G_n}^{(l)}\varphi(\omega) :=& \frac{1}{\mu_G^{(l)}(G_n)} \int_{G_n} S_g\varphi(\omega)\mu_G^{(l)}(dg) = \\ =& \frac{1}{\mu_G^{(l)}(G_n)} \int_{F_n} \int_\Gamma S'_f S_\gamma \varphi(\omega)\mu_\Gamma(d\gamma)\mu_F^{(l)}(df) = \\ =& \frac{1}{\mu_F^{(l)}(F_n)} \int_{F_n} S'_f \int_\Gamma S_\gamma\varphi(\omega)\mu_\Gamma(d\gamma)\mu_F^{(l)}(df) = \\ =& \frac{1}{\mu_F^{(l)}(F_n)} \int_{F_n} V_f \Pi_{I^\alpha(\Gamma)}^{L^\alpha}\varphi(\omega)\mu_F^{(l)}(df) \end{aligned}$$

(here $S'_f = S_g$ if $g \in f$, and V is the factor representation in $I^\alpha(\Gamma) :=$ $:= \{\psi : \psi \in L^\alpha,\ S_\gamma\psi = \psi,\ \gamma \in \Gamma\}$). Similarly,

$$\begin{aligned} S_{G_n}^{(r)}\varphi(\omega) :=& \frac{1}{\mu_G^{(r)}(G_n)} \int_{G_n} S_g\varphi(\omega)\mu_G^{(r)}(dg) = \\ =& \frac{1}{\mu_G^{(r)}(G_n)} \int_{F_n}\int_\Gamma S_\gamma S'_f\varphi(\omega)\mu_\Gamma(d\gamma)\mu_F^{(r)}(df) = \\ =& \frac{1}{\mu_F^{(r)}(F_n)} \int_{F_n} \Pi_{I^\alpha(\Gamma)}^{L^\alpha} S'_f\varphi(\omega)\mu_\Gamma(d\gamma)\mu_F^{(r)}(df) = \\ =& \frac{1}{\mu_F^{(r)}(F_n)} \int_{F_n} V_f \Pi_{I^\alpha(\Gamma)}^{L^\alpha}\varphi(\omega)\mu_F^{(r)}(df). \end{aligned}$$

These equalities imply our statement. For example, $\lim_{n\to\infty} S_{G_n}^{(l)}\varphi(\omega) =$

$$= \Pi_{I^\alpha}^{I^\alpha(\Gamma)} \Pi_{I^\alpha(\Gamma)}^{L^\alpha} \varphi(\omega) = \Pi_{I^\alpha}^{L^\alpha} \varphi(\omega) = \mathbf{M}(\varphi).$$

□

Lemma 8.2. *Let Γ be a normal subgroup of G, $F := G/\Gamma$ and let $\{F_n\}$ be a sequence of compact sets in F such that for some cross-sections $\widetilde{F}_n$ in G over F_n we have: $\{\widetilde{F}_n\Gamma_n\} \in \mathrm{MDP}_1^{(r)}(\mu_G^{(l)})$ (resp. $\{\Gamma_n\widetilde{F}_n\} \in \mathrm{MDP}_1^{(r)}(\mu_G^{(r)})$). Then $\{F_n\} \in \mathrm{MDP}_1^{(r)}(\mu_F^{(l)})$ (resp. $\{F_n\} \in$ $\in \mathrm{MDP}_1^{(r)}(\mu_F^{(r)})$). The same holds for the classes $\mathrm{MDP}_{1+0}^{(r)}$.*

The proof is almost word to word like the proof of Lemma 3.2.7.

Let F_n, K_1, K_2 be compact sets in G, $G_n := K_1F_nK_2$. Let ν_G, ν_{F_n}, ν_{K_i} be measures on G, F_n, K_i, and let ν_{G_n} be the restriction of ν_G to G_n. Below we use the following notation: $S_A\varphi = \int_A S_x\nu_A(dx)$.

Lemma 8.3. *Suppose that $\nu_{G_n} = \nu_{K_1} \times \nu_{F_n} \times \nu_{K_2}^{*)}$ $n = 1,2,\ldots$. Let $\{G_n\}$ satisfy condition (M).*

(i) If $\{F_n\}$ satisfies the conditions

(D) $\sup\limits_{1\le n<\infty} S_{F_n}\varphi \in L^\alpha(\Omega,\mathcal{F},m)$, $1<\alpha<\infty$;

$(\widetilde{P}_\alpha)$ for any $\varphi \in L^\alpha(\Omega,\mathcal{F},m)$ the limit $\lim\limits_{n\to\infty} S_{F_n}\varphi$ exists and is finite m–a.e.

for all $1<\alpha<\infty$, then $\{G_n\} \in \mathrm{MDP}_{1+}(\nu_G)$.

(ii) If $\{F_n\}$ satisfies the conditions (D) and $(\widetilde{P}_\alpha)$ with $1\le\alpha<\infty$ and if $K_1 = \{e\}$, then $\{G_n\} \in \mathrm{MDP}_1(\nu_G)$.

Proof. We have $S_{G_n}\varphi = S_{K_1}S_{F_n}S_{K_2}\varphi$ for any $\varphi \in L^\alpha(\Omega,\mathcal{F},m)$. According to Example 5.1.1, $S_{K_2}\varphi \in L^\alpha$, $\alpha>1$, and by property (D) $\sup_{1\le n<\infty} S_{F_nK_2}\varphi \in L^\alpha\Omega$, $\alpha>1$, and therefore

$$\sup_{1\le n<\infty} S_{K_1F_nK_2}\varphi \in L^\alpha(\Omega) \quad \text{for all} \quad \alpha>1.$$

Property (P_α) implies that $\lim_{n\to\infty} S_{F_nK_2}\varphi$ exists and is finite m–a.e. If $G_n = F_nK_2$ this limit coincides with $\mathbf{M}(f|S)$ since $\{G_n\}$ satisfies

---

*) More precisely: ν_{G_n} is the image of $\nu_{K_1}\times\nu_{F_n}\times\nu_{K_2}$ under the mapping $(k_1,f,k_2)\mapsto k_1fk_2$.

condition (M). So *(ii)* is proved. If $\alpha > 1$, by the Lebesgue dominated convergence theorem $\lim_{n\to\infty} S_{G_n}\varphi = \lim_{n\to\infty} \int_{K_2} S_{k_2}S_{F_n K_1}\varphi dk_2$ exists and is finite m–a.e.; it coincides with $\mathbf{M}(\varphi|S)$ since $\{G_n\}$ satisfies (M).

□

Theorem 8.4. *Let $G^{(i)}$ be subgroups of G, $i = \overline{1,m}$, and let each $G^{(i)}$ be isomorphic to either $\mathbb{R}$, $\mathbb{Z}$, or $\mathbb{K} := \mathbb{R}/\mathbb{Z}$. If $G_n^{(i)} \cong \mathbb{R}$ we set $G_n^{(i)} := [0, a_n^{(i)}]$, $a_n^{(i)} \in \mathbb{R}_+$, $a_n^{(i)} \uparrow \infty$ as $n \to \infty$; if $G^{(i)} \cong \mathbb{Z}$, then $G_n^{(i)} = [0, a_n^{(i)}] \cap \mathbb{Z}$, $a_n^{(i)} \in \mathbb{Z}_+$, $a_n^{(i)} \uparrow \infty$; and if $G^{(i)} \cong \mathbb{K}$ we set $G_n^{(i)} = = G^{(i)}$. Let $G_n = G_n^{(1)} \cdot \ldots \cdot G_n^{(m)}$ and $\mu_{G_n} := \mu_{G_n^{(1)}} \times \mu_{G_n^{(2)}} \times \ldots \times \mu_{G_n^{(m)}}$, where $\mu_{G_n^{(i)}}$ is the restriction of the Haar measure $\mu_{G^{(i)}}$ to $G_n^{(i)}$. Let S be a generalized strongly isometric measurable representation of G. We denote $S_{G_n}\varphi = \frac{1}{\mu_G(G_n)} \int_{G_n} S_g\varphi\mu_{G_n}(dg)$.*

(i) Let $1 < \alpha < \infty$. Then for any $\varphi \in L^\alpha(\Omega, \mathcal{F}, m)$,

a) $\sup\limits_{1\le n<\infty} |S_{G_n}\varphi| \in L^\alpha(\Omega)$,

b) $\lim\limits_{n\to\infty} S_{G_n}\varphi(\omega)$ exists m–a.e.;

c) $\lim\limits_{n\to\infty} S_{G_n}\varphi$ exists in $L^\alpha(\Omega)$.

(ii) If the subgroups $G^{(i)}$ commute and $a_n^{(i)} = a_n$, $1 \le i \le m$, $a_n \uparrow \infty$, then, moreover, b) holds for $\alpha = 1$ and c) holds for $\alpha \ge 1$ if $m(\Omega) < \infty$.

Proof. If $G^{(i)} \cong \mathbb{K}$, $i = \overline{1,m}$, the statement is trivial; if either all $G^{(i)} \cong \mathbb{R}$ or all $G^{(i)} \cong \mathbb{Z}$, the statements *(i)* and *(ii)* are evident consequences of the continuous and discrete versions of the Dunford–Zygmund–Fava and Dunford–Schwartz ergodic theorems, resp. (see, e.g., Dunford and Schwartz [1], Krengel [1]), since $S_{G_n} = = S_{G_n^{(1)}}S_{G_n^{(2)}}\ldots S_{G_n^{(m)}}$. In the general, "mixed" case the statement can be proved like the above mentioned theorems. They can be also proved using the trivial "periodic" extension of a representation of $\mathbb{K} = \mathbb{R}/\mathbb{Z}$ to a representation of $\mathbb{R}$ and the standard extension of a representation S: $n \to S_1^n$ of $\mathbb{Z}$ in $L^\alpha(\Omega, \mathcal{F}, m)$ to a continuous representation of $\mathbb{R}$ in some space $L^\alpha(\Omega', \mathcal{F}', m')$. We recall the construction of this standard extension. Let $W = [0,1]$; $\mathcal{W} = \mathfrak{B}(W)$; let

l be the Lebesgue measure on $\mathcal{W}$; $(\Omega', \mathcal{F}', m') = (\Omega \times W, \mathcal{F} \times \mathcal{W}, m \times l)$. In $L^\alpha(\Omega', \mathcal{F}', m')$ we consider the operator $\widetilde{S}_1 : \widetilde{S}_1\psi(\omega, w_0) = S_1\psi(\omega, w_0)$ where S_1 is our operator in $L^\alpha(\Omega, \mathcal{F}, m)$, and the operators $V_t : V_t\psi(\omega, w) = \widetilde{S}_1^{[w+t]}(\psi(\omega, \{w+t\}))$ (here $[a] = \max\{n : n \in \mathbb{Z}, n \leq a\}$, $\{a\} = a - [a]$). For $\psi(\omega, w) \equiv \varphi(\omega)$ we have $\frac{1}{n}\int_0^n V_t\psi(\omega, w)dt = \frac{1}{n}\sum_{k=0}^{n}\int_{k-w}^{k-w+1} S_1^k\varphi(\omega)dt = \frac{1}{n}\sum_{k=0}^{n} S_1^k\varphi(\omega)$. $S' : t \to S'_t$ is a measurable, hence continuous, generalized strongly isometric representation of $\mathbb{R}$. If $G^{(i)} \cong \mathbb{Z}$ and $g_0^{(i)}$ is the generator of $G^{(i)}$, we set $\overline{S}_t^{(i)}$ to be the standard extension to $\mathbb{R}$ of the representation $n \mapsto S^n_{g_0}$. If $G^{(i)} \cong \mathbb{R}$ we set $\bar{S}_t^{(i)}\psi(\omega, w_0) := S_t^{(i)}\psi(\omega, \omega_0)$, $\omega \in \Omega$, $w_0 \in W$, where $S^{(i)}$ is the restriction of S to $G^{(i)}$. If $G^{(i)} \overset{\kappa}{\cong} \mathbb{K}$ we set $\bar{S}_t^{(i)}\psi(\omega, \omega_0) = S^{(i)}_{\kappa^{-1}(\alpha)}\psi(\omega, \omega_0)$ $t \equiv \alpha (\text{mod } 1)$. Let $\widehat{S}_n^{(i)} := \frac{1}{a_n^{(i)}}\int_0^{a_n^{(i)}} \bar{S}_t^{(i)} dt$ ($a_n^{(i)} = n$ if $G^{(i)} \cong \mathbb{K}$). It is easy to check that if $\psi(\omega, w) = \varphi(\omega)$, then $S_{G_n}\varphi = \widehat{S}_n^{(1)}\widehat{S}_n^{(2)} \ldots S_n^{(m)}\psi$. Thus the "mixed" cases are reduced to the continuous parameter versions of the above mentioned theorems.

□

8.2. Construction of pointwise averaging sequences of sets on connected amenable groups. Let G be a connected amenable locally compact group with trivial compact kernel $K(G)$. Then by the Greenleaf–Emerson theorem (see App., (3.N)) G has the canonical left semidirect decomposition $G = G^{(0)} \odot G^{(1)} \odot \ldots \odot G^{(m)}$. Since $G^{(i)} \cong \mathbb{R}^1$, $i = \overline{1, m}$, we can take $G_n^{(i)} \cong [0, a_n^{(i)}]$, $G_n = G_n^{(0)} \cdot G_n^{(1)} \cdot \ldots \cdot G_n^{(m)}$, $\widetilde{G}_n = G_n^{(m)} \cdot \ldots \cdot G_n^{(0)}$. By Theorem 3.2.9 $\{G_n\} \in M^{(r)}(\mu_G^{(l)})$ and $\widetilde{G}_n^* \in M^{(r)}(\mu_G^{(r)})$. By Theorems 8.4 (*i*) and 8.3, $\{G_n\} \in \mathrm{MDP}^{(r)}_{1+0}(\mu_G^{(l)})$ and $(G_n^*) \in \mathrm{MDP}^{(r)}_{1+0}(\mu_G^{(r)})$.

If G has a non-trivial compact kernel $K(G)$ and $G' := G/K(G)$, then $K(G') = \{e\}$ and we can construct $\{G'_n\} \in \mathrm{MDP}^{(r)}_{1+0}(\mu_G^{(l)})$ and $\{G'^*_n\} \in \mathrm{MDP}^{(r)}_{1+0}(\mu_G^{(r)})$. Theorem 8.1 states that $\{\pi^{-1}(G'_n)\} \in \mathrm{MDP}^{(r)}_{1+0}(\mu_G^{(l)})$ and $\{\pi^{-1}(G'_n)\} \in \mathrm{MDP}^{(r)}_{1+0}(\mu_G^{(r)})$. We have proved the following theorem.

Theorem 8.5. *If G is a connected locally compact topological group, then* $\mathrm{MDP}^{(r)}_{1+0}(\mu^{(l)}_G) \neq \emptyset$ *and* $\mathrm{MDP}^{(r)}_{1+0}(\mu^{(r)}_G) \neq \emptyset$.

The construction of $\{G_n\} \in \mathrm{MDP}^{(l)}_{1+0}(\mu^{(l)}_G)$ was given in Greenleaf, Emerson [1].

8.3. Construction of pointwise averaging sequences of sets on connected semisimple Lie groups. Let G be a connected semisimple Lie group. We shall consider a sequence $\{G_n\} \in M(\mu_G)$ of the form $Z^{(\infty)}_n A_n N_0 K_0$ where K_0 and $Z^{(\infty)}_n$ were constructed in Lemma 4.7.5 ($K_0 = K$, $Z^{(\infty)}_n = \{e\}$ if G has finite center). We let N_0 be a compact set in N with $\mu(N_0) > 0$; since $A \cong \mathbb{R}^m$ for some $m \geq 1$, we can set $A_n \cong [0, a_n]^m$ where $a_n \uparrow \infty$; $Z^{(\infty)} \cong \mathbb{Z}^k$, $k \geq 0$, we can set $Z^{(\infty)}_n = \overline{[0, a_n]}^k$ if the center $Z(G)$ is infinite, and $Z^{(\infty)}_n = \{e\}$ if it is finite. Thus $G_n \cong \overline{[0, a_n]}^k [0, a_n]^m N_0 K_0$. Theorem 8.4 *(ii)* and Lemma 8.3 imply that $\{G_n\} \in \mathrm{MDP}_1(\mu_G)$. We have proved the following theorem.

Theorem 8.6. *If G is a connected semisimple Lie group, then* $\mathrm{MDP}_1(\mu_G) \neq \emptyset$. *If the center $Z(G)$ is finite, there exist sequences of simply connected sets in* $\mathrm{MDP}_1(\mu_G)$ *of the form* $G_n = A_n N_0 K$.

8.4. Construction of pointwise averaging sequences of sets on general groups. Let G be a connected Lie group and let $G^* = R^* \odot S^* = S^* \odot R^*$ be the right and left semidirect factorizations of the universal covering group G^* into the radical R^* and the complementary semisimple subgroup S^*. Let $R^* = R^{(0)} \odot R^{(1)} \odot \ldots \odot R^{(s)}$ be the canonical left semidirect factorization of R^* (see App., (3.N)). As in Subsection 8.2, we can set $R^{(i)}_n \cong [0, a^{(i)}_n]$, $i = \overline{1, s}$. Let Φ be the fundamental group of S (i.e. $\Phi \in S^*$ and $S^*/\Phi \cong S$); since Φ is commutative and discrete, $\Phi \cong \mathbb{Z}^l \times \Phi''$ where Φ'' is a finite subgroup of Φ and $l \in \mathbb{Z}_+$. Let $\{\Phi_n\} = \prod_{i=1}^{l} [0, d^{(i)}_n] \times \Phi'' =: \Phi'_n \times \Phi''$. If $d^{(i)}_n \uparrow \infty$, $i = \overline{1, l}$, then $\{\Phi_n\} \in M^{(r)}(\mu_\Phi)$. As in the proof of Theorem 7.9 we consider the sets $S_n = K_0 N_0 A_n Z^{(\infty)}_n$ where $A_n = \prod_{i=1}^{m} [0, b^{(i)}_n]$, $Z^{(\infty)}_n = \prod_{j=1}^{m} \overline{[0, c^{(j)}_n]}$

and the cross–sections $\widetilde{S}_n = K_0^* N_0 \Phi'_n A_n Z_n^{(\infty)}$ over S_n, and set $G_n^* = = \Phi_n \widetilde{S}_n R_n^* = \Phi'' K_0^* N_0 \Phi'_n A_n Z_n^{(\infty)} \cdot R^{(0)} R_n^{(1)} \ldots R_n^{(m)}$. A reasoning as in the proof of Theorem 4.7.9 gives $\{G_n^*\} \in M^{(r)}(\mu_{G^*}^{(l)})$. Since $R^{(0)}$ is a connected compact solvable group it is isomorphic to $\mathbb{R}^s (s \in Z_+)$. Theorem 8.4 *(i)* implies that $G_n^* = \{\Phi_n K_0^* N_0 A_n Z_n^{(\infty)} R^{(0)} \cdot R_n^{(1)} \ldots R_n^{(m)}\} \in \in \mathrm{MDP}_{1+0}^{(r)}(\mu_{G^*}^{(l)})$, $\Phi \subset \Phi(G)$; therefore Lemma 3.2.7 and Theorem 8.2 imply that $\{G_n\} := \{K_0 N_0 A_n Z_n^{(\infty)} R^{(0)} R_n^{(1)} \ldots R_n^{(m)}\} \in \mathrm{MDP}_{1+0}^{(r)}(\mu_G^{(l)})$. We can similarly prove that $\{R_n^{(m)} \ldots R_n^{(1)} R^{(0)} \cdot Z_n^{\infty} A_n N_0 K_0\} \in \in \mathrm{MDP}_{1+0}^{(r)}(\mu_G^{(r)})$.

If G is a connected locally compact topological group, there exists a compact normal subgroup Γ such that $F := G/\Gamma$ is a connected Lie group. If $F_n \in \mathrm{MDP}_{1+0}^{(r)}(\mu_F^{(l)})$ and $\{\widetilde{F}_n\} \in \mathrm{MDP}_{1+0}^{(r)}(\mu_F^{(r)})$, then by Theorem 8.1, $\{\pi^{-1}(F_n)\} \in \mathrm{MDP}_{1+0}^{(r)}(\mu_G^{(l)})$ and $\{\pi^{-1}(\widetilde{F}_n)\} \in \in \mathrm{MDP}_{1+0}^{(r)}(\mu_G^{(r)})$. Now we have achieved our main goal: to give a constructive proof of the following theorem.

Theorem 8.7. *If G is a connected locally compact group, then* $\mathrm{MDP}_{1+0}^{(r)}(\mu_G^{(l)}) \neq \emptyset$ *and* $\mathrm{MDP}_{1+0}^{(r)}(\mu_G^{(r)}) \neq \emptyset$.

§ 9. Pointwise Averaging Nets on Homogeneous Spaces

Let $Y = G/K$ be a locally compact homogeneous space; $\mathfrak{B}$ denotes the Borel σ–algebra on Y; y_0 is the K–fixed point in Y; μ is the invariant measure on $\mathfrak{B}$.

Definition 9.1. We say that a net $\{\nu_n, n \in N\}$ $(\subset \mathcal{P}(\mathfrak{B}))$ is *pointwise averaging* if for any completely measurable B–valued strictly homogeneous random field $\xi(y)$, $y \in Y$, with $E\|\xi(y)\| < \infty$, $y \in Y$, with probability *1* the following limit exists:

$$\lim_{n \in N} \int \xi(y) \nu_n(dy) = E[\xi(y) | \mathfrak{J}_\xi]$$

where $\mathfrak{J}_\xi$ is the σ-subalgebra of invariant sets; we say that a net $\{A_n, n \in N\} (\subset \mathfrak{B})$ is pointwise averaging if the net $\nu_n(A) = \frac{\mu(A_n \cap A)}{\mu(A_n)}$ has this property.

Proposition 4.1.9 on the "projections" of mean averaging nets from motion groups into homogeneous spaces remains valid in the "pointwise" case. The necessary notions of product of sets, symmetric set and convolution of measures are defined in the App., Subsect. 2.1 and 2.4. We restrict ourselves to three examples.

Example 9.1. Let Y be a metric homogeneous space (see App., Subsect. 2.1). If the net of concentric balls $\{B_r, r > 0\}$ satisfies the conditions

1) $\lim\limits_{r \to \infty} \mu(B_{r+r_0}))/\mu(B_r) = 1$, $r_0 > 0$,

and

2) $\sup\limits_{0<r<\infty} [\mu(B_{2r})/\mu(B_r)] < \infty$,

then it is pointwise averaging.

Of course, the class of homogeneous spaces with such balls is narrow. For example, the measures of the balls in the Lobachevsky space L_m grow exponentially when $r \to \infty$, so conditions *1)* and *2)* are both not satisfied.

Example 9.2. Let $Y = L_m$, $m \geq 2$. The invariant measure $\mu(A) = \int_A s_m(r) dr d\sigma$ where $s_m(r) = sh^{m-1} r$, $r = \rho(y_0, y)$ and $d\sigma$ is the element of the rotation invariant measure on the unit sphere S_1. Consider the Gaussian measure q_t with parameter $t > 0$, i.e. the probability measure with μ-density $u_t = u_t(r)$ that is the solution of the equation $\Delta u_t = \partial u_t / \partial t$ where Δ is the Laplace–Beltrami operator (its restriction to the spherically symmetric functions is given by the formula $\Delta = [s_m(r)]^{-1} \frac{d}{dr}(s_m(r) \frac{d}{dr})]$. For example, if $n = 3$, then $u_t(r) = (1/8)\pi^{-\frac{3}{2}} t^{-\frac{3}{2}} r \operatorname{sh}^{-1} r \exp(-t - r^2/4t)$ (see Karpelevich, Tutubalin, Shur [1]). We have $q_1^k = q_k$, so $\lambda_n^q := \frac{1}{n} \sum_{k=1}^n q_k$, $n = \overline{1, \infty}$, is a pointwise averaging sequence in L_m.

According to Theorem 4.7.1, the net of balls $\{B_r\}$ is mean averaging in L_m; however it seems to be unknown whether it is pointwise averaging (the majoration approach applied in Note 6.1 fails here).

The following simple consequence of Theorem 8.5 presents examples of pointwise averaging nets in symmetric homogeneous spaces.

Example 9.3. **(Shulman–Tempelman)** Let Y be a completely non–compact symmetric Riemannian uniform space or rank r; let $G =$ $=$ ANK be the Iwasawa factorization of its motion group G; denote by κ the isomorphism $\mathbb{R}^r \mapsto A$; $A_t := \kappa([-a_t, a_t]^r)$; and let N_0 be a compact subset of N. If $a_t \uparrow \infty$, then $\{G_t G := A_t N_0 y_0\}$ is a pointwise averaging net in Y. For example, let $Y = L_2$, the Lobachevsky plane. Consider the complex half–plane model of L_2 with $y_0 = i$. Then $G = SL(2, \mathbb{R})$ and $gz = \frac{\gamma_{11}z+\gamma_{12}}{\gamma_{21}z+\gamma_{22}}$ for any $g = \begin{pmatrix} \gamma_{11} & \gamma_{12} \\ \gamma_{21} & \gamma_{22} \end{pmatrix}$ and any $z = \alpha + \beta i$ with $\beta > 0$; $A = \left\{a_c = \begin{pmatrix} e^{\frac{c}{2}} & 0 \\ 0 & e^{-\frac{c}{2}} \end{pmatrix}, c \in \mathbb{R}\right\}$, $N = \left\{n_b = \begin{pmatrix} 1 & b \\ 0 & 1 \end{pmatrix}, b \in \mathbb{R}\right\}$, and thus $a_c n_b i = e^c i + e^{-c} b$. We set $t \in \mathbb{R}_+$, $c_t > 0$, $c_t \uparrow \infty$ if $t \to \infty$; $A_t := \{a_c : |c| \le c_t\}$; $N_0 := \{n_b : |b| \le b_0\}$; then $\{G_t := A_t N_0 i\}$ is a pointwise averaging net. Any set G_t is a "rectangle" bounded by the two oricycles $o_t^{\pm} := \{z : z = \alpha + \beta i,\ \alpha = e^{\pm c_t}\}$ (which are orthogonal to the straight line $l_0 := \{z : \alpha = 0\}$ and cut it at distance c_t from the point i) and the two hypercycles $h_0 := \{z : \beta = \pm b_o^{-1} \alpha\}$ (which are equidistant to l_0). In the case $Y = L_m$, $G \cong SO(m, 1)$, $m \ge 3$, A and N are described in App., Example 1.6; we put $N_0 :=$ $:= \{n : n(\mathbf{b}) = n(b_1, \ldots, b_{m-1}) : \sum b_i^2 \le 2d_0\}$; then the sets $G_t =$ $= A_t N_0 y_0$ are bounded by the two orispheres $o_t^{\pm} = \{y = \kappa(\pm a_t) n y_0,$ $n \in N_0\}$ and the hypersphere $h_0 = \{y = \kappa(a_t) n(\mathbf{b}) y_0,\ \|\mathbf{b}\| = 2d_0\}$ equidistant to the straight line $A_t y_0$. $\{G_t\}$ is a pointwise averaging net.

BIBLIOGRAPHICAL NOTES

1. The classical individual ergodic theorem was proved by G.D.Birkhoff [1] (1931) under additional assumptions; in its general form it was proved by Chintschin [1] (1932). Other proofs were given by Kolmogorov [1] (1938), Yosida [3], [4] (1940) and Riesz [6] (1945), [7] (1948). Simple proofs of this theorem, based on these works can be found in the text–books by Doob [1], Gnedenko [1], Halmos [1], and

Cornfeld, Sinai, Fomin [1]. A new simple proof of Birkhoff's theorem is given in Katznelson, Weiss [1] (1982).

2. Wiener [1] (1939) extended Birkhoff's theorem to $\mathbb{R}^m$ and $\mathbb{Z}^m$ by proving that infinitely increasing sequences of spheres and cubes form pointwise averaging nets of sets. Pitt [1] (1942) extended these results to a wide class of "regular" nets. Calderón [1] (1953) extended the pointwise ergodic theorem to a large class of unimodular amenable groups. Further generalization were given by Cotlar [1] (1956).

2. The spectral approach to the pointwise ergodic theorem (see 2) goes back to the works by Reich [1] (1981) and Blum and Reich [1]; the results of §2 are published here for the first time.

3. In the unimodular case, Theorem 3.1 and Corollary 3.2 are close to results announced in Tempelman [5] (1967); the proofs of these results and other results of §3 were given in Tempelman [7] (1972); the works by Bewley [1] (1971) and Chatard [1] (1970) are also devoted to giving proofs of the results announced in Tempelman [1]. In the non-unimodular case, Corollary 3.2 is due to Emerson [2] (1974). In Ornstein, Weiss [1] (1987) a proof of Corollary 3.2 for discrete groups is given using a quite different method. The results of § 4 generalize some statements due to Cotlar [1] (1956).

4. The martingale approach to Corollary 3.5 goes back to Pitckel, Stepin [1] (1971), who use this method to prove the pointwise ergodic theorem in the group of proper dyadic fractions.

5. Theorem 5.1 is due to Jones [1] (to appear).

6. Theorem 6.1 is contained in Tempelman [5] (1967) and in Tempelman [24] (1986).

7. The theorem stated in Note 6.5 is due to Oseledec [1] (1965).

8. The results of § 7 were published in Tempelman [8] (1972).

9. The construction of sequences $\{G_n\} \in \mathrm{MDP}^{(r)}_{1+0}(\mu^{(r)})$ on connected amenable groups (see Theorem 8.5) was given in Greenleaf, Emerson [1] (1974). The remaining results of §§ 8, 9 are published here for the first time; the proofs of Theorem 8.6 and the statements of Example 9.3 incorporate suggestions made by A. Shulman.

Let us note also some works closely connected with the material of this chapter.

1. Vershik [1] developed an approach to the point ergodic theorem on countable amenable groups based on his results on the approximation of dynamical systems with countable time by dynamical systems with finite time.

2. Gaposhkin [1]-[6] proved the pointwise ergodic theorem for wide-sense homogeneous random fields on $\mathbb{R}^m$ under very weak restrictions on the spectrum; this is a generalization and strenghtening of results due to Blanc–Lapierre and Brard [1], Blanc–Lapierre and Fortet [1], Blanc–Lapierre and Tortrat [1], Verbickaja [1] and Yurinsky [1]. Savichev [1] transferred the results of Gaposhkin [4] to the Lobachevsky space.

3. The classical strong laws of large numbers were extended to B–valued random variables in Beck [1]–[3], Beck, Warren [1], Beck, Schwartz [1], Goldsheid [1], Mourier [1], [2], and Fortet, Mourier [1].

Related results can be found in Burkholder [1], Brunel, Keane [1], Deo [1], Dunford [1], Yoshimoto [1], Conze, Dang–Ngoc [1], Olsen [1]–[3], Reich [1], Ryll–Nardzewski [1], Tempelman [8], [9], [11]–[14], and Zygmund [1].

CHAPTER 7
ERGODIC THEOREMS FOR HOMOGENEOUS RANDOM MEASURES

In this chapter G is a σ–compact locally compact topological group; e is the identity of G; $\mathfrak{B}$ is the σ–algebra of Borel sets in G and μ_l and μ_r are fixed left and right Haar measures on G, respectively; λ is a relatively invariant Borel measure on G and $\Delta_\lambda(\cdot)$ is its modular function; $\mathcal{K}$ is the field of all bounded sets in $\mathfrak{B}$; $\mathcal{K}_e := \{A : e \in A \in \mathcal{K}, \mu_l(A) > > 0\}$ (it is a direction with the following "downwards" order: $A_1 < A_2$ if $A_1 \supset A_2$); $\mathfrak{M}$ is the set of all Borel measures on G; $X = G/K$ is a left homogeneous space; μ is an invariant measure on X; $(\Omega, \mathcal{F}, P)$ is a probability space; $\widetilde{\mathbb{R}}_+ = \mathbb{R}_+ \cup \{+\infty\}$; and $\widetilde{\mathbb{R}} = \mathbb{R} \cup \{-\infty\} \cup \{+\infty\}$.

§ 1. Homogeneous and λ–Homogeneous Random Fields

Definition 1.1. A random function $\Xi(A) = \Xi(A, \omega)$, $A \in \mathfrak{B}$, $\omega \in \Omega$, is called a *random measure* (r.m.) on G if:

(i) for P–almost all ω, $\Xi(\cdot, \omega)$ is a measure on $\mathfrak{B}$;

(ii) with probability *1*, $\Xi(A < \omega) < \infty$, $A \in \mathcal{K}$;

(iii) for any $A \in \mathfrak{B}$ the function $(g\omega) \longmapsto \Xi(gA, \omega)$ is $(\mathfrak{B} \times \mathcal{F})$–measurable.

The following definition introduces the notion of a "relatively homogeneous" random measure; it generalizes the notion of relatively invariant measure.

Definition 1.2. A random measure $\Xi(\cdot)$ is called a λ*–homogeneous random measure* (λ–h.r.m.) if for any $g \in G$, $A_1, \ldots, A_n \in \mathfrak{B}$ the $\widetilde{\mathbb{R}}$–valued random vector $(\Xi(gA_1), \ldots, \Xi(gA_n))$ has the same distribution as $(\Delta_\lambda(g)\Xi(A_1), \ldots, \Delta_\lambda(g)\ \Xi(A_n))$; μ_l–homogeneous random measures will simply be called *homogeneous random measures* (h.r.m.).

Since $\Delta_{\mu_l}(\cdot) \equiv 1$, the notion of a homogeneous random measure coincides with the notion of a random function on $\mathfrak{B}$ which is homogeneous in the strict sense with respect to the group of left shifts on $\mathfrak{B}$: $A \longmapsto gA$, $g \in G$, $A \in \mathfrak{B}$ (see App., § 8). Therefore, we can generalize to such measures all notions connected with strict sense homogeneous random functions. In particular, we can consider the shift dynamical system $\{T_g,\ g \in G\}$ in the measurable space of all Borel measures $(\mathfrak{M}_+,\ S)$, where S is the σ–algebra generated by the sets $\{m : m \in \mathfrak{M}_+,\ m(A) < a\}$, $A \in \mathfrak{B}$, $a \in \mathbb{R}_+$.

There exists a simple connection between homogeneous and λ–homogeneous measures, as the following statement shows.

Lemma 1.1. *(i) If $\Xi(\cdot)$ is an h.r.m., then $\Psi(A) := \int_A \Delta_\lambda(g)\Xi(dg)$, $A \in \mathfrak{B}$, is a λ-h.r.m..*

(ii) If $\Psi(\cdot)$ is a λ-h.r.m., then $\Xi(A) := \int_A (\Delta_\lambda(g))^{-1}\Xi(g)$ is an h.r.m..

Proof. *(i).* For any $g \in G$, $A_1, \ldots, A_n \in \mathfrak{B}$, the random vector $(\Psi(gA_i),\ i = \overline{1,m}) = (\int_{gA_i} \Delta_\lambda(f)\Xi(df),\ i = \overline{1,m}) = (\Delta_\lambda(g)\int_{A_i} \Delta_\lambda(f)\Xi(df),\ i = \overline{1,m})$. Since Ξ is homogeneous, one can deduce easily that this random vector has the same distribution as $(\Delta_\lambda(g)\int_{A_i} \Delta_\lambda(f)\Xi(df),\ i = \overline{1,m})$, i.e. as $(\Delta_\lambda(g)\Psi(A_i),\ i = \overline{1,m})$

(ii) can be proved similarly.

□

Lemma 1.2. *Let $\Xi(\cdot)$ be a random measure such that $\mathbf{E}\Xi(A) < \infty$, $A \in \mathcal{K}$; then*

(i) the function $A \to M_\Xi(A) := \mathbf{E}\Xi(A)$, $A \in \mathfrak{B}$, is a measure;

(ii) for any non-negative $\varphi \in L_1(G,\ \mathfrak{B},\ M_\Xi)$ we have

$$\mathbf{E}[\int \varphi(g)\Xi(dg)] = \int \varphi(g)M_\Xi(dg).$$

Proof. *(i).* Additivity of $M_\Xi(\cdot)$ is obvious; if $A, A_1, A_2, \ldots \in \mathfrak{B}$ and $A_n \uparrow A$, then $\Xi(A_n) \uparrow \Xi(A)$ and, by Levi's theorem,

$M_\Xi(A_n) \uparrow M_\Xi(A)$. Thus $M_\Xi(\cdot)$ is a measure on $\mathfrak{B}$.

(ii) Let $\{\varphi_n, n = \overline{1,\infty}\}$ be a sequence of simple functions such that $\varphi_n(g) \uparrow \varphi(g)$, $g \in G$. The statement is clearly valid for φ_n, $n = \overline{1,\infty}$. It remains to use Levi's theorem again.

□

Lemma 1.3. *Let $\Xi(\cdot)$ be a λ-h.r.m. and $\mathbf{E}\Xi(U) < \infty$ for some $U \in \mathcal{K}$. Then $M_\Xi(A) = \alpha\lambda(A)$, $A \in \mathfrak{B}$, where $0 \leq \alpha < \infty$.*

Proof. Let $\lambda = \mu_l$. Then the measure $M_\Xi(A)$ is a left Haar measure and, consequently, $M_\Xi(A) = \alpha\mu_l(A)$ where $\alpha \geq 0$ (see, e.g., Halmos [1], §§ 52, 60, 64). If $\lambda \neq \mu_l$, then $\Xi(A) = \int_A \Delta_\lambda(g)\Psi(d\lambda)$ where $\Psi(\cdot)$ is an h.r.m. (see Lemma 1.1) and, by Lemma 1.2, $M_\Xi(A) = \int_A \Delta_\lambda(g)M_\Psi(dg) = \alpha \int_A \Delta_\lambda(g)\mu_l(dg) = \alpha\lambda(A)$, $A \in \mathfrak{B}$.

□

We give two important examples of homogeneous and λ-homogeneous random measures.

Example 1.1. Let $\xi(\cdot)$ be a non-negative completely measurable strict sense homogeneous random field and let with probability *1*, $\int_A \xi(g)\mu_l(dg) < \infty$, $A \in \mathcal{K}$ (e.g., this is fulfilled if $\mathbf{E}\xi(e) < \infty$). It is easy to check that $\Xi(A) := \int_A \xi(g)\mu_l(dg)$, $A \in \mathfrak{B}$, is a random measure on $\mathfrak{B}$. Consider a covariant (with respect to some dynamical system T) image $\widetilde{\xi}$ of ξ. Let $\widetilde{\Xi}(A) := \int_A \widetilde{\xi}(g)\mu_l(dg)$, $A \in \mathfrak{B}$; it is easy to verify that the random measure $\widetilde{\Xi}$ is a covariant (with respect to T) image of Ξ and, consequently, Ξ is homogeneous. Of course, $\Xi_\lambda(A) := \int_A \xi(g)\lambda(dg)$, $a \in \mathfrak{B}$, is a λ-h.r.m.: this follows from the equality: $\Xi_\lambda(A) = \int_A \Delta_\lambda(g)\xi(g)\mu_l(dg)$, $A \in \mathfrak{B}$, and Lemma 1.1.

Example 1.2. An important class of random measures is the class of *point random measures*, which a.s. accept only integer values. Any such measure a.s. has the following form:

$$\Xi(A,\omega) = \sum_{k=1}^{n_A(\omega)} \delta_{g_k}(\omega)(A)m_k(\omega), \qquad A \in \mathfrak{B},$$

where δ_g is the Dirac measure concentrated at the point g, $g_k(\omega) \in \in \mathbb{Z}_+$, $k = 1, 2, \ldots$; moreover, if $A \in \mathcal{K}$ then $n_A(\omega) < \infty$ a.s. If $m_k(\omega) \equiv 1$ or, in other words, $\Xi(\{g_i\}) = 1$, $i = \overline{1, \infty}$, a.s., then $M(\omega) := \{g_1(\omega), g_2(\omega), \ldots\}$ is a *random set* in G, and $\Xi(A)$ is the cardinality of the set $M(\omega) \cap A$; such point measures are called *simple point measures*. Let $\rho(\cdot) = \alpha\lambda(\cdot)$, let $\alpha > 0$. $\Xi(\cdot)$ be a random measure; let for any $A_1, \ldots, A_n \in \mathcal{K}$ the random variables $\Xi(A_1), \ldots, \Xi(A_n)$ be independent and for any $A \in \mathcal{K}$ the random variable $\Xi(A)$ have the Poisson distribution with parameter $\rho(A)$; then Ξ is a *Poisson point* λ*-h.r.m.* If G is a non-discrete group, the measure Ξ is simple (cf., e.g., Nguen and Zessin [2]).

§ 2. Local Ergodic Theorems. Lebesgue Decomposition of Homogeneous Random Measures

If G is σ-compact, then, by the well-known Lebesgue theorem, we have for any r.m. Ξ the decomposition: with probability *1* $\Xi(A, \omega) = = \Xi_{ac}(A, \omega) + \Xi_s(A, \omega)$, $A \in \mathfrak{B}$, where $\Xi_{ac}(A, \omega)$ is a.s. an absolutely continuous (with respect to μ_l) measure and $\Xi_s(A, \omega)$ is a.s. a singular measure.

Definition 2.1. An r.m. Ξ is said to be *absolutely continuous* (resp. *singular*) with respect to μ_l if for P-almost all ω the measures $\Xi(\cdot, \omega)$ are absolutely continuous (resp. singular) with respect to μ_l.

Important classes of absolutely continuous r.m. and of singular r.m. are described in Examples 1.1 and 1.2, respectively.

Definition 2.2. We say that a net of sets $\{U_n, n \in N\} \subset \mathcal{K}_e$ *converges to identity* e if for any neighborhood V of e there exists an element $n_V \in N$ such that $U_n \subset V$ for $n > n_V$.

Following Edwards and Hewitt [1] (see also Hewitt, Ross [1]) we introduce the notion of $\mathcal{D}$-sequence.

Definition 2.3. A $\mathcal{D}$*–sequence* in G is a sequence $\{U_n, n = \overline{1,\infty}\} \subset \mathcal{K}$ with the following properties:

(i) $U_1 \supset U_2 \supset \ldots$;

(ii) there exists a positive constant κ such that

$$0 < \mu_l^*(U_n U_n^{-1}) < \kappa \mu_l(U_n), \qquad n = \overline{1,\infty};$$

(iii) $\{U_n, n = 1, \infty\}$ converges to e.

$\mathcal{D}$–sequences exist in any Lie group (cf. Edwards, Hewitt [1]; in this paper some other classes of groups possessing $\mathcal{D}$–sequences of sets are also found).

Theorem 2.1. *Let Ξ be an h.r.m. on a σ–compact group G. The following statements are valid:*

1. Ξ_{ac} and Ξ_s are h.r.m.;

2. If $\{U_n, n = \overline{1,\infty}\}$ is a $\mathcal{D}$–sequence in G, then

a) for any $g \in G$ the following limit exists a.s.:

$$\lim_{n\to\infty} (\mu_l(U_n))^{-1}\Xi(gU_n) = \lim_{n\to\infty} (\mu_l(U_n))^{-1}\Xi_{ac}(gU_n) := \xi(g);$$

here $\xi(\cdot)$ is a non–negative completely measurable random field which is homogeneous in the strict sense (with respect to the left shifts) and for any $A \in \mathfrak{B}$ $\Xi_{ac}(A) = \int_A \xi(g)\mu_l(dg)$ a.s.

b) for any $g \in G$ with probability 1 the following limit exists:

$$\lim_{n\to\infty} (\mu(U_n))^{-1}\Sigma_s(gU_n) = 0.$$

Proof. Without loss of generality we can suppose that $\Omega = \mathfrak{M}_+$ and Ξ is covariant with respect to the (right) translation dynamical system $\{T_g, g \in G\}$. By the Radon–Nikodym theorem there exists a family of functions $\{\xi_1(\cdot, \omega), \omega \in \Omega\}$ such that with probability *1*, $\Xi_{ac}(A) = \int_A \xi_1(g,\omega)\mu_l(dg)$, $A \in \mathfrak{B}$. According to Theorems 4.4.18

and 4.4.19 in Hewitt, Ross [1] with probability *1*,

$$\lim_{n\to\infty}(\mu(U_n))^{-1}\Xi(fU_n) = \lim_{n\to\infty}(\mu_l(U_n))^{-1}\int_{fU_n}\xi_1(g)\mu_l(dg)+$$

$$+\lim_{n\to\infty}(\mu_l(U_n))^{-1}\Xi_s(fU_n) = \lim_{n\to\infty}(\mu_l(U_n))^{-1}\int_{fU_n}\xi(g)\mu_l(dg) = \xi_1(f)$$

for μ_l–almost all $f \in G$. It is clear, that the set $M := \{(f,\omega) : (\mu_l(U_n))^{-1}\ \Xi(fU_n,\omega)\overset{n\to\infty}{\longrightarrow}\xi_1(f,\omega)\} \in \mathfrak{B}\times\mathcal{F}$; by Fubini's theorem $(\mu_l\times P)(G\times\Omega/M) = 0$ and therefore there exists an element $g_0 \in$ $\in G$ such that $\lim_{n\to\infty}(\mu_l(U_n))^{-1}\Xi(g_0U_n) = \xi_1(g,\omega)$ a.s. The random measure Ξ is covariant; therefore if $(f,\omega)\in M$, then $(g,\ \omega T_{fg^{-1}})\in M$ and $\xi_1(f,\omega) = \xi_1(g,\ \omega T_{fg^{-1}})$, $g\in G$. This implies that

$$P\{\omega : (f,\ \omega)\in M\} = 1, \qquad f\in G. \tag{2.1}$$

Let $\xi(f,\omega) := \xi_1(g_0,\omega T_{fg_0^{-1}})$. It is clear that the r.f. $\xi(\cdot)$ is completely measurable, covariant and

$$\xi(f,\ \omega) = \xi_1(f,\ \omega), \qquad (f,\ \omega)\in M. \tag{2.2}$$

Denote

$$\Xi'_{ac}(A) := \int_A \xi(g,\ \omega)\mu_l(dg), \quad \Xi'_s(A) := \Xi(A) - \Xi'_{ac}(A), \quad A\in\mathfrak{B}.$$

In view of Example 1.1, Ξ'_{ac} and Ξ'_s are covariant random measures. Equality (2.2) implies that with probability *1*, $\Xi_{ac}(A) = \Xi'_{ac}(A)$, $\Xi'_s(A) = \Xi_s(A)$, $A\in\mathfrak{B}$. Thus Statement *1)* is proved. Statement *2)* follows from the properties of the random field $\xi(g,\ \omega)$ and equalities (2.1) and (2.2).

□

Theorem 2.1 has several simple consequences.

Corollary 2.2. *An h.r.m.* Ξ *is absolutely continuous iff there exists a completely measurable random field* $\xi(\cdot)$ *which is strict sense homogeneous (with respect to the left translations) such that with probability 1,* $\Xi(A) = \int_A \xi(g)\mu_l(dg)$, $A\in\mathfrak{B}$.

Corollary 2.3. *An h.r.m. Ξ is singular iff for any $g \in G$ and any $\mathcal{D}$-sequence $\{U_n\}$,*

$$\lim_{n\to\infty} (\mu(U_n))^{-1}\Xi(gU_n) = 0 \qquad \text{a.s.}$$

It is easy to check that the sets

$$\Lambda_{ac} := \{\omega : \Xi(A,\omega) = \Xi_{ac}(A,\omega), \quad A \in \mathfrak{B}\} \in \mathfrak{I}$$

and

$$\Lambda_s := \{\omega : \Xi(A,\omega) = \Xi_s(A,\omega), \quad A \in \mathfrak{B}\} \in \mathfrak{I}.$$

are invariant with respect to shifts of the measure Ξ (see App., Subsection 8.4). This implies the following statement.

Corollary 2.4. *If Ξ is a metrically transitive h.r.m., then only one of the following case is possible:*

(i) Ξ is completely continuous;

(ii) Ξ is singular;

(iii) with probability 1, Ξ is neither singular nor absolutely continuous.

For singular h.r.m.'s the "mean" analog of Theorem 2.1 is not true. Indeed, let Ξ be such a measure and let $0 < \mathbf{E}\Xi(A) < \infty$ for some $A \in \mathcal{K}$. Then $\mathbf{E}\Xi(A) = \alpha\mu_l(A)$, $A \in \mathcal{K}$ $(\alpha > 0)$, and therefore $\mathbf{E}[(\mu_l(A))^{-1}\Xi(A)] = \alpha > 0$, $A \in \mathcal{K}$. On the other hand, by Theorem 2.1, $\lim(\mu_l(U_n))^{-1}\Xi(U_n) = 0$ a.s. for any $\mathcal{D}$-sequence $\{U_n\}$. Thus the limit $\lim_{n\to\infty}(\mu_l(U_n))^{-1}\Xi(U_n)$ does not existsin the sense of L^1-convergence. However, for absolutely continuous h.r.m.'s the following theorem holds.

Theorem 2.5. *Let the net $\{U_n,\ n \in N\}(\subset \mathcal{K}_e)$ converge to e; let $\xi(\cdot)$ be a completely measurable random field which is strict sense homogeneous with respect to the left shifts, and let $\mathbf{E}|\xi(e)|^\alpha < \infty$ for some $\alpha \geq 1$. Then uniformly with respect to $f \in G$,*

$$\lim_{n\in N} \mathbf{E}\left|\frac{1}{\mu_l(U_n)}\int_{fU_n} \xi(g)\mu_l(dg) - \xi(f)\right|^\alpha = 0.$$

Proof. The conditions of the theorem imply that the $L^\alpha(\Lambda, \mathcal{F}, P)$–valued function $g \to \xi(g)$ is uniformly continuous (see App., Subsect. 9.2); therefore, there exists for $\varepsilon > 0$ a neighborhood V_ε of e such that $\mathbf{E}|\xi(g) - \xi(f)|^\alpha < \varepsilon$ as soon as $gf^{-1} \in V_\varepsilon$. Let $U_n \subset V_\varepsilon$ for $n > n_\varepsilon$. Using Hölder inequality and Fubini's theorem we obtain that for $n > n_\varepsilon$

$$\begin{aligned}&\mathbf{E}\left|\frac{1}{\mu_l(U_n)}\int_{fU_n} \xi(g)\mu_l(dg) - \xi(f)\right|^\alpha \leq \\ \leq &\mathbf{E}\left|\frac{1}{\mu_l(U_n)}\int_{fU_n} |\xi(g) - \xi(f)|^\alpha \mu_l(dg)\right| = \\ = &\frac{1}{\mu_l(U_n)}\int_{fU_n} \mathbf{E}|\xi(g) - \xi(f)|^\alpha \mu_l(dg) < \varepsilon.\end{aligned}$$

□

§ 3. "Global" Ergodic Theorems

Let $A \in \mathfrak{B}$, $I \in \mathcal{K}$. Denote by $C_D(A) := [\bigcup_{g\in D} Ag]$; $I_D(A)$ the interior of $\bigcap_{g\in D} Ag$; $\partial_D A := C_D(A)\backslash I_D(A)$ (the "D–boundary" of A). Let $\mathcal{A} = \{A_n,\ n \in N\}$ be a net in $\mathfrak{B}$ and $\varphi_{\mathcal{A}}(D/\lambda) := \overline{\lim}_{n\in N}\lambda(\partial_D A_n)/\lambda(A_n)$. It is clear that $\varphi_{\mathcal{A}}(D_1|\lambda) \leq \varphi_{\mathcal{A}}(D_2|\lambda)$ if $D_1 \subset D_2$.

Below we shall need the following condition of "relative smallness" of the boundaries of the sets A_n, $n \in N$.

Definition 3.1. We shall say that a net $\mathcal{A} = \{A_n,\ n \in N\}$ satisfies condition (Γ_λ) if $A_n \in \mathfrak{B}$, $n \in N$, and $\lim_{D\in\mathfrak{N}_e} \varphi_{\mathcal{A}}(D|\lambda) = 0$.

Amenable groups possess Van Hove sequences which satisfy for $\lambda = \mu_r$ even a stronger condition: $\varphi_A(D|\mu_r) = 0$, $D \in \mathcal{K}_e$ (see App., §3).

Non–amenable groups do not possess Van Hove nets, but they possess nets with property (Γ_λ).

Example 3.1. Let $G = SO_0(m, 1)$, the motion group of the Lobachevsky space L_m, $m \geq 2$, (see App., Examples 1.1 and 2.5), and consider the balls $\widetilde{B}_r = \{A : A \in SO_0(m, 1),\ a_{00} = (Ae_0, e_0) \leq \text{ch } r\}$ where $e_0 = (1, 0, \ldots, 0)$; in other words, $\widetilde{B}_r$ is the set of all motions in L_m which leave the point e_0 in the ball $B_r(e_0) \subset L_m$ of radius r and center e_0. The group $SO_0(m, 1)$ is not amenable and therefore it does not possess nets with the Van Hove property. But it follows from Example 4.1 below that $\{\widetilde{B}_r, r > 0\}$ satisfies Condition (Γ_μ).

Now we shall discuss the geometric properties of the nets $\{C_D(A_n)\}$ for various kinds of nets $\{A_n\}$. We begin with the following definition.

Definition 3.2. Let $D \in \mathcal{K}_e$. A net of measurable sets $\{A_n, n \in N\}$ is called D–regular if the net $\{C_D(A_n), n \in N\}$ is right regular with respect to μ_r.

Example 3.2. Let $G = \mathbf{R}^m$ and $D = B_r$, the ball of radius r with center 0. It is easy to verify that if a set A is convex, then its B_r–closure $C_{B_r}(A)$ is convex, too. Therefore any increasing net of convex sets is B_r–regular for any $r > 0$. Now let A be a measurable set with $0 < l(A) < \infty$, $x_0 \in \mathbf{R}^m$; $A_n := \alpha_n(A - x_0) + x_0$, $\alpha_n > 1$, $n \in N$. Then $C_{B_{\alpha_n^{-1} r}}(A) \subset C_{B_r}(A)$ and thus $C_{B_r}(A_n) \subset A'_n := \alpha_n(C_{B_r}(A) - -x_0) + x_0$. Moreover, $l(C_{B_r}(A_n))/l(A'_n) = l(C_{B_{\alpha_n^{-1} r}}(A))/l(C_{B_r}(A)) > > l(A)/l(C_{B_r}(A)) > 0$, $n \in N$. Therefore the net $\{C_{B_r}(A_n)\}$ is regular with respect to the regular net $\{A'_n\}$ and, consequently, is regular itself, i.e. $\{A_n\}$ is B_r–regular.

Lemma 3.1. *If $\{A_n, n \in N\}$ is a Van Hove net, then $\{C_D(A_n)\}$ and $\{I_D(A_n)\}$ are right ergodic (with respect to μ_r), $D \in \mathcal{K}_e$.*

Proof. The Van Hove condition implies that

$$\lim_{n\in N}[(\mu_r(A_n))^{-1}(C_D(A_n)g\triangle A_n g)] =$$
$$= \lim_{n\in N}[(\mu_r(C_D(A_n))^{-1}\mu_r(C_D(A_n)\triangle A_n)] = 0,$$
$$\lim_{n\in N}\mu_r(A_n g\triangle A_n)/\mu_r(A_n) = 0.$$

The right ergodicity of $\{C_D(A_n)\}$ follows from these relations; the right ergodicity of $\{I_D(A_n)\}$ can be proved similarly.

□

Lemma 3.2. *Let $D \in \mathcal{K}_e$. If $\{A_n,\ n \in N\}$ is a D–regular Van Hove net, then for any $D' \subset D$, $D' \in \mathcal{K}_e$, the nets $\{A_n\}$, $\{I_{D'}(A_n)\}$, $\{C_{D'}(A_n)\}$ are right regular with respect to μ_r.*

Proof. Let the net $\{C_D(A_n)\}$ be regular (with respect to μ_r) with respect to some net $\{A'_n\}$ with the properties (E_2)–(E_4) (see Ch. 5, §1); then the nets $\{A_n\}$, $\{I_D(A_n)\}$ and $\{(C_{D'}(A_n)\}$ are also regular with respect to $\{A_n\}$, since $I_D(A_n) \subset I_{D'}(A_n) \subset A_n \subset C_{D'}(A_n) \subset \subset C_D(A_n) \subset A'_n$ and $\lim_{n\in N}(\mu_r(I_D(A_n))/\mu_r(C_D(A_n)) = 1$ according to the Van Hove condition.

□

Definition 3.3. A net of sets $\{A_n,\ n \in N\}$ will be called a *D–robust mean* (resp. *pointwise*) *averaging net* with respect to the measure λ if for any $D' \subset D$, $D' \in \mathcal{K}_e$, the nets $\{C_{D'}(A_n)\}$ and $\{I_{D'}(A_n)\}$ are mean (resp. pointwise) right averaging with respect to λ.

Example 3.3. Corollary 4.2.1 and Lemma 6.3.1 imply that any Van Hove net is D–robust mean averaging with respect to μ_r for any $D \in \mathcal{K}_e$.

Example 3.4. Theorem 4.5.1 implies that on a SQM–group any net $\{A_n,\ n \in N\}$ such that $\lim_{n\in N}\mu_r(I_D(A_n)) = \infty$ is a D–robust mean averaging net with respect to μ_r.

Example 3.5. Theorems 5.4.4 and Lemmas 6.3.1 and 6.3.2 imply that any D–regular Van Hove net is D–robust pointwise averaging with respect to μ_r.

Definition 3.4. A net of sets $\{A_n,\, n \in N\} \subset \mathcal{B}$ with $0 < \lambda(A_n) < \infty$ is called an *HRM_λ–mean* (resp. *pointwise*) *averaging net* if for any λ–h.r.m. $\Xi(\cdot)$ such that $\mathbf{E}|\Xi(U)|^\alpha < \infty$ for some $1 < \alpha < \infty$ and some $U \in \mathcal{K}_e$, the limit

$$\lim_{n\in N} \Xi(A_n)/\lambda(A_n) = \mathbf{E}[\,\Xi(A)/\lambda(A)|\mathfrak{J}\,] \tag{3.1}$$

exists in $L^\alpha(\Lambda,\, \mathcal{F},\, P)$ (resp. a.s.) (for any $A \in \mathcal{K}$ the equality holds a.s.).

With an arbitrary λ–h.r.m. $\Xi(\cdot)$ and an arbitrary $D \in \mathcal{K}_e$ we can associate the random fields

$$\begin{aligned} \xi^D(g) &:= [\lambda(D)]^{-1} \int \mathcal{X}_D(g^{-1}f)\Xi(df) = [\lambda(D)]^{-1}\Xi(gD), \\ \xi_0^D(g) &:= [\Delta_\lambda(g)]^{-1}\xi^D(g). \end{aligned} \tag{3.2}$$

It is clear that these fields are strictly measurable; the field $\xi^D(\cdot)$ is "λ–homogeneous" and $\xi_0^D(\cdot)$ is homogeneous in the strict sense with respect to right shifts.

Lemma 3.3. *Let $\Xi(\cdot)$ be a λ–h.r.m. on $\mathcal{B}$ and $\mathbf{E}[\xi(D)]^\alpha < \infty$ for some $D \in \mathcal{K}_e$ and $1 \le \alpha < \infty$. Then $\mathbf{E}[\xi_0^D(e)]^\alpha < \infty$, and for any $A \in \mathcal{K}$, $\mathbf{E}[\xi_0^D(e)|\mathfrak{J}] = \mathbf{E}[\xi(A)/\lambda(A)|\mathfrak{J}]$ a.s.*

Proof. It is clear that $\mathbf{E}[\xi_0^\infty(e)]^\alpha = \mathbf{E}[\Xi(D)/\lambda(D)]^\alpha < \infty$ and $\mathbf{E}[\xi_0^D(e)|\mathfrak{J}] = \mathbf{E}[\Xi(D)/\lambda(D)|\mathfrak{J}]$ a.s.. It remains to show that for any $A \in \mathcal{K}$ we have $\mathbf{E}[\Xi(A)/\lambda(A)|\mathfrak{J}] = \mathbf{E}[\Xi(D)/\lambda(D)|\mathfrak{J}]$. For any $\Lambda \in \mathfrak{J}$,

$$\begin{aligned} &\int_\Lambda \mathbf{E}[\Xi(gA)|\mathfrak{J}]dP = \int_\Lambda \Xi(gA)dP = \\ &= \int_\Lambda \Xi(A)dP = \Delta_A(g) \int_\Lambda \mathbf{E}[\Xi(A)|\mathfrak{J}]dP, \end{aligned}$$

which implies, by arguments similar to those used in the proof of Lemma 6.1.3, that $\int_{\Lambda} \mathbf{E}[\Xi(A)|\mathfrak{J}]dP = \alpha(\Lambda)\lambda(A)$ where $0 \leq \alpha(\Lambda) < \infty$. It is clear that $\alpha(\cdot)$ is a measure on $\mathfrak{J}$ and $\alpha \ll P_{\mathfrak{J}}$, where $P_{\mathfrak{J}}$, is the restriction of P to $\mathfrak{J}$, and for any $A \in \mathcal{K}$,

$$\frac{d\alpha}{dP_{\mathfrak{J}}}(\omega) = \mathbf{E}[\Xi(A)/\lambda(A)|\mathfrak{J}] \qquad \text{a.s.}$$

□

Theorem 3.4. *If a net $\mathcal{A} = \{A_n,\ n \in N\}$ has Property (Γ_λ) and is D_0–robust mean (resp. D_0–robust pointwise) averaging with respect to λ for some $D_0 \in \mathcal{K}_e$, then it is HRM_λ–mean (resp. HRM_λ–pointwise) averaging.*

Proof. Let $D \in \mathcal{K}_e$; $\beta(\mathcal{D}) := \mu_l(D)/\lambda(D)$. Consider the field $\xi^D(\cdot)$ defined by the formula (3.2). Using Fubini's theorem we obtain

$$\int_F \xi^D(g)\mu_l(dg) = (\lambda(D))^{-1}\int \mu_l(F \cap gD)\Xi(dg), \quad F \in \mathcal{B}. \qquad (3.3)$$

Since $\{g\colon\ I_D(A_n) \cap gD \neq \emptyset\} \subset A_n$, $n \in N$, we obtain $\mu_l(I_D(A_n) \cap \cap gD) \leq \mu_l(D)$, $g \in G$, and $\mu_l(I_D(A_n) \cap gD) = 0$, $g \notin A_n$. Therefore (3.3) implies $\int_{I_D(A_n)} \xi^D(g)\mu_l(dg) \leq \beta(D)\Xi(A_n)$. Similarly, $A_n \subset \subset \{g;\ gD \subset C_D(A_n)\}$ and thus $\mu_l(C_D(A_n) \cap gD) = \mu_l(D)$, $g \in A_n$. Using equation (3.3) we obtain:

$$\int_{C_D(A_n)} \xi^D(g)\mu_l(dg) \geq$$

$$\geq (\lambda(D))^{-1}\int_{A_n} \mu_l(C_D(A_n) \cap fD)\Xi(dy) = \beta(D)\Xi(A_n).$$

Put $\varphi_n(D) : +\lambda(\partial_D A_n)/\lambda(A_n)$;

$$\gamma := \mathbf{E}([\lambda(A)]\Xi(A)|\mathfrak{J}); \quad \nu(F) := |\frac{1}{\lambda(F)}\int_F \xi^D(g)\mu(dg) - \gamma|;$$

$$\alpha_n(D) := \max\{\nu(C_D(A_n)),\ \nu(I_D(A_n))\}, \quad n \in N.$$

Without loss of generality we may assume that the h.r.m. $\Psi(A) := := \int_A [\Delta_\lambda(g)]^{-1}\, \Xi(dg)$ coincides with its coordinate representation. It is easy to check that the random field $\xi_0^{\mathcal{D}}(q)$ is covariant with respect to the shift dynamical system T associated with the random function $\Psi(\cdot)$; moreover, $\mathfrak{I} = \mathfrak{I}_T$. Since $\int_F \xi^D(g)\mu(dg) = \int \xi_0^D(g)\lambda(dg)$, $F \in \mathfrak{B}$, it follows from Lemma 3.3 and from the conditions of the theorem that for any $D \subset D_0$ in L^α (resp. a.s.), $\lim_{n\in N} \alpha_n(D) = 0$. It is easy to verify that with probability *1*,

$$\left|\frac{1}{\lambda(A_n)} \int_{I_D(A_n)} \xi^D(g)\mu(dg) - \gamma\right| \le |\gamma|\varphi_n(D) + \alpha_n(D)[1 + \varphi_n(D)],$$

$$\left|\frac{1}{\lambda(A_n)} \int_{C_D(A_n)} \xi^D(g)\mu(dg) - \gamma\right| \le |\gamma|\varphi_n(D) + \alpha_n(D)[1 + \varphi_n(D)].$$

All this implies that $|\beta(D)\Xi(A_n)/\lambda(A_n) - \gamma| \le |\gamma|\varphi_n(D) + \alpha_n(D)(1 + +\varphi_n(D))$. Therefore,

$$\beta(D) = \overline{\lim}_{n\in N} \|\Xi(A_n)/\lambda(A_n) - \gamma\| [|\beta(D) - 1| + \varphi_{\mathcal{A}}(D|\lambda)], \qquad (3.4)$$

where $\|C\|$ can be understood as the norm of the element $C \in \in L^\alpha(\Lambda, \mathcal{F}, P)$ as well as the absolute value of the number $C \in \mathbf{R}$. Recall that $[\beta(D)]^{-1} = \frac{1}{\mu(D)} \int_D \Delta_\lambda(g)\mu(dg)$; since the function $\Delta_\lambda(\cdot)$ is continuous and $\Delta_\lambda(e) = 1$, it is easy to verify that $\lim_{D\in\mathcal{K}_e} \beta(D) = 1$. It remains only to take limits of both sides of the inequality (3.4) with respect to $D \in \mathcal{K}_e$.

□

Theorem 3.4 and Examples 3.5–3.7 imply three simple consequences.

Corollary 3.5. *Any Van Hove net is HRM–mean averaging.*

Note 3.1. Garonas and Tempelman [2] have proved that equality (3.1) in $L^1(\Lambda, \mathcal{F}, P)$ holds for $\lambda = \mu_r$ if we replace the Van Hove condition by the weaker condition of right ergodicity of $\{A_n, n \in N\}$ with respect to μ_r.

Corollary 3.6. *On a SQM–group any net $\{A_n,\ n \in N\}$ satisfying condition (Γ_λ) and such that $\lim\limits_{n\in N} \mu_l(I_D(A_n)) = \infty$ for some $D \in \mathcal{K}_e$ is an HRM_{μ_l}–mean averaging net.*

Corollary 3.7. *Any D–regular Van Hove net is an HRM_{μ_r}–pointwise averaging net.*

§ 4. Homogeneous random measures on homogeneous spaces

Let $Y = G/K$ be a metric homogeneous space. We suppose that the group G is unimodular. The notion of homogeneous random measure (h.r.m) on Y is an obvious generalization of the notion of an h.r.m. on G (see Definitions 6.1.1 and 6.1.2). If Ξ is an h.r.m. on Y, then its K–uniform inverse image $\widetilde{\Xi}(\cdot)$ on G is also an h.r.m. This allows us to transfer all results of §§ 2 and 3 to h.r.m.'s on homogeneous spaces.

Let μ be the invariant Borel measure on Y (its K–uniform preimage of $\widetilde{\mu}$ is the Haar measure on G). Let us denote by $\widetilde{A}$ the preimage of $A \subset Y$ (with respect to the natural mapping $\pi : G \mapsto Y$; see App., §2). Let B_r be the ball of radius r with center $y_0 (= K)$. We denote:

$$d(y, A) := \inf\{\rho(y, x),\ x \in A\}; \quad C_r(A) := \{y : d(y, A) \le r\},$$
$$I_r(A) = \{y : d(y, Y \setminus A) > r\}, \quad \partial_r := C_r(A) \setminus I_r(A).$$

If $\mathcal{A} = \{A_n,\ n \in N\}$ is a net of Borel sets, we put $\varphi_{\mathcal{A}}(r) := := \overline{\lim}_{n\in N} \mu(\partial_r A_n)/\mu(A_n)$. We assume that the balls B_r are compact.

Definition 4.1. A net $\mathcal{A} = \{A_n,\ n \in N\}$ *satisfies condition* (Γ) if $\{A_n,\ n \in N\} \in \mathfrak{B}(Y)$ and $\lim_{r\to 0} \varphi_{\mathcal{A}}(r) = 0$. $\mathcal{A}$ is a *Van Hove net* if $\varphi_{\mathcal{A}}(r) = 0$ for every $r > 0$ (see App., §3). If Y is amenable, there exist Van Hove nets.

Example 4.1. For balls B_R in the Lobachevsky space L_m, $m \ge \ge 2$, $\mu(B_R) \sim k_m e^{(m-1)R}$ as $R \to \infty$. $I_r(B_R)$ is the interior of the ball B_{R-r}, and, if the balls are closed, $C_r(B_R) = B_{R+r}$. Therefore $\mu(\partial_r B_R) \sim k_m e^{(m-1)R}(e^{(m-1)r} - e^{-(m-1)r})$ $(R \to \infty)$, and, consequently, if $\mathcal{A} = \{B_R,\ R > 0\}$ then $\varphi_{\mathcal{A}}(r) = 2\,\mathrm{sh}\,[(m-1)r]$. Of course, $\mathcal{A}$ does not satisfy the Van Hove condition, but it satisfies the condition (Γ).

Example 4.2. We consider the Lobachevsky plain L_2 and the oricycle–hypercycle quadrangles (see Example 6.9.3) A_n with parameters $(a_n,\ b_0)$. It is easy to check that $C_r(A_n)$ and $I_r(A_n)$ are also oricycle–hypercycle quadrangles, with parameters $(a_n + r,\ b_0 + r)$ and $(a_n - r,\ b_0 - r)$, resp. Further, $\mu(A_r) = Ya_r \cdot b_0$, $\mu(C_r(A_R)) = Y(a_n + r)(b_0 + r)$, $\mu(I_r(B_n)) = Y(a_n - r)(b_0 + r)$, $\mu(\partial_r B_n) = 8a_n r + 8rb_0$, and $\varphi_{\mathcal{A}}(r) = 2b_0^{-1}r$. Thus condition (Γ) is satisfied.

Let us denote by $\widetilde{A}$ the inverse image in G of $A \subset Y$ with respect to the natural mapping π (see App., § 2).

If a net $\mathcal{A} = \{A_n,\ n \in N\} \subset \mathfrak{B}_0(Y)$ satisfies condition Γ, then the net $\widetilde{\mathcal{A}} = \{\widetilde{A}_n,\ n \in N\}$ satisfies the condition $(\Gamma_{\widetilde{\mu}})$ in G. The same applies to the Van Hove condition; this follows from the following lemma.

Lemma 4.1. *If $A \in \mathfrak{B}(Y)$, then $\widetilde{C_r(A)} = C_{\widetilde{B}_r}(\widetilde{A})$ and $\widetilde{I}_r(A) \subset I_{B_r}(\widetilde{A})$. Thus, for any net of bounded sets $\mathcal{A} \subset \mathfrak{B}(Y)$ we have $\varphi_{\mathcal{A}}(r) \geq \varphi_{\widetilde{\mathcal{A}}}(\widetilde{B}_r|\widetilde{\mu})$ (here the B_r are closed balls in Y).*

Proof. Since the set B_r is compact, $C_{\widetilde{B}_r}(\widetilde{A}) = [\widetilde{A}]\widetilde{B}_r = \widetilde{[A]}\widetilde{B}_r$. Let $g \in C_{\widetilde{B}_r}(\widetilde{A})$; this means that $g = \widetilde{g}f$ where $\widetilde{g} \in \widetilde{[A]}$, $f \in \widetilde{B}_r$, that is, $\widetilde{g}x_0 \in [A]$, $fx_0 \in B_r$. But $\rho(gx_0,\ \widetilde{g}x_0) = \rho(\widetilde{g}fx_0,\ \widetilde{g}x_0) = \rho(fx_0,\ x_0) \leq r$ and thus $gx_0 \in C_r(A)$, i.e. $g \in \widetilde{C_r(A)}$. The second statement can be proved similarly.

□

Now it is easy to transfer the main results of §§ 2 and 3 to h.r.m.'s on homogeneous spaces. We restrict ourselves to the analogs of Corollaries 3.5 and 3.6. The notion of *HRM–mean* (or *HRM–pointwise*) *averaging net* on Y used here is an obvious generalization of that of

HRM_{μ_l}–mean (resp. HRM_{μ_l}–pointwise) averaging net on G.

Theorem 4.2. *Let G be a unimodular amenable group. Then any Van Hove net on Y is an HRM–mean averaging net.*

Theorem 4.3. *On a SQM–space any net $\{A_n,\ n \in N\}$ which satisfies the condition (Γ) and for which $\lim_{n \in N} \mu(I_r A_n) = \infty$ for some $r > 0$ is an HRM–mean averaging net.*

Our next theorem gives an example of an HRM–pointwise averaging net on the Lobachevsky plain L_2. It follows from Theorem 3.4 and Example 4.2.

Theorem 4.4. *The sequence of oricycle–hypercycle quadrangles A_n with parameters a_n, b_0, $n = 1, 2, \ldots$, is an HRM–pointwise averaging sequence if $a_n \to +\infty$.*

§ 5. Homogeneous Random Signed Measures

Let G be a separable locally compact group.

Definition 5.1. A random function $\Xi(A) = \Xi(A, \omega)$, $A \in \mathfrak{B}$, on the probability space $(\Lambda, \mathcal{F}, P)$ is called a *random signed measure* if it satisfies the following conditions:

(*i*) for almost all ω, $\Sigma(\cdot, \omega)$ is a signed measure on $\mathfrak{B}$ (see, e.g., Halmos [1]);

(*ii*) with probability *1*, var $\Xi(K, \omega) < \infty$, $K \in \mathcal{K})^{*)}$;

(*iii*) for any $A \in \mathfrak{B}$ the function $(g, \omega) \longmapsto \Xi(gA, \omega)$ is $(\mathfrak{B} \times \mathcal{F})$–measurable.

The notions of homogeneous and λ–homogeneous random signed measure are defined similarly as that of h.r.m.

Let $\Xi(A, \omega)$ be a random signed measure. For almost all $\omega \in \Lambda$ we have the Jordan decomposition $\Xi(A, \omega) = \Xi^+(A, \omega) - \Xi^-(A, \omega)$, where

---

*) var $\Xi(A,\omega)$ is the variation of $\Xi(\cdot, \omega)$ on the set A.

$\Xi^+(A,\omega)$ and $\Xi^-(A,\omega)$ are measures: $\Xi^+(A,\omega) = \sup\{\Xi(B)\colon B \subset A, B \in \mathfrak{B}\}$; $\Xi^-(A,\omega) = \sup\{-\Xi(B)\colon B \subset A, B \in \mathfrak{B}\}$.

Lemma 5.1. *If Ξ is a λ-homogeneous random signed measure, then the measures Ξ^- and Ξ^+ are λ-homogeneous random signed measures.*

Proof. In view of the "signed" analog of Lemma 1.1 it suffices to consider the case when $\Xi(\cdot)$ is homogeneous. Let $U = \{B_i, i = 1, 2, \ldots\}$ be a countable family of open sets generating the σ-algebra $\mathfrak{B}$ and let $\widetilde{\mathfrak{B}}$ be the (countable) algebra generated by U. With probability *1*, for any $\varepsilon > 0$ and any $A \in \mathcal{K}$ there exists a $B \in \widetilde{\mathfrak{B}}$ such that $|\Xi(B) - \Xi(A)| < \varepsilon$. Consequently, for any $K \in \mathcal{K}$ we have with probability *1*, $\Xi^+(K) = \sup\{\Xi(B), B \subset K, B \in \widetilde{\mathfrak{B}}\}$. Similarly, $\Xi^-(K) = \sup\{-\Xi(B), B \in K, B \in \widetilde{\mathfrak{B}}\}$. This shows that $\Xi^+(K)$ and $\Xi^-(K)$ are random functions on $\mathcal{K}$. It is easy to show that these random functions are covariant with respect to the Ξ-shift dynamical system. Consequently, $\Xi^+(\cdot)$ and $\Xi^-(\cdot)$ are homogeneous random signed measures (see App. 8.9).

□

It is well known that var $\Xi(A) = \Xi^+(A) + \Xi^-(A)$. This fact and Lemma 5.1 admit to transfer easily the results of §§ 2–4 to random signed measures. We have only to replace in all statements the condition $\mathbf{E}[\Xi(A)] < \infty$ by the condition $\mathbf{E}[\mathrm{var}\,\Xi(A)] < \infty$.

BIBLIOGRAPHICAL NOTES

1. This chapter is close to Chapter 6 in Tempelman [24].

2. In Wiener [1] (1939) the local ergodic theorem (Theorem 2.1) for absolutely continuous random measures on $\mathbb{R}^m$ was proved. Akcoglu and Krengel [1] (1979), [2] (1981) have proved a local ergodic theorem for homogeneous superadditive random set functions on $\mathbb{R}^m$. For $\alpha = 1$ the assertion of Theorem 2.5 is contained in Boclé [2] (1959), [3] (1960). § 2 is textually close to § 3 of Garonas, Tempelman [2] (1984) (see also Garonas, Tempelman [1] (1981)).

3. Corollary 3.5 is a generalization of Theorem 2 in Fritz [1] (1970).

4. In Akcoglu, Krengel [1] (1979), [2] (1981), Nguyen [1] (1979), and Smythe [1] (1976), mean and pointwise "global" ergodic theorems for homogeneous super– and subadditive functions on $\mathbb{R}^m$ and $\mathbb{Z}^m$ were proved.

5. Nguyen and Zessin [1] (1976), [2] (1979) have proved ergodic theorems for homogeneous point random fields on $\mathbb{R}^m$ and homogeneous "spacial processes", i.e. homogeneous random functions on $\mathbb{R}^m$ satisfying certain additional conditions.

6. Corollary 3.7 coincides with Theorem 4.2 in Garonas, Tempelman [2] (1984), and Corollary 3.5 for $\alpha = 1$ is contained in Theorem 4.1 of this work; also § 5 is textually close to § 5 of this work (see also Garonas, Tempelman [1] (1981)). In Garonas [1] (1984), a mean ergodic theorem (similar to Corollary 3.5) for wide–sense homogeneous random fields on $\mathbb{R}^m$ was proved.

Results close to Chapter 7 are also contained in Gaposhkin [5], Krengel [1], Kubokava [1, 2], Sato [1], Terrel [1].

CHAPTER 8
SPECIFIC INFORMATIONAL AND THERMODYNAMICAL CHARACTERISTICS OF HOMOGENEOUS RANDOM FIELDS

§ 1. Main Notations and Definitions

1.1. Configuration space, shifts, invariant measures. In this section we use the following notations: G is a countable group with the identity element e; $(X, \mathfrak{X}, \mu)$ is a σ–finite measure space; $\Omega_A (A \subset G)$ coincides with the space X^A of all X–valued functions ("*configurations*") on A; $\Omega = \Omega_G$; ω_A is the restriction of $\omega \in \Omega$ to A; $\eta_A \zeta_B (A, B \subset G,\ A \cap B = \emptyset,\ \eta,\ \zeta \in \Omega)$ is the configuration $\omega_{A \cup B} (\in \Omega_{A \cup B})$ with $\omega_A = \eta_A, \omega_B = \zeta_B$; $\mathcal{F}_A$ is the σ–algebra in Ω generated by the cylindrical sets $\{\omega : \omega(g) \in B\}$, $g \in A$, $B \in \mathfrak{X}$; $\mathcal{F} \in \mathcal{F}_G$; we write $\varphi(\omega_A)$ instead of $\varphi(\omega)$ if φ is $\mathcal{F}_A$–measurable. $R_g (g \in G)$ is the right *shift* (or *translation*) in Ω by the element g: $(R_g \omega) = \omega(fg)$; $\mathcal{P}$ is the set of all probability measures on $\mathcal{F}$; $\mathcal{P}_0$ is the set of all probability measures P on $\mathcal{F}$ which are invariant under all translations R_g, i.e. satisfy the relations $P(R_g \Lambda) = P(\Lambda)$, $\Lambda \in \mathcal{F}$, $g \in G$.

If $P \in \mathcal{P}_0$, then $R = \{R_g,\ g \in G\}$ is a left dynamical system in $(\Lambda,\ \mathcal{F},\ P)$ with time G. We denote by $\mathfrak{I}_T$ the algebra $\{\Lambda : \Lambda \in \mathcal{F},\ P(R_g \Lambda \Delta \Lambda) = 0\}$.

We shall often write $\mathbf{E}_P(\varphi(\cdot)|\omega_A)$ instead of $\mathbf{E}_P(\varphi(\cdot)|\mathcal{F}_A)$.

1.2. The direction of all finite subsets of $A \in G$ (with respect to inclusion) is denoted by $\mathcal{K}_A$; $\mathcal{K} := \mathcal{K}_G$; if $\mathcal{L}$ is a subdirection of $\mathcal{K}$ and $\{b_A,\ A \in \mathcal{L}\}$ is a net of reals we set $\sum_{A \in \mathcal{L}} b_A = \lim_{A \in \mathcal{L}} \sum_{\substack{A \in \mathcal{L} \\ A \subset B}} b_A$ (provided that the limit exists); $|A| = \text{card } A$ if $A \subset G$; $\bar{A} = G \backslash A$.

1.3. Covariant random set functions. A function $\Phi(A,\omega)$, $A \in \mathcal{K}$, $\omega \in \Omega$, is called a *random set function* if $\Phi(A,\cdot)$ is $\mathcal{F}$–measurable for any $A \in \mathcal{K}$; $\Phi(A,\omega)$ is called *covariant* if $\Phi(Ag^{-1}, R_g\omega) = \Phi(A,\omega)$ for any $A \in \mathcal{K}$, $\omega \in \Omega$ and $g \in G$.

1.4. Potentials and energy. A *potential* is a real valued random set function $U(A,\omega)$, $A \in \mathcal{K}$, $\omega \in \Omega$, such that

a) $U(A,\omega) = U(A,\omega')$ if $\omega_A = \omega'_A$;

b) $\sum_{e \in A \in \mathcal{K}} |U(A,\omega)| < \infty$ for any $\omega \in \Omega$.

$\|U\| := \sum_{e \in A \in \mathcal{K}} \sup_{\omega \in \Omega} |U(A,\omega)| \leq \infty$ is the *norm* of U. The quantity $E_U(\omega_V|\omega_{\bar V}) := \sum_{A \cap V \neq \emptyset} U(A,\omega)$ $(\omega \in \Omega, V \in \mathcal{K})$ is the *(potential) energy* of the configuration ω_V on V given the environment $\omega_{\bar V}$. One can also consider the energy

$$E_U(\omega_V|\omega_{\overline{V \cup C}}\Theta_C) := \sum_{\substack{A \cap V \neq \emptyset \\ A \subset \bar C}} U(A,\omega), \qquad C \subset \bar V, \tag{1.1}$$

with formal "vacuum" on the part C of the set $\bar V$. We regard $\omega_{\overline{V \cup C}}\Theta_C$ as a configuration on $\bar V$ with values in $\widetilde{X} := X \cup \{\Theta\}$; so $E_U(\omega_V/\omega_{\bar V})$ is defined for $\omega_V \in \Omega_V$ and $\omega_{\bar V} \in \widetilde{\Omega}_{\bar V} := (X \cup \{\Theta\})^{\bar V}$. If X contains a proper *vacuum state* Θ (i.e. $U(A,\omega) = 0$ as soon as $\Theta \in \{x : x = \omega(g),\ g \in A\}$), one can treat Θ_C in (1.1) as the proper vacuum configuration on C: $\Theta_C(g) = \Theta$, $g \in C$, and set $\widetilde{X} = X$. In this case (1.1) becomes a special case of the definition of $E_U(\omega_V|\omega_{\bar V})$ for $\omega \in \Omega$ given above. Clearly, the functions $\omega \mapsto E(\omega_V|\omega_{\bar V})$, $\omega \mapsto E(\omega_V|\omega_{\overline{V \cup C}}\zeta_C)$ $(C \subset \bar V,\ \zeta_C \in \widetilde{\Omega}_C$ fixed) are $\mathcal{F}$–measurable.

1.5. Densities. Let Q be a measure on F and $B \subset G$; then Q_B denotes the restriction of Q to the σ–algebra $\mathcal{F}_B$; we shall often identify Q_B with its projection πQ_B to the σ–algebra $\mathfrak{X}^B$ $(\pi Q_B(C) = Q_B(\{\omega : \omega_B \in C\}),\ C \in \mathfrak{X}^B)$; for $A \in \mathcal{K}$ we set $\mu^A := \prod_{g \in A} \mu^g$ where $\mu^g \equiv \mu$; $\mu_A := \pi^{-1}\mu^A$; if $B \subset \bar A$, $P \in \mathcal{P}$, then $M_{A,B} := \mu^A \times P_B$.

We denote by $\mathcal{P}_\mu$ the class of all measures P $(\in \mathcal{P})$ such that $P_A \ll \mu_A$ for all $A \in \mathcal{K}$. For $P \in \mathcal{P}_\mu$ we define the *μ–density* of P_A by

$p(\omega_A) := \frac{dP_A}{d\mu_A}(\omega)$ $(A \in \mathcal{K})$; if in addition $P_{A\cup B} \ll M_{A,B}$ $(A \in \mathcal{K}$, $B \subset \bar{A})$ we say that the *conditional μ–density* $p(\omega_A|\omega_B)$ exists and define it as $p(\omega_A|\omega_B) := \frac{dP_{A\cup B}}{dM_{A,B}}$; for example, $p(\omega_A|\omega_B)$ exists if $P \in \mathcal{P}_\mu$ and $A \in \mathcal{K}$, $B \in \mathcal{K}_{\bar{A}}$. If $p(\omega_A|\omega_B)$ exists, we denote the corresponding regular conditional distribution on $\mathcal{F}_A$ by $P(\cdot|\omega_B)$, i.e.

$$P_A(\Lambda|\omega_B) := \int_\Lambda p(\omega_A|\omega_B)\mu_A(d\omega), \quad \Lambda \in \mathcal{F}_A.$$

1.6. Gibbsian measures. Let U be a potential with $\|U\| < \infty$ and let $\mu(X) < \infty$. The *Gibbsian specification corresponding* to $U$is the family of functions $\Lambda = \{\lambda_U(\omega_A|\omega_{\bar{A}}),\ A \in \mathcal{K}\}$ defined by the formulas:

$$\lambda_U(\omega_A|\omega_{\bar{A}}) = [Z_U(A|\omega_{\bar{A}})]^{-1} exp\{-E_U(\omega_A|\omega_{\bar{A}})\}, \qquad A \in \mathcal{K}, \quad (1.2)$$

where $\omega_A \in \Omega_A$, $\omega_{\bar{A}} \in \Omega_{\bar{A}}$ and

$$Z_U(A|\omega_{\bar{A}}) := \int_{\Omega_A} \exp\{-E_U(\omega_A|\omega_{\bar{A}})\}\mu_A(d\omega). \quad (1.3)$$

A measure $Q\,(\in \mathcal{P}_\mu)$ is called a *Gibbsian measure corresponding to the potential* U if the conditional densities $q(\omega_A|\omega_{\bar{A}})$ exist and with Q–probability *1*, $q(\omega_A|\omega_{\bar{A}}) = \lambda(\omega_A|\omega_{\bar{A}})$. $\mathcal{G}(U)$ is the set of all such Q. It is easy to check that $\mathcal{G}(U) \subset \mathcal{P}_0$; under certain restrictions on X and U, $\mathcal{G}(U) \neq \emptyset$ (see e.g. Dobrushin [1–3], Preston [1,2]). Formulas (1.1)–(1.3) allow us to define $\lambda_U(\omega_A|\omega_{\bar{A}})$, $A \in \mathcal{K}$, even if $\omega_A \in \widetilde{\Omega}_A := (X \cup \{\Theta\})^A$ where Θ is the "vacuum state" (see Subsect. 1.4.); we shall also assume that $q(\omega_A|\omega_{\bar{A}}) = \lambda(\omega_A|\omega_{\bar{A}})$ for all $\omega_A \in \Omega_A$, $\omega_{\bar{A}} \in \widetilde{\Omega}_{\bar{A}}$.

1.7. Invariant orders in G. A linear order $\prec$ in G is called *invariant* if $fg \prec g$ as soon as $f \prec e$; if such an order exists, G is called an *orderable* group (for example, the lexigographical order in $\mathbb{Z}^m$ is an invariant order). If G is orderable, we fix an invariant order $\prec$ and set $V_f = \{g : g \in G,\ g \prec f\}$. In an unorderable group we have to consider a family of linear orders $\{\overset{s}{\prec}, s \in S\}$ and an "ordering dynamical system" $\{S, \mathcal{S}, \lambda; \Gamma_g, g \in G\}$ (see Subsect. 1.9). This causes some complications in arguments and notations. A reader who is ready to suppose G to be orderable can neglect all marks connected with the dynamical system $\{\Gamma_g, g \in G\}$ in $\{S, \mathcal{S}, \lambda; \Gamma_g, g \in G\}$ in the formulas of § 3 and § 5, as well as omit Subsect. 1.8 and 1.9.

1.8. Random sets in G. Let $(S, \mathcal{S}, \lambda)$ be a probability space. A family $A(\cdot) = \{A(s), s \in S\}$ of subsets of G is called a *random set* in G if for any $g \in G$ the set $S_g := \{s : g \in A(s)\} \in \mathcal{S}$. This is equivalent to the requirement: for any $B \in \mathcal{K}$ and any $C \subset B$ the set $S_C^B :=$ $:= \{s : B \cap A(s) = C\} \in \mathcal{S}$; indeed, $S_C^B = \bigcap_{g \in C} S_g \cap \bigcap_{g \in B \setminus C} \bar{S}_g$. Note that if B is fixed, the sets S_C^B, $C \subset B$, are disjoint and $\bigcup_{C \subset B} S_C^B = S$. Clearly, if $A(\cdot)$ is a random set, then $g^{-1}A(\cdot)$, $A(\cdot)g^{-1}$, $A(\cdot) \cup B$ are also random sets for any $g \in G$, $B \subset G$.

1.9. Homogeneous random orders in G. A dynamical system $\{\Gamma_g, g \in G\}$ in $(S, \mathcal{S}, \lambda)$ is called an *ordering dynamical system* if $\mathcal{S}_{\Gamma,\lambda} =$ $= \{\emptyset, S\}$ (λ–mod 0) and there exists a family of linear orders $\{\overset{s}{\prec}, s \in$ $\in S\}$ in G with the following properties:

(*i*) $\{s : s \in S, g \overset{s}{\prec} e\} \in \mathcal{S}$ for any $g \in G$;

(*ii*) $fg \overset{s}{\prec} g$ if and only if $f \overset{\Gamma_g s}{\prec} e$ $(f, g \in G,\ s \in S)$.

Such a family $\{\overset{s}{\prec}, s \in S\}$ is called a *homogeneous random order* in G; $V_g(s) := \{f : f \in G, f \overset{s}{\prec} g\}$. Property (*i*) means that $S_g :=$ $:= \{s : g \in V(s)\} \in \mathcal{S}$, i.e. $V(\cdot)$ is a random set (see Subsect. 1.8); property (*ii*) means that $V_g(s)g^{-1} = V_e(\Gamma_g s)$. Homogeneous random orders exist in any countable group (see Kieffer [1], Stepin [1]). If G is orderable, we can take the ordering system to be trivial:

$$S = \{s_0\}, \qquad \Gamma_g s_0 = s_0 (g \in G), \quad \lambda(\{s_0\}) = 1, \quad \overset{s_0}{\prec} = \prec,$$

and avoid any references to S and λ.

§ 2. Generalized Ergodic Theorems

In the sequel we shall need some generalized versions of the mean and individual ergodic theorems (see Chapters 4 and 6). Let $\{T_g, g \in G\}$ be a dynamical system in a probability space $(W, \mathcal{W}, P)$.

Theorem 2.1. *Suppose $\{A_n\}$ is a (Følner) sequence of sets in G; φ, φ_A $(e \in A \in \mathcal{K})$ are functions on W, and*

1) $\varphi, \varphi_A \in L^1(W, \mathcal{W}, P)$, $e \in A \in \mathcal{K}$;

2) $\lim\limits_{e \in A \in \mathcal{K}} \varphi_A = \varphi$ in $L^1(W, \mathcal{W}, P)$;

3) $\sup\limits_{e \in A \in \mathcal{K}} \mathbf{E}_P|\varphi_A(w)| < \infty$.

Then in $L^1(W, \mathcal{W}, P)$ the following limit exists

$$\lim_{n \to \infty} |A_n|^{-1} \sum_{g \in A_n} \varphi_{A_n g^{-1}}(T_g w) = \mathbf{E}_P[\varphi(w)|\mathfrak{I}_T] \tag{2.1}$$

(equality holds with P-probability 1).

Theorem 2.2. *Suppose $\{A_n\}$ is a regular left-ergodic (Følner) sequence of sets in G and*

1) $\varphi, \varphi_A \in L^1(W, \mathcal{W}, P)$, $e \in A \in \mathcal{K}$;

2) $\lim\limits_{e \in A \in \mathcal{K}} \varphi_A(w) = \varphi(w)$ with P-probability 1;

3) $\mathbf{E}_P[\sup\limits_{e \in A \in \mathcal{K}} |\varphi_A(w)|] < \infty$.

Then the limit in (2.1) exists with P-probability 1.

Note 2.1. If $\varphi_A = \varphi$ for all $A \in \mathcal{K}$, these theorems coincide with the usual form of the mean and pointwise ergodic theorems for dynamical systems with group time (see Theorem 4.2.1 and Corollary 6.3.2). In the proofs of Theorems 2.1 and 2.2 we shall use the following two statements.

Lemma 2.3. *Let $\mathcal{A} = \{A_n\}$ be a Følner sequence; $D \in \mathcal{K}$; $C_n := := \bigcap\limits_{g \in D} g^{-1} A_n$. Then*

a) $A_n g^{-1} \supset D$ for any $g \in C_n$;

b) $\{C_n\}$ *is a Følner sequence;*
c) $\lim_{n\to\infty} |A_n|^{-1}|A_n \Delta C_n| = 0$*;*
d) $\lim_{n\to\infty} \frac{|C_n|}{|A_n|} = 1$*;*
e) if $\{A_n\}$ *is regular, then* $\{C_n\}$ *is regular.*

Proof. *a)* for any $d \in D$, $c \in C_n$ there exists an element $a\,(a \in A_n)$ such that $c = d^{-1}a$ and thus $d = ac^{-1} \in A_n C_n^{-1}$.
b)–d) follow from the inequality $|A_n \Delta C_n| \le \sum_{g\in D} |A_n \Delta g^{-1} A_n|$.
e) Let $\{A_n\}$ be regular and let $\widetilde{C}_n := \bigcup_{k=1}^{n} C_n$. Then $\widetilde{C}_n = \bigcap_{g\in D} g^{-1}\widetilde{A}_n$, and hence $|\widetilde{C}_n \cdot C_n^{-1}| \le |d_0^{-1}\widetilde{A}_n \cdot A_n^{-1}| \le |\widetilde{A}_n \cdot A_n^{-1}| \le \gamma(\mathcal{A})|A_n| \le \gamma_1 |C_n|$ by *d)* $(d_0 \in D)$.

□

Lemma 2.4. *Suppose* $\{A_n\}$ *is a Følner sequence of sets in* G, $\{a_A,$ $e \in A \in \mathcal{K}\}$ *is a net of reals and* $\lim_{e\in A\in\mathcal{K}} a_A = a$. *If either*
1) $\sup_{e\in A\in\mathcal{K}} |a_A| < \infty$, *or*
2) $a = -\infty$ *and* $\sup_{e\in A\in\mathcal{K}} a_A < +\infty$, *or*
3) the net $\{a_A,\ e \in A \in \mathcal{K}\}$ *is nonincreasing and* $\sup_{e\in A\in\mathcal{K}} a_A < \infty$,

then $\lim_{n\to\infty} |A_n|^{-1} \sum_{g\in A_n} a_{A_n g^{-1}} = a$.

Proof. In case *1)*, for any number $\varepsilon > 0$ there exists a set $D\,(e \in D \in$ $\in \mathcal{K})$ such that $|a_A - a| < \varepsilon$ if $A \supset D$. Let $C_n = \bigcap_{g\in D} g^{-1} A_n$. According to Lemma 2.3

$$\lim_{n\to\infty} \Big| |A_n|^{-1} \sum_{g\in A_n} a_{A_n g^{-1}} - a \Big| \le \overline{\lim_{n\to\infty}} |A_n| \sum_{g\in C_n \cap A_n} |a_{A_n g^{-1}} - a| +$$
$$+ \lim_{n\to\infty} \frac{|A_n \setminus C_n|}{|A_n|} \sup_{e\in A\in\mathcal{K}} |a_A| \le \varepsilon,$$

and since ε is chosen arbitrarily our statement is proved. Case *2)* can be considered similarly. Case *3)* is a simple consequence of *1)* and *2)*.

□

Proof of Theorem 2.1. Let $a_A := \|\varphi_A - \varphi\|_{L_1}$. Then

$$\| \sum_{g \in A_n} \varphi_{A_n g^{-1}}(T_g w) - \sum_{g \in A_n} \varphi(T_g w) \| \le \sum_{g \in A_n} a_{A_n g^{-1}},$$

and our statement follows at once from Lemma 2.4 and the mean ergodic theorem 4.2.1.

□

Proof of Theorem 2.2. Let us fix a set D, $e \in D \in \mathcal{K}$, and set $C_n = \bigcap_{g \in D} g^{-1} A_n$. By applying Corollary 6.3.2 with the "averaging sequences" $\{A_n\}$ and $\{C_n\}$ to the dynamical system $\{T_g,\, g \in G\}$ (see Lemma 2.3) we obtain with P–probability *1*:

$$\lim_{n \to \infty} |A_n|^{-1} \sum_{g \in A_n \setminus C_n} \sup_{e \in A \in \mathcal{K}} |\varphi_A(T_g w) - \varphi(T_g w)| = 0$$

and thus

$$\begin{aligned} \overline{\lim_{n \to \infty}} |A_n|^{-1} \sum_{g \in A_n} |\varphi_{A_n g^{-1}}(T_g w) - \varphi(T_g w)| \le \\ \le \lim_{n \to \infty} |A_n|^{-1} \sum_{g \in C_n} \sup_{D \subset A \in \mathcal{K}} |\varphi_A(T_g w) - \varphi(T_g w)| = \\ = \mathbf{E}_P\{ \sup_{D \subset A \in \mathcal{K}} |\varphi_A(w) - \varphi(w)| | \mathfrak{I}_T \}. \end{aligned}$$

Since the set D can be chosen arbitrarily large,

$$\lim_{n \to \infty} |A_n|^{-1} \sum_{g \in A_n} |\varphi_{A_n g^{-1}}(T_g w) - \varphi(T_g w)| = 0.$$

It remains to apply Corollary 6.3.2 once more.

§ 3. Specific Energy

The main result of this section is the following theorem.

Theorem 3.1. *Let $U(A,\omega)$ be a potential, $P \in \mathcal{P}_0$ and let $\{A_n\}$ be an ergodic (resp. regular ergodic) sequence of sets.*

(i) If $\mathbf{E}_P[\sum_{e\in A\in\mathcal{K}} |A|^{-1}|U(A,\omega)|] < \infty$, then in $L^1(\Omega, \mathcal{F}, P)$ (resp. with P-probability 1) the limit

$$\lim_{n\to\infty} |A_n|^{-1} E_U(\omega_{A_n}|\Theta_{\bar{A}_n}) := e_{P,U}(\omega)$$

exists, where

$$e_{P,U}(\omega) = \mathbf{E}_P[\sum_{e\in A\in\mathcal{K}} |A|^{-1}U(A,\omega)|\mathfrak{I}_R].$$

(ii) If $\mathbf{E}_P[\sum_{e\in A\in\mathcal{K}} |U(A,\omega)|] < \infty$, then in $L^1(\Omega, \mathcal{F}, P)$ (resp. with P-probability 1),

$$\lim_{n\to\infty} |A_n|^{-1} E_U(\omega_{A_n}|\omega_{\bar{A}_n}) = e_{P,U}(\omega).$$

(iii) If $\|U\| < \infty$, then for any $\xi_n \in \widetilde{\Omega}_{\bar{A}_n}$ (see Subsect. 1.5) we have in $L^1(\Omega,\mathcal{F},P)$ (resp. with P-probability 1),

$$\lim_{n\to\infty} |A_n|^{-1} E_U(\omega_{A_n}|\xi_n) = e_{P,U}(\omega).$$

Here $e_{P,U}(\omega)$ is the specific potential energy of the configuration ω; $e_{P,U} := \mathbf{E}_P(e_{P,U}(\omega)) = \mathbf{E}_P\{\sum_{e\in A\in\mathcal{K}} |A_n|^{-1}U(A,\omega)\}$ is the specific potential energy of the measure P (under the potential U).

The proof of the theorem is preceeded by two simple auxiliary statements.

Lemma 3.2. *Let $\Phi(A,\omega)$, $A \in \mathcal{K}$, $\omega \in \Omega$, be a random set function and $\sum_{e \in A \in B} |\Phi(A,\omega)| < \infty$; let $B(\cdot)$ be a random set in G. Then the function $\psi(\omega, s) := \sum_{e \in A \in \mathcal{K}} \Phi(A,\omega)$ is $(\mathcal{F} \times \mathcal{S})$–measurable.*

Proof. This statement follows immediately from the equality $\psi(\omega, s) := \sum_{e \in A \in \mathcal{K}} \chi_{S_A^A}(s)\Phi(A,\omega)$ where $S_A^A = \{s : A \subset B(s)\}$ (see Subsect. 1.8).

□

Lemma 3.3. *Suppose $\Phi(A,\omega)$ is a covariant random set function, $(S, \mathcal{S}, \{\Gamma_g,\ g \in G\})$ is a G–ordering dynamical system acting in $(S, \mathcal{S}, \lambda)$ and $P \in \mathcal{P}_0$. If either $\Phi(A,\omega) \geq 0$ or $\mathbf{E}_P\{\sum_{e \in A \in \mathcal{K}} |A|^{-1}|\Phi(A,\omega)|\} < \infty$, then*

$$\mathbf{E}_{P\times\lambda}\Big[\sum_{\substack{e \in A \in \mathcal{K} \\ A \subset V_e(s)}} \Phi(A,\omega)\Big] = \mathbf{E}_P\Big[\sum_{e \in A \in \mathcal{K}} |A_n|^{-1}\Phi(A,\omega)\Big].$$

Proof. The following relations can be justified by Levi's monotone convergence theorem and Lebesgue's dominated convergence theorem:

$$\begin{aligned}
\mathbf{E}_P\Big\{\sum_{e \in A \in \mathcal{K}} |A|^{-1}\Phi(A,\omega)\Big\} &= \\
&= \mathbf{E}_{P\times\lambda}\Big\{\sum_{g \in G}\sum_{e,g \in A \in \mathcal{K}} |A|^{-1}\Phi(A,\omega)\Big\} = \\
&= \sum_{g \in G} \mathbf{E}_{P\times\lambda}\Big\{\sum_{\substack{e,g \in A \in \mathcal{K} \\ A \subset V_g(s)\cup\{g\}}} |A|^{-1}\Phi(Ag^{-1}, R_g\omega)\Big\} = \\
&= \sum_{g \in G} \mathbf{E}_{P\times\lambda}\Big\{\sum_{\substack{e,g^{-1} \in D \in \mathcal{K} \\ D \subset V_e(\Gamma_g s)\cup\{e\}}} |D|^{-1}\Phi(D, R_g\omega)\Big\} =
\end{aligned}$$

$$= \mathbf{E}_{P\times\lambda}\{\sum_{g\in G} \sum_{\substack{e,g\in D\in\mathcal{K}\\ D\subset V_e(s)\cup\{e\}}} |D|^{-1}\Phi(D,\omega)\} =$$

$$= \mathbf{E}_{P\times\lambda}\{ \sum_{\substack{e\in D\in\mathcal{K}\\ D\subset V_e\cup\{e\}}} \Phi(D,\omega)\}.$$

□

Proof of theorem 3.1. Clearly, for any $\omega \in \Omega$, $A \in \mathcal{K}$,

$$E_U(\omega_A|\omega_{\bar{A}}) = \sum_{g\in A} \sum_{\substack{g\in B\in\mathcal{K}\\ A\cap B\subset V_g(s)\cup\{g\}}} U(B,\omega) =$$
$$= \sum_{g\in A} E_U(\omega_g|\omega_{V_g(s)\cup\bar{A}}\Theta_{A\setminus(V_g(s)\cup\{g\})}). \qquad (3.1)$$

Thus,

$$E_U(\omega_{A_n}|\Theta_{\bar{A}_n}) = \sum_{g\in A} \varphi^{(1)}_{A_n g^{-1}}(R_g\omega, \Gamma_g s),$$
$$E_U(\omega_{A_n}|\omega_{\bar{A}_n}) = \sum_{g\in A_n}^{n} \varphi^{(2)}_{A_n g^{-1}}(R_g\omega, \Gamma_g s),$$

where

$$\varphi^{(1)}_A(\omega, s) = E_U(\omega_e|\omega_{A\cap V_e(s)}\Theta_{\overline{A\cap V_e(s)}\setminus\{e\}}) = \sum_{\substack{e\in B\in\mathcal{K}\\ B\subset A\cap V_e(s)\cup\{e\}}} U(B,\omega),$$
$$\varphi^{(2)}_A(\omega, s) = E_U(\omega_e|\omega_{V_e(s)\cup\bar{A}}\Theta_{A\setminus(V_e(s)\cup\{e\}}) = \sum_{\substack{e\in B\in\mathcal{K}\\ B\subset \bar{A}\cup V_e(s)\cup\{e\}}} U(B,\omega).$$

Denote

$$\widetilde{P} := P\otimes\lambda; \qquad \varphi(\omega, s) := E_U(\omega_e|\omega_{V_e(s)}\Theta_{\overline{V_e(s)}\setminus\{e\}}).$$

Clearly, $|\varphi_A^{(i)}(\omega,s)-\varphi(\omega,s)| \le \sum_{\substack{e\in B\in\mathcal{K}\\ B\not\subset A}} |U(B,\omega)|$, $i=1,2$, and therefore, for all $(\omega,s)\in\Omega\times S$,

$$\lim_{e\in A\in\mathcal{K}} \varphi_A^{(i)}(\omega,s) = \varphi(\omega,s), \qquad i=1,2. \tag{3.2}$$

Moreover,

$$\begin{aligned}\mathbf{E}_{\widetilde{P}}\{\sup_{e\in A\in\mathcal{K}} |\varphi_A^{(1)}(\omega,s)|\} \le \mathbf{E}_{\widetilde{P}}\{ &\sum_{\substack{e\in B\in\mathcal{K}\\ B\subset V_e(s)\cup\{e\}}} |U(B,\omega)|\} = \\ &= \mathbf{E}_P\{\sum_{e\in A\in\mathcal{K}} |A|^{-1}|U(A,\omega)|\}\end{aligned}$$

(see Lemma 3.3) and

$$\mathbf{E}_{\widetilde{P}}\{\sup_{e\in A\in\mathcal{K}} |\varphi_A^{(2)}(\omega,s)|\} \le \mathbf{E}_P\{\sum_{e\in A\in\mathcal{K}} |U(A,\omega)|\}.$$

So (3.2) holds in $L^1(\widetilde{P})$ too. Now we apply Theorems 2.1 and 2.2 to the nets $\{\varphi_A^{(1)},\ e\in A\in\mathcal{K}\}$ and $\{\varphi_A^{(2)},\ e\in A\in\mathcal{K}\}$ and the dynamical system $\{\widetilde{R}_g,\ g\in G\}$ in $(\Omega\times S,\ \mathcal{F}\times\mathcal{S},\ \widetilde{P})$, where $\widetilde{R}_g(\omega,s)=(R_g\omega,\Gamma_g s)$. Under the corresponding conditions of Statements *(i)* and *(ii)* we obtain that in $L^1_{\widetilde{P}}$ ($\widetilde{P}$–almost everywhere)

$$\begin{aligned}\lim_{n\to\infty} |A_n|^{-1}E_U(\omega_{A_n}|\Theta_{A_n}) &= \mathbf{E}_{\widetilde{P}}\{\varphi(\omega,s)|\mathfrak{I}_{\widetilde{R}}\} = \\ &= \mathbf{E}_{\widetilde{P}}\{\mathbf{E}_{\widetilde{P}}[\varphi(\omega,s)|\mathfrak{I}_{\widetilde{R}}]|\mathfrak{I}_R\times\{\emptyset,S\}\} = \\ &= \mathbf{E}_{\widetilde{P}}[\varphi(\omega,s)|\mathfrak{I}_R\times\{\emptyset,S\}]\end{aligned}$$

and

$$\lim_{n\to\infty} |A_n|^{-1}E_U(\omega_{A_n}|\omega_{\bar{A}_n}) = \mathbf{E}_{\widetilde{P}}\{\varphi(\omega)|\mathfrak{I}_R\times\{\emptyset,S\}\}.$$

Now let $\psi(\omega,s)=\psi(\omega)$ be a $(\mathfrak{I}_R\times\{\emptyset,S\})$–measurable function. Lemma 3.3 with $\Phi(A,\omega)=\psi(\omega)U(A,\omega)$ gives:

$$\mathbf{E}_{\widetilde{P}}\{\psi(\omega)\varphi(\omega,s)\} = \mathbf{E}_P\{\sum_{e\in A\in\mathcal{K}} |A|^{-1}U(A,\omega)\psi(\omega)\}.$$

Thus statements *(i)* and *(ii)* are proved.

To prove statement *(iii)* we introduce the function

$$\widetilde{\varphi}_A(\omega,s) := \sup_{\zeta_{\bar{A}}\in\Omega_{\bar{A}}} |\varphi_A^{(2)}(\omega_A\zeta_{\bar{A}},s) - \varphi(\omega,s)|.$$

Clearly, for any $A(e\in A\in\mathcal{K})$, $\omega\in\Omega$, $s\in S$,

$$\widetilde{\varphi}_A(\omega,s) \le \sup_{\zeta_{\bar{A}}\in\Omega_{\bar{A}}} \sum_{\substack{e\in B\in\mathcal{K}\\ B\not\subset A}} |U(B,\omega_A\zeta_{\bar{A}})| \le 2\|U\|$$

and, consequently, $\lim_{e\in A\in\mathcal{K}}\widetilde{\varphi}_A(\omega,s)=0$. We have

$$\begin{aligned}\sup_{\zeta_n\in\widetilde{\Omega}_{\bar{A}_n}} &\left||A_n|^{-1}E_U(\omega_{A_n}|\zeta_n) - \mathbf{E}_{\widetilde{P}}[\varphi(\omega,s)|\mathfrak{I}_{\widetilde{R}}]\right| \le \\ &\le \left||A_n|^{-1}\sum_{g\in A_n}\widetilde{\varphi}_{A_ng^{-1}}(\widetilde{R}_g(\omega,s))\right| + \\ &+ \left||A_n|^{-1}\sum_{g\in A_n}\varphi(\widetilde{R}_g(\omega,s)) - \mathbf{E}_{\widetilde{P}}[\varphi(\omega,s)|\mathfrak{I}_{\widetilde{R}}]\right|.\end{aligned}$$

It remains to apply Theorems 2.1 and 2.2.

□

§ 4. Relative Entropy and Associated Convergence Theorems

4.1. Definitions and main properties. In this section $(W,\mathcal{W})$ is an arbitrary measurable space, Σ is the direction of all σ–subalgebras of $\mathcal{W}$ ordered by inclusion; P (resp. M) is a probability (resp. a σ–finite) measure on $\mathcal{W}$; $P_{\mathcal{C}}$ (resp. $M_{\mathcal{C}}$) is the restriction of P (resp. M) to $\mathcal{C}$ ($\mathcal{C}\in\Sigma$). We denote $\Psi_{\mathcal{C}}(w) := \frac{dP_{\mathcal{C}}}{dM_{\mathcal{C}}}(w)$ (the Radon–Nikodym derivative) and $\rho_{\mathcal{C}}(w) := -\log\Psi_{\mathcal{C}}(w)$.

If $M_{\mathcal{C}}$ is σ–finite, the *relative entropy* $H_{P:M}(\mathcal{C})$ is defined in the following way: $H_{P:M}(\mathcal{C}) = -\mathbf{E}_P\rho_{\mathcal{C}}$ if $P_{\mathcal{C}} \ll M_{\mathcal{C}}$ (provided the expectation $\mathbf{E}_P$ exists) and otherwise $H_{P:M}(\mathcal{C}) = -\infty$; if $M_{\mathcal{C}}$ is not σ–finite or $\mathbf{E}_P$ does not exist, $H_{P:M}(\mathcal{C})$ remains undefined. $\Sigma_{P:M}$ will denote the subdirection of Σ consisting of all $\mathcal{C}$ for which $H_{P:M}(\mathcal{C})$ is defined.

Let $\mathcal{C} \subset \mathcal{C}' \in \Sigma$. With the aid of Jensen's inequality it can be shown that

a) if $H_{P:M}(\mathcal{C}) < \infty$, then $\mathcal{C}' \in \Sigma_{P:M}$ and $H_{P:M}(\mathcal{C}') < \infty$;

b) if $H_{P:M}(\mathcal{C}') > -\infty$ and $M_{\mathcal{C}}$ is σ–finite, then $\mathcal{C} \in \Sigma_{P:M}$ and $H_{P:M}(\mathcal{C}) > -\infty$;

c) the net $\{\rho_{\mathcal{C}},\, \mathcal{C} \in \Sigma_{P:M}\}$ is a P–supermartingale, i.e.

$$\int_\Lambda \rho_{\mathcal{C}'} dP \leq \int_\Lambda \rho_{\mathcal{C}} dP, \quad \mathcal{C}, \mathcal{C}' \in \Sigma_{P:M}, \quad \mathcal{C} \subset \mathcal{C}', \quad \Lambda \in \mathcal{C}. \tag{4.1}$$

Especially,

$$H_{P:M}(\mathcal{C}') \leq H_{P:M}(\mathcal{C}), \quad \mathcal{C}, \mathcal{C}' \in \Sigma_{P:M}, \quad \mathcal{C} \subset \mathcal{C}'; \tag{4.2}$$

equality holds if $\Psi_{\mathcal{C}} = \Psi_{\mathcal{C}'}$ P–a.e. (see Csiszár [1], Fritz [1], Nguyen, Zessin [1]).

Note, by the way, that the net $\{\Psi_{\mathcal{C}},\, \mathcal{C} \in \Sigma\}$ is an M–martingale: $\int_\Lambda \Psi_{\mathcal{C}} dM = \int_\Lambda \Psi_{\mathcal{C}'} dM (= P(\Lambda))$, $\mathcal{C} \subset \mathcal{C}'$, $\Lambda \in \mathcal{C}$.

If $M(W) < \infty$, then $H_{P:M}(\{\emptyset, W\}) = \log M(W) < \infty$ and thus $\Sigma_{P:M} = \Sigma$ and $H_{P:M}(\mathcal{C}) \leq \log M(W) < \infty$, $\mathcal{C} \in \Sigma$ (equality holds iff $\Psi_{\mathcal{C}} = [M(W)]^{-1}$ P–a.e.). If $\int_W \Psi_{\mathcal{C}}\, dP \leq C < \infty$, Jensen's inequality shows that $H_{P:M}(\mathcal{C}) \geq -\log C$ (for example, this is the case when M is purely atomic and $M(A) > \frac{1}{C}$, $A \in \mathcal{C}$).

4.2. Convergence theorems. We state here two convergence theorems for relative entropies on σ–algebras forming increasing or decreasing nets; these theorems are an important item in the study of specific entropy (see § 5).

Theorem 4.1. **(Perez–Fritz)** *Let A be a direction, let $\{\mathcal{C}_\alpha,\, \alpha \in A\}$ be an increasing net of σ–subalgebras of $\mathcal{W}$ and let $\mathcal{C}^* = \vee_{\alpha \in A} \mathcal{C}_\alpha$. We suppose also that all $H_{P:M}(\mathcal{C}_\alpha)$ exist. The following statements are true:*

(i) $\lim_{\alpha \in A} H_{P:M}(\mathcal{C}_\alpha) = \inf_{\alpha \in A} H_{P:M}(\mathcal{C}_\alpha)$ *exists.*

(ii) *If* $\inf_{\alpha \in A} H_{P:M}(\mathcal{C}_\alpha) < \infty$, *then* $\lim_{\alpha \in A} H_{P:M}(\mathcal{C}_\alpha) = H_{P:M}(\mathcal{C}^*)$.

(iii) *If* $-\infty < \inf_{\alpha \in A} H_{P:M}(\mathcal{C}_\alpha) < \infty$, *then*

a) $P_{\mathcal{C}^*} \ll M_{\mathcal{C}^*}$;

b) $(L^1_P)\text{-}\lim_{\alpha \in A} \rho_{\mathcal{C}_\alpha}(w) = \rho_{\mathcal{C}^*}(w)$.

Proof. Statement *(i)* is an evident consequence of (4.2); statement *(ii)* follows from (4.2) and *(iii)*. The proof of statement *(iii)* is given in Fritz [1]; a short proof in the case when $M(W) = 1$ was given by Nguyen and Zessin [1].

□

The next theorem is a "backward" version of Theorem 4.1; for simplicity we consider only "*separable*" directions, i.e. directions with a cofinal subsequence $\alpha_1 \prec \alpha_2 \prec \ldots$.

Theorem 4.2. *Let A be a separable direction, let $\{\mathcal{C}_\alpha,\ \alpha \in A\}$ be a decreasing net of σ–subalgebras of $\mathcal{F}$ and let $\mathcal{C}_* := \bigwedge_{\alpha \in A} \mathcal{C}_\alpha$. Suppose that $M_{\mathcal{C}_*}$ is σ–finite and the entropies $H_{P:M}(\mathcal{C}_\alpha)$ exist. Then*

(i) $\lim_{\alpha \in A} H_{P:M}(\mathcal{C}_\alpha) = \sup_{\alpha \in A} H_{P:M}(\mathcal{C}_\alpha)$;

(ii) *if* $\sup_{\alpha \in A} H_{P:M}(\mathcal{C}_\alpha) > -\infty$, *then* $\lim_{\alpha \in A} H_{P:M}(\mathcal{C}_\alpha) = H_{P:M}(\mathcal{C}_*)$;

(iii) *if* $-\infty < \sup_{\alpha \in A} H_{P:M}(\mathcal{C}_\alpha) < \infty$, *then the following limit exists in* L^1_P:

$$\lim_{\alpha \in A} \rho_{\mathcal{C}_\alpha}(w) = \rho_*(w).$$

Proof. As in Theorem 4.1, *(i)* and *(ii)* follow at once from (4.2) and *(iii)*. So let us prove statement *(iii)*. Let at first $A = \{1, 2, \ldots\}$ with the natural order. Since $M_{\mathcal{C}_*}$ is σ–finite, the martingale backward convergence theorem shows that M–a.e. the limit $\lim_{\alpha \to \infty} \psi_{\mathcal{C}_\alpha} =: \psi_*$ exists. The equality $\int_\Lambda \psi_{\mathcal{C}_\alpha} dM = \int_\Lambda \psi_{\mathcal{C}_1} dM$, $\Lambda \in \mathcal{C}_\alpha$, implies that the functions ψ_α are uniformly integrable with respect to M and thus $P(\Lambda) = \int_\Lambda \psi_{\mathcal{C}_1} dM = \int_\Lambda \psi_* dM$, $\Lambda \in \mathcal{C}_*$; consequently, $\psi_* = \frac{dP_{\mathcal{C}_*}}{dM_{\mathcal{C}_*}}$ M– (and

P–) a.e.. This equality and the supermartingale backward convergence theorem show that (L^1_P)–$\lim_{\alpha\to\infty} \rho_{C_\alpha} = \rho_{C_*}$. If A is an arbitrary separable direction we have (L^1_P)–$\lim_{i\to\infty} \rho_{C_{\alpha_i}} = \rho_{C'_*}$ with $C'_* := \wedge_{i=1}^{\infty} C_{\alpha_i} = C_*$ for any increasing cofinal subsequence $\{\alpha_i,\ i = 1, 2, \ldots\}$ in A, and thus (L^1_P)–$\lim_{\alpha\in A} \rho_{C_\alpha} = \rho_{C_*}$.

□

§ 5. Generalizations of the McMilan Theorem. Specific Entropy

5.1. Definition and properties of μ–entropy. Let $P \in \mathcal{P}_0$, $A \in \mathcal{K}$, $B \subset \bar{A}$. As in Subsect. 1.5 we consider the σ–finite measure $M_{A,B} = \mu^A \otimes P_B$ on the σ–algebra $\mathcal{F}_{A\cup B}$. The *μ-entropy* $H_P(A)$ and the *conditional μ-entropies* $H_P(A|B)$ and $H_P(A|\omega_B)$ are defined as follows (see also Fritz [1]).

$$H_P(A) := H_{P_A:\mu_A}(\mathcal{F}_A) = -\mathbf{E}\log p(\omega_A),$$

$$H_P(A|B) :=$$

$$:= H_{P:M_{A,\bar{A}}}(\mathcal{F}_{A\cup B}) = \begin{cases} -\mathbf{E}_P \log p(\omega_A|\omega_B) & \text{if } P_{A\cup B} \ll M_{A,B}, \\ -\infty & \text{if } P_{A\cup B} \not\ll M_{A,B}; \end{cases}$$

$$H_P(A|\omega_B) = \begin{cases} -\mathbf{E}_P[\log p(\omega_A|\omega_B)|\omega_B] & \text{if } P_{A\cup B} \ll M_{A,B}, \\ -\infty & \text{if } P_{A\cup B} \not\ll M_{A,B}; \end{cases}$$

provided $\mathbf{E}_P$ exists. If $P_{A\cup B} \ll M_{A,B}$ but $\mathbf{E}_P$ does not exist, the corresponding entropy remains undefined. We set

$$H_P(A|\emptyset) = H_P(A|\omega_\emptyset) = H_P(A); \quad p(\omega_A|\omega_\emptyset) = p(\omega_A).$$

If G is an unorderable group we also have to consider the following generalized version $H_{\widetilde{P}}(A|B(\cdot))$ of the conditional entropy

$H_P(A|B)$*). Let $(S, \mathcal{S}, \lambda)$ be a probability space and let $B(\cdot) = \{B(s), s \in S\}$ be a random subset of $\bar{A}(A \in \mathcal{K})$. We denote: $\widetilde{P} := P \otimes \lambda$, $\widetilde{M}_{A,\bar{A}} := M_{A,\bar{A}} \otimes \lambda$, $\mathcal{C}(s) = \mathcal{F}_{A \cup B(s)}$. Note that the measure $\widetilde{M}_{A,\bar{A}}$ is σ–finite, as are its restrictions $(\widetilde{M}_{A,\bar{A}})_{\mathcal{F}_{A\cup B}} = \widetilde{M}_{A,B}$, $B \subset \bar{A}$. Let us define $H_{\widetilde{P}}(A|B(\cdot)) := H_{\widetilde{P}:\widetilde{M}_{A,\bar{A}}}(\widetilde{\mathcal{C}})$ provided the entropy $H_{\widetilde{P}:\widetilde{M}_{A,\bar{A}}}(\widetilde{\mathcal{C}})$ is well–defined.

The following properties of μ–entropies can be easily deduced from the definitions and properties of the relative entropy (see Subsect. 4.1 and App., § 11).

(A) $H_P(Ag) = H_P(A)$, $H_P(Ag|Bg) = H_P(A|B)$, $g \in G$, $B \subset \bar{A}$.

(B) If $B(s) = B$ with λ–probability *1*, then $H_{\widetilde{P}}(A|B(\cdot)) = H_P(A|B)$.

(C) If $H_{\widetilde{P}}(A|B(\cdot)) > -\infty$, then the densities $p(\omega_A|\omega_{B(s)})$ exist for almost all $s \in S$, and the function $(\omega, s) \to p(\omega_A|\omega_{B(s)})$ has an $(\mathcal{F} \times \mathcal{S})$–measurable standard modification; moreover $H_{\widetilde{P}}(A|B(\cdot)) = -\mathbf{E}_{\widetilde{P}} \log p(\omega_A|\omega_{B(\cdot)})$.

(D) If $-\infty < H_{\widetilde{P}}(A|B(\cdot)) < \infty$, then $H_{\widetilde{P}}(A|B(\cdot)) = \mathbf{E}_\lambda H_P(A|B(s))$.

(E) If $H_P(A) < \infty$ (or $H_P(A|\bar{A}) > -\infty$), then $H_{\widetilde{P}}(A|B(\cdot))$ exists for any random subset $B(\cdot)$ of $\bar{A}(A \in \mathcal{K})$, and $H_{\widetilde{P}}(A|B(\cdot)) < \infty$ (resp. $H_P(A|B(\cdot)) > -\infty$).

(F) If $B_1(\cdot)$ and $B_2(\cdot)$ are random subsets of $\bar{A}(A \in \mathcal{K})$, $B_2(s) \subset B_1(s)$ with probability *1* and the entropies $H_{\widetilde{P}}(A|B_i(\cdot))$, $i = 1, 2$, exist, then $H_{\widetilde{P}}(A|B_1(\cdot)) \leq H_{\widetilde{P}}(A|B_2(\cdot))$.

Let $\{\Gamma_g,\ g \in G\}$ be a G–ordering dynamical system acting in $(S, \mathcal{S}, \lambda)$. If $P \in \mathcal{P}_\mu$, $A \in \mathcal{K}$, $B \in \mathcal{K}_{\bar{A}}$, then

$$\begin{aligned} p(\omega_A) &= \prod_{g \in A} p(\omega_g|\omega_{A \cap V_g(s)}) = \\ &= \prod_{g \in A} p((R_g\omega)_e|(R_g\omega)_{Ag^{-1} \cap V_e(\Gamma_g s)}) \end{aligned} \tag{5.1}$$

*) We recall that for G orderable we can omit all references to the ordering dynamical system $\{\Gamma_g, g \in G\}$ in $(S, \mathcal{S}, \lambda)$ and confine ourselves to the consideration of P, $M_{A,\bar{A}}$ and $H_P(A|B)$ instead of $\widetilde{P}$, $\widetilde{M}_{A,\bar{A}}$ and $H_P(A|B(\cdot))$ (see Subsect. 1.9).

and

$$p(\omega_A|\omega_B) = \prod_{g\in A} p(\omega_g|\omega_{(A\cap V_g(s))\cup B}) =$$
$$= \prod_{g\in A} p((R_g\omega)_e|(R_g\omega)_{(Ag^{-1}\cap V_e(\Gamma_g s))\cup B}). \qquad (5.2)$$

These formulas immediately imply the following two properties:

(G) If $H_P(e) < \infty$, then for any $A \in \mathcal{K}$, $B \in \mathcal{K}_{\bar{A}}$ we have $H_P(A) < \infty$, $H_P(A|B) > \infty$ and

$$H_P(A) = \sum_{g\in A} H_{\widetilde{P}}(e|Ag^{-1} \cap V_e(\cdot)), \qquad (5.3)$$

$$H_P(A|B) = \sum_{g\in A} H_{\widetilde{P}}(e|(Ag^{-1} \cap V_e(\cdot)) \cup Bg^{-1}). \qquad (5.4)$$

(H) If $H_P(e|\{e\}) > -\infty$, then $H_P(A) > -\infty$, $H_P(A|B) > -\infty$ and the equalities (5.1)–(5.4) hold for any $A \in \mathcal{K}$, $B \subset \bar{A}$.

(I) If $H_P(e) = +\infty$ and $H_P(e|\overline{\{e\}}) > -\infty$, then $H_P(A) = +\infty$ for any $A \in \mathcal{K}$. Indeed, from (5.1) we obtain

$$H_P(A) = \sum_{g\in A} H_P(g|A \cap V_g(s))$$

for any $s \in S$, where

$$H_P(g|A \cap V_g(s)) \geq H_P(g|\overline{\{g\}}) = H_P(e|\overline{\{e\}}) > -\infty \quad (g \in G,\ s \in S);$$

if g_0 is the s_0–least element in A, then $H_P(g_0|\omega_{A\cap V_g(s_0)}) = H_P(g_0) =$
$= H_P(e) = +\infty$ and thus $H_P(A) = +\infty$.

(J) If $H_P(e|\overline{\{e\}}) = -\infty$ and $H_P(e) < +\infty$,then $H_P(A|\bar{A}) = -\infty$ for any $A \in \mathcal{K}$.

The proof is similar to the proof of (I).

5.2. Convergence of conditional densities and entropies. Let us consider two random σ–algebras in Ω:

$$\{\mathcal{C}_A(s) = \mathcal{F}_{A\cap V_e(s)},\ e \in A \in \mathcal{K}\}$$

and

$$\{D_A(s) = \mathcal{F}_{\bar{A}\cap V_e(s)},\ e \in A \in \mathcal{K}\}.$$

Note that the first net is increasing while the second one is decreasing.

$$\mathcal{C}^*(s) := \vee_{e\in A\in\mathcal{K}}\mathcal{C}_A(s) = \mathcal{F}_{V_e(s)}$$

and

$$D_*(s) = \wedge_{e\in A\in\mathcal{K}} D_A(s) =: \mathcal{F}_{V_e(s),\infty}$$

(clearly, $\mathcal{F}_{V_e(s)} \vee \mathcal{F}_\infty \subset \mathcal{F}_{V_e(s),\infty}$ where $\mathcal{F}_\infty := \wedge_{e\in A\in\mathcal{K}}\mathcal{F}_{\bar{A}}$). Let $M = M_{\{e\},\overline{\{e\}}}$; $\widetilde{M} = M \otimes \lambda$. We recall that

$$H_{\widetilde{P}}(e|A\cap V_e(\cdot)) = H_{\widetilde{P}:\widetilde{M}}(\widetilde{\mathcal{C}}_A),\ H_{\widetilde{P}}(e|V_e(\cdot)) = H_{\widetilde{P}:\widetilde{M}}(\mathcal{C}^*),$$

while if $\widetilde{P}_{\widetilde{\mathcal{C}}_A} \ll \widetilde{M}_{\widetilde{\mathcal{C}}_A}$, then $\frac{d\widetilde{P}_{\widetilde{\mathcal{C}}_A}}{d\widetilde{M}_{\widetilde{\mathcal{C}}_A}}(\omega) = p(\omega_e|\omega_{A\cap V_e(s)})$. For $\widetilde{P}_{\widetilde{\mathcal{D}}_*} \ll$ $\ll \widetilde{M}_{\widetilde{\mathcal{D}}_*}$ we put

$$p(\omega_e|\omega_{V_e(\cdot),\infty} := \frac{dP_{\widetilde{\underline{D}}_*}}{dM_{\widetilde{\underline{D}}_*}}(\omega, s);$$

$$H_{\widetilde{P}}(e|V_e(\cdot),\infty) := H_{\widetilde{P}_{\widetilde{\mathcal{D}}_*}:\widetilde{M}_{\widetilde{\mathcal{D}}_*}}(\omega).$$

We also consider

$$H_P(e|V_e(\cdot),\infty) := -\mathbf{E}_P \log p(\omega_e|\omega_{V_e(s),\infty})$$

if $P_{\underline{D}_*(s)} \ll M_{\underline{D}_*(s)}$ and $\mathbf{E}_P$ exists. Theorems 4.1 and 4.2 imply, respectively, the following two statements.

Lemma 5.1. *Let the entropies $H_{\widetilde{P}}(e|A\cap V_e(\cdot))$, $e \in A \in \mathcal{K}$, exist. Then*

(*i*) $\lim_{e\in A\in\mathcal{K}} H_{\widetilde{P}}(e|A\cap V_e(\cdot)) = \inf_{e\in A\in\mathcal{K}} H_{\widetilde{P}}(e|A\cap V(\cdot))$ *exists;*

(*ii*) *if* $\inf_{e\in A\in\mathcal{K}} H_{\widetilde{P}}(e|A\cap V_e(\cdot)) < \infty$, *then*

$$\lim_{e\in A\in\mathcal{K}} H_{\widetilde{P}}(e|A\cap V_e(\cdot)) = H_{\widetilde{P}}(e|V_e(\cdot));$$

(iii) if $-\infty < \inf_{e\in A\in\mathcal{K}} H_{\widetilde{P}}(e|V_e(\cdot)) < \infty$, *then*

(a) $H_{\widetilde{P}}(e|V_e(\cdot)) = \mathbf{E}_\lambda H_P(e|V_e(s)$;

(b) $(L^1_{\widetilde{P}})$- $\lim_{e\in A\in\mathcal{K}} \log p(\omega_e|\omega_{A\cap V_e(s)}) = \log p(\omega_e|\omega_{V_e(s)})$.

Lemma 5.2. *Let the entropies* $H_{\widetilde{P}}(e|\bar{A}\cup V_e(\cdot))$, $e\in A\in\mathcal{K}$, *exist. Then*

(i) $\lim_{e\in A\in\mathcal{K}} H_{\widetilde{P}}(e|\bar{A}\cup V_e(\cdot)) = \sup_{e\in A\in\mathcal{K}} H_{\widetilde{P}}(e|\bar{A}\cup V_e(s))$ *exists;*

(ii) if $\sup_{e\in A\in\mathcal{K}} H_{\widetilde{P}}(e|A\cup V_e(\cdot)) > -\infty$, *then*

$$\lim_{e\in A\in\mathcal{K}} H_{\widetilde{P}}(e|\bar{A}\cup V_e(\cdot)) = H_{\widetilde{P}}(e|V_e(\cdot),\infty);$$

(iii) If $-\infty < \sup_{e\in A\in\mathcal{K}} H_{\widetilde{P}}(e|\bar{A}\cup V_e(\cdot)) < +\infty$, *then*

(a) $H_{\widetilde{P}}(e|V_e(\cdot),\infty) = \mathbf{E}_\lambda H_P(e|V_e(s),\infty)$;

(b) (L^1_P)- $\lim_{e\in A\in\mathcal{K}} \log p(\omega_e|\omega_{\bar{A}\cup V_e(s)}) = \log p(\omega_e|\omega_{V_e(s),\infty})$.

5.3. Specific entropy. Generalized McMillan theorem. We suppose here again that $P\in\mathcal{P}_\mu$. Let $\{A_n\}$ be an ergodic (Følner) sequence of sets. If either $H_P(e) < \infty$ or $H_P(e|\overline{\{e\}}) > -\infty$, the entropies $H_P(A)$ and $H_P(A|\bar{A})$, $A\in\mathcal{K}$, exist; we can then define the *upper specific entropy* $\bar{h}_P := \inf_{1\le n\le\infty} |A_n|^{-1} H_P(A_n)$ and the *lower specific entropy* $\underline{h}_P := \sup_{1\le n<\infty} |\bar{A}_n|^{-1} H_P(A_n|\bar{A}_n)$. Obviously, $\underline{h}_P \le \bar{h}_P$. If $H_P(e|\overline{\{e\}})$ exists, then $\underline{h}_P \ge H_P(e|\overline{\{e\}})$, and if $H_P(e)$ exists, then $\bar{h}_P \le H_P(e)$.

The following two theorems show that $\bar{h}_P$ and $\underline{h}_P$ do not depend on the choice of the Følner sequence $\{A_n\}$ and can indeed be considered as the "upper" and "lower" specific entropies of the measure P. These theorems also include generalized versions of the McMillan theorem; the first one is a generalization of the well–known results of Föllmer [1], Kieffer [1], Pitskel [1], Pitskel, Stepin [1], Thouvenot [1]. It is also related to the version of McMillan's theorem due to Fritz [1] (see also

Nguyen, Zessin [1]) and to the results of Stepin [1] and Tagi–zade [1–3] concerning the topological entropy of group actions. Below we consider the following conditions:

(E) is an ergodic sequence of sets in G;

(H) either $H_P(e) < \infty$ or $H_P(e|\overline{\{e\}}) > -\infty$.

Theorem 5.3. *Under the conditions $P \in \mathcal{P}_\mu$, (E) and (H),*

(i) $\bar{h}_P = \lim_{n\to\infty} |A_n|^{-1} H_P(A_n)$;

(ii) $\bar{h}_P \leq H_P(e)$; $\bar{h}_P = \infty$ *iff* $h_P(e) = \infty$; *if* $\bar{H}_P < \infty$, *then* $\bar{H}_P = = H_{\widetilde{P}}(e|V_e(\cdot))$;

(iii) if $|\bar{h}_P| < \infty$, *then*

(a) $\bar{h}_P = \mathbf{E}_\lambda H_P(e|V_e(s))$;

(b) the following limit exists:

$$
\begin{aligned}
(L^1_P)-\lim_{n\to\infty} |A_n|^{-1}[-\log p(\omega_{A_n})] =\\
= -\mathbf{E}_\lambda \mathbf{E}_P\{\log p(\omega_e|\omega_{V_e(s)}|\mathfrak{J}_R\} =: \bar{h}_P(\omega);
\end{aligned}
$$

(c) $\mathbf{E}_P \bar{h}_P(\omega) = \bar{h}_P$.

Proof. Let us denote

$$
\begin{aligned}
a_A &:= H_{\widetilde{P}}(e|A \cap V_e(\cdot));\\
\varphi_A(\omega, s) &:= -\log p(\omega_{A\cap V_e(s)}).
\end{aligned}
$$

According to (5.3) and (5.1), $H_P(A_n) = \sum_{g\subset A_n} a_{A_n g^{-1}}$ and $-\log p(\omega_{A_n}) = \sum_{g\in A_n} \varphi_{A_n g^{-1}}(\widetilde{R}_g(\omega, s))$ where $\widetilde{R}_g(\omega, s) = (R_g\omega, \Gamma_g s)$.

(i)–(ii). If $H_P(e) = +\infty$ and $H_P(e|\overline{\{e\}}) > -\infty$, then $H_P(A_n) = +\infty$ according to (I) (Subsect. 5.1) ,and thus $\bar{h}_P = \lim_{n\to\infty} |A_n|^{-1} H_P(A_n) = = +\infty$. Let $H_P(e) < \infty$. Since $\{a_A,\ e \in A \in \mathcal{K}\}$ is nonincreasing and $\sup_{e\in A\in\mathcal{K}} a_A \leq H_P(e) < \infty$, Lemmas 2.4 and 5.1 show that the following limit exists:

$$
\lim_{n\to\infty} |A_n|^{-1} H_P(A_n) = \inf_{e\in A\in\mathcal{K}} H_{\widetilde{P}}(e|A \cap V_e(\cdot)).
$$

On the other hand,

$$\lim_{n\to\infty} |A_n|^{-1} H_P(A_n) \geq \inf_{1\leq n\leq\infty} |A_n|^{-1} H_P(A_n) \geq$$
$$\geq \inf_{e\in A\in\mathcal{K}} H_{\widetilde{P}}(e|A\cap V_e(\cdot)).$$

Thus statements *(i)–(ii)* are proved.

(iii) Statement *(a)* is a consequence of *(ii)* and property (D) (Subsect. 5.1). In view of (5.1), Lemma 5.1 and Theorem 2.1 imply the existence of

$$(L_P^1) - \lim_{n\to\infty} [|A_n|^{-1} \log p(\omega_{A_n})] = \mathbf{E}_{\widetilde{P}}[\log p(\omega_e|\omega_{V_e(s)})\mathfrak{I}_{wtR,\widetilde{P}}] =$$
$$= \mathbf{E}_{\widetilde{P}}[\log p(\omega_e|\omega_{V_e(s)})|\mathfrak{I}_{R,P} \times \mathfrak{I}_{\Gamma,\lambda}] = \mathbf{E}_\lambda \mathbf{E}_P[\log p(\omega_e|\omega_{V_e(s)})|\mathfrak{I}_{R,P}].$$

The equality $\mathbf{E}_P\bar{h}_P(\omega) = \bar{h}_P$ is a simple consequence of these relations.

□

The following statement is a "conditional" version of Theorem 5.3.

Theorem 5.4. *Let $P \in \mathcal{P}_\mu$; under the conditions (F) and (H)*

(i) $\underline{h}_P = \lim_{n\to\infty} |A_n|^{-1} H_P(A_n|\bar{A}_n)$;

(ii) $\underline{h}_P \geq H_P(e|\overline{\{e\}})$; $\underline{h}_P = -\infty$ *iff* $H_P(e|\overline{\{e\}}) = -\infty$; *if* $\underline{h}_P < \infty$, *then* $\underline{h}_P = H_{\widetilde{P}}(e|V_e(\cdot),\infty)$;

(iii) if $|\underline{h}_P| < \infty$, *then*

a) $\underline{h}_P = \mathbf{E}_\lambda H_P(e|V_e(s),\infty)$;

b) *the following limit exists:*

$$(L_P^1) \lim_{n\to\infty} |A_n|^{-1}[-\log p(\omega_{A_n}|\omega_{\bar{A}_n})] =$$
$$= -\mathbf{E}_\lambda \mathbf{E}_P\{\log p(\omega_e|\omega_{V_e(s),\infty})|\mathfrak{I}_{R,P}\} := \underline{h}_P(\omega);$$

c) $\underline{h}_p(\omega) = (L_P^1) \lim_{n\to\infty} |A_n|^{-1} H_P(A_n|\omega_{\bar{A}_n})$;

d) $\mathbf{E}_P[\underline{h}_P(\omega)] = \underline{h}_P$.

Proof. Let us denote

$$a_A := H_{\widetilde{P}}(e|\bar{A} \cup V_e(\cdot)); \qquad \varphi_A(\omega, s) := -\log p(\omega_e|\omega_{\bar{A}\cup V_e(s)}).$$

According to (5.4) and (5.2),

$$H_P(A_n|\bar{A}_n) = \sum_{g \in A_n} a_{A_n g^{-1}}$$

and

$$-\log p(\omega_{A_n}) = \sum_{g \in A_n} \varphi_{A_n g^{-1}}(R_g\omega, \Gamma_g s).$$

If $H_P(e|\overline{\{e\}}) = -\infty$ and $H_P(e) < \infty$, then $H_P(A|\bar{A}) = -\infty$ for any $A \in \in \mathcal{K}$ (see (I) in Subsect. 5.1) and thus $\underline{h}_P = \lim_{n\to\infty} |A_n|^{-1} H_P(A_n|\bar{A}_n) = = -\infty$. Now suppose that $H_P(e|\overline{\{e\}}) > -\infty$; then $\inf_{e\in A\in\mathcal{K}} a_A \geq \geq H_P(e|\overline{\{e\}}) > -\infty$, and since the net $\{a_A,\ e \in A \in \mathcal{K}\}$ is nondecreasing, Lemmas 2.4 and 5.2 imply the existence of

$$\lim_{n\to\infty} |A_n|^{-1} H_P(A_n|\bar{A}_n) = \sup_{e\in A\in\mathcal{K}} H_{\widetilde{P}}(e|\bar{A} \cup V_e(\cdot)) = H_P(e|V_e(\cdot), \infty).$$

According to equality (5.4),

$$\lim_{n\to\infty} |A_n|^{-1} H_P(A_n|\bar{A}_n) \leq \sup_{1\leq n<\infty} |A_n|^{-1} H_P(A_n|\bar{A}_n) \leq$$
$$\leq \sup_{e\in A\in\mathcal{K}} H_{\widetilde{P}}(e|A \cap V_e(\cdot)).$$

Thus statements *(i)* and *(ii)* are proved.

The assertions *(iiia)*,*(iiib)* and *(iiid)* can be proved as *(iiia)*– *(iiic)* of Theorem 5.3 with the aid of Lemma 5.2 and Theorem 2.1. Let us prove *(iiic)*. For any $\Lambda \in \mathfrak{I}_R$ and $\varepsilon > 0$ there exist nets $B_\varepsilon \in \mathcal{K}$ and $\Lambda_\varepsilon \in \mathcal{F}_{B_\varepsilon}$ such that $P(\Lambda \Delta \Lambda_\varepsilon) < \varepsilon$; if $A \in \mathcal{K}$ and $g \notin A^{-1}B_\varepsilon$, then $B_\varepsilon g^{-1} \in \bar{A}$, and since $\Lambda (\in \mathfrak{I}_R)$ and $\varepsilon > 0$ are chosen arbitrarily, $\mathfrak{I}_R \subset \mathcal{F}^*_{\bar{A}}$ (P–completion of $\mathcal{F}_{\bar{A}}$) for any $A \in \mathcal{K}$. Thus $\underline{h}_P(\cdot)$ is measurable with respect to $\mathcal{F}_{\bar{A}_n}$; we have the following inequality:

$$\mathbf{E}_P\big||A_n|^{-1}H_P(A_n|\omega_{\bar{A}_n}) - \underline{h}_P(\omega)\big| \le$$
$$\le \mathbf{E}_P\big||A_n|^{-1}\log p(\omega_{A_n}|\omega_{\bar{A}_n}) + \underline{h}_P(\omega)\big| \qquad (n = 1, 2, \ldots),$$

which implies *(iiic)* in view of statement *(iiib)*.

□

In "regular" cases $\bar{h}_P = \underline{h}_P := h_P$, and the common value h_P may be called the *specific entropy* of the measure P. For example, Theorem 7.2 below shows that $\bar{h}_P = \underline{h}_P$ if $\mu(X) < \infty$ and the measure P admits a Gibbsian representation with finite norm potential*).

On the other hand, Gurevich [1] has constructed a homogeneous random measure P on $\mathbb{Z}^1$ with $|X| < \infty$, $\mathcal{F}_\infty := \wedge_{0\in A\in\mathcal{K}}\mathcal{F}_{\bar{A}} = \mathcal{F}$ (P–mod 0) and $\wedge_{x\in\mathbb{Z}^1}\mathcal{F}(-\infty, x) = \{\emptyset, \Omega\}$ (P–mod 0); obviously, in this case $H_P(A|\bar{A}) = 0$, $A \in \mathcal{K}$, and thus $\underline{h}_P = 0$; $\bar{h}_P$ coincides with Kolmogorov–Sinai entropy of the K–automorphism R_1 in Ω and hence $\bar{h}_P > 0$ (see Rokhlin [1]). It also follows from results of Föllmer and Snell [1] that $\bar{h}_P$ and $\underline{h}_P$ may be different in the case of a Markov field on a homogeneous graph.

§ 6. Specific Free Energy

Let us suppose that $P \in \mathcal{P}_\mu$, and $U(A,\omega)$, $A \in \mathcal{K}$, $\omega \in \Omega$, is a potential. The *individual free energy* $F_{P,U}(\omega_A|\omega_{\bar{A}})$ of the configuration ω_A on $A \in \mathcal{K}$ given the environment $\omega_{\bar{A}}$ is defined by the equality

$$F_{P,U}(\omega_A|\omega_{\bar{A}}) := E_U(\omega_A|\omega_{\bar{A}}) + \log p(\omega_A|\omega_{\bar{A}}), \tag{6.1}$$

---

*) This is the case when $G{=}\mathbb{Z}^m$ and the conditional densities $p(\omega_A|\omega_{\bar{A}})$, $A\in\mathcal{K}$, have versions such that $m_A \le p(\omega_A|\omega_{\bar{A}}) \le M_A$, $\omega\in\Omega_A$, for some $0<m_A<M_A<\infty$, and the weak Markov property holds: $\sum_{r=1}^{\infty} r^{m-1}\varphi(r)<\infty$ where

$$\varphi(r) := \sup_{\omega,\omega':\omega_{S_r}=\omega'_{S_r}} |P(\omega_0|\omega_{\overline{\{0\}}}) - P(\omega_0|\omega'_{\overline{\{0\}}})|, \quad S_r = \{t : |t| \le r\}$$

(see Kozlov [1]; related questions are also considered in Sullivan [1]).

provided that $p(\omega_A|\omega_{\bar{A}})$ exists.

If

$$\mathbf{E}_P[|E_U(\omega_A|\omega_{\bar{A}})|] < \infty$$

and either $H_P(e) < \infty$ or $H_P(e|\overline{\{e\}}) > -\infty$, we can define the *mean free energy on* $A: F_{P,U}(A|\bar{A}) := \mathbf{E}_P[F_{P,U}(\omega_A|\omega_{\bar{A}})]$, and the *mean free energy on A given $\omega_{\bar{A}}$:*

$$\begin{aligned} F_{P,U}(A|\bar{A}) &:= \mathbf{E}_P\{F_{P,U}(\omega_A|\omega_{\bar{A}})|\omega_{\bar{A}}\} = \\ &= \mathbf{E}_P\{E_U(\omega_A|\omega_{\bar{A}})|\omega_{\bar{A}}\} - H_P(A|\omega_{\bar{A}}). \end{aligned}$$

From Theorems 3.1 and 5.2 we obtain the following general L^1–version of the well–known Lee–Yang and Van Hove theorems concerning the thermodynamical limit of free energy.

Theorem 6.1. *Let $\{A_n\}$ be a Følner sequence. Suppose $P \in \mathcal{P}_\mu$ and $\mathbf{E}_P\{\sum_{e\in A\in\mathcal{K}} |U(A,\omega)|\} < \infty$.*

(i) If either $H_P(e|\overline{\{e\}}) > -\infty$ or $H_P(e) < \infty$, then the specific free energy of the measure $f_{P,U} := \lim_{n\to\infty} |A_n|^{-1} F_{P,U}(A_n|\bar{A}_n)$ exists.

(ii) If $|\underline{h}_P| < \infty$, then

$$f_{P,U}(\omega) := (L^1_P) \lim_{n\to\infty} F_{P,U}(A_n|\omega_{\bar{A}_n}) = (L^1_P) \lim_{n\to\infty} F_{P,U}(\omega_{A_n}|\omega_{\bar{A}_n})$$

exists. Clearly,

$$e_{P,U} = \underline{h}_P + f_{P,U}; \quad \mathbf{E}_P f_{P,U}(\omega) = f_{P,U},$$

and

$$e_{P,U}(\omega) = \underline{h}_P(\omega) + f_{P,U}(\omega).$$

§ 7. Convergence Theorems for Gibbsian Measures. Specific Relative Entropy

In this section we assume that $\mu(X) < \infty$, U is a finite–norm potential, i.e. $\|U\| < \infty$, and $\mathcal{G}(U) \neq \emptyset$ (see Subsect. 1.6). *The Gibbsian free energy* corresponding to the potential U is:

$$F_U(A|\omega_{\bar{A}}) := -\log Z_U(A|\omega_{\bar{A}})^{*)}.$$

Clearly, if $Q \in \mathcal{G}(U)$, then

$$F_{Q,U}(\omega_A|\omega_{\bar{A}}) = F_{Q,U}(A|\omega_{\bar{A}}) = F_U(A|\omega_{\bar{A}}).$$

The following lemma states some useful properties of Gibbsian measures. We denote the densities corresponding to the measure Q by $q(\omega_A)$ and $q(\omega_A|\omega_B)$ $(A \in \mathcal{K},\ B \subset \bar{A})$.

Lemma 7.1. *For any $A \in \mathcal{K}$, $B \subset \bar{A}$,*

1) $\sup\limits_{\substack{\omega_A \in \Omega_A \\ \omega_{\bar{A}} \in \tilde{\Omega}_{\bar{A}}}} |E_U(\omega_A|\omega_{\bar{A}})| \le |A|\|U\|$;

2) $\sup\limits_{\substack{\omega_A \in \Omega_A \\ \eta_{\bar{A}}, \zeta_{\bar{A}} \in \tilde{\Omega}_{\bar{A}}}} |E_U(\omega_A|\eta_B\zeta_{\bar{A}\setminus B}) - E_U(\omega_A|\eta_{\bar{A}})| \le \sum_{g\in A} \delta_{(A\cup B)g^{-1}}$

where $0 \le \delta_D \le 2 \sum\limits_{\substack{e \in C \in \mathcal{K} \\ C \not\subset D}} \sup\limits_{\omega \in \Omega} |U(C,\omega)|$,

and thus $\lim\limits_{D\in\mathcal{K}} \delta_D = 0$;

3) $|A|(-\|U\| + \log\mu(X)) \le \log Z_U(A|\omega_{\bar{A}}) \le |A|(\|U\| + \log\mu(X))$;

4) $|A|(-2\|U\| + \log\mu(X)) \le \log\lambda_U(\omega_A|\omega_{\bar{A}}) \le |A|(2\|U\| + \log\mu(X))$.

If $Q \in \mathcal{G}(U)$, then with Q-probability 1,

5) $|A|(-2\|U\| + \log\mu(X)) \le \log q(\omega_A|\omega_B) \le |A|(2\|U\| + \log\mu(X))$;

6) $|A|(-2\|U\| + \log\mu(X)) \le \log q(\omega_A) \le |A|(2\|U\| + \log\mu(X))$;

7) $Q_A \sim \mu_A$ *and thus* $P_A \ll Q_A$ *if* $P \in \mathcal{P}_\mu$.

Proof. Statements *1)* and *2)* are simple consequences of the equality (3.1). Each of the remaining statements can be easily deduced from the previous ones.

□

This lemma shows that the densities $q(\omega_A)$ and $q(\omega_A|\omega_B)$ $(A \in \mathcal{K},\ B \in \mathcal{K}_{\bar{A}})$ are defined Q–almost everywhere. We recall that the

---

*) Here and in the sequel we suppose that $\omega_{\bar{A}} \in \tilde{\Omega}_{\bar{A}}$ (see Subsect. 1.4).

density $q(\omega_A|\omega_{\bar{A}})$ is supposed to be equal to $\lambda_U(\omega_A|\omega_{\bar{A}})$ everywhere; the conditional measure $Q_A(\cdot|\omega_{\bar{A}})$ on $\mathcal{F}_A$ is defined by

$$Q_A(\Lambda|\omega_{\bar{A}}) := \int_\Lambda q(\omega_A|\omega_{\bar{A}})Q_A(d\omega_A), \quad \Lambda \in \mathcal{F}_A, \quad \omega_{\bar{A}} \in \widetilde{\Omega}_{\bar{A}}.$$

Lemma 7.1 allows us to prove stronger versions of the theorems of § 5 and § 6 for Gibbsian measures. We begin with a generalized version of the Lee–Yang convergence Theorem; this is a refinement of Theorem 6.1.

Theorem 7.2. *(i) If $Q \in \mathcal{G}(U)$, then $f_{Q,U}(\omega) \equiv f_{Q,U}$.*
(ii) The quantity $f_{Q,U}$ does not depend on the choice of $Q \in \mathcal{G}(U)$: $f_{Q,U} =: f_U$.
(iii) For any Følner sequence $\{A_n\}$,

$$\lim_{n\to\infty} \sup_{\zeta_{\bar{A}_n} \in \widetilde{\Omega}_{\bar{A}_m}} ||A_n|^{-1} F_U(A_n|\zeta_{\bar{A}_n}) - f_U| = 0.$$

□

Proof. Let us consider the quantities

$$r_n := \sup_{\omega_{\bar{A}_n}, \zeta_{\bar{A}_n} \in \widetilde{\Omega}_{\bar{A}_n}} \frac{Z_U(A_n|\omega_{\bar{A}_n})}{Z_U(A_n|\zeta_{\bar{A}_n})} =$$

$$= \sup_{\omega_{\bar{A}_n}, \zeta_{\bar{A}_n} \in \widetilde{\Omega}_{\bar{A}_n}} \frac{\int_{\Omega_{A_n}} \psi_n(\omega, \zeta_{\bar{A}_n}) \exp\{-E_U(\omega_{A_n}|\zeta_{\bar{A}_n})\}\mu_{A_n}(d\omega_{A_n})}{\int_{\Omega_{A_n}} \exp\{-E_U(\omega_{A_n}|\zeta_{\bar{A}_n})\}\mu_{A_n}(d\omega_{A_n})},$$

where $\psi_n(\omega, \zeta_{\bar{A}_n}) = \exp[E_U(\omega_{A_n}|\zeta_{\bar{A}_n}) - E_U(\omega_{A_n}|\omega_{\bar{A}_n})]$.
Using statement *2)* of Lemma 7.1 with $B = \emptyset$ we obtain

$$|\log r_n| \leq \rho_n, \tag{7.1}$$

where $\rho_n = \sum_{g \in A_n} \delta_{A_n g^{-1}}$. Since $\lim_{e \in A \in \mathcal{K}} \delta_A = 0$ Lemma 2.4 immediately gives us $\lim_{n\to\infty} |A_n|^{-1} \log r_n = 0$. Now if $Q \in \mathcal{G}(U)$,

$$\overline{\lim_{n\to\infty}} \sup_{\zeta_{\bar{A}_n} \in \widetilde{\Omega}_{\bar{A}_n}} ||A_n|^{-1} F_U(A_n|\zeta_{\bar{A}_n}) - f_U| \leq$$

$$\leq \lim_{n\to\infty} |A_n|^{-1} \log r_n + \lim_{n\to\infty} \mathbf{E}_Q||A_n|^{-1} F_{Q,U}(A|\omega_{A_n}) - f_{Q,U}(\omega)| = 0,$$

in view of Theorem 6.1. This implies all statements *(i)–(iii)*.

□

The quantity $f_U = f_{Q,U}$ for $Q \in \mathcal{G}(U)$ is called the *specific Gibbsian free energy* corresponding to the potential U.

Let $h_{P,U}(\omega) := e_{P,U}(\omega) - f_U(\omega)$, $h_{P,U} := \mathbf{E}_P h_{P,U}(\omega) = e_{P,U} - f_U$. Thus, if $P \in \mathcal{G}(U)$, then $h_{P,U}(\omega) = \underline{h}_P(\omega)$ ($\underline{h}_P(\cdot)$ and $\underline{h}_P$ were defined in § 6).

Theorem 7.3. *If $P \in \mathcal{P}_\mu$ and $Q \in \mathcal{G}(U)$, then for any regular ergodic sequence $\{A_n\}$ in G, with P-probability 1,*

1) $\lim_{n\to\infty} \sup_{\zeta_{\bar{A}} \in \widetilde{\Omega}_{\bar{A}}} |h_{P,U}(\omega) + |A_n|^{-1} \log q(\omega_{A_n}|\zeta_{\bar{A}_n})| = 0;$

2) $\lim_{n\to\infty} [-|A_n|^{-1} \log q(\omega_{A_n})] = h_{P,U}(\omega);$

3) $\lim_{n\to\infty} |A_n|^{-1} \mathbf{E}_P[-\log q(\omega_{A_n})] = h_{P,U};$

4) $h_{P,U}(\omega) = \bar{h}_P(\omega) = \underline{h}_P(\omega)$; $h_{P,U} = \bar{h}_P = \underline{h}_P$.

Proof. Statement *1)* follows immediately from Theorems 3.1 and 7.2 in view of (6.1). Using (7.1) we deduce from statement *2)* of Lemma 7.1 that

$$e^{-2\rho_n} \leq \frac{q(\omega_{A_n}|\omega_{\bar{A}_n})}{q(\omega_{A_n}|\zeta_{\bar{A}_n})} \leq e^{2\rho_n},$$

where $\rho_n = \sum_{g\in A_n} \delta_{A_n g^{-1}}$. By integration with respect to $Q_{\bar{A}_n}$ we obtain

$$e^{-2\rho_n} \leq \frac{q(\omega_{A_n})}{q(\omega_{A_n}|\zeta_{\bar{A}_n})} \leq e^{2\rho_n}$$

for Q–almost P–almost all $\omega_{A_n} \in \Omega_{A_n}$. Since $|A_n|^{-1}\rho_n \to 0$, statement *2)* is proved. Statement *3)* is a consequence of statement *2)* and Lemma 7.1 (6). Statement *4)* follows from statements *1)–3)*.

□

Statement *2)* of Theorem 7.3 is an extension of Breiman's "almost sure" version of the McMillan theorem. Note that we proved it under certain restrictions on the measure Q (see the footnote on page 286); however, we have gained a.e.–convergence with respect to any measure $P \in \mathcal{P}_\mu$. Using different method, Ornstein and Weiss [1] have extended the usual version of Breiman's theorem to general amenable groups for all measures $Q \in \mathcal{P}_\mu$.

Let $P \in \mathcal{P}_\mu$, $A \in \mathcal{K}$, $B \subset \bar{A}$; since $x \log x \geq -e^{-1}$ for $x > 0$, we have

$$H_P(A|B) \leq H_P(A) \leq e^{-1}\mu(X^A) = e^{-1}[\mu(X)]^{|A|} < \infty.$$

Taking in account Lemma 7.1 ((5), (6)), we can introduce the *relative entropy* $H_{P:Q}(A) := -\mathbf{E}_P \log \frac{p(\omega_A)}{q(\omega_A)}$ (≤ 0) (see Subsect. 4.1), as well as the *conditional relative entropy*: $H_{P:Q}(A|B) := -\mathbf{E}_P \log \frac{p(\omega_A|\omega_B)}{q(\omega_A|\omega_B)}$ ($B \in \mathcal{K}_{\bar{A}}$ or $B = \bar{A}$) if $p(\omega_A|\omega_B)$ exists, and $:= -\infty$ in the opposite case. Clearly, $H_{P:Q}(A|B) = \mathbf{E}_P[H_{P_A(\cdot|\omega_B):Q(\cdot|\omega_B)}] \leq 0$ and equality holds iff $p(\omega_A|\omega_B) = q(\omega_A|\omega_B)$ P–almost everywhere (see Subsect. 4.1).

The following theorem is an obvious consequence of Theorems 5.1, 5.2 and 7.3.

Theorem 7.4. *Let $P \in \mathcal{P}_\mu$, $Q \in \mathcal{G}(U)$ and let $\{A_n\}$ be an ergodic sequence of sets. Then the upper and the lower specific relative entropy exist:*

$$\bar{h}_{P:Q} := \lim_{n\to\infty} |A_n|^{-1} H_{P:Q}(A_n),$$
$$\underline{h}_{P:Q} := \lim_{n\to\infty} |A_n|^{-1} H_{P:Q}(A_n|\bar{A}_n) = f_U - f_{P,U},$$

and $\underline{h}_{P:Q} \leq \bar{h}_{P:Q} \leq 0$.

§ 8. The Variational Principle

As in § 7, let $\|U\| < \infty$, $\mu(X) < \infty$, $P \in \mathcal{P}_\mu$, and $Q \in \mathcal{G}(U)$. According to (6.1), for any $A \in \mathcal{K}$,

$$F_U(A|\omega_{\bar{A}}) = E_U(\omega_A|\omega_{\bar{A}}) + \log q(\omega_A|\omega_{\bar{A}});$$

hence

$F_U(A|\omega_{\bar{A}}) = \mathbf{E}_P[E_U(\omega_A|\omega_{\bar{A}})|\omega_{\bar{A}}] + \mathbf{E}_P[\log q(\omega_A|\omega_{\bar{A}})|\omega_{\bar{A}}]$. Comparing this with (6.1) we obtain

$$F_{P,U}(A|\omega_{\bar{A}}) - F_U(A|\omega_{\bar{A}}) = -H_{P(\cdot|\omega_{\bar{A}}):Q(\cdot|\omega_{\bar{A}})} \geq 0$$

with P–probability 1. Thus $F_{P,U}(A|\omega_{\bar{A}}) \geq F_U(A|\omega_{\bar{A}})$ with equality only in the case when $p(\omega_A|\omega_{\bar{A}} = \lambda(\omega_A|\omega_{\bar{A}})$ with P–probability 1. This simple statement is a "local variational principle". It immediately implies that $f_{P,U} \geq f_U$. We know that $f_{P,U} = f_U$ if $P \in \mathcal{G}(U)$. In this section we shall show that the converse is also true: if $f_{P,U} = f_U$, then $P \in \mathcal{G}(U)$.

We begin with a simple lemma.

Lemma 8.1. *If $A \in \mathcal{K}$, $Q \in \mathcal{G}(U)$ and $P \in \mathcal{P}_\mu$, then*

a) $\lim_{V \in \mathcal{K}_{\bar{A}}} (P) - \operatorname{ess\,sup}_{\omega \in \Omega} |\log q(\omega_A|\omega_V) - \log q(\omega_A|\omega_{\bar{A}})| = 0;$

b) $\lim_{V \in \mathcal{K}_{\bar{A}}} \mathbf{E}_P \log q(\omega_A|\omega_V) = \mathbf{E}_P \log q(\omega_A|\omega_{\bar{A}});$

c) $\lim_{V \in \mathcal{K}_{\bar{A}}} H_{P:Q}(A|V) = H_{P:Q}(A|\bar{A}).$

Proof. As in the proof of Theorem 7.2, we obtain by integration with respect to $Q_{\bar{A}\setminus V}(\cdot|\omega_V)$ that with $Q_{A\cup V}$– and $P_{A\cup V}$–probability 1,

$$e^{-2\widetilde{\rho}_V} \leq q(\omega_A|\omega_V)/q(\omega_A|\omega_{\bar{A}}) \leq e^{2\widetilde{\rho}_V}$$

where $\widetilde{\rho}_V = \sum_{g \in A} \delta_{(A\cup V)g^{-1}}$. Since $\lim_{V \in \mathcal{K}_{\bar{A}}} \delta_{(A\cup V)g^{-1}} = 0$ statement a) is proved, statement b) follows from a), and statement c) follows from b) and Theorem 4.1.

$\square$

Theorem 8.2. *Let G be an amenable group; let $P \in \mathcal{P}_\mu$ and $\mathcal{G}(U) \neq \neq \emptyset$. If $f_{P,U} = f_U$, then $P \in \mathcal{G}(U)$.*

Proof. According to Theorem 7.4, the condition $f_{P,U} = f_U$ is equivalent to $\underline{h}_{P:Q} = \bar{h}_{P:Q} = 0$ where $Q \in \mathcal{G}(U)$. Let us choose a set $B\,(B \in \mathcal{K})$ and then a set $Y\,(Y \in \mathcal{K}_{\bar{B}})$; let $\widetilde{Y} = B \cup Y$. Let $\{A_n\}$ be a Følner sequence in G. According to Lemma 5.2.3, we can find a family of disjoint sets $\widetilde{Y}g_1, \ldots, \widetilde{Y}g_n$ such that $g_1, \ldots, g_{l_n} \in A_n$ and $l_n \geq \alpha_Y|A_n|$, where $\alpha_{\widetilde{Y}} = |\widetilde{Y}^{-1}\widetilde{Y}|^{-1}$. Let m_n be the number of $\widetilde{Y}g_1, \ldots, \widetilde{Y}g_{l_n}$ which lie in A_n; we change the indexes if necessary and assume that $\widetilde{Y}_{g_1}, \ldots, \widetilde{Y}_{g_{m_n}} \in A_n$. If $g \in C_n := \bigcap_{h \in \widetilde{Y}} h^{-1}A_n$, then $\widetilde{Y}g \subset A_n$ (see Lemma 2.3). Thus $0 \leq l_n - m_n \leq |A_n \setminus C_n|$ and

$\lim_{n\to\infty} \frac{|l_n - m_n|}{|A_n|} = 0$. Hence $\lim_{n\to\infty} m_n|A_n|^{-1} \geq \alpha_{\widetilde{Y}} > 0$. Let $\{W_1^{(n)}, \dots, W_{m_n}^{(n)}\}$ be a partition of A_n such that $W_k^{(n)} \supset \widetilde{Y}g_k$ ($k = 1, \dots, m_n$). We denote

$$C_{k-1}^{(n)} = \bigcup_{i=1}^{k} W_i^{(n)} \setminus Bg_k,$$

$$D_k^{(n)} = \bigcup_{i=1}^{k} W_i^{(n)} = C_{k-1}^{(n)} \cup Bg_k,$$

$k = 1, \dots, m_n$ (all these sets depend on the choice of the sets B, Y). Clearly, $C_{k-1}^{(n)} g_k^{-1} \supset (W_k^{(n)} \setminus Bg_k)g_k^{-1} \supset (\widetilde{Y}g_k \setminus Bg_k)g_k^{-1} = Y$, $C_{k-1}^{(n)} \subset D_k^{(n)} \subset C_k^{(n)}$ for $k = 1, \dots, m_n$. It is easy to see that if $E_1 \subset E_2 \subset \dots \subset E_{n-1} \subset E_n$, then $H_{P:Q}(E_n) = H_{P:Q}(E_1) + \sum_{r=2}^{n} H_{P:Q}(E_r \setminus E_{r-1}|E_{r-1})$. Let us set $E_{2k-1} = C_{k-1}^{(n)}$ and $E_{2k} = D_k^{(n)}$, $k = 1, 2, \dots$. We have

$$H_{P:Q}(A_n) = H_{P:Q}(C_0^{(n)}) + \sum_{k=2}^{m_n} H_{P:Q}(Bg_k|C_{k-1}^{(n)}) +$$

$$+ \sum_{k=1}^{m_n} H_{P:Q}(C_k^{(n)} \setminus D_k^{(n)}|D_k^{(n)}) \leq \sum_{k=2}^{m_n} H_{P:Q}(Bg_k|C_{k-1}^{(n)});$$

since the fields P and Q are homogeneous,

$$H_{P:Q}(A_n) \leq \sum_{k=2}^{m_n} H_{P:Q}(B|C_{k-1}^{(n)} g_k^{-1}).$$

According to Lemma 8.1, $H_{P:Q}(A_n) \leq (m_n - 1)(H_{P:Q}(B|\bar{B}) + \varepsilon)$ for any $\varepsilon > 0$ as soon as Y is sufficiently large ($Y \in \mathcal{K}_{\bar{B}}$). Thus $0 = \bar{h}_{P:Q} \leq \alpha_{\widetilde{Y}}(H_{P:Q}(B|\bar{B}) + \varepsilon)$. Recalling that $H_{P:Q}(B|\bar{B}) \leq 0$, we obtain $H_{P:Q}(B|\bar{B}) = 0$ and therefore $p(\omega_B|\omega_{\bar{B}}) = q(\omega_B|\omega_{\bar{B}})$ with P–probability *1*. Since B is an arbitrary finite set in G, the theorem is proved.

□

BIBLIOGRAPHICAL NOTES

1. This Chapter is close to Chapter 7 in Tempelman [24] and to Tempelman [22] (1984); the results were announced in Tempelman [17] (1977), [18] (1980), [20] (1977).

2. The definition of random fields by the means of specifications and the notion of a Gibbsian random field were introduced and studied in Dobrushin [1], [2] (1968), [3] (1969) and in Lanford, Ruelle [1] (1968).

3. Homogeneous random orders in countable groups were introduced in Kieffer [1] (1975) where they were used in extending the McMillan theorem to groups; similar arguments were earlier used in Pitckel, Stepin [1] (1971) in the case of the group of proper fractions with the natural addition mod 1.

4. Ergodic theorems of the type of Theorems 2.1 and 2.2 for set functions were usually used as a tool in proving the existence of various "summatory" characteristics of stationary random processes and homogeneous fields, begining McMillan [1] (1953) and Breiman [1] (1957). They are close to the ergodic theorems for "spatial processes" in $\mathbb{R}^m$ due to Nguyen, Zessin [2] (1979).

5. Theorem 3.1 generalizes and partly strengthens the results in Föllmer [1] (1973), Nguyen, Zessin [2] (1979), Preston [1] (1976).

6. Properties of the relative entropy (Subsect.4.1) were investigated in Csiszar [1] (1969), Fritz [1] (1970) and Perez [1] (1956). Theorem 4.1 is due to Fritz [1] (1970); it is a generalization of a theorem in Perez [1] (1956).

7. In 1970 Fritz [1] extended McMillan's theorem to $\mathbb{R}^m$; for a different approach see also Nguyen, Zessin [2] (1979). Further generalizations were obtained in Conze [1] (1973), Kieffer [1] (1975), Pitckel [1] (1971), [2] (1975), [3] (1978), Pitckel, Stepin [1] (1971), Stepin [1] (1978), Thouvenot [1] (1972). Tagi–zade [1], [3] (1978), [4] (1979) has defined and investigated the topological entropy of actions of amenable groups in compact metric spaces.

Theorems 5.3 and 5.4 were announced in Tempelman [17] (1977), [18] (1980). The notion of the lower entropy introduced in theorem 5.4 coincides with the notion of "inner entropy" defined earlier in Föllmer, Snell [1] (1973) for homogeneous random fields on graphs.

8. Theorem 6.1 is an L^1–analog of the classical limit theorems of Lie–Yang and Van Hove (see, e.g., Ruelle [1, 2] and Ulenbeck, Ford [1]).

9. The "uniform" versions of the assertions of §§ 3, 5 and § 6 proved in § 7 generalize results of Föllmer [1] (1973). The statement 2 of Theorem 7.3 is also a generalization of the Breiman theorem to Gibbsian random fields with a bounded potential. In Ornstein, Weiss [1] (1987), the Breiman's theorem was extended to arbitrary homogeneous random fields on discrete amenable groups.

10. Theorem 8.2 is a generalization of the variational principle due to Lanford and Ruelle [1] (1968), Spitzer [1] (1971) and of similar results in Föllmer [1] (1977) and Preston [1] (1976) related to $\mathbb{Z}^m$ and of results of Moulin-Ollagnier and Pinchon [1] (1976) and Föllmer and Snell [1] (1977) related to more general groups and graphs. In Stepin, Tagi–zade [1] (1980) the variational principle for topological pressure is proved in relation to a continuous action of a countable amenable group in a compact topological space. Theorem 8.2 was announced in Tempelman [20] (1977), [21] (1979). The proof was published in [18] (1980), [22] (1984).

11. In Künsch [1] (1981), Lebowitz, Presutti [1] (1976) and Pirlot [1] (1980) statements similar to the theorems of §§ 7, 8 are proved in the case $G = \mathbb{Z}^m$ under essentially weaker restrictions on the potential; under different conditions the variational principle is proved in Gurevich [1] (1978).

APPENDIX

§ 1. Groups and Semigroups

This section is devoted mainly to some non–standard notions connected with topological semigroups. Some examples of groups and semigroups widely used in this book are given for the reader's convenience. The terminology connected with topological groups coincides with that of Pontryagin [1]; the only exception is the use of the term "compact" instead of "bicompact" and "almost simple group" instead of "simple group".

1.1. Definitions and examples. A set X is called a *semigroup* if to any ordered pair of elements $(x, y) \in X \times X$ corresponds a "product" $x \cdot y = xy \in X$, with the following property: $(xy)z = x(yz)$, $x, y, z \in X$; the mapping $(x, y) \mapsto xy$ is called *multiplication*, or *composition*. If $xy = yx$, $x, y \in X$, the semigroup is called *commutative*. If a semigroup X has an element e for which $xe = ex = x$, $x \in X$, then e is called the *identity* of X; a semigroup X has at most one identity. For $x \in X$, $y \in X$ we set $Ax^{-1} = \{y : y \in X,\ yx = A\}$, $x^{-1}A = \{y : xy \in A\}$. A semigroup X with identity e is called a *group* if to any $x \in X$ there corresponds an "inverse" element x^{-1}, such that $xx^{-1} = x^{-1}x = e$; $x \to x^{-1}$ is a one–to–one correspondence; in this case $Ax^{-1} = \{y : y = ax^{-1},\ a \in A\}$ and $x^{-1}A = \{y : y = x^{-1}a,\ a \in A\}$, $x \in X$. If a semigroup X is also a topological (resp. measurable) space, it is said that the multiplication in X is *separately continuous* (resp. *separately measurable*) if for any $x_0 \in X$ the mappings $x \mapsto xx_0$ and $x \mapsto x_0x$ are continuous (resp. measurable); the multiplication is *jointly continuous* (resp. *jointly measurable*) if the mapping $(x, y) \mapsto xy$ is continuous (resp. measurable). A *semitopological* (resp. *topological*) semigroup is a Hausdorff topological space which is a semigroup in which multiplication is separately (resp. jointly) continuous. A *semimeasurable* semigroup and a *measurable* semigroup are defined similarly. In a topological (resp. measurable) group the mapping $x \mapsto x^{-1}$ is also required to be continuous (measurable). A semitopological semigroup can be considered as a semimeasurable semigroup with respect to the Borel σ–algebra $\mathfrak{B}$; a separable topological semigroup is a measurable semigroup.

A compact semitopological semigroup, which is a group, is a topological group (see DeLeeuw and Glicksberg [1]).

We give several examples of topological groups and semigroups.

Example 1.1. *a)* The m–dimensional linear space $\mathbb{R}^m$, with the natural topology, and the integer lattice $\mathbb{Z}^m$ are commutative locally compact topological groups with respect to addition of vectors. $\mathbb{R}^m_+ :=$ $:= \{x : x = (x_1, \dots, x_m),\ x_1 \geq 0, \dots, x_m \geq 0\}$ and $\mathbb{Z}^m \bigcap \mathbb{R}^m_+$ are commutative locally compact topological semigroups.

b) $GL(m, \mathbb{R})$, the group of all invertible linear operators in $\mathbb{R}^m$, $m \equiv 1$; such operators are uniquely defined by their (non–singular) matrices of order $m \times m$ with respect to the basis $e_1 = (1, 0, \dots, 0), \dots,$ $e_m = (0, \dots, 0, 1)$; this gives a one–to–one mapping of $GL(m, \mathbb{R})$ into an open set in $\mathbb{R}^{m^2}$ and this mapping induces a natural topology in $GL(m, \mathbb{R})$. Clearly, $GL(m, \mathbb{R})$ is a locally compact topological group; it consists of two components, and the component of the identity, $GL_0(m, \mathbb{R})$, is formed by all "proper" operators that preserve the orientation of $\mathbb{R}^m$, i.e. of operators A whose matrices a have positive determinants ($\det(a) > 0$).

c) $SL(m, \mathbb{R})$, $m \geq 2$, is the subgroup of $GL(m, \mathbb{R})$ consisting of all operators A with $\det(a) = 1$.

d) $SO(m)$, $m \geq 2$, is the group of all proper rotations of $\mathbb{R}^m$, i.e. the group of all operators $A \in GL_0(m, \mathbb{R})$ which preserve the quadratic form $Q_m(x) := \sum_{i=1}^m x_i^2$ (i.e. $Q_m(Ax) = Q_m(x)$); in "matrix" terms, $SO(m)$ is the group of all orthogonal matrices of order $m \times m$ with $\det(a) = 1$. It is a connected compact group (cf., e.g., Chevalley [1], v. 1, Ch. 2, § 4).

e) $SO(m, 1)$ is the subgroup of $SL(m), \mathbb{R})$ consisting of all operators that preserve the quadratic form $Q_{m,1}(x) := \sum_{i=1}^m x_i^2 - x_0^2$, $x = (x_0, x_1, \dots, x_m)$. It is a locally compact topological group consisting of two components. The identity component is the group $SO_0(m, 1)$

of all operators $A \in SO(m,1)$ with $A_{00} = (Ae_0)_0 > 0$ (here $e_0 = (1,0,\dots,0) \in \mathbf{R}^{m+1}$; a_{00} is the left upper element of the matrix a of A).

f) The subgroup of $GL(m,\mathbf{R})$ consisting of all matrices of the form $\left(\begin{smallmatrix} a & b \\ 0 & 1 \end{smallmatrix}\right)$, $a \in \mathbf{R}_{++}$, $b \in \mathbf{R}$. This group is isomorphic to the group $\mathbf{R} \odot \mathbf{R}^{\times}_{++}$ all pairs of reals (b,a), $b \in \mathbf{R}$, $a > 0$, with multiplication $(b_1,a_1)(b_2,a_2) = (a_1b_2+b_1, a_1a_2)$ ($\mathbf{R} \times \mathbf{R}^{\times}_{++}$ is the group of affine transformation of the real line; see also Subsect. 2.2).

With any measurable semigroup $(X,\mathfrak{B})$ and any Banach space B we can associate the Banach space $F_B = F_B(X,\mathfrak{B})$ of all B-valued measurable functions f with norm $\|f\|_{F_B} = \sup_{x\in X}\|f(x)\|_B$; with any topological semigroup X we associate the subspace $C_B(X)$ of F_B consisting of all bounded B-valued continuous functions. Moreover, on a topological semigroup we consider the subspaces $UC_B^{(r)}(X)$ and $UC_B^{(l)}(X)$ of all right- (resp., left-) uniformly continuous bounded functions. We recall that a B-valued function f on a topological semigroup X is called *right uniformly continuous* if for any $\varepsilon > 0$ and any $x \in X$ there exists a neighborhood $V_{\varepsilon,x}$ of x such that $\|f(xz) - f(yz)\|_B < \varepsilon$ for all $z \in X$ and $y \in V_{\varepsilon,x}$. If X is a group, this is equivalent to the property $\|f(x) - f(y)\| < \varepsilon$ as soon as $xy^{-1} \in V_{\varepsilon,e}$.

1.2. Semigroups, subsemigroups, full subsemigroups. A subsemigroup (a subgroup) of a topological semigroup (resp. of a group) X is a closed set $Y \subset X$ which is a subsemigroup (resp. subgroup) of X (in the algebraic sense). Certainly, Y is itself a topological semigroup (or topological group) with respect to the topology induced by X. A subsemigroup Y is said to be *full* with respect to the right shifts (or, simply, *r-full*) if $Yx^{-1} \subset Y$, $x \in Y$. An l-full subsemigroup is defined similarly. For example, the subsemigroup $\mathbb{Z}_+^{(e)}$ of all even non-negative integers is full in $\mathbb{Z}_+$, but the subsemigroup $\mathbb{Z}_+^{(e)}\setminus\{0,2\}$ is not. It is easy to verify that: *1)* any subgroup of a group is r-full and l-full; *2)* if X has identity e, then any r- and l-full subsemigroup contains e, too; *3)* a semigroup is an r- and l-full subsemigroup of itself; *4)* the intersection of all r-full subsemigroups containing some set A is also an r-full subsemigroup; we say that it is the r-full subsemigroup *generated* by the set A.

1.3. Open semigroups; the interior of a semigroup. Let X be a topological semigroup. A point $x \in X$ is said to be a *proper interior point* of X if for any open neighbourhood U of x and any $z \in X$ the sets zU and Uz are open in X; the set of all proper interior points of X is called the *proper interior* of X and is denoted by $\mathrm{Int}_p(X)$; it is an open set in X and a subsemigroup in the algebraic sense. For example, $\mathrm{Int}_p(\mathbb{R}^m_+)$ coincides with the topological interior of $\mathbb{R}^m_+$ in $\mathbb{R}^m$. A semigroup X is said to be *open* if $\mathrm{Int}_p(X) = X$. It is clear that any semitopological group is open. The semigroup $\mathbb{R}^m_{++} := \{x : x \in \mathbb{R}^m, x = (x_1, \ldots, x_m), x_i > 0, i = \overline{1,m}\}$ is open but $\mathbb{R}^m_+$ is not. If X is a subsemigroup (in the algebraic sense) of a topological group G, we can consider the topological interior $\mathrm{Int}\,(X|G)$. We have $\mathrm{Int}_p(X) \subset \subset \mathrm{Int}\,(X|G)$; these interiors not always coincide; for example, $\mathrm{Int}_p\mathbb{Z} = = \mathbb{Z}$, but $\mathrm{Int}\,(\mathbb{Z}|\mathbb{R}) = \emptyset$.

1.4. Left and right transformation semigroups. Let T be an arbitrary set and let $\mathcal{T}(T)$ be the class of all transformations of T. In $\mathcal{T}(T)$ we define the *left composition* $(\tau_1\tau_2)(t) = \tau_1(\tau_2(t))$, $\tau_1, \tau_2 \in \in \mathcal{T}(T)$, $t \in T$, and the *right composition* $(\tau_1\tau_2)(t) = \tau_2(\tau_1(t))$, and convert $\mathcal{T}(T)$ into the *left semigroup* $\mathcal{T}_l(T)$ or into the *right* semigroup $\mathcal{T}_r(T)$, respectively. It is convenient to denote $\tau(t)$ by τt if we consider τ as an element of $\mathcal{T}_l(T)$, and by $t\tau$ if we consider τ as an element of $\mathcal{T}_r(T)$ – in both cases the "brackets omitting rule" holds: $(\tau_1\tau_2)(t) = \tau_1(\tau_2(t)) = \tau_1\tau_2 t$ if $\tau_1, \tau_2 \in \mathcal{T}_l(T)$, and $(t)(\tau_1\tau_2) = = ((t)\tau_1)\tau_2 = t\tau_1\tau_2$ if $\tau_1, \tau_2 \in \mathcal{T}_r(T)$.

Subsemigroups of $\mathcal{T}_l(T)$ (resp. of $\mathcal{T}_r(T)$) are called *left* (resp. *right*) transformation semigroups in T. In particular, if T is a topological space, it is natural to consider $\mathcal{H}o_l(T)$ and $\mathcal{H}o_r(T)$, the left and right groups of homeomorphisms of T.

Example 1.2. If $T = B$ is a Banach space, then in $\mathcal{T}_l(B)$ it is natural to consider the (left) semigroups of all bounded linear operators, $\mathcal{L}_l(B)$; in $\mathcal{L}_l(B)$, in its turn, two important subsemigroups can be considered: the semigroup of all contractions, $C_l(B) = \{A,\ A \in \mathcal{L}_l(B),\ \|A\| \le 1\}$, and the semigroup of isometric operators, $\mathfrak{I}_l(B) = \{A : A \in \mathcal{L}_l(B), \|Ab\| = \|b\|, b \in B\}$. If $T = H$ (a Hilbert space), then in $\mathfrak{I}_l(H)$ the left subgroup $U_l(H)$ of all unitary operators is considered, i.e. the subgroup of all invertible isometric operators.

If we equip $\mathcal{L}_l(B)$ and $\mathfrak{I}_l(B)$ with the strong (resp. weak) operator topology, they turn into topological (resp. semitopological) semigroups. On $U_l(H)$ both topologies coincide: $U_l(H)$ is a topological group. The corresponding right semigroups are defined similarly.

A homomorphism $\gamma : x \to \gamma_x$ from a topological semigroup (or group) X into a topological subgroup (group) X_0 of $\mathcal{T}_l(T)$ (resp. of $\mathcal{T}_r(T)$) is called a *left* (resp. *right*) *action of the semigroup* (resp. *group*) X in T by transformations of X_0; it is often convenient to write $x(t)$ instead of $\gamma_x(t)$ and treat x as a transformation of T. Any left action γ of X in T induces the conjugate right action γ^* of X in $\mathbb{C}^T$, the space of all functions on T: $(\varphi\gamma_x^*)(t) = \varphi(\gamma_x t)$, $\varphi \in \mathbb{C}^T$, $x \in X$, $t \in T$. If X is a semigroup of $\mathcal{T}(T)$, then speaking about its action in T we have in mind the "direct" action, i.e. γ is the identity mapping of X onto X.

An action of a topological semigroup X in a topological space T is said to be *continuous* if for any $t \in T$ the mapping from T into T, $x \mapsto x(t)$, is continuous; an action is said to be *completely continuous* if $(x, t) \mapsto x(t)$ is a continuous mapping from the topological space $X \times T$ into T.

Let γ_i be a continuous action of a topological semigroup X_i in a topological space T_i, $i = 1, 2$. Two actions γ_1 and γ_2 are said to be *equivalent* if there exist a homomorphism $\tau : T_1 \to T_2$ and an isomorphism $i : X_1 \mapsto X_2$ such that $\tau(\gamma_1(x_1)t_1) = \gamma_2(i(x_1)\tau(t_1))$, $x_1 \in X_1$, $t_1 \in T_1$; in this case we shall also say that the action of $g_1 \in G_1$ in T_1 is *equivalent* to the action of $g_2 = \tau(g_1) \in G_2$ in T_2.

Example 1.3. If $T = X$ is a topological semigroup, we can consider the left action l of X in itself by left translations ($l_x y = xy$, x, $y \in X$) and the right action r by right translations ($yr_x = yx$, x, $y \in X$). The conjugate actions L and R in C^x are: $(\varphi L_x)(y) = \varphi(xy)$ and $R_x\varphi)(y) = \varphi(yx)$, $x, y \in X$. These actions are completely continuous. If X is a group, then the actions r and l are equivalent:

we can set $\tau(x) = x^{-1}$ and $i(x) = x^{-1}$.

1.5. Simple, almost simple, and semisimple groups. Perfect groups. Any group G has two trivial normal subgroups: $\{e\}$ and G. If G has no other normal subgroups it is called a *simple* group. A topological group G is said to be *semisimple* if its center Z is discrete and the group G/Z is a direct composition of simple groups $G_1, \ldots, G_m$ $(1 \leq m < \infty)$; if, moreover, $m = 1$, G is said to be *almost simple*. A Lie group is almost simple (resp. semisimple) if its Lie algebra is simple (resp. semisimple). We shall say that a semisimple group G is of *(non–)compact type* iff all subgroups G_i are (non–)compact.

(1.A). (H. Weyl) *Any connected semisimple Lie group of compact type is compact* (see, e.g., Naimark [1], Ch. 11, § 7).

(1.B). 1. *Any simply connected semisimple Lie group G is a direct composition of simply connected almost simple groups $\widetilde{G}_i$ ($\widetilde{G}_i$ is the universal covering group of G_i); this factorization is unique* (see, e.g., Naimark [1], Ch. 10, § 7 and Ch. 11, § 4).

2. *Any simply connected semisimple Lie group is a direct composition of a simply connected Lie group of non–compact type and a simply connected compact group; this factorization is unique.*

Example 1.4. *a)* The group $SO(m)$, $m > 2$ is simple if m is odd; if $m > 6$ is even, $SO(m)$ is almost simple: its center $Z = \{\mathbb{I}, -\mathbb{I}\}$ where $\mathbb{I}$ is the identity matrix in $\mathbb{R}^m$ (see, e.g., Pontryagin [1], §§ 51, 65).

b) The group $SO_0(m, 1)$, $m \geq 2$, is simple: its Lie algebra is simple (see, e.g., Chevalley [1] or Goto and Grosshans [1]) and the center $Z = \{\mathbb{I}\}$; this follows from the previous example, since it is well–known that Z is contained in the center of the maximal compact subgroup $\{\left(\begin{smallmatrix} 1 & 0 \\ 0 & A \end{smallmatrix}\right), \quad A \in SO(m)\}$ and $\left(\begin{smallmatrix} 1 & 0 \\ 0 & \mathbb{I}_m \end{smallmatrix}\right) \notin Z$.

(1.C). (K. Iwasawa) *Any connected non–compact semisimple Lie group $G = NAK = KAN$ where N, A and K are subgroups of G with the following properties: $K \cap A = K \cap N = A \cap N = \{e\}$, $A \cong \mathbb{R}^m$,*

$m \geq 1$, N is nilpotent, K contains the center Z of G, and K/Z is compact (thus K is compact iff Z is finite).

Example 1.5. If $G = SL(m, \mathbb{R})$, $m \geq 2$, then $G = ANK$ where $K = SO_0(m, \mathbb{R})$, A is the group of all diagonal matrices and N is the subgroup of all triangular matrices with units on the diagonal.

Example 1.6. Let $G = SO(m, 1)$. We identify G with the matrices in the natural basis in $\mathbb{R}^{m+1}$. Besides the natural coordinates $(x_0, \ldots, x_m)$ of a point $x \in \mathbb{R}^{m+1}$ we shall consider "isotropic" coordinates $(y_0, \ldots, y_m)$, where $y_0 = \frac{x_0 + x_m}{\sqrt{2}}$, $y_1 = x_1, \ldots, y_{m-1}$, $y_m = \frac{x_m - x_0}{\sqrt{2}}$. Then the hyperbolic quadratic form Q_m can be written as follows: $Q_m(y) = 2y_0 y_m + y_1^2 + \ldots + y_{m-1}^2$. The group of matrices $\widetilde{G}$ which preserve this quadratic form is isomorphic to G; the isomorphism α is given by the formula $g = \alpha(\widetilde{g}) = g_0^{-1} \widetilde{g} g_0$, where

$$g_0 = \begin{pmatrix} \frac{1}{\sqrt{2}} & 0 & \ldots & 0 & \frac{1}{\sqrt{2}} \\ 0 & 1 & \ldots & 0 & 0 \\ \vdots & \vdots & \ddots & \vdots & \vdots \\ 0 & 0 & \ldots & 1 & 0 \\ -\frac{1}{\sqrt{2}} & 0 & \ldots & 0 & \frac{1}{\sqrt{2}} \end{pmatrix}.$$

$\widetilde{G} = \widetilde{A}\widetilde{N}\widetilde{K}$ where $\widetilde{K} = g_0 \left(\begin{smallmatrix} 1 & 0 \\ 0 & SO(m) \end{smallmatrix}\right) g_0^{-1}$,

$$\widetilde{A} = \left\{ \begin{pmatrix} a & & & & \\ & 1 & & & \\ & & \ddots & & \\ & & & 1 & \\ & & & & a^{-1} \end{pmatrix}, \quad a \in \mathbb{R}_{++} \right\}$$

and

$$\widetilde{N} = \left\{ \begin{pmatrix} 1 & b_1 & b_2 & \ldots & b_{m-1} & -r \\ 0 & 1 & 0 & \ldots & 0 & -b_1 \\ \vdots & \vdots & \vdots & \ddots & \vdots & \vdots \\ 0 & 0 & 0 & \ldots & 1 & -b_{m-1} \\ 0 & 0 & 0 & \ldots & 0 & 1 \end{pmatrix}, \ (b_1, \ldots, b_{m-1}) \in \mathbb{R}^{m-1} \right\}$$

where $r = \frac{1}{2}\sum_{i=1}^{m-1} b_i^2$. Of course, $G = ANK$ where $A = \alpha(\widetilde{A})$, $N = \alpha(\widetilde{N})$, $K = \begin{pmatrix} 1 & 0 \\ 0 & SO(m) \end{pmatrix}$.

(1.D). (E. Levi). *Any simply connected Lie group $G = R \odot Q = Q \odot R$ (respectively, the right and the left semidirect product*)) where Q is a simply connected semisimple subgroup of G and R is a simply connected solvable subgroup of G (the radical of G).*

We denote by G' the closure of the set $\{gf - fg, f, g \in G\}$. G' is a normal subgroup of G and the group G/G' is commutative. G is said to be *perfect* if $G = G'$. Of course, any semisimple group is perfect.

1.6. Relatively semiinvariant measures. Let X be a locally compact topological semigroup and $\mathfrak{B}$ the σ–algebra of Borel sets in X. *The right modular function* of a measure on $\mathfrak{B}$ is the function $\Delta_\mu^{(r)}(x) = \inf\{\mu(D)/\mu(Dx^{-1}), D \in \mathfrak{B}\}$**). If Δ_μ is strictly positive on X, we say that μ is a *right relatively semiinvariant* measure on X. Evidently,

$$\Delta_\mu^{(r)}(x)\mu(Dx^{-1}) \le \mu(D), \quad x \in X, \quad D \in \mathfrak{B}, \tag{1.1}$$

and, consequently,

$$\Delta_\mu^{(r)}(x)\mu(D) \le \mu(Dx), \quad x \in X, \quad D \in \mathfrak{B}. \tag{1.2}$$

If

$$\Delta_\mu^{(r)}(x)\mu(Dx^{-1}) = \mu(D), \quad x \in X, \quad D \in \mathfrak{B}, \tag{1.3}$$

μ is called a *right relatively invariant measure*. If $\Delta_\mu^{(r)} \equiv 1$ and (1.3) (resp. (1.1)) holds, the measure μ is said to be *right invariant* (resp. *right semiinvariant*).

---

*) The definitions of a right and a left semidirect product of a subgroup and a normal subgroup of a group is given below, in Subsect. 2.2.

**) We set here $\frac{a}{0}=+\infty$, $\frac{a}{+\infty}=0$, $a \in \mathbb{R}_+$, and do not consider the sets D with $\mu(D)=\mu(Dx^{-1})=+\infty$.

The corresponding "left" notions are defined similarly.

A measure μ is said to be *strictly positive* if $\mu(U) > 0$ as soon as U is open, $U \neq \emptyset$ and $U \subset \mathrm{Int}_p(X)$. A semigroup X is said to be *right regular* if there exists a strictly positive regular right semiinvariant measure.

We shall make some remarks on relatively invariant and relatively semiinvariant measures.

1. *If μ is a right (relatively) semiinvariant measure, then its "scalar multiples" $\alpha\mu$, $\alpha \in (0, \infty)$, have the same properties.*

If $\mu(X) = \infty$, we shall identify scalar multiples $\alpha\mu$ on X; if $\mu(X) < \infty$, then we consider the normalized version with $\mu(X) = 1$.

2. *Any right invariant measure is right semiinvariant.*

3. *Any right semiinvariant measure on a group is right invariant.*

4. *On any locally compact topological group there exist a unique left invariant strictly positive regular Borel measure μ_l ("the left Haar measure") and a unique right invariant strictly positive regular Borel measure μ_r ("the right Haar measure")*; see, e.g., Hewitt and Ross [1].

If $\mu_l = \mu_r$, i.e. $\Delta_{\mu_l}^{(r)} \equiv \Delta_{\mu_r}^{(l)} \equiv 1$, the group is called *unimodular*.

We say that a topological semigroup X is *imbeddable* if $\mathrm{Int}_p(X) \neq \emptyset$ and there exists an algebraic and topological isomorphism from X into a set X' contained in some topological group G (of course, X' is a subsemigroup of G in the algebraic sense and a topological semigroup with respect to the induced topology).

5. *An imbeddable locally compact semigroup is left and right regular.*

6. *If G is a locally compact topological group, then the right Haar measure is the unique right invariant and, consequently, the unique right semiinvariant Borel measure on G.*

Quite different is the situation with semiinvariant measures on semigroups. For example, on $\mathbb{R}_+$ any Borel measure $\mu(E) := \int_E f(x)dx$ is semiinvariant if f is a positive non–decreasing function.

(1.E). *Let X be a locally compact topological semigroup and let μ be a right relatively (semi)invariant Borel measure μ. If f is a measurable function, then for any $y \in X$ the function $R_y f(x) := f(xy)$ is* *measurable; if, moreover, $f \geq 0$, then*

$$\Delta_\mu(y) \int_D f(xy)\mu(dx) \overset{(\leq)}{=} \int_{Dy} f(x)\mu(dx)$$

for any $y \in X$, $D \in \mathfrak{B}$. If $f \in L_B^\alpha$ $(\alpha \geq 1)$, then $R_y f \in L_B^\alpha$, $y \in X$.

We shall say that a function $\varphi(x)$, $x \in X$, is *(semi)multiplicative* if it is strictly positive and $\varphi(xy) = (\geq)\varphi(x)\varphi(y)$; if X has identity e, then we, moreover, require that $\varphi(e) = 1$. For example, on $GL(m, \mathbb{R})$ the functions $\varphi(a) = \|a\|^{-\alpha}$, $\alpha > 0$, are semimultiplicative.

Relations (1.1) and (1.3) imply the following property of modular functions.

(1.F). *If μ is a right relatively (semi)invariant measure, then the function $\Delta_\mu^{(r)}$ is (semi)multiplicative.*

Using (1.E) and (1.F), it is easy to prove the following statement.

(1.G). *Let X have identity element e. If ν is a right relatively semiinvariant measure and φ is a continuous semimultiplicative function on X, then the measure $\mu(A) := \int_A \varphi(x)\nu(dx)$, $A \in \mathfrak{B}$, is a right semiinvariant measure, and $\Delta_\mu^{(r)}(x) = \varphi(x)\Delta_\mu^{(r)}(x)$. If, moreover, ν is right relatively invariant and φ is multiplicative, then μ is right relatively invariant.*

(1.H). *If X is a group, then a left relatively invariant measure ν is right relatively invariant and, conversely, any right relatively invariant measure is left relatively invariant; moreover,*

$$\nu(A) = \int_A \Delta_\nu^{(r)}(x)\mu_r(dx) = \int_A \Delta_\nu^{(l)}(x)\mu_l(dx),$$

and thus

$$\Delta_\nu^{(r)}(x) = \Delta_\nu^{(l)}(x)/\Delta\mu_r^{(l)}(x);$$

in particular, $\Delta_{\mu_l}^{(r)}(x) = (\Delta_{\mu_r}^{(l)}(x))^{-1}$ (cf. Halmos [1], Ch. 11, § 60).

1.7. Convolution of probability measures. Let X be a locally compact semitopological semigroup and let μ, ν be Borel probability measures on X. If $f \in C_0(X)$, then the function $y \mapsto \int f(xy)\mu(dx)$ is bounded and continuous and, consequently, the following functional $f \to I(f)$ is well defined:

$$I(f) := \int \left(\int f(xy)\mu(dx) \right) \nu(dy) = \int \left(\int f(xy)\nu(dy) \right) \mu(dx).$$

It is a continuous linear positive functional and $I(\mathbf{1}) = 1$; by the Riesz C_0^*–representation theorem, there exists a probability measure λ on X such that $I(f) = \int f(x)\lambda(dx)$. The measure λ is called the *convolution* of μ and ν and is denoted by $\lambda = \mu * \nu$. If X is a separable locally compact topological semigroup, λ is measurable and we have for any measurable bounded function $f(x)$, $x \in X$:

$$\int f(x)(\mu * \nu)(dx) = \int \left(\int f(xy)\mu(dx) \right) \nu(dy) =$$
$$= \int \left(\int f(xy)\nu(dy) \right) \mu(dx);$$

consequently,

$$(\mu * \nu)(E) = \int \mu(Ey^{-1})\nu(dy) =$$
$$= \int \nu(x^{-1}E)\mu(dx), \quad E \in \mathfrak{B};$$

The last two formulas can be considered as the definition and the main property of $\mu * \nu$ on any measurable semigroup.

1.8. Cross–sections over factor groups. Let G be a locally compact topological group, Γ be a normal subgroup and $F := G/\Gamma$ (the left factor group). Let π be the natural homomorphism of G onto F. A *cross–section in* G *over a set* $A \subset F$ is an arbitrary set $\widetilde{A} \subset G$ such that the restriction of π to $\widetilde{A}$ is a one–to–one mapping from $\widetilde{A}$ onto A. Let $p_A(a)$ $(a \in A)$ be the unique element of the set $\pi^{-1}(a) \cap \widetilde{A} \subset G$; it is clear that $p_A : a \mapsto p_A(a)$ is a one-to-one mapping from A onto $\widetilde{A}$. A cross-section $\widetilde{A}$ is called *continuous* if the mapping p_A is continuous; $\widetilde{A}$ is called *piecewise continuous* if there exist σ-compact sets A_i, $i = \overline{1,m}$ $(m \leq \infty)$, such that $A_i \cap A_j = \emptyset$ if $i \neq j$, $\bigcup_{i=1}^{m} A_i = A$ and $\widetilde{A}_i = p_A(A_i)$ are continuous cross-sections over A_i, $i = \overline{1,\infty}$. For example, the cross-section $[0,1)$ in $\mathbb{R}$ over $\mathbb{R}/\mathbb{Z}$ is not continuous, but it is piecewise continuous since it can be divided into two continuous "pieces": $[0,1] = [0,a) \cup [a,1)$ $(0 < a < 1)$. It is clear that:

(1.I). *Any piecewise continuous cross-section $\widetilde{A}$ over $A \subset F$ is σ-compact, and therefore it is a Borel set in G; if C is a σ-compact set in G, then $\widetilde{A} \cdot C$ and $C \cdot \widetilde{A}$ are σ-compact.*

If G is a Lie group, then over some compact neighbourhood A of e_F, the identity of F, there exists a continuous cross-section $\widetilde{A}$ (see, e.g., Chevalley [1], v. 1, Ch. 4, § 5). This implies the following statement.

(1.J). *If G is a Lie group, then over any σ-compact set $A \subset F$ there exists a piecewise continuous cross-section $\widetilde{A}$.*

The following relation between integrals with respect to the left Haar measures μ_G, μ_F and μ_Γ is well known (see, e.g. Bourbaki [1], Ch. 7, § 2).

(1.K). *Let B be a Banach space and $\varphi \in L^1_B\ (G, \mathfrak{B}_G, \mu_G)$. Then 1) the function $\varphi(g) = \int_\Gamma \varphi(g\gamma)\mu_\Gamma(d\gamma)$ is constant on the cosets with respect to Γ and can be considered as a function $\varphi(f)$ on F; 2) $\varphi(f) \in$ $\in L^1_B\ (F, \mathfrak{B}_F, \mu_F)$ and*

$$\int_G \varphi(g)\mu_G(dg) = \int_F \Big(\int_\Gamma \varphi(f\gamma)\mu_\Gamma(d\gamma) \Big) \mu_F(df). \tag{1.4}$$

Let us consider a section $\widetilde{A}$ over $A \subset \mathfrak{B}_F$, the σ-algebra $\widetilde{\mathfrak{B}}_{\widetilde{A}}$ on

$\widetilde{A}$ induced by $\mathfrak{B}_A$ and the mapping π, and the measure $\widetilde{\mu}_F^{(\widetilde{A})}$ on $\mathfrak{B}_{\widetilde{A}}$ induced by π: $\widetilde{\mu}_F^{(\widetilde{A})}(\widetilde{E}) = \mu_F(\pi(\widetilde{E}))$, $\widetilde{E} \in \widetilde{\mathfrak{B}}_{\widetilde{A}}$ (if $\widetilde{A} \in \mathfrak{B}(G)$, then $\widetilde{\mathfrak{B}}_{\widetilde{A}} = \mathfrak{B}_{\widetilde{A}}$, the restriction of $\mathfrak{B}(G)$ to $\widetilde{A}$).
Any function $\psi(f) \in L_B^1\ (A, \mathfrak{B}_A, \mu_F)$ can be considered as a function $\psi(\widetilde{f}) \in L^1\ (\widetilde{A}, \widetilde{\mathfrak{B}}_{\widetilde{A}}, \widetilde{\mu}_F^{(\widetilde{A})})$, and therefore (1.4) can be rewritten in the following form: if $C \in \mathfrak{B}_\Gamma$, $\widetilde{A} \cdot C \in \mathfrak{B}_G$ (see (1.I)) and $\varphi \in$ $\in L_B^1\ (\widetilde{A}C, \mathfrak{B}_{\widetilde{A}C}, \mu_G)$, then

$$\int_{\widetilde{A}C} \varphi(g)\mu_G(dg) = \int_{\widetilde{A}}\left(\int_C \varphi(\widetilde{f}\gamma)\mu_\Gamma(d\gamma)\right)\widetilde{\mu}_F^{(\widetilde{A})}(d\widetilde{f}); \tag{1.5}$$

in particular,

$$\mu_G(\widetilde{A}C) = \mu_F^{(\widetilde{A})}(\widetilde{A})\mu_\Gamma(C) = \mu_F(A)\mu_\Gamma(C). \tag{1.6}$$

1.9. Factorization of the Haar measure. Let G be a σ–compact group and $G = G_1 \cdot G_2 \cdot \ldots \cdot G_n$ where the G_i's are subgroups of G. Denote $G^{(i)} := G_{n-i+1} \cdot \ldots \cdot G_n$, $i = 1, \ldots, n$, ($G^{(n)} = G$, $G^{(1)} =$ $= G_n$). Suppose that the $G^{(i)}$ are groups and all groups $G_{n-i} \cap G^{(i)}$ are compact, $i = 1, \ldots n-1$*). The following two statements are simple consequences of Proposition 13 in Bourbaki [1, Ch. 7, § 2]. For any $(g_1, \ldots, g_n) \in G_1 \times \ldots \times G_n$ we set $j(g_1, \ldots, g_n) = g_1 \cdot \ldots \cdot g_n\ (\in G)$.

(1.L). *If $G^{(i)}$ is a normal subgroup of $G^{(i+1)}$**) $(i = 1, \ldots, n-1)$, then a function $\varphi \in L_B^1\ (G, \mathfrak{B}_G, \mu_G^{(l)})$ iff the function $\varphi \cdot j \in L_B^{(1)}\ (G_1 \times \ldots \times G_n,\ \ \mathfrak{B}_{G_1} \times \ldots \times \mathfrak{B}_{G_n}, \mu_{G_1}^{(l)} \times \ldots \times \mu_{G_n}^{(l)})$; in this case*

$$\int_G \varphi(g)\mu_G^{(l)}(dg) = \int_{G_1 \times \ldots \times G_n} \varphi(g_1 \cdot \ldots \cdot g_n)\mu_{G_1}^{(l)}(dg_1) \ldots \mu_{G_n}^{(l)}(dg_n),$$

*) We shall call this "*left*" *factorization* of G.

**) In this case we say that $G=G_1 \cdot \ldots \cdot G_n$ is an "*almost semidirect*" left factorization.

i.e. j is a measure preserving mapping from $(G_1 \times \ldots \times G_n, \mathfrak{B}_{G_1} \times \ldots \times \mathfrak{B}_{G_n}, \mu^{(l)}_{G_1} \times \ldots \times \mu^{(l)}_{G_n})$ *onto* $(G, \mathfrak{B}_G, \mu^{(l)}_G)$.

(1.M). *If all groups $G^{(i)}$, $i = 1, \ldots, n$, are unimodular, then the conclusion of the previous statement is true.*

Various combinations of Statements (1.L) and (1.K) are possible. Here is an important example.

(1.N). *Let the group G be unimodular and $G = ANK$ where K is a unimodular subgroup, A, N and AN are subgroups, N is a normal subgroup of AN and the subgroups $AN \cap K$ and $A \cap N$ are compact. Then for any $\varphi \in L^1$ $(G, \mathfrak{B}_G, \mu_G)$ we have*

$$\int_G \varphi(g)\mu_G(dg) = \int_{A\times N\times K} \varphi(ank)\mu^{(l)}_A(da)\mu^{(l)}_N(dn)\mu_K(dk).$$

In order to prove this statement we have to use (1.L) with $G = AN$, $G_1 = A$, $G_2 = N$, and (1.M) with $G = ANK$, $G_1 = AN$, $G_2 = K$.

We can also consider the inverse factorization of G: $G = KNA$; then we get

$$\int_G \varphi(g)\mu_G(dg) = \int_{K\times N\times A} \varphi(kna)\mu^{(r)}_A(dk)\mu^{(r)}_N(dn)\mu_K(da).$$

§ 2. Homogeneous and Group–Type Homogeneous Spaces

2.1. Homogeneous spaces. Let Y be a topological space, X a left toplogical semigroup of continuous left transformation of the space Y, and for any $y_0 \in Y$ let the mapping $\pi_{y_0} : x \mapsto xy_0$ be a continuous mapping from X into Y; it is then said that Y is a left X–space. Let $X = G$ be a group; then a left G–space Y is said to be a *left* *homogeneous space* (with respect to G) if G acts *transitively* on Y, i.e. if for $y_0 \in Y$ the transformation π_{y_0} is a mapping *on* Y. In this case the elements of G are called *motions* of Y. The set $K_{y_0} = \{g : g \in G, gy_0 = y_0\}$ is a subgroup of G; it is called the *stationary subgroup* of the point y_0. The stationary subgroup of some other point $y = gy_0$ is related to K_{y_0} in the following way: $K_y = gK_{y_0}g^{-1}$. Right G–spaces and right homogeneous spaces are defined similarly.

In this book, for simplicity reasons we consider only homogeneous spaces having the following additional properties:

a) *π_y is an open mapping*, $y \in G$;

b) *the stationary subgroups K_y are compact*, $y \in Y$.

In the sequel we fix some $y_0 \in Y$ and put $K := K_{y_0}$.

Example 2.1. A topological group G is a left (right) homogeneous space with the group G_l (G_r) of all left (right) translations as group of motions (see Example 1.3); the stationary subgroup of an arbitrary point consists of the identity translation l_e (resp. r_e).

Example 2.2. Let K be a compact subgroup of a topological group G. The space of left cosets G/F (see Pontryagin [1], § 19) is a left homogeneous space with respect to the group of left translations l_g : $l_g(fF) = gfF$, $g, f \in G$; the space of right cosets $F\backslash G$ is a right homogeneous space with respect to the right translations $r_g : (Ff)r_g = = Ffg$. The stationary subgroup of the point $F \in G/F$ is the subgroup F itself.

A G_1–space Y_1 and G_2–space Y_2 are *equivalent* if the direct actions of G_1 on Y_1 and G_2 on Y_2 are equivalent (see Subsect. 1.4), i.e. if there exist a homomorphism $\tau : Y_1 \mapsto Y_2$ and a homomorphism $i : G_1 \mapsto G_2$ such that $\tau(g_1 y_1) = i(g_1)\tau(y_1)$, $g_1 \in G_1$, $y_1 \in Y_1$, The restriction of i to K_{y_0} is a homomorphism $K_{y_0} \mapsto K_{\tau(y_0)}$.

(2.A). *Any left homogeneous G–space Y is equivalent to the left space $G\backslash K$ where K is the stationary subgroup of some point $y_0 \in Y$; here $i(g) = l_g$, and if $y = gy_0$ then $\tau(y) = gK$* (see, e.g. Naimark [1], Ch. 3, § 2).

It is easy to verify that the following statement is true.

(2.B). *If i is an isomorphism of a group G onto a group G', K is a compact subgroup of G and $K' := i(K)$, then the homogeneous spaces G/K and G'/K' are equivalent.*

In this book we identify equivalent homogeneous spaces, and therefore any homogeneous G–space with a stationary subgroup $K_{y_0} = K$ is denoted by G/K; in the situation described in (2.B) we identify G/K and G'/K' (e.g., we write $G/\{e\}$ instead of $G_l/\{l_e\}$). This does not cause misunderstandings, because all the notions and properties we consider are preserved if we replace a homogeneous space by another one.

(2.C). *A homogeneous space G/K is locally compact iff the group G is locally compact* (cf. Pontryagin [1], § 19).

(2.D). *On a locally compact homogeneous space G/K there exists an invariant Borel measure μ, and $\mu(A) < \infty$ if A is compact; invariancy means that $\mu(gA) = \mu(A)$, $A \in \mathfrak{B}(G/K)$, $g \in G$* (see, e.g., Weil [1], Hewitt and Ross [1]).

Any set of the form $\{ky, k \in K\}$ is called a *sphere* with center y_0; a set A with the property $kA = A$, $k \in K$, is said to be *spherically symmetric*. If V is a neighborhood of the point y_0, then the set $V' = \bigcup_{k \in K} kV$ is a spherically symmetric neighbourhood of y_0. If V is a spherically symmetric neighbourhood of y_0 and $y \in Y$, then for all $g \in G$ such that $y = gy_0$ we obtain the same neighbourhood gV of the point y; we denote it by yV.

If $A, B \subset Y$ and B is spherically symmetric, then we set $A \cdot B := \pi(\pi'(A) \cdot \pi^{-1}(B))$ and $B^{-1} := \pi\{[\pi^{-1}(B)]^{-1}\}$. Then we have $\pi^{-1}(A \cdot B) = \pi^{-1}(A) \cdot \pi^{-1}(B)$ and $\pi^{-1}(B^{-1}) = [\pi^{-1}(B)]^{-1}$.

We say that Y is a *metric homogeneous* space if there is a metric ρ on Y which induces the initial topology in Y and which is G–invariant: $\rho(gy_1, gy_2) = \rho(y_1, y_2)$, $g \in G$, $y_1, y_2 \in Y$. It is easy to verify that the balls $B_r := \{y : \rho(y_1 y_0) \le r\}$ are spherically symmetric and also inversely symmetric, i.e. $B_r^{-1} = B_r$; moreover, $B_{r_1} \cdot B_{r_2} \subset B_{r_2+r_2}$.

2.2. Semidirect products of groups and group–type homogeneous spaces. In many situations it is reasonable to extend the action l of a group A on itself by the left translations (see Examples 1.3 and 2.1), "joining" to l an action in A of some group Σ of automorphisms of A. Such an "extension" of A and of its action is performed in the following way.

Let A and Σ be topological groups, Aut $_l(A)$ the left group of all automorphisms of A, and let γ be a completely continuous (left) action of Σ in A by automorphisms of some subgroup Σ_0 ($\subset$ Aut $_l(A)$). We shall write $\sigma(a)$ instead of $\gamma_\sigma(a)$, $\sigma \in \Sigma$, $a \in A$. We define the joint left action $\tilde{\gamma}$ in A of a pair $(a, \sigma) \in A \times \Sigma$ as follows: $(a, \sigma)(a') = a\sigma(a')$. Then $(a_1, \sigma_1)(a_2, \sigma_2)(a') = (a_1, \sigma_1)(a_2, \sigma_2(a')) = a_1 \cdot \sigma_1(a_2) \cdot \sigma_1\sigma_2(a')$. If we introduce the multiplication $(a_1, \sigma_1)(a_2\sigma_2) = (a_1\sigma_1(a_2), \sigma_1\sigma_2)$ in the topological space $A \times \Sigma$. This space turns into a topological group, the *right semidirect product* of the groups A and Σ (with respect to the homomorphism γ); we denote it by $A \overset{\gamma}{\odot} \Sigma$. The formula $(a, \sigma)(a') =$ $= a\sigma(a')$ defines a completely continuous action of $A \overset{\gamma}{\odot} \Sigma$ on A.

We shall consider three special cases of this construction, corresponding to various methods of choosing Σ_1, Σ_0 and γ.

Case I. Σ is an arbitrary topological group, $\Sigma_0 = \{e_{Aut_l(A)}\}$, i.e. $\sigma(a) \equiv a$, $\sigma \in \sigma$. In this case $A \overset{\gamma}{\odot} \Sigma = A \times \Sigma$, the direct product of A and Σ.

Case II. $\Sigma = \Sigma_0$ and γ is the identity mapping; in this case we shall write $A \odot \Sigma$ instead of $A \overset{\gamma}{\odot} \Sigma$. A is a homogeneous space with motion group $A \odot \Sigma$, the stationary subgroup of the identity e_A is the subgroup $\Sigma' = \{(e_A, \sigma), \sigma \in \Sigma\}$; we shall identify Σ' and Σ (this is quite natural in view of statement (2.E) presented below). Homogeneous spaces of the form $A \odot \Sigma/\Sigma$ will be called *group–type homogeneous spaces*. If $\Sigma = \{e\}$, then $A \odot \Sigma$ can be identified with A and the space $A \odot \Sigma/\Sigma$ coincides with the A–translation space A considered in Example 2.1.

Example 2.3. The group $\mathbb{R}^m \odot GL(m)$ is called the affine group, or the group of affine transformations, of the space $\mathbb{R}^m$ (if $m = 1$ this is the group $\mathbb{R} \odot R^{\times}_{++}$ considered in Example 1.1).

Example 2.4. The group $\mathbb{R}^m \odot SO(m)$ is the group of proper Euclidean motions of $\mathbb{R}^m$; the metric $\rho(x,y) = (\sum_{i=1}^m (x_i - y_i)^2)^{\frac{1}{2}}$, $x = (x_1, \ldots, x_m)$, $y = (y_1, \ldots, y_m)$ is $\mathbb{R}^m \odot SO(m)$–invariant. The space $E_m := (\mathbb{R}^m \odot SO(m))/SO(m)$ is called the *Euclidean homogeneous space*.

Now we shall give an example of a homogeneous but not group–type space.

Example 2.5. Let us consider in $\mathbb{R}^{m+1}$ the bilinear form $A(x,y) = \sum_{i=1}^m x_i y_i - x_0 y_0$, $x = (x_0, x_1, \ldots, x_m)$, $y = (y_0, y_1, \ldots, y_m)$, and the corresponding quadratic form $Q(x) = \sum_{i=1}^m x_i^2 - x_0^2$. The set $S^+_{m,1} := \{x : Q(x) = -1, \ x_0 \geq 1\}$ is invariant with respect to the group $SO_0(m,1)$ (see Example 1.1) and is a homogeneous $SO_0(m,1)$–space; the stationary subgroup of the point $(1, 0, \ldots, 0)$ is isomorphic to the group $SO(m)$. Recall that the space $S^+_{m,1}$ is the base of one of the models of the m–dimensional Lobachevsky space. The metric $\rho(x,y) = \operatorname{arch}|A(x,y)|$ on $S^*_{m,1}$ is $SO_0(m,1)$–invariant. The space $L_m := SO_0(m,1)/SO(m)$ is called the m–dimensional homogeneous *Lobachevsky space*. The group $SO_0(m,1)$ can not be decomposed as $SO_0(m,1) = A \odot SO(m)$, and thus L_m is not an group–type space.

Let $G = A \odot \Sigma$. It is easy to verify that the following statements are true.

(2.E). *The mapping $i_\Sigma : \sigma \to (e_A, \sigma)$ is an isomorphism $\Sigma \mapsto \Sigma' := \{(e_A, \sigma), \sigma \in \Sigma\} \subset G$ and $i_A := a \mapsto (a, e_\Sigma)$ is an isomorphism $A \mapsto A' = \{(a, e_\Sigma), \ a \in A\}$. We have $(a, e_\Sigma)(a_0) = aa_0$ and $(e_A, \sigma)(a_0) = \sigma(a_0)$, i.e. the translation actions of the elements $a' = (a, e_\Sigma) \in A'$ and $\sigma' = (e_A, \sigma) \in \Sigma'$ are equivalent to the translation action of a and the automorphism action of σ on A.*

(2.F). *The subgroups A' and Σ' have the following properties:*

a) *A' is a normal subgroup of G;*

b) *the restriction of the natural homomorphism $\pi : G \mapsto A'\backslash G$ (the right factor group) to the subgroup Σ' is an isomorphism: $\Sigma' \cong G/A'$;*

c) *$A' \cap \Sigma' = \{e\}$;*

d) *$G = A'\Sigma'$.*

The union of the properties a), b) *is equivalent to the union of* a), c), d).

(2.G). *The action of an automorphism $\sigma \in \Sigma$ on A is equivalent to the action on A' of the element $\sigma' = i_\Sigma(\sigma) = (e_A, \sigma) \in \Sigma'$ by the inner automorphism $\alpha_{\sigma'}$ defined by σ'; indeed, $\alpha_{\sigma'}(a, e_\Sigma) = (e_A, \sigma)(e_A, e_\Sigma) =$ $= (\sigma(a), e_\Sigma)$.*
The translation action of an element $a \in A$ on A is equivalent to the translation action of $a' = (a, e_\Sigma)$ on A'.

Case III. G is a topological group, A is a normal subgroup, Σ is a subgroup of G, and $\gamma_\sigma = a_\sigma$ is the restriction to A of the inner automorphism defined by the element $\sigma \in \Sigma$, i.e. $\alpha_\sigma(a) = \sigma a \sigma^{-1}$, $a \in A$. In this case we can consider the "inner" semidirect product $A\overset{\alpha}{\odot}\Sigma$.

(2.H). *If the subgroups A and Σ of G have the properties* a)–d) *of statement* (2.F)*, then the mapping $a\sigma \mapsto (a, \sigma)$ is an isomorphism $G \mapsto$ $\mapsto A\overset{\alpha}{\odot}\Sigma$. Conversely, if $G \cong A\overset{\alpha}{\odot}\Sigma$ and $a \mapsto (a, e_\Sigma)$, $\sigma \mapsto (e_A, \sigma)$, $fa \in A$, $\sigma \in \Sigma$, then A and Σ have the properties* a)–d).

If A and Σ have the properties a) and b), we say that G is the *right semidirect product of its subgroups A and Σ* (or that G is decomposed into the right semidirect product of A and Σ) and write $G = A \odot \Sigma$.

We can generalize the notion of semidirect product to the case of several subgroups. Let $G_1, \dots, G_n$ be subgroups of a topological group G and $G = G_1 \cdot \ldots \cdot G_n$. Denote $G^{(i)} = G_n \dots G_{n-i+1}$, $i = 1, \dots, n$ ($G^{(1)} = G_1$, $G^{(n)} = G$). Suppose that: *1)* all $G^{(i)}$ are groups, *2)* $G^{(i)}$ is a normal subgroup of $G^{(i+1)}$, and *3)* $G^{(i+1)}/G^{(i)} \cong G_{n-i}$, $i - 1, \dots, n-1$; then we say that G is a *right semidirect product of the subgroups* $G_1, \dots, G_n$ and write $G = G_n \odot \dots \odot G_1$ or $G = \prod_{i=1}^{n} G_i$ (condition *3)* can be replaced by the condition *3')* $G_{i+1} \bigcap G^{(i)} = \{e\}$). This *semidirect factorization of a group* is a special case of a right almost semidirect factorization of G (see Subsect. 1.9).

2.3. Simple homogeneous spaces. Now we shall extend to homogeneous spaces the notion of simple group introduced in Subsect. 1.5.

Let G/K be a homogeneous space and Φ a normal subgroup of G. We shall say that Φ is *transitive* if $\Phi K = G$, i.e. if Φ acts transitively on G/K; if $\Phi K \neq G$, we shall say that Φ is *intransitive*. We shall say that the factor group G/Φ is *small* (*large*) if Φ is transitive (resp. intransitive).

A homogeneous space G/K is called *simple* if $\{e\}$ is the only intransitive normal subgroup of G. It is obvious that: *1)* a group G regarded as a homogeneous space (see Example 2.1) is simple iff it is a simple group; and *2)* if a group G is simple, then the homogeneous space is simple. In particular, the Lobachevsky space L_n is simple.

In the case of a group–type homogeneous space, simplicity admits of a descriptive characterization. We shall say that a group of automorphisms K of a group A is *γ–irreducible* (on A) if A has no K–invariant subgroups besides the two trivial ones: $\{e\}$ and A.

(2.I). *Let* $G = A \odot K$.

a) If K is γ–irreducible on A, then the homogeneous space G/K is simple.

b) If A is commutative, then γ–irreducibility of K is also necessary in order that G/K be simple.

Proof. *a)* According to statements (2.E) and (2.F), $G = A' \odot K'$ where $K' = \{(e_A, k),\ k \in K\}$ and $A' = \{(a, e_K),\ a \in A\}$. Let N be a normal subgroup of G; then $A_0 := NK' \cap A'$ is a subgroup of A'. For any $k' \in K'$ we have

$$k' A_0 (k')^{-!} = k' N K' (k')^{-1} \cap k' A' (k')^{-1} =$$
$$= k' N (k')^{-1} k' K' (k')^{-1} \cap A' = NK' \cap A' = A_0'.$$

The subgroup A'_0 of A' is K'–invariant and, according to (2.G), the subgroup $A_0 := i_A^{-1}(A'_0)$ of A is K–invariant. Therefore either $A'_0 = \{e\}$, or $A'_0 = A'$. Let us consider the first case: $A'_0 = \{e\}$. Suppose that some $(a,k) \in N\backslash K'$. Then $(a,k)(e_A,k^{-1}) = (a,e_K) \in A'$. Thus $(a,e_K) \in A'_0$; this is a contradiction since $a \neq e_A$. Therefore $N \subset K'$. If $k \neq e_K$, there exists an element $a_0 \in A$ such that $k(a_0) \neq a_0$, and

$$(a_0,e_k)(e_A,e_k)(a_0,e_k)^{-1} = (k(a_0^{-1})a_0,k) = \left((k(a_0))^{-1}a_0,k\right) \notin K'.$$

Thus, if $A_0 = \{e\}$, then $N = \{e\}$. Now we shall consider the case $A'_0 = A'$. Then $NK' \supset A'$ and $NK' \supset NK' \cdot K' \supset A'K' = G$, i.e. N is transitive, and assertion a) is proved.

b) If a commutative and non–trivial subgroup A_0 of A is K–invariant, then, by virtue of (2.F,d) and (2.G), A_0 is an intransitive normal subgroup of G; this is impossible if G/K is simple.

□

If $A = \mathbb{R}^m$ and $K \subset GL(m,\mathbb{R})$, it is natural to compare the property of γ–irreducibility of K in $\mathbb{R}^m$ with the well–known property of (*linear*) *irreducibility*: $\mathbb{R}^m$ does not contain non–trivial (i.e. different from $\{0\}$ and $\mathbb{R}^m$) K–invariant subspaces. Of course, *γ–irreducibility of K implies irreducibility*. The converse is not always true, as the following example shows.

Example 2.6. Let $G = SL(2,\mathbb{Z})$ be the subgroup of $SL(2,\mathbb{R})$, consisting of all operators with integer–valued matrices. If l is a one–dimensional subspace of $\mathbb{R}^2$, then the set $G(l) \cap l$ is countable and thus G is irreducible on $\mathbb{R}^2$. But it is not γ–irreducible: $G(\mathbb{Z}^2) = \mathbb{Z}^2$.

(2.J). *If a connected subgroup K of $GL(m,\mathbb{R})$ is irreducible on $\mathbb{R}^m$, then it is γ–irreducible.*

Proof. Let K be an irreducible connected subgroup of $GL(m,\mathbb{R})$ and A_0 a non–trivial K–invariant subgroup of $\mathbb{R}^m$. We can choose a basis in $\mathbb{R}^m$ such that $A_0 = \mathbb{R}^n + \mathbb{Z}^l$, $n + l \leq m$ (see, e.g., Bourbaki [2], Ch. 7, § 1). If $0 < n < m$, then $\mathbb{R}^n$ is a K–invariant subspace of $\mathbb{R}^m$ (indeed, $k(\mathbb{R}^n)$ is connected and therefore $k(\mathbb{R}^n) \subset \mathbb{R}^n$, $k \in \in K$); this is impossible. If $n = 0$, then (A_0, e_K) is a discrete normal subgroup of the group $G := \mathbb{R}^m \odot K$. Consequently (see Pontryagin [1], § 22) the elements of (A_0, e_K) commute with all elements of G. But $(e_A, k)(a, e_K) = (k(a), k)$, $(a, e_K)(e_A, k) = (a, k)$, $a \in A_0$, $k \in K$, and thus $(e_A, k)(a, e_K) \neq (a, e_K)(e_A, k)$ if $k(a) \neq a$. If there exists an $a_0 \in \mathbb{R}^m$ such that $k(a_0) = a_0$, $k \in K$, then the whole space $\{\lambda a_0, \lambda \in \mathbb{R}\}$ is K–invariant and this contradicts irreducibility.

□

Example 2.7. The group $SO(m)$ is connected and irreducible in $\mathbb{R}^m$. Consequently, it is γ–irreducible. According to statement (2.J), the Euclidean space $E_m = (\mathbb{R}^m \odot SO(m))/SO(m)$ is simple.

2.4. On measures on homogeneous spaces. Let $Y = G/K$ be a locally compact homogeneous space; $\mathfrak{B}(G)$ and $\mathfrak{B}(Y)$ the σ–algebras of Borel sets on G and Y, respectively. A measure μ on $\mathfrak{B}(Y)$ is called *spherically symmetric* if $\mu(kA) = \mu(A)$, $k \in K$, $A \in \in \mathfrak{B}(Y)$. The notions of modular function and relatively (semi)invariant measure, introduced in Subsect. 1.6 for groups, are naturally extended homogeneous spaces; the properties of such measures remain true (with obvious changes). It is easy to verify that any relatively invariant measure ν on $\mathfrak{B}(Y)$ is spherically symmetric.

Let π be the natural mapping of G onto Y, and $\pi^{-1}(A)$ the complete inverse image of A with respect to π. If the measures ν on $\mathfrak{B}(Y)$ and $\widetilde{\nu}$ on $\mathfrak{B}(G)$ are connected by the relation $\widetilde{\nu}(\pi^{-1}A) = \nu(A)$, $A \in \mathfrak{B}(Y)$, then we say that $\widetilde{\nu}$ is a *prototype*, or *preimage*, of ν, or that ν is the *projection* of $\widetilde{\nu}$ (notation: $\nu = \widetilde{\nu} \circ \pi^{-1}$). Denote $\widetilde{f} := f \circ \pi$, the prototype of $f(y)$, $y \in Y$, on G. It is easy to verify that the following statement is true.

(2.K). *Let $\widetilde{\nu}$ be a prototype of a measure ν. Then*

a) $f \in L^1(Y, \mathfrak{B}(Y), \nu)$ *iff* $\widetilde{f} \in L^1(G, \mathfrak{B}(G), \widetilde{\nu})$;

b) $\int_Y f\, d\nu = \int_G \widetilde{f}\, d\widetilde{\nu}$, $f \in L^1_B(Y, \mathfrak{B}(Y), \nu)$.

Let μ_K be the normalized Haar measure on K. It is well known (see, e.g., Hewitt and Ross [1], Ch. 4, § 15) that for any $\widetilde{A} \in \mathfrak{B}(G)$ the function $g \mapsto \int_K \chi_{\widetilde{A}}(gk)\mu_K(dk)$ is constant on the left K–cosets. Therefore we can put $\int_K \chi_{\widetilde{A}}(yk)\mu_K(dk) : \int_K \chi_{\widetilde{A}}(gk)\mu_K(dk)$, if $y = gy_0$. To any measure ν on $\mathfrak{B}(Y)$ we can associate a measure $\widetilde{\nu}$ on G in the following way: $\widetilde{\nu}(\widetilde{A}) = \int_Y \left(\int \chi_{\widetilde{A}}(yk)\mu_K(dk)\right)\nu(dy)$, $\widetilde{A} \in \mathfrak{B}(G)$. It is clear that $\widetilde{\nu}$ is a prototype of ν with the property $\widetilde{\nu}(\widetilde{A}k) = \widetilde{\nu}(\widetilde{A})$, $\widetilde{A} \in \mathfrak{B}(G)$, $k \in K$; we shall call this measure the *K–uniform prototype of ν* . The K–uniform prototype of the invariant measure μ is the left Haar measure μ_l on G; the right Haar measure μ_r is the K–uniform prototype of some relatively invariant measure on Y; $\mu_r \circ \pi^{-1}(A) = \mu(A)$ if A is spherically symmetric and $A = A^{-1}$. The K–uniform prototype of the measure $\nu_A(C) = \frac{\mu(A \cap C)}{\mu(A)}$ is the measure $\nu_{\widetilde{A}}(\widetilde{C}) := \mu_l(\widetilde{A} \cap \widetilde{C})/\mu_l(\widetilde{A})$, $A, C \in \mathfrak{B}(Y)$, $\widetilde{A} = \pi^{-1}(A)$, $\widetilde{C} \in \mathfrak{B}(G)$.

Let ν_1, ν_2 be probability measures on $\mathfrak{B}(Y)$ and suppose that ν_2 is spherically symmetric. Then for any $A \in \mathfrak{B}(Y)$ the function $g \mapsto \nu_2(g^{-!}A)$ is constant on the left K–cosets of G; we can set $\nu_2(y^{-1}A) := \nu_2(g^{-1}A)$ if $y = gy_0$, and define the convolution $\nu_1 * \nu_2 := \int_Y \nu_2(y^{-1}A)\nu_1(dy)$. It is easy to check that the K–uniform prototypes of ν_1, ν_2 and $\nu_1 * \nu_2$ are connected by the following relation: $\widetilde{\nu_1 * \nu_2} = \widetilde{\nu}_1 * \widetilde{\nu}_2$.

§ 3. Amenable Semigroups and Ergodic Nets

3.1. Definition and main classes of amenable semigroups. Let X be a semigroup and $\Phi = \Phi(X)$ the Banach space of all bounded complex–valued functions on X with the norm $\|f\|_\Phi := \sup_{x \in X} |f(x)|$. Let us consider the right–translation operators on Φ, i.e. the operators $R_x : (R_x f)(y) = f(yx)$, $x, y \in X$. Let Q be a subspace of $\Phi(X)$; suppose it is R–invariant (i.e. $R_x(Q) \subset Q$, $x \in X$) and $*$–invariant (i.e., $f \in Q \Rightarrow \bar{f} \in Q$) and $\mathbf{I} \in Q$ ($\mathbf{I}(x) \equiv 1$). A *right–invariant mean* on Q is a continuous linear functional $m^{(r)}$ on Q with the following properties: *1)* $m^{(r)}(R_x f) = m^{(r)}(f)$; *2)* $m^{(r)}(\bar{f}) = \overline{m^{(r)}(f)}$, $f \in Q$; *3)* $m^{(r)}(f) \geq 0$ if $f \geq 0$; and *4)* $m^{(r)}(\mathbf{I}) = 1$. A subgroup X is called Q–

right–amenable if there exists at least one right–invariant mean. A left–invariant mean and left amenability are defined similarly. A semigroup X is *Q–amenable* if it is both Q–left– and Q–right–amenable. Any Q–right– (or Q–left–) amenable *group* is Q–amenable (cf. Greenleaf [1]). Clearly, if Q_1 and Q_2 are R– and $*$–invariant subspaces of $\Phi(X)$, and $\mathbf{I} \in Q_1 \subset Q_2$, then Q_2–amenability implies Q_1–amenability. A discrete (resp. measurable, semitopological) semigroup is called *right–amenable* if it is Φ– (resp. F–, C–) right–amenable (recall that F is the subspace of all measurable functions and C is the subspace of all continuous functions on Φ). Left amenability and amenability of discrete, measurable and semitopological subgroups is defined similarly. It is shown in Chapter 1 that on any semigroup X in Φ there exists a large subspace of "right–averageable" functions $A^{(r)}$, and that X is $A^{(r)}$–right–amenable.

(3.A). *Any subgroup of an amenable locally compact topological group is amenable* (see Greenleaf [1], § 2.3).

(3.B). *Any commutative semigroup is amenable* (see Greenleaf [1], § 1.2).

This statement and theorem 2.3.3 in Greenleaf [1] imply:

(3.C). *Any solvable locally compact topological group is amenable.*

Let X be a semitopological semigroup. A non–empty closed set $D \subset X$ is called a *two–sided ideal* of X if $XDX \subset D$. The intersection $K(X)$ of all two–sided ideals of X is called the *kernel* of X. If X is a group, then $K(X) = X$. If X is a compact semigroup, then $K(X) \neq \emptyset$ and it is the least two–sided ideal of X.

(3.D). *A compact semigroup X is amenable iff $K(X)$ is a subgroup of X. Then $m^{(r)}$ and $m^{(l)}$ are unique and $m^{(r)}(f) = m^{(l)}(f) = \int_{K(X)} f d\mu$, $f \in C(X)$, where μ is the normalized Haar measure on $K(X)$* (DeLeeuw and Glicksberg [1]).

(3.E). *A non-compact semisimple Lie group is not amenable.*

This is a simple consequence of a theorem of Furstenberg [1] and of the fixed point theorem due to Day [3,4] (see also Greenleaf [1], § 3.3). Theorems (3.C) and (1.D) imply the following statement.

(3.F). *A simply connected Lie group is amenable iff it is a semidirect product of a solvable subgroup and a compact semisimple subgroup.*

3.2. Ergodic nets of measures, functions and sets. Let $(X, \mathfrak{B})$ be a measurable semigroup. A net of measures $\{\nu_n, n \in N\} (\subset \subset P\mathfrak{B})\}$ is called a *weakly right-ergodic net* (resp. a *right-ergodic net*) if $\lim_{n\in N}(\nu_n(E) - \nu_n(Ex^{-1})) = 0$, $E \in \mathfrak{B}$, $x \in X$ (resp. $\lim_{n\in N} \operatorname{var}(\nu_n(E) - -\nu_n(Ex^{-1})) = 0$, $x \in X$).

Let μ be a measure on $\mathfrak{B}$; *a (probability) density with respect to* μ is a non-negative function $p(\cdot)$ on X such that $\int_X p\, d\mu = 1$. A set of μ-densities $\{p_n, n \in N\}$ is said to be a (*weakly*) *right-ergodic net* (with respect to μ) if the measures $\nu_n(A) := \int_A p_A\, d\mu$ have the corresponding properties. If X is a group, right ergodicity of p_n is equivalent to the property: $\lim_{n\in N} \int_x |p_n(y) - p_n(yx)| \mu(dy) = 0$, $x \in X$. Finally, a net of measurable sets $\{A_n\}$ is a *weakly right-ergodic net* (with respect to μ) if $0 < \mu(A_n) < \infty$ and $\lim_{n\in N}[(\mu(A_n \cap E) - \mu(A_n) \cap Ex^{-1})/\mu(A_n)] = 0$ and $\{A_n\}$ is a *right-ergodic net* if $\lim_{n\in N} \operatorname{var}_{E\in\mathfrak{B}}(\mu(A_n \cap E) - \mu(A_n \cap \cap Ex^{-1}))/\mu(A_n) = 0$, $x \in X$.

If X is a locally compact group, then right ergodicity of $\{A_n\}$ with respect to the right Haar measure is equivalent to the "Følner property":

(F) $\lim_{n\in N} \frac{\mu_r(A_n \Delta A_n x)}{\mu_r(A_n)} = 0$, $x \in X$.

Note that if there exists a net $\{A_n\}$ with the property (F) with respect to the *left* Haar measure μ_l then the group X is unimodular, i.e. $\mu_r = \mu_l$. Indeed, then $\overline{\lim_{n\in N}} |1 - \mu_l(A_n x)/\mu_l(A_n)| \le \lim_{n\in N} \frac{\mu_l(A_n \Delta A_n x)}{\mu_l(A_n)} = 0$ and the modular function $\Delta_{\mu_l}^{(r)}(x) := \frac{\mu_l(A_n x)}{\mu_l(A_n)} \equiv 1$.

(3.G). **(Day [1,2])** *A semigroup X is right amenable iff on X there* *exists at least one weakly right ergodic net of measures.*

(3.H). (Følner–Hulanicki) *A locally compact topological group is amenable iff there exist right–ergodic nets of compact Baire sets on it* (see Greenleaf [1], Theorem 3.6.2).

(3.I). (Emerson and Greenleaf [1]) *Let X be an amenable locally compact topological group, $\varepsilon > 0$, K a compact subset of X, and $e \in K$. Then there exists a compact set $F (\subset X)$ such that $\frac{\mu_r(FK\setminus F)}{\mu_r(F)} < \varepsilon$.*

(3.I) implies the following statement.

(3.J). *On any amenable σ–compact topological group there exists a right–ergodic sequence of compact sets.*

The "left" analogues of the notions of (weakly) right ergodic net can be introduced easily and the "left" analogues of Theorems (3.G)–(3.J) are true.

In conclusion, we give several examples of right ergodic nets of sets.

Example 3.1. (Day [1]) Let $X \in \mathbf{R}^m$. Denote by $r(A)$ the least upper bound of the radii of all balls contained in the set $A \subset \mathbf{R}^m$. Any net of bounded convex sets $\{A_n,\ n \in N\}$ with $\lim_{n \in N} r(A_n) = \infty$ is ergodic.

Example 3.2. In $\mathbf{R}^m$ any net of sets $\{A_n,\ n \in N\}$ which are similar to a measurable set with Lebesgue measure $0 < l(A) < \infty$ is ergodic if the similarity coefficients $\lambda(A_n : A) \to \infty$.

Example 3.3. Let us consider the semisimple product $\mathbf{R} \odot \mathbf{R}^{\times}_{++}$, the affine group of the real line (see Example 1.1); recall that this group consists of all pairs of reals (β, α) where $\alpha > 0$; $(\beta_1, \alpha_1)(\beta_2, \alpha_2) = (\alpha_1\beta_2 + \beta_1, \alpha_1\alpha_2)$; $\mu_r(d\beta d\alpha) = \alpha^{-1} d\beta d\alpha$. It is easy to check that the sequence of "rectangles" $C_n = [-b_n, b_n] \times [\alpha_n^{-1}, \alpha_n]$ is right ergodic with respect to μ_r iff $\lim_{n\to\infty} \alpha_n = +\infty$, $\lim_{n\to\infty} \frac{b_n}{a_n} = \infty$. Of course, then the sequence $\{C_n^{-1}\}$ is left ergodic. Note that in $\mathbf{R} \times \mathbf{R}^{\times}_{++}$ there are no left ergodic sequences of "rectangles" $A_n \times B_n$, $A_n \in \mathbf{R}$, $B_n \in \mathbf{R}_+$.

3.3 Van Hove nets. Let G be a locally compact topological group, $\mathfrak{B}$ the σ–algebra of Borel sets in G, $\mathcal{K}$ the ring of all bounded subsets of $\mathfrak{B}$; $\mathfrak{N}_e := \{A : e \in A \in \mathcal{K},\ \mu_r(A) > 0\}$.

Let $A \in \mathfrak{B}$, $D \in \mathcal{K}$. We denote $C_D(A) := [\bigcup_{g \in D} Ag] = [AD]$; $I_D(A) := \operatorname{Int}(\bigcap_{g \in D} Ag) = A \backslash C_D(G \backslash A)$; $\partial_D A := C_D(A) \backslash I_D(A)$, the "$D$–boundary" of A. If $\mathcal{A} = \{A_n,\ n \in N\} \subset \mathfrak{B}$, then $\varphi_{\mathcal{A}}(D|\mu_r) := := \overline{\lim_{n \in N}} \frac{\mu_r(\partial_D A_n)}{\mu_r(A_n)}$. Of course, $\varphi_{\mathcal{A}}(D_1|\mu_r) \le \varphi_{\mathcal{A}}(D_2|\mu_r)$ if $D_1 \subset D_2$.

It is said that a net $\mathcal{A}$ satisfies *the (right) Van Hove condition* (or *is a right Van Hove net*) if $\varphi_{\mathcal{A}}(D|\mu_r) = 0$, $D \in \mathfrak{N}_e$.

It is obvious that if $\{A_n,\ n \in N\}$ is a Van Hove net, then for any $g_n \in G$ the net $\{g_n A_n,\ n \in N\}$ is a Van Hove net. If $g \in D$, then $Ag \Delta A \subset \partial_D A$; therefore any Van Hove net is ergodic.

Let the topology in G be induced by an invariant metric ρ, i.e. $\rho(g_1, g_2) = \rho(fg_1, fg_2)$, $g_1, g_2, f \in G$. Then we can consider in $\mathfrak{N}_e$ the set $D = B_R := \{\rho(e,x) \le R\}$, the ball of radius R; we suppose that all B_r are compact. If A is a closed subset of G, then $C_R(A) := C_{B_R}(A) = = \{g : \inf_{g \in A} \rho(g, f) \le R\}$, the "$R$–closure" of A, $I_R(A) := I_{B_R}(A) = = \{g : \inf_{f \in G \backslash A} \rho(g, f) > R\}$, the "$R$–interior" of A, and $\partial_R A := \partial_{B_R} A$ is the "boundary layer of thickness $2R$". The Van Hove condition is equivalent to the following:

$$\overline{\lim_{n \in N}} \frac{\mu_r(\partial_R A_n)}{\mu_r(A_n)} = 0, \quad R > 0. \tag{3.1}$$

Example 3.4. Let $G = \mathbf{R}^m$ and let $\{A_n,\ n \in N\}$ be a net of convex sets with $\lim_{n \in N} r(A_n) = \infty$ (see Example 3.1). Then $\{A_n\}$ is a Van Hove net. In order to prove this, let us consider a sequence of balls $B_{r_n}(a_n) := \{x : |x - a_n| < r_n\} \subset A_n$ with $r_n \to \infty$; then $B_{r_n} = = B_{r_n}^{(0)} \in A'_n := A_n - a_n$. Let us choose some $r > 0$ and consider the set $C_r(A'_n) = [A'_n + B_r]$. If $x \in A'_n + B_r$, then $x = x'_1 + x'' = x'_1 + \frac{r}{r_n} x''_1$ where $x'_1 \in A'_n$, $x'' \in B_r$ and $x''_1 \in B_{r_n}$; thus $x = \frac{r_n + r}{r_n}\left(\frac{r_n}{r_n + r} x'_1 + + \frac{r}{r_n + r} x''_1\right) \subset \frac{r_n + r}{r_n} A'_n$ since x'_1, $x''_1 \in A'_n$, and for any $\varepsilon > 0$ $C_r(A'_n) \subset \subset \left(\frac{r_n + r + \varepsilon}{r}\right) A'_n$ (cf. Day [1]). Now we suppose that $r_n > r$; let

$x_0 \in \partial A'_n$, the boundary of A'_n. We consider the cone $F_n := \{x : x = (1-\lambda)x_1 + \lambda x_0,\ x_1 \in B_{r_n},\ \rho \leq \lambda \leq 1\}$. It is easy to verify that if $0 \leq \lambda < \frac{r_n - r}{r_n}$, then $\lambda x_0 \in I_r(F_n) \subset I_r(A'_n)$. Thus $\frac{r_n-r-\varepsilon}{r_n}A'_n \subset I_r(A'_n)$, and $\partial_r(A'_n) \subset \frac{r_n+r+\varepsilon}{r_n}A'_n - \frac{r_n-r-\varepsilon}{r_n}A'_n$. Therefore $\lim\limits_{n\in N} \frac{l(\partial_r(A'_n))}{l(A'_n)} = \lim\limits_{n\in N}\left[\left(\frac{r_n+r+\varepsilon}{r_n}\right)^m - \left(\frac{r_n-r-\varepsilon}{r_n}\right)^m\right] = 0$ (l is Lebesgue measure), and $\{A'_n\}$ is a Van Hove net; consequently, $\{A_n\}$ is a Van Hove net. If we remove the "rational" points from all A_n, the new net is still ergodic, but does not satisfy the Van Hove condition.

Example 3.5. Let A be a Jordan measurable subset of $\mathbb{R}^m$, i.e. $l(\partial A) = 0$. Then any net of sets $\{A_n,\ n \in N\}$ which are similar to A and with $\gamma_n = \gamma(A_n : A) \to \infty$ satisfies the Van Hove condition. Indeed, $\lim\limits_{n\to\infty} \frac{l(\partial_r A_n)}{l(A_n)} = \lim\limits_{n\to\infty} \frac{l(\partial_{\gamma_n^{-1} r} A)}{l(A)} = 0$ (compare with Example 3.2).

(3.K). *A net $A = \{A_n,\ n \in N\}$ satisfies the right Van Hove condition iff it is right ergodic and $\varphi_A^{(r)}(D_0|\mu_r) = 0$ for some open set $D_0 \in \mathfrak{N}_e$.*

Proof. Necessity is obvious, so let us show that the condition is sufficient. Let $D \in \mathfrak{N}_e$. Consider a finite set $\{g_1, \ldots, g_k\} \subset G$ such that $D \subset \bigcup_{i=1}^k D_0 g_i$. It is easy to verify that $\partial_D A_n \in \in \bigcup_{i=1}^k (\partial_{D_0} A_n) g_i \bigcup \bigcup_{i=1}^k (A_n \Delta A_n g_i)$. Thus

$$\varphi_A(D|\mu_r) \leq k \lim_{n\in N} \frac{\mu_r(\partial_{D_0} A_n)}{\mu_r(A_n)} + \sum_{i=1}^k \lim_{n\in N} \frac{\mu_r(A_n \Delta A_n g_i)}{\mu_r(A_n)} = 0.$$

□

(3.L). *Let G be an amenable σ-compact, locally compact topological group and K_0 a compact set of $\mathfrak{N}_e$. There exists a right Van Hove sequence $A = \{A_n\}$ of compact sets on G of the form: $A_n = F_n K_0$ where the F_n are compact.*

Proof. Let us fix an open set $V_0 \in \mathfrak{N}_e$ and some compact set $D \in \mathfrak{N}_e$. Let K'_n be an increasing sequence of compact sets such that $K'_1 =$

$= \{e\}$ and $\bigcup_{n=1}^{\infty} K'_n = G$; $K_n := [V_0]D^{-1}K_0^{-1}K_0DK'_n$. According to Theorem (3.I), there exists a sequence of compact sets $\{F_n\}$ such that

$$\frac{\mu_r(F_nK_n\Delta F_n)}{\mu_r(F_n)} < \frac{1}{n}, \quad n = 1, 2, \ldots. \tag{3.2}$$

Let us set $A_n := F_n[V_0]D^{-1}K_0^{-1}$. Then $\bigcap_{g\in K_0D} A_ng = F_n[V_0] = [F_nV_0]$, and thus $F_n \subset F_nV_0 \subset I_{K_0D}(A_n)$. Further, $C_{K_0D}(A_n) = A_nK_0D = F_nK_1$, and (3.2) implies the relations $\lim\limits_{n\to\infty} \frac{\mu_r(C_{K_0D}(A_n)\setminus F_n)}{\mu_r(F_n)} = 0$, $\lim\limits_{n\to\infty} \frac{\mu_r(A_n\setminus F_n)}{\mu_r(F_n)} = 0$ and $\lim\limits_{n\to\infty} \frac{\mu_r(A_n)}{\mu_r(F_n)} = 1$.
Now it is easy to verify that $\varphi_{\mathcal{A}}(K_0D|\mu_r) = 0$;
moreover, $\lim\limits_{n\to\infty} \frac{\mu(A_nK'_n\setminus A_n)}{\mu(A_n)} = 0$ and thus $\{A_n\}$ is right ergodic. It remains to recall (3.K).

□

Let $Y = G/K$ be a metric homogeneous space (see § 2) and let G be unimodular. Assume that the balls $B_R(y_0)$ with center $y_0 = K$ are compact. We can introduce $\partial_R A$, the "boundary layer of thickness $2R$" of a set A, as in the case of a metric group, and define Van Hove nets on Y by (3.1), considering the invariant measure μ in Y instead of μ_r. It is obvious that if $\{F_nK, n \in N\}$ is a Van Hove net of compact sets on G, then the net $\{\pi(F_n), n \in N\}$ has the same property on Y. Therefore Theorem (3.L) can be reformulated in the following way.

(3.L'). *On any amenable metric homogeneous space with a unimodular motion group there exists a Van Hove sequence of compact sets.*

In fact there are a lot of such sequences: recall that if $\{A_n\}$ is a Van Hove sequence on Y and $g_n \in G$, then $\{g_nA_n\}$ is also a Van Hove sequence.

3.4. Canonical factorizations. We shall say that a left semidirect factorization $G = G^{(0)} \odot G^{(1)} \odot \ldots \odot G^{(m)}$ is *canonical* if the subgroup G^0 is compact and $G^{(i)} \cong \mathbb{R}$ if $1 \le i \le n$. $G = G^{(m)} \odot \ldots \odot G^{(1)} \odot G^{(0)}$ denotes a right canonical factorization.

(3.M). *Any simply connected solvable Lie group admits a canonical semidirect factorization* (see, e.g., Naimark [1]).

(3.N). (Greenleaf, Emerson [1]) *Let G be a connected locally compact amenable group. Denote by $K(G)$ a maximal compact normal subgroup of G (all such subgroups are isomorphic)*

(i) If $K(G) = \{e\}$, then G admits a canonical semidirect factorization.

(ii) If $K(G) \neq \{e\}$, then the group $G' = G/K(G)$ is a connected amenable Lie group with $K(G') = \{e\}$; thus G' admits a canonical semidirect factorization.

§ 4. Positive Definite Functions

4.1. Positive definite functions and reproducing kernel Hilbert spaces. Let T be an arbitrary set. A function $R(s,t)$, $s,t \in T$, is said to be *positive definite* if $\sum_{i,j=1}^{m} \alpha_i \bar{\alpha}_j R(t_i,t_j) \geq 0$, $\alpha_1,\ldots,\alpha_m \in \mathbb{C}$, $t_1,\ldots,t_m \in T$, $m = \overline{1,\infty}$. Suppose now that T is a left G–space. A function $R(s,t)$ on $T \times T$ is called G–invariant if $R(gs,gt) = R(s,t)$, $s,t \in T$, $g \in G$.

Let H be a Hilbert space of complex valued functions defined on T with the scalar product $\langle \psi, \varphi \rangle_H$. Let $R(s,t), (s,t) \in T \times T$, be a function with the following properties:

a) $R(\cdot,t) \in H$, $t \in T$.

b) $\langle R(\cdot,t), \varphi \rangle = \varphi(t)$, $t \in T$, $\varphi \in H$.

Then R is called a *reproducing kernel* of the Hilbert space H; it is also said that H is a *reproducing kernel Hilbert space* with kernel R; we shall denote such spaces by $H(R)$. This terminology is justified by the fact that the Hilbert space and its reproducing kernel R determine each other in a unique manner. *A function R on $T \times T$ is the reproducing kernel of some Hilbert space H of functions on T iff R is positive definite. The Hilbert space $H(R)$ is generated by the family of functions $\{R(\cdot,t),\ t \in T\}$*

These statements are proved in Aronszajn [1].

4.2. Positive definite functions on groups. Let $T = G$ be a topological group; we shall consider G as a homogeneous space with respect to the group of left shifts G_l (see Example 2.1). If $R(g,f)$ is a G_l–invariant function on $G \times G$, then $R(g,f) = R(l_{g^{-1}}g, l_{g^{-1}}f) = R(e, g^{-1}f) = R_0(g^{-1}f)$ where $R_0(g) = R(e,g)$. We shall say that $R_0(g)$ is a *positive definite function* if $R(g,f)$ is such; this means that $\sum_{i,j=1}^{m} \alpha_i \bar{\alpha}_j R_0(g_i^{-1}g_j) \geq 0$, $\alpha_1, \ldots, \alpha_m \in \mathbb{C}$, $g_1, \ldots, g_m \in G$, $m = \overline{1, \infty}$. Thus $R \Leftrightarrow R_0$ is a one–to–one correspondence between the invariant positive definite functions on $G \times G$ and the positive definite functions on G. The class of all continuous positive definite functions on G will be denoted by $\mathbf{P}(G)$.

If R_0 is a positive definite function on G, we shall denote by $H(R_0)$ the Hilbert space $H(R)$ where $R(g,f) = R_0(g^{-1}f)$.
A function $\varphi \in \mathbf{P}(G)$ is said to be *elementary* if $\varphi(e) = 1$ and if a decomposition $\varphi = \varphi_1 + \varphi_2$, with φ_1, $\varphi_2 \in \mathbf{P}(G)$, is possible only if $\varphi_1 = \alpha_1\varphi$, $\varphi = \alpha_2\varphi$ for some α_1, $\alpha_2 \in \mathbb{C}$. We shall denote the class of all elementary positive definite continuous functions on G by $\mathbf{EP}$ or $\mathbf{EP}(G)$.

Let G be a locally compact group. We shall provide $\mathbf{EP}$ with the topology of uniform convergence on compact sets. Let $\mathfrak{B}(\mathbf{EP})$ be the σ–algebra of Borel sets on $\mathbf{EP}$. It is easy to verify that $(\varphi, g) \mapsto \varphi(g)$ is a continuous mapping of $\mathbf{EP} \times G$ into $\mathbb{C}$.

If G is commutative, $\mathbf{EP}$ coincides with topological space $\widehat{G}$ of all characters; a *character* φ on G is a continuous function with the following properties: $|\varphi(g)| \equiv 1$ and $\varphi(gf) = \varphi(g)\varphi(f)$). Any $\varphi \in \mathbf{EP}$ determines a unitary representation $g \mapsto U_g$ of G in the one–dimensional space by the formula: $U_g c = \varphi(g)c$, $c \in \mathbb{C}$, $g \in G$.

The following statement is a generalization of the well–known theorem of Bochner on the spectral decomposition of positive definite functions on $\mathbb{R}$.

(4.A). *If G is separable or commutative, then any function $R \in \mathbf{P}(G)$ can be represented in the form $R(g) = \int_{\mathbf{EP}} \varphi(g)F(d\varphi)$ where F is a finite measure on $\mathfrak{B}(\mathbf{EP})$* (see Dixmier [1], 13.6.8).

The measure F is called a *spectral measure* of R; in the commutative case F is determined uniquely by R.

4.3. Positive definite functions on homogeneous spaces. Let $Y = G/K$ be a homogeneous space. To any function R on $Y \times Y$ there corresponds a function R_1 on $G \times G$; it is defined as follows: $R_1(g, f) := R(x, y)$ if $x = gK$, $y = fK$. If R is G–invariant, then we have for any $k_1, k_2 \in K$: $R_1(g_1k_1, g_2k_2) = R_1(e, k_1^{-1}g_1^{-1}g_2k_2) =: R_0(k_1^{-1}g_1^{-1}g_2k_2)$; moreover, $R_1(k_1, gk_2) = R_1(e, g)$, and therefore $R_0(k_1^{-1}gk_2) = R_0(g)$. Thus, to any G–invariant function R we have associated a function R_0 on G which is constant on the two–sided cosets of G with respect to K. It is clear that if R is positive definite, then so is R_0; in this case it is said that R_0 is a *spherical positive definite* function on G.

We denote by $\mathbf{SP}(G, K)$ the class of all spherical positive definite continuous functions. Any function $\widetilde{R}_0 \in \mathbf{SP}(G, K)$ can be considered as a function $R_0(y)$ on Y with the property $R_0(ky) = R_0(y)$, $k \in K$, $y \in Y$. In this sense we may identify the class $\mathbf{SP}(Y)$ of all such *spherical positive definite functions* on Y with $\mathbf{SP}(G, K)$.

It is easy to see that the set $\mathbf{ESP}(G, K) := \mathbf{EP}(G) \cap \mathbf{SP}(G)$ is closed in $\mathbf{EP}(G)$. Therefore $\mathfrak{B}(\mathbf{ESP})$, the restriction of $\mathfrak{B}(\mathbf{EP})$ to $\mathbf{ESP}$, is the σ–algebra of Borel sets in $\mathbf{ESP}$.

(4.B). *If $R \in \mathbf{SP}(G)$, then there exists a measure F on $\mathfrak{B}(\mathbf{ESP})$ such that $R(g) = \int_{\mathbf{ESP}} \varphi(g)F(d\varphi)$.*

4.4. Positive definite generalized functions on $\mathbb{R}^m$. Let $K = K(\mathbb{R}^m)$ be the Schwartz space on $\mathbb{R}^m$ (see, e.g., Gel'fand and Shilov [1]). It is a G–space with respect to the group G of all translations L_x, $(L_x\varphi)(y) = \varphi(x + y)$, $\varphi \in K$, $x, y \in \mathbb{R}^m$.

(4.C). *To any G–invariant positive definite bilinear function $R(\varphi, \psi)$ on $K \times K$ we can associate a generalized function R_0 such that $R(\varphi, \psi) = R_0(\varphi * \bar{\psi})$, where $\varphi * \bar{\psi}$ is the convolution of the functions φ and $\bar{\psi}$; the function R_0 is called a positive definite generalized function on $\mathbb{R}^m$; it admits the spectral representation $R_0(\varphi) = \int_{\mathbb{R}^m} \widetilde{\varphi}(\lambda)F(d\lambda)$, $\varphi \in K$, where $\widetilde{\varphi}$ is Fourier transformation of φ and F is a Borel measure on $\mathbb{R}^m$ such that $\int_{\mathbb{R}^m}(1 + \|\lambda\|^2)^{-p}F(d\lambda) < \infty$ for some $p \geq 0$ (the* *spectral measure of the function R_0). If p is the least integer with this property, it is said that R_0 is of class $\mathfrak{S}_p$. The class $\mathfrak{S}_0$ coincides with the class $\mathbf{P}(\mathbb{R}^m)$.*

§ 5. Representations of Semigroups in Banach Spaces

A left (right) representation of a semigroup (group) X in a Banach space B is a left (right) action $S : x \mapsto S_x$ in B by operators of $\mathcal{L}_l(B)$ (resp. $\mathcal{L}_r(B)$) (see Example 1.2). Thus, e.g., a left representation of X in B is a mapping $x \mapsto S_x$ where $x \in X$ and S_x are bounded linear operators, and $S_{x_1} S_{x_2} = S_{x_1 x_2}$; if X is a group, $S_e = \mathbf{I}$ (the identity operator). In this section we shall consider only left representations and omit the word "left"; all notions and results can be easily transferred to right representations.

Let S be a representation of X in B. If $b \in B$, then the set $O(b) := \{S_x b,\ x \in X\}$ is called the *orbit of b with respect to S*, or simply the *S-orbit* of b; the B-valued function $\varphi_b(x) := S_x b$, $x \in X$, is called the *orbital function of b*. An element $b \in B$ is said to be *cyclic* (with respect to S) if its S-orbit spans the space B. A representation S is said to be *isometric (contractive)* if all S_x, $x \in X$, are isometric operators (resp. contractions); a representation U is a Hilbert space H is *unitary* iff all U_x are unitary operators. A representation S in B is *uniformly bounded* if $\sup_{x \in X} \|S_x\| < \infty$, and is *bilaterally uniformly bounded* if there exist constants $c > 0$ and $C > 0$ such that $c\|b\| \leq \|S_x b\| \leq C\|b\|$, $x \in X$, $b \in B$.

Let $(\Omega, \mathcal{F}, m)$ be a measure space. A representation S: $x \mapsto S_x$ in $L^1(\Omega, \mathcal{F}, m)$ is *strongly isometric* (resp., *strongly contractive*) if there exists a family of representations $S^{(\alpha)} : x \mapsto S_x^{(\alpha)}$ in $L^\alpha(\Omega, \mathcal{F}, m)$, $1 \leq \alpha \leq \infty$, such that: *1)* all the representations $S^{(\alpha)}$ are isometric (resp., contractive); *2)* $S_x^{(1)} = S_x$; and *3)* $S_x^{(\alpha_1)} f = S_x^{(\alpha_2)} f$ if $f \in L^{\alpha_1} \cap L^{\alpha_2}$ $(1 \leq \alpha_1, \alpha_2 \leq \infty)$.

A representation S of a semitopological (semimeasurable) subgroup in B is said to be *continuous* (resp. *measurable*) if all orbital functions (with respect to S) are continuous (resp. measurable in the sense of Bochner with respect to any measure $\mu \in \mathfrak{M}(B)$). *Throughout this book it is assumed that a representation of a semitopological semigroup is continuous.*

Example 5.1. Let $\Phi_B = \Phi_B(X)$ be the Banach space of B–valued functions on X with the norm $\|\varphi\|_{\Phi_B} = \sup_{x \in X} \|\varphi(x)\|_B$, $\varphi \in \Phi_B$. To any $x \in X$ we can associate two operators: $R_x : (R_x\varphi)(y) = \varphi(yx)$, the right translation operator, and $L_x : (L_x\varphi)(y) = \varphi(xy)$, the left translation operator. The action $R : x \mapsto R_x$ (resp. $L : x \mapsto L_x$) is a left (resp. right) isometric representation of X in Φ_B. If X is a semitopological (resp. semimeasurable) semigroup, we can consider the restrictions of the operators R_x and L_x to the subspace C_B (resp. F_B) of all continuous (resp. measurable) B–valued functions.

Let S be a representation of X in B. A subspace $B' \subset B$ is said to be *S–invariant* if $S(B') \subset B'$. S is said to be *irreducible* if there are no nontrivial (i.e. different from $\{0\}$ and B) S–invariant subspaces. If B' is an S–invariant subspace of B and the operators S'_x are the restrictions of S_x to B', then $S' : x \mapsto S'_x$ is a representation of X in B'; it is said that S' is a *subrepresentation* of S in B', or that S contains S'.

Let K be a subgroup of G, S a representation of G in B; we say that an element $b_0 \in B$ is *K–invariant*, or *K–fixed*, if $S_K b_0 = b_0$, $k \in K$. The set of all K–invariant elements is a subspace I_K of B. A representation S of X in B is called a *class–1 representation* (with respect to K) if the set $\bigcup_{b \in I_K} O(b) = \{S_x b,\ x \in X,\ b \in I_K\}$ spans the space B.

Example 5.2. Let G be a locally compact semitopological group; μ_r its right Haar measure; K a compact subgroup of G; and V a bounded compact neighbourhood of the identity e of G. The *regular representation* R of G, i.e. the representation by the right translations R_g, $g \in G$, in $H = L^2(G, \mathfrak{B}(G), \mu_r)$, is a unitary (left) representation of G. The function $\mathcal{X}_{VK}(\cdot) \in H$ is K–invariant; the restriction of R to the subspace H_K generated by the set of functions $\{R_g \mathcal{X}_{VK},\ g \in G\}$ is a class–1 representation of G in H_K.

Two representations of a semigroup X, $S^{(1)} : x \mapsto S_x^{(1)}$ in $B^{(1)}$ and $S^{(2)} : x \mapsto S_x^{(2)}$ in $B^{(2)}$ are said to be *(unitary) equivalent* if there exists an isometric isomorphism i: $B^{(1)} \mapsto B^{(2)}$ such that $S_x^{(2)} = iS_x^{(1)}i^{-1}$, $x \in X$.

If $U : g \to U_g$ is a unitary representation of a group G in a Hilbert space H, then any function $\varphi(g) = \langle U_g h_1, h_2 \rangle_H$, h_1, $h_2 \in H$, is called a *matrix coefficient* of this representation. The "diagonal" coefficients with $h_1 = h_2$ are of special interest.

(5.A). *Let U be a unitary representation of G in H and $h \in H$; then $g \to \varphi_h(g) := \langle U_g h, h \rangle$ is a continuous positive definite function on G.*

Let us consider some $R \in \mathbf{P}(G)$ and the reproducing kernel Hilbert space $H(R)$; it is easy to verify that the *translation representation* $\widehat{U}$ in $H(R)$ defined by the formula $(\widehat{U}_g\psi)(f) = \psi(fg)$, $\psi \in H(R)$, is unitary and $R(\cdot)$ is a cyclic element of $\widehat{U}$; moreover, $\langle U_g R, R \rangle_{H(R)} = R(g)$. Thus we have proved the following statement, which is a converse to (4.A).

(5.B). *(**Gel'fand – Raikov**) If $R \in \mathbf{P}(G)$ then there exists a unitary representation U of G in a Hilbert space and a cyclic element $h \in H$ such that*

$$R(g) = \langle U_g h, h \rangle. \tag{5.1}$$

The following statement shows that the unitary representation U is defined by R uniquely up to equivalence; we shall say that U is *associated* with R.

(5.C). *Let U be a unitary representation of G in H and let h be a cyclic element of H with respect to U; $\varphi_h := \langle U_g h, h \rangle_H$. Then U is unitary equivalent to the translation representation $\widehat{U}$ in $H(\varphi_h)$.*

The equivalence isomorphism $i : H \mapsto H(\varphi_h)$ is defined as follows: $\widetilde{h} \mapsto \psi(g) := \langle U_g \widetilde{h}, h \rangle_H$.

(5.D). *$R \in \mathbf{SP}(G, K)$ iff R can be represented in the form (5.1) where U is a class–1 unitary representation in H with respect to K and h is a K–invariant cyclic element of H.*

(5.E). *A function $\varphi \in \mathbf{EP}(G)$ ($\varphi \in \mathbf{ESP}(G/K)$) iff $\varphi \in \mathbf{P}(G)$ (resp. $\varphi \in \mathbf{SP}(G, K)$) and the associated unitary representation is irreducible.*

Let $(\Lambda, \mathcal{D})$ be a measurable space and F a measure on $\mathcal{D}$. We denote by $L^2_H(\Lambda, \mathcal{D}, F)$ the Hilbert space of all H–valued F–measurable (in the sense of Bochner; see, e.g., Dunford and Schwartz [1]) functions φ on Λ with $\int_\Lambda \|\varphi(\lambda)\|^2 F(d\lambda)$; the scalar product is $\langle\varphi, \psi\rangle :=$ $:= \int_\Lambda \langle\varphi(\lambda), \psi(\lambda)\rangle_H F(d\lambda)$, where φ, $\psi \in L^2_H$ and $\langle\cdot,\cdot\rangle_H$ is the scalar product in H. Let $\Lambda = \bigcup_{k=1}^\infty \Lambda_k \bigcup \Lambda_\infty$ be a decomposition of Λ in disjoint measurable sets $\Lambda_1, \Lambda_2, \ldots, \Lambda_\infty$; let $\mathbf{C}^k$ be the standard k–dimensional Hilbert spaces and $\mathbf{C}^\infty$ the standard infinite–dimensional Hilbert–space. $L^2_{1,\infty}(\Lambda, \mathcal{D}, F) := \sum_{1\le k\le\infty} \oplus L^2_{\mathbf{C}^K}(\Lambda_k, \mathcal{D}_k, F_k)$ where $\mathcal{D}_k$ and F_k are the restrictions of $\mathcal{D}$ and F to Λ_k.

The space $L^2_{1,\infty}(\Lambda, \mathcal{D}, \mu)$ consists of the functions φ on Λ whose restrictions to Λ_k belong to $L^2_{\mathbf{C}^k}$, $k = 1, 2, \ldots, \infty$, and $\|\varphi\|^2 :=$ $:= \sum_{k=1}^\infty \|\varphi_k\|^2_{L^2_{\mathbf{C}^k}} < \infty$. It is a Hilbert space with the scalar product

$$\langle\varphi, \psi\rangle = \sum_{k=1}^\infty \langle\varphi_k, \psi_k\rangle^2_{L^2_{\mathbf{C}^k}(F_k)} + \langle\varphi_\infty, \psi_\infty\rangle_{L^2_{\mathbf{C}^\infty}(F_\infty)}.$$

$L^2_{1,\infty}(\Lambda, \mathcal{D}, F)$ is often called a "direct integral" and denoted by $\int_\Lambda \oplus H(\lambda)F(d\lambda)$, and the subspace of all functions vanishing F–a.e. outside some set $M \in \mathcal{D}$ is denoted by $\int_M \oplus H(\lambda)F(dx)$.

Let us fix some topological group G. An *IUR–function on Λ* is a mapping $\lambda \mapsto U(\lambda)$ where the $U(\lambda)$ are irreducible unitary representations of G in a Hilbert space $H(\lambda) = \mathbf{C}^{d(\lambda)}$; an IUR–function $\lambda \mapsto U(\lambda)$ is said to be *measurable* if for any k ($1 \le k \le \infty$), any measure F on $\mathcal{D}$, any measure μ on $\mathfrak{B}$ and any $h \in L^2_{\mathbf{C}^k}(F)$ the function $(g, \lambda) \mapsto$ $\mapsto U_g(\lambda)h(\lambda)$ is $(\mu \times F)$–measurable. Let $U(\cdot)$ be a measurable IUR–function. $(\Lambda, \mathcal{D}, U(\cdot))$ is called a *spectral system of the group G* if for any unitary representation U of G (in some Hilbert space H) there exists a measure F on $\mathcal{D}$ such that U is equivalent to the unitary representation $\widetilde{U}$ in $L^2_{1,\infty}(\Lambda,\mathcal{D},F)$ defined by the formula $(\widetilde{U}_g\varphi)(\lambda) := := U_g(\lambda)\varphi(\lambda)$, $\lambda\in\Lambda$, $\varphi\in L^2_{1,\infty}(\Lambda,\mathcal{D},F)$. The measure F is called the *spectral measure* of the representation U. U is defined by F up to equivalence (but, in general, F is not defined uniquely by U). H is often identified with $L^2_{1,\infty}(\Lambda,\mathcal{D},F) = \int_\Lambda \oplus H(\lambda)F(d\lambda)$, and then U is written in the "spectral" form $U = \int_\Lambda \oplus U(\lambda)F(d\lambda)$. Clearly, for any $M\,(\in\mathcal{D})$ the subspace $\int_M \oplus H(\lambda)F(d\lambda)$ is U–invariant and the restriction $U^{(M)}$ of U to this subspace has the spectral form $\int_M \oplus U(\lambda)F(d\lambda) = L_{1,\infty}(M,\mathcal{D}^{(M)},F^{(M)})$ where $\mathcal{D}^{(M)}$ and $F^{(M)}$ are the restrictions of $\mathcal{D}$ and F to M.

(5.F). (Adel'son–Vel'ski – Kolmogorov – Mautner) *A separable locally compact topological group has at least one spectral system.*

Such a system can be constructed as follows. Recall that the Hilbert space of any unitary representation U can be decomposed into a direct sum of k_U subspaces $H^{(i)}$ in such a way that the restriction $U^{(i)}$ of U to $H^{(i)}$ has a cyclic vector h_i (we shall suppose that k_U is the least possible number in such decompositions; of course, k_U is the same for all equivalent representations); let $k = \sup k(U)$ where the supremum is taken over all classes of equivalent unitary representations of G (k is a natural number or ∞). Let $(\Lambda_i,\mathcal{D}_i)$, $i=1,\dots,k$, be copies of the measurable space $(\mathbf{EP},\mathfrak{B}(\mathbf{EP}))$ and let $(\Lambda,\mathcal{D}) := \bigcup_{\Lambda_i,\mathcal{D}_i}$; if the point $\lambda(\in\Lambda)$ corresponds to $\psi\in\mathbf{EP}$, we shall write φ_λ instead of ψ. Let $U(\lambda)$ denote the (irreducible) translation representation of G in $H(\varphi_\lambda)$. If a representation U in H has a cyclic element h (we may suppose that $\|h\|=1$) and $\varphi_h(\cdot) = \langle U.h,h\rangle$ has a spectral measure F_h on $\mathbf{EP}$, then this measure is also a spectral measure of U on $\mathbf{EP} = \Lambda_1 \subset \Lambda$. In the general case we consider on Λ_i the spectral measure $F^{(i)}$ of $U^{(i)}$ in $H^{(i)}$, $i=1,\dots,k_U$; the direct sum F of these measures is a measure on $\bigcup_{i=1}^{k_U}(\Lambda_i,\mathcal{D}_i) \subset (\Lambda,\mathcal{D})$ and is a spectral measure of U.

Let us note that if G is a *group*, then, according to the definition, for any representation S of G in B we have $S_e = I$, the identity operator in B, and $S_{g^{-1}} = S_g^{-1}$. But sometimes it is convenient to consider $g\mapsto S_g$ as a semigroup action, and then S_e need not be the identity operator.

Let S be an isometric (or unitary) representation of G in B and M an S–invariant subspace of B. Then $S' : g \mapsto S'_g = \Pi^B_M S_g = S_g \Pi^B_M = \Pi^B_M s_g \Pi^B_M$ is a semigroup action of G in B with $S'_e = \Pi^B_M$. Following Emerson and Greenleaf [1] we shall call such actions *generalized isometric* (resp. *unitary*) *representations* of G in B. It is evident that the restriction of S to M is an ordinary isometric (unitary) representation.

§ 6. Weakly Almost Periodic Elements and Functions

6.1. Almost periodic and weakly almost periodic elements. Let $S : x \mapsto S_x$ be a uniformly bounded (left) representation of a semigroup X in a Banach space B. An element $b \in B$ is said to be *(weakly) almost periodic* with respect to S if its orbit $O(b) = \{S_x b, x \in X\}$ is a relatively (weakly) compact subset of B. We denote by $W(B|S)$ the almost periodic elements (with respect to S). It is easy to verify that

(6.A). *$W(B|S)$ is an S–invariant subspace of B.*

(6.B). *If B is reflexive, then $W(B|S) = B$ for any B.*

Statement (6.B) follows from the well–known fact that in a reflexive Banach space any set A with $\sup(\|a\|, a \in A) < \infty$ is relatively compact (see Dunford and Schwartz [1], Ch. 4, § 4). Denote by $K(b)$ the closed convex hull of the set $O(b)$. The well–known theorem due to Krein and Shmulian (see, e.g., Dunford and Schwartz [1], Ch. 5, § 6) implies the following statement:

(6.C). *If $b \in W(B|S)$, then $K(b)$ is weakly compact.*

6.2. Weakly almost periodic functions. Let X be a semi-topological semigroup; $C(X)$ the Banach space of all complex–valued bounded continuous functions on X; let $R : x \to R_x$ and $L : x \to L_x$ be the representations of X by right and left translations in $C(X)$.

A function $f\,(\in C(X))$ is called a *(weakly) almost periodic function* if it is a (weakly) almost periodic element of C with respect to R. This "right" weak almost periodicity is equivalent to "left" weak almost periodicity with respect to L (the proof concerning almost periodic functions can be found in Loomis [1]; for weakly almost periodic functions this fact was proved by Grothendieck [1]).

The subspace of all weakly almost periodic functions in $C(X)$ will be denoted by $W(X)$, or simply by W; $AP(X) = AP$ is the subspace of all almost periodic functions in $C(X)$; it is obvious that $AP \subset W$.

(6.D). **(Eberlein [1])** *If X is a topological group, then any $f \in W(X)$ is right and left uniformly continuous.*

(6.E). *If a representation $S : x \mapsto S_x$ is continuous, then for any $b \in$ $\in W(B|S)$ there exists a functional $l \in B^*$ such that the function $x \mapsto \langle S_x b, l\rangle$ belongs to $W(X)$* (see Greenleaf [1], Lemma 3.8.2).

We shall recall several important classes of weakly almost periodic functions. Statements (6.B) and (6.E) imply:

(6.F). *The matrix coefficients of a unitary representation of a group G are weakly almost periodic functions; in particular,* $\mathbf{P} \subset W$.

(6.G). *On a compact semitopological semigroup $AP = C$* (see DeLeeuw and Glicksberg [1]).

Let X be a locally compact semitopological semigroup and $C_a(X)$ the subspace of all $\varphi \in (C(X))$ for which $\lim_{y\to\infty} \varphi(y)$ exists. Since $C_a(X) = C(X_\infty)$, where X_∞ is the one–point compactification of X $(x \cdot \infty = \infty \cdot x = \infty \cdot \infty = \infty)$, (6.G) implies the following statement.

(6.H). **(Eberlein [1]**, see also DeLeeuw and Glicksberg [1]) $C_a \subset W$.

6.3. Almost periodic compactifications. Let G be a topological group. We let $\mathcal{L}_l(B,s)$ (resp. $\mathcal{L}_l(B,w)$) be the left semigroup (with respect to the usual multiplication) of all bounded linear operators in a Banach space B with the strong (resp. weak) operator topology; note that both $\mathcal{L}_l(B,w)$ and $\mathcal{L}_l(B,s)$ are *semi*topological semigroups (in $\mathcal{L}_l(B,s)$ the multiplication is jointly continuous on any set $D_c := \{A : \|A\| < C\}$, $0 < C < \infty$). The set of right translation operators $\{R_g,\ g \in G\}$ is relatively compact in $\mathcal{L}_l(\mathbf{AP},s)$ and in $\mathcal{L}_l(W,w)$ (note that $\{R_x,\ x \in G\} \subset D_1$). Let us denote by $\widetilde{G}^{(a)}$ and $\widetilde{G}^{(w)}$ the closures of the sets $\{R_g,\ g \in G\}$ in $\mathcal{L}_l(\mathbf{AP},s)$ and $\mathcal{L}_l(W,w)$, respectively. $\widetilde{G}^{(a)}$ is a compact topological group; $\widetilde{G}^{(w)}$ is a compact semitopological semigroup with identity. $G^{(a)}$ is called the *almost periodic*, or *Bohr, compactification of* G; $G^{(w)}$ is the *weak almost periodic compactification of* G. In DeLeeuw and Glicksberg [1], the following statements are proved.

(6.I). 1. *The mapping $R : g \to R_g$ is a (continuous) semigroup homomorphism from G onto dense subsets in $\widetilde{G}^{(a)}$ and $\widetilde{G}^{(w)}$.*

2. *The mapping $\widetilde{R} : (\widetilde{R}f)(g) = f(R_g)$ is an isometric isomorphism from $C(\widetilde{G}^{(a)})$ onto $AP(G)$ and from $C(\widetilde{G}^{(w)})$ onto $W(G)$. $\widetilde{R}^{-1}(\varphi)$ is the continuous extension of $\varphi \in W$ (resp. AP) onto $\widetilde{G}^{(w)}$ (resp. $\widetilde{G}^{(a)}$).*

The results of DeLeeuw and Glicksberg [1] and the theorem of Ryll–Nardzewski on the averageability of weak almost periodic functions (see Ch. 1, § 4) imply the following statement.

(6.J). *The semigroup $\widetilde{G}^{(w)}$ is amenable.*

Theorem (6.I) implies:

(6.K). *If $f, g \in W$, $\alpha, \beta \in \mathbf{C}$, $\gamma \in \mathbf{R}_+$, then $\bar{f} \in W$, $\alpha f + \beta g \in W$, $fg \in W$, $|f| \in W$, $f^\gamma \in W$.*

(6.L). *Let $\widetilde{m}$ be an invariant mean on $C(\widetilde{G}^{(w)})$ (see (3.D)); then the functional $m(\varphi) = \widetilde{m}(\widetilde{R}^{-1}\varphi)$ is the only invariant mean on $W(G)$.*

According to Theorem 1.2.9, the mean value $\mathbf{M}(\varphi) = m(\varphi)$, $\varphi \in W$.

Example 6.1. It is obvious that if the group G is locally compact, then $m_a(\varphi) := \lim_{g\to\infty} \varphi(g)$ is the (unique) invariant mean on C_a.

Note that $\lim_{g\to a} \varphi(g) = a$ means that $\varphi \in C_a$ and $\mathbf{M}|\varphi(\cdot) - a| = 0$. If $\varphi(\cdot) \in W$ and $\mathbf{M}|\varphi(\cdot) - a| = 0$, we say that the *strong Cesaro limit exists and is equal to* a, and write: $(sC) \lim \varphi(g) = a$. It is obvious that if $\lim_{g\to\infty} \varphi(g) = a$, then $(sC) \lim \varphi(g) = a$.

Theorem (3.D) gives the construction of $\widetilde{m}$ on $\widetilde{G}^{(w)}$; this theorem combined with Theorem (6.L) imply the following statement.

(6.M). *Let* $f \in W$; $M(|f|^\gamma) = 0$ *for some* γ $(0 < \gamma < \infty)$ *iff the function* $\widetilde{R}^{-1}(f) = 0$, $f \in K(\widetilde{G}^{(w)})$; *in this case* $M(|f|^{\gamma'}) = 0$ *for any* $0 < \gamma' < \infty$.

Since $C_a \subset W$, Theorem (6.I) implies:

(6.N). *If the group G is locally compact, the homomorphism* $g \mapsto R_g$ *is a one-to-one mapping of G onto a dense subset of* $\widetilde{G}^{(w)}$.

Quite different is the situation with the homomorphism $g \to R_g$ of G into $\widetilde{G}^{(a)}$: it is one–to–one only for a special class of locally compact groups. Following A. Weil [1], we say that a group G is *representable in a compact group* if there exists a one–to–one homomorphism of G in a dense subset of some compact group $\widetilde{G}$; in such a case we can take $\widetilde{G}^{(a)}$ as $\widetilde{G}$.

(6.O). **(Freudenthal–Weil)** *A connected locally compact topological group is representable in a compact group iff it is isomorphic to a group of the form* $\mathbb{R}^m \times K$, *where* $m \geq 0$ *and* K *is a compact group.*

The proof is given in Weil [1].

§ 7. Dynamical systems

Let $(\Omega, \mathcal{F}, m)$ be a σ–finite measure space and let X be a subgroup. A (left) *dynamical system in the phase space Ω with time X* is a family $T = \{T_x, x \in X\}$ of measurable transformations of Ω with the following properties: *1)* $T_{x_1}T_{x_2} = T_{x_1x_2}$, $x_1, x_2 \in X$; and *2)* T_x are measure preserving transformations, i.e. $m(T_x^{-1}\Lambda) = m(\Lambda)$, $x \in X$, $\Lambda \in \mathcal{F}$, where $T_x^{-1}\Lambda = \{\omega : T_x\omega \in \Lambda\}$; if X is a group we require, moreover, *3)* $T_e = I$, the identity transformation. We can also consider a left dynamical system as a left action of X in Ω by a measure preserving transformation. Right dynamical systems are defined similarly.

A set $\Lambda \in \mathcal{F}$ is said to be *invariant* with respect to a dynamical system $T = \{T_x, x \in X\}$ if $m(T_x^{-1}\Lambda\Delta\Lambda) = 0$, $x \in X$. A B–valued measurable function $f(\omega)$, $\omega \in \Omega$, is said to be *invariant* with respect to T if $f(T_x\omega) = f(\omega)$ m–a.e., $x \in X$. It is easy to verify the following properties of invariant (with respect to a fixed T) functions and sets.

(7.A). *A set Λ is invariant iff its indicator $\chi_A(\omega)$ is invariant.*

(7.B). *A B–valued function f on Ω is invariant iff for any $A \in \mathfrak{B}_B$ the set $\{\omega : f(\omega) \in A\}$ is invariant.*

(7.C). *The family $\mathfrak{I}$ of all invariant sets is a σ–subalgebra of $\mathcal{F}$.*

(7.D). *The family I_B^α if all invariant functions f in $L_B^\alpha(\Omega, \mathcal{F}, m)$ is a subspace of L_B^α $(1 \le \alpha \le \infty)$.*

A left dynamical system $T = \{T_x, x \in X\}$ in $(\Omega, \mathcal{F}, P)$ can also be considered as a right measure preserving action $\widehat{T}$ of X in $\mathcal{F}$: $\Lambda\widehat{T}_x = T_x^{-1}\Lambda$. Therefore we say that two dynamical systems $\{T^{(1)}$, $x \in X\}$ in $(\Omega^{(1)}, \mathcal{F}^{(1)}, P^{(1)})$ and $\{T^{(2)}, x \in X\}$ in $(\Omega^{(2)}, \mathcal{F}^{(2)}, P^{(2)})$ are *equivalent* if there exists a one–to–one mapping $i : \mathcal{F}^{(1)} \mapsto \mathcal{F}^{(2)}$ such that $P^{(2)}(i(\Lambda)) = P^{(1)}(\Lambda)$ and $(i(\Lambda))\widehat{T}_x^{(2)} = \Lambda\widehat{T}_x^{(1)}$. It is easy to check that if, e.g., $T^{(1)}$ is ergodic, mixing, or quasimixing, then $T^{(2)}$ has the corresponding property.

To any *left* dynamical system $\{T_x, x \in X\}$ there corresponds the *associated right* isometric representation $U_x^{(\alpha)}$ in L_B^α $(1 \leq \alpha \leq \infty)$: $(fU_x^{(\alpha)})(\omega) = f(T_x\omega)$ ($U^{(2)}$ is unitary if X is a group). It is evident that I_B^α is the subspace of all U^α–invariant functions of L_B^α. To any $f\,(\in L_B^\alpha)$ corresponds a L_B^α–valued *orbital function* $\varphi_f(\cdot)$: $\varphi_f(x) = f(T_x\omega) = (fU_x^\alpha)(\omega)$. Moreover, to any B–valued function $f(\omega)$, $\omega \in \Omega$, and $\omega \in \Omega$ there corresponds the *realization*: $\psi_{f,\omega}(x) = f(T_x\omega)$.

Let $X = G$ be a group and K a compact subgroup of K. We shall say that a dynamical system $\{T_g,\ g \in G\}$ in $(\Lambda,\ \mathcal{F},\ m)$ is a *class–1* dynamical system if the associated unitary representation $U^{(2)}$ is of class *1*.

§ 8. Homogeneous Random Functions

8.1. Random functions. Let $(\Omega, \mathcal{F}, P)$ be a probability space, B a Banach space.

Any B–valued P–measurable (in the sense of Bochner) function ξ, $\xi = \xi(\omega)$, $\omega \in \Omega$, is called a *B–valued random variable* (*r.v.*) on $(\Omega,\ \mathcal{F},\ P)$. If $\xi \in L_B^1(\Omega,\ \mathcal{F},\ P)$, i.e. $\mathbf{E}\|\xi\| < \infty$, then the element $\mathbf{E}\xi := \int_\Omega \xi(\omega)P(d\omega)$ (the Bochner integral) is the *expectation* of ξ. Let Y be an arbitrary set. A *B–valued random function (r.f.)* on Y is a family of random variables $\xi(y)$, $y \in Y$. If $\omega_0\ (\in \Omega)$ is fixed, then the function $y \to \xi(y, \omega_0)$ is a *realization* of $\xi(\cdot)$.

If $Y = \{1, \ldots, k\}$, the family $\{\xi_1, \ldots, \xi_k\}$ is called a *random vector*; if $Y = \{1, 2, \ldots\}$, $\xi = \xi(\cdot)$ is a *random sequence*; and if $Y = (a, b) \subset \mathbb{R}^1$, then $\xi(y)$ is a *random process* on (a, b). Finally, if Y is a semigroup or homogeneous space (in particular, if Y is a group), we say that $\xi(\cdot)$ is a *random field* on Y.

The notion of a real Gaussian vector is well-known (see, e.g., Doob [1] or Shiryaev [1]). A complex random vector $(\xi_1, \ldots, \xi_m) = (\eta_1 + i\xi_1, \ldots, \eta_m + i\xi_m)$ is called *Gaussian* if the vector $(\eta_1, \ldots, \eta_m, \xi_1, \ldots, \xi_m)$ is Gaussian and $\mathbf{E}\xi_k\xi_l = \mathbf{E}\xi_k\mathbf{E}\xi_l$, $k, l = \overline{1, m}$. A scalar (i.e. either real or complex) r.f. $\xi(\cdot)$ is called *Gaussian* if any random vector $(\xi(y_1), \ldots, \xi(y_k))$ is Gaussian $(y_1, \ldots, y_k \in Y,\ k = \overline{1, \infty})$. An $\mathbb{R}^m$–valued r.f. $\xi(y) = (\xi_1(y), \ldots, \xi_m(y))$ is called *Gaussian* if for any $y \in Y$, $k = \overline{1, m}$, the r.f. $(y, k) \to \xi_k(y)$ is Gaussian.

If $\xi(y)$, $y \in Y$, is a r.f. with values in a Hilbert space $\mathcal{H}$ and $\mathbf{E}\|\xi(y)\|^2 < \infty$, $y \in Y$. We can define its *scalar correlation function* $\widetilde{R}(y,z) = \mathbf{E}\langle\xi(y), \xi(z)\rangle$ and the *operator correlation function* $\widetilde{R}(y,z)$: $\langle\widetilde{R}(y,z)a, b\rangle = \mathbf{E}\langle\xi(y), a\rangle\langle\xi(z), b\rangle$, $a, b \in \mathcal{H}$; we can also consider the *covariance function* $B(y,z) = \mathbf{E}\langle(\xi(y) - m(y)),\ (\xi(z) - m(z))\rangle = R(y,z) - \langle m(y), m(z)\rangle$ where $m(y) = \mathbf{E}[\xi(y)]$ and R and B are positive definite function. And, conversely, any positive definite function is the correlation function of some Gaussian r.f. with $m(y) \equiv 0$ (see, e.g., Doob, Ch. 2, § 3).

8.2. Images of random functions and covariance with respect to semigroup actions. Let $\xi(y)$, $y \in Y$, be a r.f on $(\Omega, \mathcal{F}, P)$ and $\mathcal{F}_\xi$ the σ–algebra generated by $\xi(\cdot)$ (i.e. the minimal σ–algebra with respect to which all r.v. $\xi(y)$, $y \in Y$, are measurable). Let $(\widetilde{\Omega}, \widetilde{\mathcal{F}})$ be a measurable space. A $\xi(\cdot)$–*admissible reduction* of $(\Omega, \mathcal{F}, P)$ into $(\widetilde{\Omega}, \widetilde{\mathcal{F}})$ is a mapping r of some $\mathcal{F}$–measurable set Ω' with $P(\Omega') = 1$ into $\widetilde{\Omega}$ such that the induced mapping $\mathcal{F}'_\xi \to \widetilde{\mathcal{F}}$ is one–to–one (here $\mathcal{F}'_\xi$ is the restriction of $\mathcal{F}$ to Ω'). We can consider the probability space $(\widetilde{\Omega}, \widetilde{\mathcal{F}}, \widetilde{P})$ with $\widetilde{P} = P \circ r^{-1}$.

The following statement is a simple generalization of a theorem due to Doob (Doob [1], App., § 3).

(8.A). *Let $\xi(y)$, $y \in Y$, be a B–valued r.f on $(\Omega, \mathcal{F}, P)$, r an admissible reduction of Ω into $\widetilde{\Omega}$, and $\eta(\omega)$, $\omega \in \Omega$, an $\mathcal{F}_\xi$–measurable B_1–valued r.v. Then there exists a B_1–valued r.v. $\widetilde{\eta}$ on $(\widetilde{\Omega}, \widetilde{\mathcal{F}}, \widetilde{P})$ such that $\widetilde{\eta}(r(\omega)) = \eta(\omega)$, $\omega \in \Omega_1$. If $\widetilde{\eta}_1$ is another such r.v., then $\widetilde{P}\{\widetilde{\omega} : \widetilde{\eta}(\widetilde{\omega}) = \widetilde{\eta}_1(\widetilde{\omega})\} = 1$.*

The r.v. $\widetilde{\eta}$ is called the *image* of the r.v. η on $(\widetilde{\Omega}, \widetilde{\mathcal{F}}, \widetilde{P})$ (with respect to the reduction r). If $\eta(y)$, $y \in Y$, is a r.f. and $\widetilde{\eta}(y)$ is the image of $\eta(y)$ with respect to r, then we say that the r.f. $\widetilde{\eta}(\cdot)$ is the *image* of $\eta(\cdot)$. It is easy to verify that $P\{\eta(y_1) \in \Gamma_1, \ldots, \eta(y_n) \in \Gamma_n\} = \widetilde{P}\{\widetilde{\eta}(y_1) \in \Gamma_1, \ldots, \widetilde{\eta}(y_n) \in \Gamma_n\}$ where $\Gamma_i \in \mathfrak{B} = \mathfrak{B}(B)$.

Let Y be a left X–space and $T : x \to T_x$ some right action of the semigroup X in $(\widetilde{\Omega}, \widetilde{\mathcal{F}})$ by measurable transformations.

We say that a r.f. $\xi(\cdot)$ on $(\Omega, \mathcal{F}, P)$ is *covariant* with respect to T if there exists a ξ–admissible reduction r: $\Omega \mapsto \widetilde{\Omega}$ such that $\widetilde{\xi}(xy, \widetilde{\omega}) = \widetilde{\xi}(y, \widetilde{\omega}T_x)$, $x \in X$, $y \in Y$, $\widetilde{\omega} \in \widetilde{\Omega}$.

Example 8.1. Let $\xi(y)$, $y \in Y$, be a B–valued r.f. on $(\Omega, \mathcal{F}, P)$, $\mathfrak{B} = \mathfrak{B}(B)$, and let $(B^Y, \mathfrak{B}^Y)$ be the measurable space of all B–valued functions, where $\mathfrak{B}^Y$ is the σ–algebra generated by the cylinder sets $\{\widetilde{\omega} = \widetilde{\omega}(\cdot) \in B^Y;\ \widetilde{\omega}(y) \in \Gamma\}$, $y \in Y$, $\Gamma \in \mathfrak{B}$. Let $\widetilde{\Omega}^\xi \subset B^T$, $\Omega_1 :=$ $:= \{\omega : \xi(\cdot, \omega) \in \widetilde{\Omega}^\xi\}$; suppose that $\Omega_1 \in \mathcal{F}$ and $P(\Omega_1) = 1$. Let $\widetilde{\mathcal{F}}^\xi$ be the restriction of the σ–algebra $\mathfrak{B}^Y$ to $\widetilde{\Omega}^\xi$; $r_\xi(\omega) = \xi(\cdot, \omega)$, $\omega \in \Omega_1$ (in many cases it is convenient to set $\widetilde{\Omega}^\xi = B^Y$; then $\Omega_1 = \Omega$, $\widetilde{\mathcal{F}}^\xi = \mathfrak{B}^Y$). Since $r_\xi^{-1}(\widetilde{\omega} : \widetilde{\omega}(t) \in \Gamma) = \{\omega : \xi(t, \omega) \in \Gamma\} \cap \Omega_1$, $\Gamma \in \mathfrak{B}$, we have $r_\xi^{-1}(\widetilde{\mathcal{F}}^\xi) = \mathcal{F}_{1,\xi}$ and thus r_ξ is an admissible reduction of Ω onto $\widetilde{\Omega}^\xi$. The r.f. $\widetilde{\xi}(t, \widetilde{\omega}) := (\widetilde{\omega}(t))$ is called the *coordinate image* of the r.f. $\xi(\cdot)$; r_ξ is called the *coordinate reduction* of Ω (with respect to the r.f. $\xi(\cdot)$). If Y is an X–space, the r.f. $\xi(\cdot)$ is covariant with respect to the (right) action by left translations $x \to L_x$ of X in $\widetilde{\Omega}^\xi : \widetilde{\omega} \to (\widetilde{\omega}L_x)(y) = \widetilde{\omega}(xy)$.

8.3. Homogeneous random functions. Let X be a semigroup and Y a left X–space. A B–valued r.f. $\xi(y)$, $y \in Y$, is said to be *strict sense homogeneous* if for any $y_1, \ldots, y_k \in Y$, $A_1, \ldots, A_k \in \mathfrak{B}(B)$ the probability $P\{\omega : \xi(xy_1) \in A_1, \ldots, \xi(xy_k) \in A_k\}$ does not depend on $x \in X$ $(k = \overline{1, \infty})$.

Let $\mathcal{H}$ be a Hilbert space and $\langle \cdot, \cdot \rangle$ the scalar product in $\mathcal{H}$. An $\mathcal{H}$–valued r.f. $\xi(t)$, $t \in T$, is said to be *wide sense homogeneous* if: *1)* $\mathbf{E}\|\xi(y)\|^2 < \infty$; *2)* $\mathbf{E}\xi(xy) = \mathbf{E}\xi(y)$; and *3)* $R(xy_1, xy_2) =$ $= R(y_1, y_2)$, $y_1, y_2 \in T$, $x \in X$. $\xi(\cdot)$ is wide sense strongly homogeneous if conditions *1)* and *2)* are fulfilled as well as *3′)* $\widetilde{R}(xy_1, xy_2) = \widetilde{R}(y_1, y_2)$. Certainly, if an $\mathcal{H}$–valued r.f. is strict sense homogeneous and $\mathbf{E}\|\xi(y)\|^2 < \infty$, $y \in Y$, then it is strongly wide sense homogeneous. For a Gaussian r.f. the notions of wide sense strong homogeneity and strict sense homogeneity coincide.

(8.B). *If a r.f. $\xi(\cdot)$ is covariant with respect to some dynamical system, then $\xi(\cdot)$ is strict sense homogeneous. And, conversely, if a strict sense* *homogeneous random function $\xi(x)$, $x \in X$, is covariant with respect to a measurable action T of X in a probability space $(\Omega', \mathcal{F}', P')$, then T is a right dynamical system; in particular, the left translation action of X in B^X is a right dynamical system.*

The dynamical system $\{L_x,\ x \in X\}$ in $\widetilde{\Omega}^\xi$ defined in subsection 8.2 is called the *translation* (or *shift*) *dynamical system.*

It is easy to verify the following:

(8.C). *All dynamical systems with respect to which some homogeneous r.f. is covariant are equivalent to each other.*

8.4. The translation representation. If Y is a left X–space and $\xi(y)$, $y \in Y$, is a wide sense homogeneous r.f. over $(\Omega, \mathcal{F}, P)$ with values in a Hilbert space $\mathcal{H}$, we can consider the Hilbert space H_ξ, the subspace of $L^2_{\mathcal{H}}\ (\Omega, \mathcal{F}, P)$, generated by the subset $H^0_\xi = \{\xi(y),\ y \in Y\}$. In H^0_ξ we consider the operators $U_x : U_x(\xi(y)) = \xi(xy)$, $x \in X$; they can be extended to isometric operators on the whole of H_ξ. The mapping U: $x \to U_x$ is an isometric (unitary, if X is a group) left representation of X in H_ξ; we call it the *translation representation* of X defined by the r.f. $\xi(\cdot)$.

8.5. Homogeneous random fields on semigroups and homogeneous spaces. A semigroup X is a left (right) X–space with respect to the left (right) translations; this allows us to speak of random fields on X which are *homogeneous* with respect to the left (or right) translations, or simply of *l–homogeneous* (or *r–homogeneous*) random fields.

The correlation function $R(x, y)$ and covariance function $B(x, y)$ of a wide sense homogeneous random field are X–invariant. Therefore, if $X = G$ is a group and a field $\xi(\cdot)$ is homogeneous with respect to the left (right) translations then $R(x, y) = R_0(x^{-1}y)$ (resp. $R(x, y) = R_0(xy^{-1})$); similarly, $B(x, y) = B_0(x^{-1}y)$ (resp. $B(x, y) = B_0(xy^{-1})$). The functions $R_0(\cdot)$ and $B_0(\cdot)$ are also called the *correlation* and the *covariance function* of $\xi(\cdot)$. The spectral measure F of the function R (see App., § 4) is called the *(mean square) spectral measure* of $\xi(\cdot)$.

Let X be a semigroup with identity e and $T = \{T_x,\ x \in X\}$ a right dynamical system in $(\Omega, \mathcal{F}, P)$. Covariance of a r.f. $\xi(x) = \xi(x,\omega)$, $x \in X$, $\omega \in \Omega$, with respect to T is equivalent to the relation: $\xi(x,\omega) = \xi(e, \omega T_x) = \eta(\omega T_x)$ where $\eta = \xi(e)$. Therefore statement (8.B) implies the following assertion.

(8.D). *If there exist a dynamical system T in $(\Omega, \mathcal{F}, P)$ and a random variable η such that $\xi(x,\omega) = \eta(\omega T_x)$, then the random field $\xi(\cdot)$ is strict sense l–homogeneous; any l–homogeneous random field is related with its left translation system L in the following way: $\xi(x,\omega) = \eta(r_\xi(\omega) L_x)$.*

A wide sense homogeneous random field $\xi(\cdot)$ in a left homogeneous space G/K can be considered as a wide sense homogeneous random field $\widehat{\xi}$ on G that is constant on the left cosets with respect to K. The family of such fields $\widehat{\xi}$ on G can be described in the following way.

(8.E). *The correlation function R of an l–homogeneous r.f. $\widehat{\xi}$ on G which is constant on the left cosets with respect to K is a positive definite spherical function.*

(8.F). *The translation representation U of a wide sense homogeneous random field on $Y = G/K$ is a (left) unitary K–class–1 (left) representation: if y_0 is the K–fixed point in G/K, then $\xi(y_0)$ is a K–invariant element of H_ξ and H_ξ is generated by the U–orbit of $\xi(y_0)$ (i.e. by the set $\{\xi(y),\ y \in Y\}$).*

8.6. Homogeneous generalized random fields. A *generalized random field* on $\mathbb{R}^m$ over $(\Omega, \mathcal{F}, P)$ is a continuous linear mapping $\varphi \to \xi(\varphi)$ from the Schwartz space $K(\mathbb{R}^m)$ into the Hilbert space $L^2\ (\Omega, \mathcal{F}, P)$ (see Gel'fand and Shilov [1], Ch. 3). Like in Subsection 4.4, we regard $K(\mathbb{R}^m)$ as a G–space with respect to the translations in $\mathbb{R}^m$. This allows us to introduce a *wide sense homogeneous* generalized random field. The correlation function of such a field is invariant; its spectral measure (see Subsect. 4.4) is the (*mean square*) *spectral mea-*

sure of $\xi(\cdot)$. The field ξ is of class $\mathfrak{S}_p$ if R_0 is of class $\mathfrak{S}_p$. In particular, a field ξ of class $\mathfrak{S}_0$ has the following form: $\xi(\varphi) = \int \xi_0(x)\varphi(x)dx$ where $\xi(x)$ is a continuous homogeneous random field on $\mathbf{R}^m$.

(8.G). (I. Gel'fand) *Any wide sense homogeneous generalized random field $\xi(\varphi)$ admits the spectral decomposition $\xi(\varphi) = \int \widetilde{\varphi}(\lambda)Z(d\lambda)$ where $Z(d\lambda)$ is a σ-additive random set function on $\mathfrak{B}(\mathbf{R}^m)$ such that* $\mathbf{E}Z(\Lambda_1)\overline{Z(\Lambda_2)} = F(\Lambda_1 \cap \Lambda_2)$ (see Gel'fand and Vilenkin [1]).

§ 9. Measurability and Continuity of Representations, Dynamical Systems and Homogeneous Random Fields

Let X be a locally compact topological semigroup and $\mathfrak{B} = B(X)$ the σ-algebra of Borel sets in X. If λ is a measure on $\mathfrak{B}$, *λ-measurability* of a function $f(x)$, $x \in X$, taking values in a Banach space B, means λ-measurability in the sense of Bochner.

If B is separable, we say that f is $\mathfrak{B}$-measurable if $\{x : f(x) \in A\} \in \mathfrak{B}$ for any Borel set A in B; in this case f is λ-measurable with respect to any σ-finite measure λ on $\mathfrak{B}$ (see Dunford and Schwartz [1], Ch. 3). The same terms will be used if f is defined on an arbitrary measurable space.

9.1. Measurable and continuous representations. We shall say that a representation $x \mapsto S_x$ of a semigroup X in a Banach space is *continuous* (*uniformly continuous*, *measurable*) if all orbital functions $\varphi_b(x) = S_x b$ possess the corresponding property.

(9.A). *If $x \mapsto S_x$ is a continuous uniformly bounded left representation of a semigroup X, then any orbital function $\varphi_b \in UC_B^{(l)}$.*

Proof. The assertion follows from the relations: $\|\varphi_b(zx) - \varphi_b(zy)\|_B = \|S_{zx}b - S_{zy}b\| \le \sup_{z\in X} \|S_z\| \|\varphi_b(x) - \varphi_b(y)\|_B$.

□

Consider a compact set K in X, a measure μ on $\mathfrak{B}$ and a function f on X. We let $\mathfrak{B}_K$, μ_K and f_K be the restrictions of $\mathfrak{B}$, μ and f to K. The notation $f \in L^2_{B,loc}(X,\mathfrak{B}, \mu)$ means that $f_K \subset \subset L^\alpha_B (K, \mathfrak{B}_K, \mu_K)$ for any compact set $K \subset X$.

We shall need the following auxiliary statement.

(9.B). *Let μ be a regular Borel right semiinvariant measure on X and $f \in L^\alpha_{B,loc} (X, \mathfrak{B}, \mu)$ for some $\alpha \geq 1$; then for any compact set $K \subset X$ with $\mu(K) > 0$ and any $x \in \mathrm{Int}_p(X)$, the function $\psi^{(f)}_{K,x}(y) := := \int_K \|f(zx) - f(zy)\|^\alpha_B \mu(dz)$ is continuous on X.*

Proof. Let us fix some $\varepsilon > 0$ and a bounded neighbourhood V of x. Denote $K_1 := K \cdot [V]$. Since μ is regular, the B-valued continuous functions on K_1 form a dense set $C_B(K_1)$ in $L^\alpha_B(K_1, \mathfrak{B}_{K_1}, \mu_{K_1})$ (see, e.g., Hewitt and Ross [1], Ch. 3, Th. 12.10). Therefore we can choose a function f_ε $(\in C_B(K_1))$ such that $\int_{K_1} \|f - f_\varepsilon\|^\alpha d\mu < \varepsilon$. But now, for all $y \in V$ we have

$$\int_K \|f(zy) - f_\varepsilon(zy)\|^\alpha_B \mu(dz) \leq \int_{K_1} \|f(z) - f_\varepsilon(z)\|^\alpha_B \mu(dz) < \varepsilon.$$

Since f_ε is continuous on K_1, it is uniformly continuous, i.e. for any $\varepsilon > 0$ there is a neighborhood V_ε of x such that $V_\varepsilon \subset V$ and $\|f_\varepsilon(zx) - f_\varepsilon(zy)\|_B < \varepsilon$, $z \in K$, $y \in V_\varepsilon$. This implies that for all $y \in V_\varepsilon$ we have $|\psi^{(f)}_{K,x}(x) - \psi^{(f)}_{K,x}(y)| < (2 + \mu(K))\varepsilon$ and, since $\varepsilon > 0$ is chosen arbitrarily, our statement is proved.

□

(9.C). *If X is a right-regular locally compact topological semigroup and μ is a regular Borel right semiinvariant measure on X, then any μ-measurable isometric representation of X in a Banach space B is left uniformly continuous on the set $\mathrm{Int}_p(X)$.*

Proof. Let $x \mapsto U_x$ be a measurable isometric representation of X in B. For any $b \in B$, $x, y, z \in X$ we have $\|U_x b - U_y b\| = \|U_{zx} b - U_{zy} b\|_B$. This implies that for any compact set K with $\mu(K) > 0$ we have

$\|U_x b - U_y b\| = \frac{1}{\mu(K)} \int_K \|U_{zx} b - U_{zy} b\| \mu(dz)$. Now we have to use statements (9.A) and (9.B).

□

Special cases of statement (9.C) can be found in Hewitt and Ross [1], (Subsect. 22.20) and in Hille and Phillips [1] (Theorems 10.2.3. and 10.10.1). In the case when X is a group similar results are contained in Moore [1].

The following, in some sense converse, statement is also true.

(9.D). *If X is a separable topological semigroup, then any continuous bounded representation S of it in a Banach space B is measurable with respect to any Borel measure on X.*

Proof. The subspace B_φ generated by an orbital function φ_b of S is separable. Also it is obvious that for $l \in B^*$ the scalar function $l \circ \varphi_b$ on X is continuous. By the well-known theorem due to Pettis (see Dunford and Schwartz [1], Ch 3, § 6, Th. 11) this implies the measurability of φ_b.

□

9.2. Measurable and continuous dynamical systems. Measurability and continuity of dynamical systems can also be defined in terms of orbital functions: namely, we shall say that a dynamical system $\{T_x,\ x \in X\}$ with phase space $(\Omega, \mathcal{F}, m)$ is *measurable* (*continuous*, or *uniformly continuous*) if all its orbital functions in the spaces $L^\alpha_B\,(\Omega, \mathcal{F}, m)$, $\alpha \geq 1$, have the corresponding property.

(9.E). *If X is a left–regular locally compact topological semigroup and $T = \{T_x,\ x \in X\}$ is a measurable (left) dynamical system, then T is right uniformly continuous on the set* $\mathrm{Int}_p(X)$.

9.3. Integration. Completely measurable dynamical systems and representations in L^α. Consider a σ–finite measure space $(\Omega, \mathcal{F}, m)$. Let $S : x \mapsto S_x$ be a uniformly bounded representation of X in $L^1(\Omega, \mathcal{F}, m)$; $C := \sup_{x \in X} \|S_x\|$.

(9.F). *If S is measurable with respect to some $\lambda \in \mathcal{P}(X)$ and $f \in$ $\in L^1\ (\Omega, \mathcal{F}, m)$, then*
a) the Bochner integral $\int_X S_x f \lambda(dx) \in L^1(\Omega)$ exists;
b) the integral $\int_\Lambda S_x f m(d\omega)$, $x \in X$, $\Lambda \in \mathcal{F}$ exists, and

$$\int_X \left\{ \int_\Lambda S_x f(\omega) m(d\omega) \right\} \lambda(dx) = \int_\Lambda \left\{ \int_X (S_x f(\omega) \lambda(dx) \right\} m(d\omega).$$

Proof. Any orbital function $\varphi_f(x) = S_x f$ is λ-measurable, and

$$\|\varphi_f(x)\|_{L^1(\Omega)} = \int_\Omega |S_x f(\omega)| m(d\omega) \leq C \int_\Omega |f(\omega)| m(d\omega) < \infty.$$

Consequently, $\varphi_f \in L^1_{L^1(\Omega)}(\lambda)$, and a) is proved. Let us consider the linear functional $L^1(\Omega) \xrightarrow{l} \mathbb{C} : l(g) = \int_\Lambda g\, dm$, $g \in L^1(\Omega)$. Assertion b) is implied by the theorem on transferring an operator under the Bochner integral sign (see Dunford and Schwartz [1], Ch. 3, § 2, Th. 19), which we have to apply to l.

The notion of measurability of dynamical systems introduced in Subsect. 9.2 does not imply measurability of realizations of dynamical systems. This circumstance compels one to introduce another notion of measurability of dynamical systems.

A dynamical system $T = \{T_x,\ x \in X\}$ with measurable time $(X, \mathfrak{B})$ and acting in the space $(\Omega,\ \mathcal{F},\ m)$ is called *completely measurable* if for any $\Lambda (\in \mathcal{F})$ the set $\{(x, \omega) : (x, \omega) \in X \times \Omega,\ T_x\omega \in \Lambda\} \in \mathfrak{B} \times \mathcal{F}$.

(9.G). *Let λ be a σ-finite measure on $\mathfrak{B}$; then any completely measurable left dynamical system has the following properties:*

(i) if $f(\omega)$, $\omega \in \Omega$, is an m-measurable B-valued function, then the function $\psi(x, \omega) = f(T_x\omega)$ is $(\lambda \times m)$-measurable for any σ-finite Borel measure λ; consequently, m-almost all realizations $\varphi_{\omega,f}(x) = f(T_x\omega)$ are λ-measurable;

(ii) if $f \in L^\alpha_B\ (\Omega, \mathcal{F}, m)$ $(\alpha \geq 1)$, then for any β, $1 \leq \beta \leq \alpha$, and $A \in \mathfrak{B}$ with $\lambda(A) < \infty$ we have for almost all $\omega \in \Omega$, $\int_A \|f(T_x\omega)\|^\beta_B \lambda(dx) < \infty$;

in particular, the integral $\int_A f(T_x\omega)\lambda(dx)$ *exists* m*–a.e.;*

(iii) if $A \in \mathfrak{B}$, $\lambda(A) < \infty$, *then the operator* $U, (fU)(\omega) = \int_A f(T_x\omega)\lambda(dx)$ *is a bounded linear operator in* L^α_B *and* $\|U\| = (\lambda(A))^{\frac{1}{\alpha}}$; *U is isometric if* $\lambda(A) = 1$;

(iv) the dynamical system T is λ*–measurable; for any* L^α_B*–valued orbital function* $\varphi_f(x) = f(T_x\omega)$ *and any* $A \in B$ *with* $\lambda(A) < \infty$, *the Bochner integral* $\int_A f(T_x\omega)\lambda(dx)$ *coincides* m*–a.e. with the integral of realizations defined in (ii)* $(\alpha \geq 1)$;

(v) if X is a left–regular locally compact topological semigroup, then any L^α_B*–valued orbital function is right uniformly continuous on the set* $\mathrm{Int}_p(X)$, $\alpha \geq 1$.

Proof. Statement *(i)* follows from the definition of complete measurability. Statements *(ii)* and *(iii)* follow from the relations

$$\int_\Omega \left\{ \int_A \|f(T_x\omega)\|^\alpha \lambda(dx) \right\} m(d\omega) =$$
$$= \int_A \left\{ \int_\Omega \|f(T_x\omega)\|^\alpha m(d\omega) \right\} \lambda(dx) = \lambda(A) \int_\Omega \|f(\omega)\|^\alpha m(d\omega)$$

and (for $1 \leq \beta \leq \alpha$)

$$\int_A \|f(T_x\omega)\|^\beta \lambda(dx) \leq \left\{ \int_A \|f(T_x\omega)\|^\alpha \lambda(dx) \right\}^{\frac{\beta}{\alpha}} [\lambda(A)]^{1-(\beta/\alpha)}.$$

It is very easy to verify that $f(T_x\cdot) \in L^\alpha_B\ (\Omega, \mathcal{F}, m)$ if $f(\cdot) \in L^\alpha_B\ (\Omega, \mathcal{F}, m)$; therefore *(iv)* is a simple consequence of Lemma 3.11.16 in Dunford and Schwartz [1]. Statement *(v)* follows from Statement *(iv)* and Theorem (9.C).

□

The same problem arises in connection with "realizations" of $\varphi_\omega(x) = (S_x f)(\omega)$ for a representation S in L^α.

We say that a representation S in L^α is *completely measurable* if for any $f \in L^\alpha\ (\Omega, \mathcal{F}, m)$ the function $(x, \omega) \to S_x f(\omega)$ is $(\mathfrak{B} \times \mathcal{F})$–measurable. Of course, if T is a completely measurable dynamical system, then the associated representation in L^α is also completely measurable.

We rephrase Statement (9.G) in the following way.

(9.G′). *Let S be a uniformly bounded completely measurable representation of X in L^α and λ be as in (9.G).*

(i) The functions $\varphi_{\omega,f} := (S_x f)(\omega)$ are $\mathfrak{B}$-measurable for m-almost all $\omega \in \Omega$;

(ii) if $f \in L^\alpha\ (\Omega, \mathcal{F}, m)$ $(\alpha \geq 1)$, then for any β, $1 \leq \beta \leq \alpha$, and A with $\lambda(A) < \infty$, we have for m-almost all $\omega \in \Omega$,

$$\int_A \|S_x f(\omega)\|^\beta \lambda(dx) < \infty;$$

in particular, the integral $\int_A (S_x f)(\omega)\lambda(dx)$ exists m-a.e.;

(iii) if $A \in \mathfrak{B}$ and $\lambda(A) < \infty$, then the operator $\widehat{S}$, $(\widehat{S}f)(\omega) = = \int_A (S_x f)(\omega)\lambda(dx)$ is a bounded linear operator;

(iv) S is λ-measurable and if $\lambda(A) < \infty$, then the Bochner integral $\int_A S_x \lambda(dx)$ coincides for m-almost all ω with the integral $\int_A (S_x f)(\omega)\lambda(dx)$;

(v) if X is a right-regular locally compact semigroup, then any orbital function $x \mapsto S_x f$ is left uniformly continuous on $\mathrm{Int}_p(X)$.

9.4. Measurable and continuous homogeneous random fields. Let B be a separable Banach space. A B-valued random field $\xi(\cdot)$ is said to be *completely measurable* if the function $(x, \omega) \mapsto \xi(x, \omega)$ is $(\mathfrak{B} \times \mathcal{F})$-measurable. It is evident that if a field $\xi(\cdot)$ is completely measurable, then all its realizations are $\mathcal{F}$-measurable. A field $\xi(\cdot)$is said to be *λ-measurable* if it is λ-measurable as an L^α_B-valued function on X as soon as $\mathbf{E}\|\xi(x)\|^\alpha < \infty$, $x \in X$ $(\alpha \geq 1)$. Similarly, *continuity* of a random field on a topological semigroup can be defined. In the sequel, when speaking about a wide sense homogeneous random field we have in mind that $B = \mathcal{H}$ (a Hilbert space) and $\alpha = 2$. The following statements can be easily deduced from the theorems of Subsect. 9.1 and 9.2; they are true for fields which are right homogeneous either in the wide or strict sense (the necessary changes for the "left" versions are obvious).

(9.H). *A continuous r–homogeneous random field is right uniformly continuous.*

(9.I). *Any λ–measurable r–homogeneous random field on a left–regular locally compact semigroup is right uniformly continuous on the set* $\mathrm{Int}_p(X)$.

(9.J). *Let λ be a σ–finite Borel measure on X. A completely measurable B–valued r–homogeneous random field $\xi(\cdot)$ with $\mathbf{E}\|\xi(x)\|^\alpha < < \infty$, $x \in X$, has the following properties:*

(i) if $A \in \mathfrak{B}$ and $\lambda(A) < \infty$, then with probability 1,

$$\int_A \|\xi(x)\|^\beta \lambda(dx) < \infty \text{ for } \beta \le \alpha$$

and, in particular, the integral of the realizations $\int_A \xi(x)\lambda(dx)$ is defined a.s.;

(ii) $\xi(\cdot)$ is λ–measurable and, if $\lambda(A) < \infty$, the Bochner integral $\int_A \xi(x)\lambda(dx)$ coincides with the integral of the realizations P–a.e.;

(iii) if X is a left–regular locally compact topological semigroup, then the field $\xi(\cdot)$ is right–uniformly continuous on $\mathrm{Int}_p(X)$.

§ 10. Flight–Functions and FD–Continuous Spectrum

10.1. Flight–functions and representations with FD–continuous spectrum. Let G be a topological group. A function f ($\in W$) is called a *flight–function* if $\mathrm{M}(|f|) = 0$; this is equivalent to the following property: the continuous "extension" of f to the semigroup $\widetilde{G}^{(W)}$, i.e. the function $\widetilde{R}^{-1}(f)$, vanishes on the kernel $K(\widetilde{G}^{(W)})$ of $\widetilde{G}^{(W)}$.

The set of all flight–functions is denoted by W_0; it is clear that it is a subspace of W and that $f \in W_0 \Rightarrow f^\alpha \in W_0$, $\alpha > 1$.

Example 10.1. If G is locally compact, then, in view of Example 6.1, $C_0 \subset W_0$.

(10.A). **(DeLeeuw and Glicksberg [1])** $W = W_0 \oplus AP$, *i.e. any*

$f \in W$ has a unique decomposition $f = f_d + f_c$ where $f_d \in AP$, $f_c \in W_0$.

This decomposition is closely connected with the theorem concerning decomposition of unitary representations due to Godement [1] and Maak [1] (see also Jacobs [1]), which we state below. Let $U : x \to U_x$ be a unitary representation of G in a Hilbert space H. We write $U = \oplus_{\gamma \in \Gamma} U^\gamma$ if $U^{(\gamma)}$ are subrepresentations of U in subspaces $H^{(\gamma)}$ of H, and $H = \oplus_{\gamma \in \Gamma} H^\gamma$.

(10.B). *I. Any unitary representation U admits a unique decomposition $U = U^{(d)} \oplus U^{(c)}$ where $U^{(d)}$ is a subrepresentation with almost periodic matrix coefficients and $U^{(c)}$ is a subrepresentation, all matrix coefficients of which are flight–functions.*

II. There is a unique decomposition: $U^{(d)} = \oplus_{\gamma \in \Gamma} U^{(\gamma)}$ where $U^{(\gamma)}$ are finite–dimensional irreducible subrepresentations of U.

III. $U^{(c)}$ does not contain finite–dimensional representations.

This theorem has two important corollaries concerning positive definite functions.

(10.C). *A continuous elementary positive definite function is a flight–function (resp. an almost periodic function) iff the associated irreducible unitary representation is infinite–dimensional (resp. finite–dimensional).*

(10.D). **(Godement [1])** *I. The components R_c and R_d of a function $R \in \mathbf{P}(G)$ belong to $\mathbf{P}(G)$. If $R(g) = \langle U_g h, h\rangle$, then $R_c(g) = \langle U_g^{(c)} h^{(c)}, h^{(c)}\rangle$ and $R_d(g) = \langle U_g^{(d)} h^{(d)}, h^{(d)}\rangle$ where $h^{(c)}$ and $H^{(d)}$ are the projections of h to the subspaces $H^{(c)}$ and $H^{(d)}$ in which $U^{(c)}$ and $U^{(d)}$ act.*

II. $R_d(g) = \sum_{i=1}^{k} \sigma_i \varphi_i(g) = \sum_{i=1}^{k} S_p[U_g^{(i)} F^{(i)}]$ where $1 \le k < \infty$, $\sigma_i > 0$, φ_i are continuous almost periodic elementary positive definite functions, $U^{(i)}$ are associated finite–dimensional irreducible unitary representations in $H^{(i)}$, and $F^{(i)}$ are positive definite Hermitian operators *in* $H^{(i)}$.

III. The function $R_c(g)-\sigma\varphi(g)$ *is not positive definite for any* $\varphi \in \in \mathbf{P} \cap AP$ *and* $\sigma > 0$.

Now it is easy to prove the following statement.

(10.E). *If* U *is a class–1 unitary representation with respect to a subgroup* K *of* G, *then the representations* $U^{(c)}$, $U^{(d)}$, $U^{(\gamma)}$ *which appear in Statement* (10.*B*) *are class–1 unitary representations. If* $R \in \mathbf{SP}(G,K)$, *then* $R_c \in \mathbf{SP}(G,K)$ *and* $R_d \in \mathbf{SP}(G,K)$.

If $U = U^{(d)}$ (resp. $U = U^{(c)}$), we shall say that U has *finite-dimensionally discrete spectrum*, or, for short, *FD–discrete spectrum* (resp. *FD–continuous spectrum*). A function $R \in \mathbf{P}$ is said to have *FD*–discrete (resp. *FD*–continuous) spectrum if it is an almost periodic function (resp. a flight–function).

Thus, the decompositions $U = U^{(d)} \oplus U^{(c)}$ and $R = R_d + R_c$ are decompositions into components with *FD*–discrete and *FD*–continuous spectrum.

This "spectral" terminology has an obvious background. Let G have a spectral system $(\Lambda, \mathcal{D}, U(\cdot))$ (see § 5). We shall say that a measure F on Λ has a *finite–dimensional (infinite–dimensional) atom* if $F(\{\lambda\}) \neq \neq 0$ and the Hilbert space $H(\lambda)$ is finite–dimensional (resp. infinite–dimensional); a measure F on **EP** has *finite–dimensional (infinite–dimensional) atom at* $\varphi(\in \mathbf{EP})$ if $F(\{\varphi\}) \neq 0$ and the associated irreducible unitary representation is finite–dimensional (resp. infinite–dimensional). A measure F on $\mathcal{D}$ is *discrete* if there is a countable set $D_0 \subset \Lambda$ such that $\mu(D) = \sum_{\lambda \in D_0 \cap D} F(\{\lambda\})$, $D \in \mathcal{D}$.

(10.F). *A* **P**–*function* R, *or a unitary representation* U, *has FD–discrete spectrum iff it has a discrete spectral measure whose atoms are finite–dimensional;* R, *or* U, *has FD–continuous spectrum iff any spectral measure of it has no finite–dimensional atoms on* $\Lambda \backslash \{\lambda_1\}$.

On a compact topological group any continuous function is almost periodic (see, e.g., DeLeeuw and Glicksberg [1], Loomis [1]). This implies the following statement.

(10.G). *On a compact topological group any* **P**-*function and any unitary representation have FD–discrete spectrum.*

All irreducible unitary representations of a commutative group are one–dimensional; thus we have a simple consequence of (10.F).

(10.H). *If G is commutative, then any* **P**-*function and any unitary representation have FD–continuous (FD–discrete) spectrum iff the spectral measure F is continuous (discrete).*

10.2. Homogeneous random functions with FD–continuous spectrum. Let G be a group and Y a left G–space. Let $\xi(\cdot)$ be a wide sense homogeneous random function on Y; $\xi_0(y) := \xi(y) - \mathbf{E}\xi(y)$, $y \in Y$; we denote by U the translation representation of $\xi_0(\cdot)$ in H_{ξ_0} and let $\varphi_{x,y}(g) := \mathbf{E}\xi_0(gx)\overline{\xi_0(xy)} = \langle U_g\xi_0(x), \xi_0(y)\rangle$. Since $\varphi_{x,y}$ is a matrix coefficient of U, $\varphi_{x,y} \in W$, $x, y \in Y$. We shall say that the random function $\xi(\cdot)$ *has FD–continuous (FD–discrete) spectrum* if all $\varphi_{x,y}$, $x, y \in Y$, are flight–functions (almost periodic functions). The following statement can be easily deduced from the results of Subsect. 10.1.

(10.I). *A homogeneous r.f. $\xi(\cdot)$ has FD–continuous (resp. FD–discrete) spectrum iff the spectrum of U is of the corresponding type.*

(10.J). *Any continuous homogeneous r.f. $\xi(\cdot)$ has a unique decomposition $\xi(\cdot) = \xi_c(\cdot) + \xi_d(\cdot)$ where $\xi_c(\cdot)$ and $\xi_d(\cdot)$ are homogeneous random functions with FD–continuous and FD–discrete spectrum, respectively, and $\mathbf{E}\xi_c(x)\overline{\xi_d(y)} = 0$, $\mathbf{E}\xi_d(y) \equiv 0$, $x, y \in Y$.*

Example 10.2. The decomposition $\xi = \xi_c + \xi_d$ of a homogeneous random field $\xi(\cdot)$ on a group G corresponds to the decomposition of its correlation function $R(g) = R_c(g) + R_d(g)$ into components with FD–continuous and FD–discrete spectrum. Thus, $\xi(\cdot)$ has FD–continuous (FD–discrete) spectrum iff the spectrum of its correlation function is of the corresponding kind.

Example 10.3. The decomposition $\xi = \xi_c + \xi_d$ of a homogeneous continuous random field $\xi(\cdot)$ on a homogeneous space $Y = G/K$ corresponds to the decomposition of its correlation function $R_0(y), R_0(y) = R_{0c}(y) + R_{0d}(y)$, where R_{0c} and R_{0d} are continuous positive definite spherical functions on G with F–continuous and F–discrete spectrum, respectively.

Example 10.4. A homogeneous generalized random field $\xi(\varphi)$, $\varphi \in K(\mathbb{R}^m)$, has F–continuous (F–discrete) spectrum iff its spectral measure is continuous (discrete).

§ 11. Relative Entropy on Random σ–Algebras

11.1. Random σ–algebras and their graphs. Let $(W, \mathcal{W})$ be a measurable space and M a measure on $\mathcal{W}$. Let $\mathcal{C}$ be a σ–subalgebra of $\mathcal{W}$ and let ρ_M be the usual M–semimetrics in $\mathcal{C}$: $\rho_M(A, B) = \arctan M(A \Delta B)$, $A, B \in \mathcal{C}$. If the metric space $(\mathcal{C}, \rho_M)$ is separable, $\mathcal{C}$ is said to be *countably generated* (with respect to M); if a class of sets Ψ_0 is dense in $(\mathcal{C}, \rho_M)$, we say that Ψ_0 *generates* $\mathcal{C}$ or that Ψ_0 is a *generating class* of $\mathcal{C}$ (with respect to M).

Let $(S, \mathcal{S})$ be a measurable space. For any $\Lambda \in \mathcal{W} \times \mathcal{S}$ and $s \in S$ we put $\Lambda_s := \{w : w \in W, (w, s) \in \Lambda\}$. Let $\mathcal{C}(\cdot) := \{\mathcal{C}(s), s \in S\}$ be a family of σ–subalgebras of $\mathcal{W}$. The σ–algebra $\widetilde{\mathcal{C}}$ consisting of all sets Λ such that $\Lambda \in \mathcal{W} \times \mathcal{S}$ and $\Lambda_s \in \mathcal{C}(s)$, $s \in S$, is called the *graph* of $\mathcal{C}(\cdot)$. A class of sets $\Psi \in \mathcal{W} \times \mathcal{S}$ will be called a *generating class* of $\mathcal{C}(\cdot)$ (with respect to M) if for any $s \in S$ the class $\{\Lambda_s, \Lambda \in \Psi\}$ generates (with respect to M) the σ–algebra $\mathcal{C}(s)$; obviously, $\Psi \subset \widetilde{\mathcal{C}}$. The family $\mathcal{C}(\cdot)$ is called a *random σ–algebra* in $\mathcal{W}$ (with respect to M) if it has a countable generating class Ψ (clearly, Ψ can be assumed to be an algebra). If $\mathcal{C}_i(\cdot)$, $i = 1, 2$, are random σ–algebras and $\mathcal{C}_1(s) \subset \mathcal{C}_2(s)$, $s \in S$, then $\widetilde{\mathcal{C}}_1 \subset \widetilde{\mathcal{C}}_2$. It is easy to check that if $\mathcal{C}_\alpha(\cdot)$, $\alpha \in A$, is a countable set of random σ–algebras, then the families $\mathcal{C}^*(\cdot) = \vee_{\alpha \in A} \mathcal{C}(\cdot)$

and $\mathcal{C}_*(\cdot) = \wedge_{\alpha\in A}\mathcal{C}_\alpha(\cdot)$ are random σ–algebras, and $\widetilde{\mathcal{C}}^* = \vee_\alpha\widetilde{\mathcal{C}}_\alpha$, $\widetilde{\mathcal{C}}_* = \wedge_{\alpha\in A}\widetilde{\mathcal{C}}_\alpha$.

Let $\mathcal{C}^i$ be countably generated σ–subalgebras of $\mathcal{W}$, $S^i \in \mathcal{S}$, $i = 1,\dots,m$, $m < \infty$, $\bigcup_{i=1}^m S^i = S$, $S^i \cap S^j = \emptyset$ if $i \neq j$. The family $\mathcal{C}(s) := \mathcal{C}^i$ if $s \in S^i$ $(i = 1,\dots,m)$ is a "*simple*" random σ–algebra. If Ψ_0^i is a countable generating class for $\mathcal{C}^i$, then

$$\Psi := \{\Lambda : \Lambda = \bigcup_{i=1}^m (D^i \times S^i), \quad D^i \in \Psi_0^i, \quad i = 1,\dots,m\}$$

is a countable generating class for $\mathcal{C}(\cdot)$; the graph $\widetilde{\mathcal{C}}$ can be described as follows: $\widetilde{\mathcal{C}}_{W\times S^i} = \mathcal{C}^i \times \mathcal{S}_{S^i}$ where $\widetilde{\mathcal{C}}_{W\times S^i}$ and $\mathcal{S}_{S^i}$ are the restrictions of $\widetilde{\mathcal{C}}$ and $\mathcal{S}$ to $W \times S^i$ and S^i, respectively.

11.2. Radon–Nikodym derivatives and relative entropy on random σ–algebras and graphs. Let $(S, \mathcal{S}, \lambda)$ be a probability space, let $\{\mathcal{C}(s),\ s \in S\}$ be a random σ–algebra in $\mathcal{W}$ with respect to M, and let $\widetilde{\mathcal{C}}$ be its graph; $\widetilde{M} := M \otimes \lambda$. We suppose that $\widetilde{M}_{\widetilde{\mathcal{C}}}$, the restriction of $\widetilde{M}$ to $\widetilde{\mathcal{C}}$, is σ–finite. Let P be a probability measure on $\mathcal{W}$; $\widetilde{P} := P \times \lambda$.

Two functions on $W \times S$, ψ and φ, will be called *standard modifications* of each other if for λ–almost all $s \in S$ we have $\varphi(\cdot, s) = \psi(\cdot, s)$ M–almost everywhere. In the sequel we do not differ between standard modifications, and we shall consider $(\mathcal{W}\times\mathcal{S})$–measurable standard modifications only, if such exist.

(11.A). *$\widetilde{P}_{\widetilde{\mathcal{C}}} \ll \widetilde{M}_{\widetilde{\mathcal{C}}}$ iff $P_{\mathcal{C}(s)} \ll M_{\mathcal{C}(s)}$ for λ-almost all $s \in \mathcal{S}$; in this case the function $\psi(\omega, s) := \frac{dP_{\mathcal{C}(s)}}{dM_{\mathcal{C}(s)}}(\omega)$ has a $\widetilde{\mathcal{C}}$-measurable standard modification and*

$$\frac{dP_{\mathcal{C}(s)}}{dM_{\mathcal{C}(s)}}(\omega) = \frac{d\widetilde{P}_{\widetilde{\mathcal{C}}}}{d\widetilde{M}_{\widetilde{\mathcal{C}}}}(\omega, s) \tag{11.1}$$

for $\widetilde{M}$-almost all $(\omega, s) \in W \times S$.

Proof. Sufficiency. Let $P_{\mathcal{C}(s)} \ll M_{\mathcal{C}(s)}$ for λ–almost all $s \in S$. Let $\Lambda \in \widetilde{\mathcal{C}}$ and $\widetilde{M}(\Lambda) = 0$; according to Fubini's theorem, $M(\Lambda_s) = 0$ and thus $P(\Lambda_s) = 0$ for almost all $s \in S$. It follows at once that $P(\Lambda) = 0$; therefore $\widetilde{P}_{\widetilde{\mathcal{C}}} \ll \widetilde{M}_{\widetilde{\mathcal{C}}}$.

Necessity. Let $\widetilde{P}_{\widetilde{\mathcal{C}}} \ll \widetilde{M}_{\widetilde{\mathcal{C}}}$, $\Gamma \in \mathcal{S}$, and $\Lambda \in \Psi$ where Ψ is a countable algebra generating $\mathcal{C}(\cdot)$; then $\Lambda(\Gamma) := \Lambda \cap (\Omega \times \Gamma) \in \widetilde{\mathcal{C}}$ and $\widetilde{P}(\Lambda(\Gamma)) = = \int_\Gamma P(\Lambda_s)\lambda(ds)$. On the other hand,

$$\widetilde{P}(\Lambda(\Gamma)) = \int_{\Lambda(\Gamma)} \frac{d\widetilde{P}_{\widetilde{\mathcal{C}}}}{d\widetilde{M}_{\widetilde{\mathcal{C}}}} d\widetilde{M} = \int_\Gamma \left(\int_{\Lambda_s} \frac{d\widetilde{P}_{\widetilde{\mathcal{C}}}}{dM_{\widetilde{\mathcal{C}}}} dM \right) d\lambda.$$

Since Γ is chosen arbitrarily,

$$P(\Lambda_s) = \int_{\Lambda_s} \frac{d\widetilde{P}_{\widetilde{\mathcal{C}}}}{dM_{\widetilde{\mathcal{C}}}}(w,s) M(dw)$$

for $s \in S \setminus \Gamma_\Lambda$ with $\lambda(\Gamma_\Lambda) = 0$. If Λ runs over Ψ then Λ_s runs over an algebra generating $\mathcal{C}(s)$ and we obtain for $s \in S \setminus \bigcup_{\Lambda \in \Psi} \Gamma_\Lambda$ the relation $P(D) = \int_D \frac{d\widetilde{P}_{\widetilde{\mathcal{C}}}}{dM_{\widetilde{\mathcal{C}}}}(w,s) M(dw)$ for all $D \in \mathcal{C}(s)$, and thus $P_{\mathcal{C}(s)} \ll M_{\mathcal{C}(s)}$ and (11.1) holds.

□

From (11.A) and Fubini's theorem we immediately obtain the following statement, concerning the relative entropy (see Ch. 8, § 4).

(11.B). *If* $-\infty < H_{\widetilde{P}:\widetilde{M}}(\widetilde{\mathcal{C}}) < \infty$, *then* $H_{\widetilde{P}:\widetilde{M}}(\widetilde{\mathcal{C}}) = \mathbf{E}_\lambda H_{P:M}(\mathcal{C}(s))$.

With the aid of Statements (11.A) and (11. B), Theorems 8.4.1 and 8.4.2 imply obvious generalizations to monotone countable nets of random σ–algebras; $H_{P:M}(\mathcal{C}_\alpha)$, $H_{P:M}(\mathcal{C}^*)$, $H_{P:M}(\mathcal{C}_*)$ are to be replaced, respectively, by $H_{\widetilde{P}:\widetilde{M}}(\widetilde{\mathcal{C}}_\alpha)$, $H_{\widetilde{P}:\widetilde{M}}(\widetilde{\mathcal{C}}^*)$ and $H_{\widetilde{P}:\widetilde{M}}(\widetilde{\mathcal{C}}_*)$; the functions $\rho_{\mathcal{C}_\alpha}(s)(w)$, $\rho_{\mathcal{C}^*(s)}(w)$ and $\rho_{\mathcal{C}_*(s)}(w)$ figure instead of $\rho_{\mathcal{C}_\alpha}(w)$, $\rho_{\mathcal{C}^*}(w)$ and $\rho_{\mathcal{C}_*}(w)$.

§ 12. The Banach Convergence Principle

Let $(\Omega, \mathcal{F}, m)$ and $(\Omega_1, \mathcal{F}_1, m_1)$ be σ–finite measure spaces. Consider an arbitrary Banach space B and the space $\widetilde{F}_B = \widetilde{F}_B(\Omega, \mathcal{F}, m)$ of all B–valued functions on Ω with the semimetric

$$\rho(\varphi, \psi) = \inf_{\gamma>0} \arctan[\gamma + m(\{\omega : \|\varphi(\omega) - \psi(\omega)\| > \gamma\})]$$

(convergence in $\widetilde{F}_B$ coincides with convergence with respect to m).

Let $\{T_n\}$ be a sequence of operators from $L^{\alpha}_{B}\ (\Omega_1, \mathcal{F}_1, m_1)$ $(\alpha \geq 1)$ into $\widetilde{F}_B(\Omega, \mathcal{F}, m)$.

(12.A). *Suppose that the following conditions are fulfilled:*
a) *all T_n are continuous;*
b) *all T_n are semiadditive, i.e.*

$$\|T_n(\varphi + \psi)\|_B \leq \|T_n(\varphi)\|_B + \|T_n(\psi)\|_B, \quad \varphi, \psi \in L^{\alpha}_{B}(\Omega_1, \mathcal{F}, m_1);$$

c) *all T_n are semihomogeneous, i.e.*

$$\|T_n(\alpha\varphi)\|_B = |\alpha| \|T_n(\varphi)\|_B;$$

d) $\sup_n \|T_n\varphi(\omega)\|_B < \infty \quad m\text{–}a.e., \quad \varphi \in L^{\alpha}_{B}\ (\Omega_1, \mathcal{F}_1, m_1);$

e) $\lim_{n\to\infty} T_n\varphi(\omega)$ *exists* (12.1)

for any φ from a subset $S \subset L^{\alpha}_{B}\ (\Omega_1, \mathcal{F}_1, m_1)$.
Then (12.1) *is true for any $\varphi \in [S]_{L_B}$ (the subspace of $L^{\alpha}_{B}\ (\Omega_1, \mathcal{F}_1, m_1)$ spanned by S).*

This is a version of the well–known Banach Convergence Principle; its proof almost literally follows the proof of Theorem 4.11.3 in Dunford and Schwartz [1].

§ 13. Directions and Nets

13.1 Directions, nets and limits. A partially ordered set N is called a *direction* (or *directed set*) if any two elements n_1 and n_2 of N have a common majorant n_3: $n_1 \le n_3$, $n_2 \le n_3$. Let D be some set; any subset $\{d_n, n \in N\}$ ($\subset D$) is called a *net* in D. If D is a topological space and $\{d_n, n \in N\}$ is a net in D and $d \in D$, the notation $\lim_{n \in N} d_n = d$ means that for any neighbourhood U of d there exists an element $n_U \in N$ such that $d_n \in U$ if $n > n_U$; d is the *limit* of the net $\{d_n, n \in N\}$; any net has at most one limit.

13.2. Separable and cofinally separable nets of sets. Let $(X, \mathfrak{B})$ be a measurable semigroup and μ a σ–finite measure on $\mathfrak{B}$. Let $\mathcal{A} = \{A_n, n \in N\}$ be a net of nonnull integrable sets. Consider the mapping $\sigma_{\mathcal{A}} : n \mapsto A_n$ from N into the space $\mathfrak{S}$ of all nonnull integrable sets in $\mathfrak{B}$ with the metric $\rho(A, B) = \mu(A \Delta B)$ and the induced topology $\mathcal{T}_{\mathcal{A}}$ in N. Note that the mapping $n \mapsto \mu(A_n)$ is continuous with respect to this topology, since $|\mu(A_n) - \mu(A_m)| \le$ $\le \mu(A_n \Delta A_m)$.

We say that the net $\mathcal{A}$ is *separable* if N contains a countable dense subset N_0. Any such N_0 is called a *separability set*. We say that $\mathcal{A}$ is *cofinally separable* if there exist a separability set N_0 such that for any $n \in N$ the set $N_0 \cap \{k : k > n\}$ is dense in $\{k : k > n\}$; in this case N_0 is called a *c–separability set*.

Of course, if N is countable, any net $\{A_n, n \in N\}$ is cofinally separable.

A cofinally separable set is separable. The converse is also true, if any set $\{k : k > n\}$ is a $\mathcal{T}_{\mathcal{A}}$–open set.

We state several simple propositions giving wide conditions under which a net $\mathcal{A}$ is separable or cofinally separable.

(13.A). *Let N be a topological space and N_0 a countable dense subset of N. If the mapping $\sigma_{\mathcal{A}}$ is continuous, then the net $\mathcal{A}$ is separable and N_0 is a separability set. If, moreover, all subsets $\{k : k > n\}$, $k \in N$, are open, then $\mathcal{A}$ is cofinally separable and N_0 is a c–separability set.*

The proof is evident.

(13.B). *Let X be a second countable topological semigroup and $\mathfrak{B}$ the σ–algebra of Borel sets. Then every net $\mathcal{A}$ is separable.*

Proof. This follows from the separability of the space $\mathfrak{B}$.

□

(13.C). *Every monotone net $\mathcal{A}$ is cofinally separable.*

Proof. Suppose $\mathcal{A}$ is increasing. Consider the set $D = \{\mu(A_n),\ n \in N\}$ in $\mathbb{R}_+$ with the natural order. It is clear that the mapping $\varphi : n \mapsto \mu(A_n)$ from N to D is order preserving and that the φ–induced topology coincides with $\mathcal{T}_{\mathcal{A}}$. Therefore, if a set D_0 is countable and dense in D, then $N_0 = \varphi^{-1}(D_0)$ is a c–separability set in N.

□

The introduction of the notions: "separable" and "cofinally separable" nets is justified by the following useful statement.

Recall that for any completely measurable uniformly bounded representation S of X in $L^\alpha(\Omega, \mathcal{F}, m)$ and for any $f \in L^\alpha(\Omega, \mathcal{F}, m)$ we denote

$$\widehat{f}_n(\omega) = \frac{1}{\mu(A_n)} \int_{A_n} (fS_x)(\omega)\mu(dx), \quad f^*(\omega) = \sup_{n \in N} |\widehat{f}_n(\omega)|.$$

(13.D). *(i) Suppose that $\mathcal{A}$ is a separable net and let N_0 be some separability set in N. Then for any S and f $(\in L^\alpha(\Omega))$, $f^*(\omega) = \sup_{n \in N_0} |\widehat{f}_n(\omega)|$ and therefore the function $f^*(\omega)$ is $\mathcal{F}$–measurable.*

(ii) Let $\mathcal{A}$ be cofinally separable and let N_0 be a c–separability set in N. If $\lim_{n \in N_0} \widehat{f}_n(\omega)$ exists m–a.e., then $\lim_{n \in N} \widehat{f}_n(\omega)$ exists m–a.e., and both limits coincide a.e.

Proof.

(*i*) let Λ be the set of all $\omega (\in \Omega)$ such that $\int_{A_n} |(fS_x)(\omega)|\mu(dx) < \infty$ for all $n \in N_0$. According to App. (9.G'), we have $m(\Omega \backslash \Lambda) =$

$= 0$. It remains to note that since for any $\omega \in \Lambda$ the mappings $A \mapsto \int_A f S_x(\omega)\mu(dx)$ and $A \mapsto \mu(A)$ are continuous, the mapping $n \mapsto \widehat{f}_n(\omega)$ is also continuous, and therefore the set $\{\widehat{f}_n(\omega),\, n \in N_0\}$ is dense in $\{\widehat{f}_n(\omega),\, n \in N\}$.

(*ii*) Let Λ' be the subset in Ω for which $\lim_{n\in N_0} \widehat{f}_n(\omega) =: \widehat{f}(\omega)$ exists; we have $m(\Omega\backslash\Lambda') = 0$. Fix some $\varepsilon > 0$. Let $\omega \in \Lambda'$ and let $|\widehat{f}_n(\omega) - -\widehat{f}(\omega)| < \frac{\varepsilon}{2}$ for all $n\,(\in N_0)$, $n > n_\varepsilon$; for any $n' > n_\varepsilon$ there exists an $n \in N_0$, $n > n_\varepsilon$, such that $|\widehat{f}_n(\omega) - \widehat{f}_{n'}(\omega)| < \frac{\varepsilon}{2}$.

Therefore $|\widehat{f}_{n'}(\omega) - \widehat{f}(\omega)| < \varepsilon$.

□

13.3. Uniformly separable families of nets. Consider a family of nets $\mathcal{A}^{(\gamma)} = \{A_n^{(\gamma)},\, n \in N^{(\gamma)}\}$, $\gamma \in \Gamma$, where $N^{(\gamma)}$ are subdirections of a direction in N with topology $\mathcal{T}$ generated by the sets $\bigcap_{\gamma\in\Gamma} O^{(\gamma)}$, $O^{(\gamma)} \in \mathcal{T}_{\mathcal{A}^{(\gamma)}}$.

Replace in the definition of separable nets (Subsect. 13.2) the topology $\mathcal{T}_{\mathcal{A}}$ by $\mathcal{T}$. Then we get the definitions of a *uniformly separable* family of nets $\mathcal{A}^{(\gamma)}$, of a *uniform cofinally separable* family $\mathcal{A}^{(\gamma)}$, and of a *uniform c-separability set* N_0.

We also present "uniform" versions of Propositions (13.A) and (13.D). The first one is trivial.

(13.E). *Let N be a topological space and N_0 a countable dense subset of N. If the mappings $\sigma_{\mathcal{A}^{(\gamma)}}$ are uniformly continuous (with respect to $\gamma \in \Gamma$), then the family $\mathcal{A}^{(\gamma)}$, $\gamma \in \Gamma$, is uniformly separable and N_0 is a uniform separability set; if all subsets $\{k : k > n\}$, $n \in N$, are open, then $\mathcal{A}^{(\gamma)}$ is uniformly cofinally separable and N_0 is a uniform c-separability set.*

Let S be a completely measurable representation of X in $L^\alpha(\Omega, \mathcal{F}, m)$; for any f we denote

$$\widehat{f_n^{(\gamma)}}(\omega) = \frac{1}{\mu(A_n^{(\gamma)})} \int_{A_n^{(\gamma)}} (fS_x)(\omega)\mu(dx); \quad f_n^*(\omega) := \sup_{\gamma \in \Gamma} |\widehat{f_n^{(\gamma)}}(\omega)|;$$

$$f^{**}(\omega) := \sup_{\substack{n \in N \\ \gamma \in \Gamma}} |\widehat{f_n^{(\gamma)}}(\omega)| = \sup_{n \in N} f_n^*(\omega).$$

(13.F). *Suppose $\mathcal{A}^{(\gamma)}$ is a uniformly separable family of nets and let N_0 be some uniform separability set in N.*

(i) For any $f \in L^\alpha$, we have $f^{**}(\omega) = \sup_{n \in N_0} f_n^*(\omega)$, and thus, if all functions f_n^* are measurable, then so is f^{**}.

Let, moreover, N_0 be a uniform c–separability set. Then

(ii) if $\lim_{n \in N_0} f_n^*(\omega)$ exists m–a.e., then $\lim_{n \in N} f_n^*(\omega)$ also exists m–a.e., and both limits coincide.

(iii) Suppose there exist functions $\widehat{f^{(\gamma)}}(\omega)$ such that

$$\lim_{n \in N_0} \sup_{\gamma \in \Gamma} |\widehat{f_n^{(\gamma)}}(\omega) - \widehat{f^{(\gamma)}}(\omega)| = 0 \qquad m\text{–}a.e.$$

Then $\lim_{n \in N} \sup_{\gamma \in \Gamma} |\widehat{f_n^{(\gamma)}}(\omega) - \widehat{f^{(\gamma)}}(\omega)| = 0$ m–a.e., too.

Proof. Reasoning as in the proof of Proposition (13.D) *(i)* we deduce that the mappings $K^{(\gamma)} : n \mapsto \widehat{f_n^{(\gamma)}}(\omega)$ from N into $\mathbb{C}$ are uniformly continuous with respect to γ. Therefore, if $\varepsilon > 0$ and n_1 and n_2 are close enough, $|f_{n_1}^*(\omega) - f_{n_2}^*(\omega)| \le \sup_{\gamma \in \Gamma} |\widehat{f_{n_1}^{(\gamma)}}(\omega) - \widehat{f_{n_2}^{(\gamma)}}(\omega)| < \varepsilon$ for all $\omega \in \Lambda$ with $m(\Omega \backslash \Lambda) = 0$. This implies that the set $\{f_n^*(\omega),\, n \in N_0\}$ is dense in $\{f_n^*(\omega),\, n \in N\}$, $\omega \in \Lambda$. So statement *(i)* is proved. To prove *(ii)* we repeat the proof of Proposition (13.*D*) *(ii)* with f_n^* instead of f_n; the proof of *(iii)* is similar.

□

Let X be a locally compact topological semigroup and let μ be a Borel measure on X.

We shall again consider the set $\mathcal{S}$ of all nonnull integrable Borel sets in X with the semimetric

$$\rho'(A,B) = \int \left| \frac{\mathcal{X}_A(x)}{\mu(A)} - \frac{\mathcal{X}_B(x)}{\mu(B)} \right| \mu(dx).$$

Note that $\rho'(A,B) \leq \frac{\mu(A \triangle B)}{\max(\mu(A),\mu(B))}$.

Let $\mathcal{A} = \{A_n, n \in N\}$ be a net in $\mathcal{S}$. We say that $\mathcal{A}' = \{A_n, n \in N'\}$ is an *ε–subnet* of $\mathcal{A}$ if N' is a subdirection of N and for any $n \in N$ there exists an $n' \in N'$, $n' > n$, such that $\rho'(A_n, A_{n'}) < \varepsilon$.

Consider a net $\{A_n, n \in N\}$ and a completely measurable dynamical system $T = \{T_x, \in X\}$ in $(\Omega, \mathcal{F}, m)$. For $f \in L^\alpha(\Omega, \mathcal{F}, m)$ we denote $\widehat{f}_n(\omega) := \frac{1}{\mu(A_n)} \int_{A_n} f(T_x\omega)\mu(dx)$.

(13.G). *Let $\mathcal{A} \in MAX_\alpha(T|\mu)$ for all T. Suppose that $\mathcal{A}$ is cofinally countable and that for any $\varepsilon > 0$ there exists a pointwise averaging ε–subnet $\mathcal{A}^{(\varepsilon)}$ of $\mathcal{A}$. Then $\mathcal{A}$ is also pointwise averaging.*

Proof. Let $f \in L^\alpha$ and $\sup_{\omega \in \Omega} |f(\omega)| =: C < \infty$; then

$$|\widehat{f}_n(\omega) - \widehat{f}_m(\omega)| \leq C\rho'(A_n, A_m).$$

Let us consider the subdirection $N^{(\varepsilon_k)}$ where $\varepsilon_k = \frac{1}{2Ck}$, and let Λ_k be the set of all $\omega \in \Omega$ for which

$$\lim_{n \in N^{(\varepsilon_k)}} \widehat{f}_n(\omega) = \mathbf{M}(f|T).$$

Let $\omega \in \Lambda_k$; there exists an $n_k \in N^{(\varepsilon_k)}$ such that

$$|\widehat{f}_m(\omega) - \mathbf{M}(f|T)| < \frac{1}{2k} \quad \text{if } m \geq n_k,\ m \in N^{(\varepsilon_k)}. \tag{13.1}$$

Since $\mathcal{A}^{(\varepsilon_k)}$ is a $\frac{1}{2Ck}$–subnet, for $n \geq n_k$ there exists an $m \in N^{(\varepsilon_k)}$, $m > n_k$, such that

$$|\widehat{f}_n(\omega) - \widehat{f}_m(\omega)| < \frac{1}{2k}. \tag{13.2}$$

Inequalities (13.1) and (13.2) show that $|\widehat{f}_n(\omega) - \mathbf{M}(f|T)| < \frac{1}{k}$ for all $n > n_k$, $n \in N$. Thus

$$\lim_{n \in N} \widehat{f}_n(\omega) = \mathbf{M}(f|T) \tag{13.3}$$

for any $\omega \in \Lambda_0 = \bigcap_{k=1}^{\infty} \Lambda_k$. Consequently, $m(\Omega \setminus \Lambda_0) = 0$.

We have proved that (13.3) holds a.e. for all bounded functions in L^α. Since such functions form a dense subset in L^α, it remains to apply Theorem (12.A) to some c-separability subnet $N_0 \subset N$ and then use (13.D).

□

§ 14. Correspondence between "Left" and "Right" Objects and Conditions

The following table shows the way in which the "left" and "right" versions of various notions (objects) and conditions considered in the book are logically connected and combined in statements. We devide all such versions into two classes (called the "left" and "right" classes according to the type of dynamical sytems contained in each of them). The "left" and "right" versions of various objects are usually combined within each of these classes. An example is: to any left dynamical system a right isomteric representation in L^α is assoriated ("left–class" objects are combined). Another example: the orbital functions of a contracting left representation of a right–amenable semigroup in a uniformly convex Banach space are right averageable (this is the "right" version of Theorem 1.5.6 where the "right–class" objects are involved; the "left" version of this theorem can be obtained by replacing the objects of the "left class" by the corresponding objects of the "right class").

"Left" class of objects and conditions	*"Right" class of objects and conditions*
left dynamical systems	*right* dynamical systems
right representations of semigroups	*left* representations of semigroups
left translations in a semigroup	*right* translations in a semigroup
left translations of functions on a semigroup	*right* translations of functions on a semigroup
homogeneous random functions	homogeneous random functions

on a *right* X–space	on a *left* X–space
random fields homogeneous with respect to *right* translations	random fields homogeneous with respect to *left* translations
left amenable semigroups	*right* amenable semigroups
left averageable functions	*right* averageable functions
left ergodic (Følner) nets of sets or measures	*right* ergodic (Følner) nets of sets or measures
left mean value	*right* mean value
left averaging nets of measures or sets	*right* averaging nets of measures or sets
left regular nets of sets	*right* regular nets of sets
right relatively semiinvariant measures on semigroups	*left* relatively semiinvariant measures on semigroups
left version of condition (A_0)	*right* version of condition (A_0)
right uniformaly continuous functions	*leftt* uniformaly continuous functions
left–regular semigroup	*right*–regular semigroup

References

Adams J.F.

1. *Lectures on Lie groups.* W.A.Benjamin, New York, Amsterdam, 1969.

1a. *Lectures on Lie Groups.* – Moscow: Nauka, 1979, p.144 (Russian translation).

Akhiezer N.I., Glazman I.M.

1. *Theory of Linear Operators in Hilbert Space.*
Vol.I. Frederick Ungar Publishing Co., New York, 1961.
Vol.II. Frederick Ungar Publishing Co., New York, 1963.

Akilov G.P., Kantorovich L.V.

1. *Functional Analysis.* Pergamon Press, Oxford–Elmsford, New York, 1982.

Akcoglu M.A., Krengel U.

1. *Ergodic theorems for superadditive process.* – C.R. Math. Acad., Soc. R. Can., v.2, p.175–179.
2. *Ergodic theorems for superadditive process.* – J. Reine Angew. Math., 1981, v.323, p.53–67.

Akcoglu M., del Junco A.

1. *Convergence of averages of point transformations.* – Proc. Amer. Math. Soc., 1975, v.49, p.265–266.

Akcoglu M., Sucheston L.

1. *On operator convergence in Hilbert space and in Lebesgue space.* – Period. Math. Hungar., 1972, v.2, p.235–244.
2. *On weak and strong convergence of positive contractions in L_p spaces.* – Bull. Amer. Math. Soc., 1975, v.81, p.105–106.

Alaoglu L., Birkhoff G.

1. *General ergodic theorems.* – Proc. Math. Acad. Sci. USA, 1939, v.25, p.628–630.
2. *General ergodic theorems.* – Ann. of Math., 1940, v.41, p.293–309.

Aleksandrian R.A., Mirzahanian E.A.

1. *General Topology.* Moscow, 1979, Vysshaja shkola (in Russian).

Alfsen E.M.
1. *Some Covering Theorems of Vitali Type.* – Math. Annalen, 1965, v.159, No.3, p.203–215.

Aribaud F.
1. *Un théorème ergodique pour les espaces L^1.* – J. Funct. Analysis, 1970, v.7, p.395–411.

Arnold V.I., Krylov A.L.
1. *Uniform distribution of points on a sphere and some ergodic properties of solutions of linear ordinary differential equations in the complex region.* – Sov. Math., Doklady, 1963, v.4, No.1, p.1–5.

Aronszajn N.
1. *Theory of reproducing kernels.* – Trans. of Amer. Math. Soc., 1950, v.68, No.3, p.337–404.

Asplund E. – see Namioka I.

Auslander L., Green L.
1. *G-induced flows.* – Amer. J. Math., 1966, v.88, p.43–60.

Auslander L., Green L., Hahn F.
Flows on homogeneous spaces. Annals of Mathematics Studies, No.53. – Princeton University Press, Princeton, N. J., 1963.

Bagget L., Taylor K.
1. *Riemann–Lebesgue subsets of R^n and representations which vanish at infinity.* – J. Funct. Analysis, 1978, v.28, p.168–181.

Baxter G.
1. *An ergodic theorem with weighted averages.* – J. Math. Mech., 1964, v.13, No.3.

Beck A.
1. *Une loi forte des grandes nombres dans des espaces de Banach uniformément convexes.* – Ann. Inst. H. Poincaré, 1958, v.16, p.35–45.
2. *A convexity condition in Banach spaces and the strong law of large numbers.* – Proc. Amer. Math. Soc., 1962, v.13.
3. *On the strong law of large numbers. Ergodic Theory,* ed. by F.B.Wright, N.Y.–London Academic Press, 1963, p.21–54.

Beck A., Warren P.
1. *Strong laws of large numbers for weakly orthogonal sequences of Banach space-valued random variables.* – Ann. Probab., 1979, v.2,

No.5, p.918–925.

Beck A., Schwartz J.T.

1. *A vector-valued random ergodic theorem.* – Proc. Amer. Math. Soc., 1957, v.8, p.1049–1059.

Bellow A., Jones R., Rosenblatt J.

1. *Convergence for moving averages.* – Ergodic theory and dynamical systems, 1990, v.10, p.43–62

Bellow A., Losert V.

1. *On sequences of density zero in ergodic theory.* – Contemporary Mathematics, 1984, v.28, p.49–60.

Bergelson V., Rosenblatt J.

1. *Joint ergodicity for group actions.* – Ergodic Theory and Dynam. Systems, 1988, v.8, p.351–364.
2. *Mixing actions of groups.* – 1988, Ill. J. Math., v.32, No.1, p.65–80.

Bewley T.

1. *Extension of the Birkhoff and von Neumann ergodic theorems to semigroup actions.* – Ann. Inst. H. Poincaré, 1971, v.7, No.4, p.283–291.

Birkhoff G.D.

1. Proof of the ergodic theorem. – Proc. Nat. Acad. Sci. USA , 1931, v.17, p.656–660.

Birkhoff G.

1. *An ergodic theorem for general semi-groups.* – Proc. Nat. Acad. Sci. USA, 1939, v.25, p.625–627.
2. *The mean ergodic theorem.* – Duke Math. J., 1939, v.5, p.19–20.

Birkhoff G., Kampé de Fériet J.

1. *Kinematics of homogeneous turbulence.* – J. Math. Mech., 1962, v.11, No.3, p.319–340.

Blanc–Lapierre A., Brard R.

La loi forte des grands nombres pour les fonctions aléatoires stationnaires continues. C. R. Acad. Sci. Paris, 1945, v.220, p.134–136.

Blanc–Lapierre A., Fortet R.

1. *Les functions aléatoires stationnaires et la loi des grands nombres.* – Bull. Soc. Math. France, 1946, v.74, p.102–115.

Blanc–Lapierre A., Tortrat A.

1. *Sur la loi forte des grands nombres.* – C.R. Acad. Sci. Paris, 1968, v.267A, p.740–743.

Blum J.R., Cogburn R.

1. *On ergodic sequences of measures.* – Proc. Amer. Math. Soc., 1975, v.51, p.359–365.

Blum J.R., Eisenberg B.

1. *Generalized summing sequences and the mean ergodic theorem.* – Proc. Amer. Math. Soc., 1974, v.42, p.423–429.
2. *Conditions for metric transitivity for stationary Gaussian processes on groups.* – The Ann. of Math. Statistics, 1972, v.43, No.5, p.1737–1741.

Blum J.R., Eisenberg B., Hahn L.–S.

1. *Ergodic theory and the measure of sets in the Bohr group.* – Acta Scientiarum Mathem., 1973, v.34, p.17–24.

Blum J.R., Hanson D.L.

1. *On the mean ergodic theorem for subsequences.* – Bull. Amer. Math. Soc., 1960, v.66, p.308–311.

Blum J.R., Reich J.

1. *Pointwise ergodic theorems in l.c.a. groups.* – Pac. J. M., 1982, v.103, p.301-306.

Boclé J.

1. *Sur l'existence d'une measure invariante par un groupe de transformations.* – C.R. Acad. Sci., Paris, 1958, v.247, p.798–800.
2. *Théorèmes de dérivation globale dans les groupes topologiques localement compacts.* – C.R. Ac. Sci., Paris, 1959, v.248, p.2063–2065.
3. *Sur la théorie ergodique.* – Ann. Inst. Fourier, 1960, v.10, p.1–45.

Borel A., Walach N.R.

1. *Continuous cohomology discrete subgroups and representations of reductive groups.* – Ann. of Math. Studies, v.94, Princeton University Press, Princeton, N.J., University of Tokyo Press, Tokyo, 1980.

Bourbaki N.

1. *Èlèments de mathèmatique. Premiére Partie.* Livre VI, Hermann, Paris.

2. *Éléments de mathématique. Première Partie.* Livre III, *Topologie Générale.* Hermann, Paris.

Bourgain J.

1. *Averages in the plane over convex curves and maximal operators.* – J. d'Analyse Math., 1986, v.47, p.69–85.

Brard R. – see Blanc–Lapierre A.

Bray G.

1. *Théorèmes ergodiques de convergence en moyenne.* – C.R. Acad. Sci., 1968, v.267, No.12, p.418–420.

Breiman L.

1. *The individual ergodic theorem of information theory.* – Ann. Math. Stat., 1960, v.21, p.809–810

Brunel A.

1. *Théorème ergodique punctuel pour un semigroupe commutative finiment engendré de contraction de L.* – Ann. Inst. H. Poincarè, Sec.B, 1973, v.9, p.327–343

Brunel A., Keane M.

1. *Ergodic theorems for operator sequences.* – Z. Wahrscheinlichkeitstheorie und verw. Geb., 1969, v.12, No.3, p.231–241.

Buldygin V.V.

Convergence of Random Elements in Topological Spaces. – Kiev, Naukova dumka, 1980, p.240 (in Russian).

Burkholder D.L.

1. *Maximal inequalities as necessary conditions for almost everywhere convergence.* – Z. Wahrscheinlichkeitstheorie und verw. Geb., 1964, v.3, p.75–88.

Calderón A.P.

1. *A general ergodic theorem.* – Ann. of Math., 1953, v.58, No.1, p.182–191.

2. *Ergodic theory and translation–invariant operators.* – Proc. Nat. Akad. Sci., 1968, v.59, p.349–353

Chacon R.V.

1. *Identification of the limit of operator averages.* – J. Math. and Mech., 1962, v.11, p.961–968.

2. *Convergence of operator averages.* – In: *Ergodic Theory*, ed. by F.B. Wright, New York–London Academic Press, 1963, p.89–120.

3. *Ordinary means imply recurrent means.* – Bull. Amer. Math. Soc., 1964, v.70, No.6, p.796–797.

Chacon R.V., McGrath S.A.

Estimates of positive contractions. – Pacif. J. Math., 1969, v.30, No.3, p.609–620.

Chacon R.V., Olsen J.

1. *Dominated estimates of positive contractions.* – Proc. Amer. Math. Soc., 1969, v.20, No.1 , p.266–271.

Chacon R.V., Ornstein D.S.

1. *A general ergodic theorem.* – Ill. Math. J., 1960, v.4, p.153–160.

Chatard J.

1. *Application des propriétés de moyenne d'un groupe localement compact á la théorie ergodique.* – Ann. Inst. Poincaré, 1970, VI.

Chatterji S.D.

1. *Vector–valued martingales and their applications.* – In: *Probability in Banach Spaces.* ed. by Dold A., Eckmann B. Lecture notes in Math., 1976, v.526.

Chevalley C.

1. *Théorie des Groupes de Lie.* – Tome I, 1948, Tome II, 1951, Tome III, 1958, Hermann, Paris.

Chintschin A.

1. *Zu Birkhoffs Lösung des Ergoden–problems.* –Math. Ann., 1932, v.107, p.485–488.

Chou C.

1. *Weakly almost periodic functions with zero mean.* – Bull. of the Amer. Math. Soc., 1974, v.80, No.2, p.297–299.

Cogburn R. – see Blum J.R.

Conze J.P.

1. *Entropie d'un groupe abélien de transformations.* – Z. Wahrscheinlichkeitstheorie und verw. Geb., 1972, v.25, p.11–30.

Conze J.P., Dang–Ngoc N.

1. *Ergodic theorem for non-commutative dynamical systems.* –Inventiones Math., 1978, v.46, No.1, p.1–15.

Cordoba A.

1. *On the Vitali covering properties of a differential base.* Appendix to Guzmán [1].

Cornfeld I.P., Fomin S.V., Sinai Ya.G.

1. *Ergodic theory.* – Grundlehren der Mathematischen Wissenschaften, v.245, Berlin–Heidelberg–New York, Springer, 1982.

Cotlar M.

1. *A unified theory of Hilbert transforms and ergodic theorems.* – Rev. Math. Cuyana, 1956, v.1, p.105–167.

Csiszar I.

1. *On generalized entropy.* – Studia Math. Sci. Hung., 1969, v.4, p.401–415.

Dang–Ngoc N. – see Conze J.P.

Day M.M.

1. *Ergodic theorems for Abelian semigroups.* – Trans. Amer. Math. Soc., 1942, v.51, No.2, p.399–412.
2. *Means for bounded functions and ergodicity of the bounded representations of semigroups.* – Trans. Amer. Math. Soc., 1950, v.69, p.276–291.
3. *Amenable semigroups.* – Ill. J. Math., 1957, v.1, p.509–544.
4. *Fixed-point theorems for compact convex sets.* – Illinois J.of Math., 1961, v.5, p.585–590.

Davis H.W.

1. *On the mean value of measurable almost periodic functions.* – Duke Math. J., 1967, v.37, p.201–214.
2. *Generalized almost periodicity in groups.* – Trans. Amer. Math. Soc., 1974, v.191, p.329–352.

DeLeeuw K., Glicksberg I.

1. *Applications of almost periodic compactifications.* – Acta Math., 1961, v.105, p.63–97.

Deo C.M.

1. *Strong laws of large numbers for weakly stationary random fields.* – Sankhya, 1978, v.40, Ser A, No.1, p.19–27.

Derriennic Y.

1. *Entropy and boundary for random walks on locally compact groups.* – Trans. of the Tenth Prague Conference on Inform. Theory, Statist. Decision Functions Random Processes (Prague, July 7–11, 1986), Prague, 1988, p.269–272.

Derriennic Y., Krengel U.
1. *Subadditive mean ergodic theorems.* – Ergodic theory and dynamical systems, 1981, v.1, p.33-48.

Derriennic Y., Lin M.
1. *Convergence of iterates of averages of certain operator representations and of convolution powers.* – J. Funct. Anal., 1989, v.85, No.1, p.86–102.

Diestel J.
1. *Geometry of Banach spaces.* – Lecture Notes in Math., Springer, Berlin, 1975, p.485.

Dixmier J.
1. *Les C^*-algèbres et leur representations.* Paris, Gautheir – Villars editeur, 1969.

Dobrushin R.L.
1. *The description of the random field by its conditional distributions and its regularity conditions.*–Teoriya Veroyatnostei i ee Primeneniya, 1968, v.13, No.2, p.201–229 (in Russian).
2. *Gibbsian random fields for lattice systems with pair interaction.* – Funkcionalny Analiz i ego Primeneniya, 1968, v.2, p.281–301 (in Russian).
3. *Gibbsian fields. General case.* – Funkcionalny Analiz i ego Primeneniya, 1969, v.3, p.27–35 (in Russian).

Doob J.L.
1. *Stochastic processes.* New–York – John Wiley, Sons London – Chapman, 1953.
2. *A ratio operator limit theorem.* – Z. Wahrscheinlichkeitstheorie verw. Geb., 1963, v.1, p.288–294.

Doubrovine B., Novikov S., Fomenko A.
1. *Geometrie contemporaine.* Méthodes et Applications. Mir, Moscow, 1982, p.371.

Dye H.A.
1. *On the mixing theorem.* – Trans. Amer. Math. Soc., 1965, v.118, No.6, p.123–130.

Duncan R.
1. *Pointwise convergence theorems for self- adjoint and unitary contractions.* – Ann. Probab., 1977, v.5, No.4, p.622–626.

Dunford N.

1. *An ergodic theorem for n-parameter groups.* – Proc. Nat. Acad. Sci. U.S.A., 1939, v.25, p.195–196.
2. *A mean ergodic theorem.* – Duke Math. J., 1939, v.5, p.635–646.
3. *An individual ergodic theorem for non-commutative transformations.* – Acta Sci. Math. Szeged, 1951, v.14, p.1–4.

Dunford N., Schwartz J.T.

1. *Linear operators. Part 1. General theory.* Interscience Publishers, New York, London, 1958.

Eberlein W.F.

1. *Abstract ergodic theorems and weak almost periodic functions.* – Trans. Amer. Math. Soc., 1949, v.67, p.217–240.

Edwards R.E.

1. *Functional analysis.* Theory and Applications. Australian National University. Holt, Rinehart and Winston. New York, Chicago, San Francisco, Toronto, London, 1965.

Edwards R.E., Hewitt E.

1. *Pointwise limits for sequences of convolution operators.* – Acta Math., 1965, v.113, No.3–4, p.181–218.

Eisenberg B.

1. *A note on metric transitivity for stationary Gaussian processes on groups.* – Ann. Math. Statist., 1972, v.43, No.2, p.683–687.

Ellis R.S.

1. *Entropy, large deviations and statistical mechanics.* Springer–Verlag, New York, Berlin, Heidelberg, Tokyo, 1985.

Emerson W.R. (see also Greenleaf F.P.)

1. *Ratio properties in locally compact amenable groups.* – Trans. Amer. Math. Soc., 1968, v.133, p.179–204.
2. *The pointwise ergodic theorem for amenable groups.* – American Journal of Mathematics, 1974, v.96, No.3, p.472–478.

Emerson W.R., Greenleaf F.P.

1. *Covering properties and Fölner conditions for locally compact groups.* – Math. Zeitschrift, 1967, v.102, p.370–384.

England J.W. – see Martin F.G.

Feller W.

1. *An introduction to probability theory and its applications.* Vol.1,2.

John Wiley and Sons, New York, London, Sydney, 1966.

Fleischmann K.

1. *Mixing properties of cluster-invariant distributions.* – Lietuvos Matematikos Rinkinys, 1978, v.18, No.3, p.191–199.

Föllmer H.

1. *On entropy and information gain in random fields.* – Z. Wahrscheinlichkeitstheorie verw. Geb., 1973, v.26, No.3, p.207–217.

Föllmer H., Snell J.L.

1. *An "inner" variational principle for Markov fields on a graph.* – Z. Wahrscheinlichkeitstheorie verw. Geb., 1977, v.39, p.187–195.

Fomenko A.T. – see Doubrovine B.

Fomin S.V. (see also Kolmogorov A.W., Naimark M.A.)

1. *To the theory of dynamical systems with continuous spectrum.* – Doklady Akademii Nauk, 1949, v.67, p.435–437 (in Russian).

Fong H.

1. *Weak convergence of semigroups implies strong convergence of weighted averages.* – Proc. Amer. Math. Soc., 1976, v.56, p.157–161.

Fong H., Sucheston L.

1. *On a mixing property of operators in L_p spaces.* – Z. Wahrscheinlichkeitstheorie verw. Geb., 1974, v.28, p.165–171.

Ford J. – see Ulenbek J.

Fortet R. – see Blanc–Lapierre A.

Fortet R., Mourier E.

1. *Loi des grands nombres et théorie ergodique.* Comptes Rendus Acad. Sci., 1952, v.234, No.7, p.699–700.
2. *Resultats complementaires sur les éléments aléatoires dans une espace de Banach.* – Bull. Sci. Math., 1954, v.78.
3. *Les fonctions aléatoires dans les espaces de Banach.* – Studia Math., 1955, v.15.

Fritz J.

1. *Generalization of McMillan's theorem to random set functions.* – Studia Scientiarum Mathematicarum Hungarica, 1970, v.5, p.369–394.

Gaposhkin V.F.

1. *On the strong law of large numbers for wide-sense stationary*

processes and sequences. – Teoriya Veroyatn. i ee Primen., 1973, v.18, No.2, p.388–392 (in Russian).

2. *Criteria of the strong law of large numbers for some classes of stationary processes and homogeneous random fields.* – Doklady Akademii Nauk SSSR, 1975, v.223, No.5, p.1009–1013 (in Russian).
3. *Explicit estimates of the rate of convergence in the strong law of large numbers for some classes of wide-sense stationary sequences and processes.* – Uspekhi Matem. Nauk, 1976, v.31, p.233–234 (in Russian).
4. *Criteria of the strong law of large numbers for some classes of stationary processes and homogeneous random fields.* Teoriya Veroyatn. i ee Primenen., 1977, v.22, p.295–319 (in Russian).
5. *The local ergodic theorem for groups of unitary operators and second-order stationary processes.* Matem. Sbornik, 1980, v.111, p.249–265 (in Russian).
6. *On the individual ergodic theorem for normal operators.* – Funkc. Analiz i ego Prilozh., 1981, v.15, No.1, p.18–22 (in Russian).

Garonas E.A.

1. *Ergodic theorem for homogeneous in the wide sense random measures.* Lietuvos Matematikos Rinkinys, 1984, v.24, No.1, p.35–43.

Garonas E.A., Tempelman A.A.

1. *Ergodic Theorems for Homogeneous Random Measures and Signed Measures on Groups.* In: III international Vilnius conference on probability theory and mathematical statistics. Summaries. – Vilnius, 1981.
2. *Ergodic theorems for homogeneous random fields on groups.*– Lietuvos Matematikos Rinkinys, 1984, v.24, No.1, p.19–34. English c/c translation, p.11–21

Garsia A.M.

1. *A simple proof of E. Hopf's maximal ergodic theorem.* – J. Math. Mech., 1965, v.14, p.381–382.

Gel'fand I.M., Fomin S.

1. *Geodesic flows on manifolds of constant negative curvature.* Uspekhi Matemat. Nauk, 1952, v.7, p.118–137.

Gel'fand I.M., Shilov G.E.

1. *Generalized Functions.* v.1.*Properties and operations.* Academic

Press, New York – London, 1964.

Gelfand I.M., Vilenkin N.Ya.

1. *Generalized Functions.* v.4. *Applications of harmonic analysis.* Academic Press, New York – London, 1964.

Glazman I.M. – see Akhiezer N.I.

Glieksberg I. (see also DeLeeuw K.)

1. *Weak compactness and separate continuity.* – Pacific J. Math., 1961, v.11, p.205–214.

Gnedenko B.

1. *The Theory of Probability.* "Mir", Moscow, 1988.

Godement R.

1. *Les fonctions de type positif et la théorie des groupes.* – Trans. Amer. Math. Soc., 1948, v.63, p.1–84.

Goldsheid I.Ya.

1. *The law of large numbers in some functional spaces.* – Uspekhi Matemat. Nauk, 1976, v.31. No.2(188), p.211–212.

Goldstein S., Penrose O.

1. *A nonequilibrium entropy for dynamical systems.* – J. Statist. Physics, 1981, v.24, p.325–343.

Gorbis A.Z., Tempelman A.A.

1. *Averaging of almost periodic functions and of finite dimensional unitary representations on free groups.* – Lietuvos Matematikos Rinkinys, 1988, v.28. No.4, p.662–668.

Goto M., Grosshans F.D.

1. *Semisimple Lie Algebras.* Marcel Dekker, New York and Basel, 1978.

Green L.(see also Auslender L.)

1. *Spectra of nilflows.* – Bull. Amer. Math. Soc., 1966, v.67, p.414–415.

Greenleaf F.P.(see also Emerson W.R.)

1. *Invariant Means on Topological Groups and their Applications.* – Van Nostrand, 1969.

1a. *Invariant Means on Topological Groups.* – Moscow, Mir, 1973 (Russian translation).

2. *Concrete methods for summing almost periodic functions and their relation to uniform distribution of semigroup actions.* – Colloq.

Math., 1979, v.61, p.105–116.

3. *Ergodic theorems and construction of summing sequences in amenable locally compact groups.* – Comm. Pure Appl. Math., 1973, v.26, No.1, p.29–46.

Greenleaf F.P., Emerson W.R.

1. *Group structure and the pointwise ergodic theorem for connected amenable groups.* – Advances in Math., 1974, v.14, p.153–172.

Grenander U.

1. *Stochastic processes and statistical inference.* – Arkiv for Matematik., 1950, v. 1, No.17, p.195–277.
2. *Probabilities on Algebraic Structures.* – Almqvist and Wiksel, Wiley and Sons, 1963.

Grosshans F. – see Goto M.

Grothendieck A.

1. *Critères de compacité dans les espaces fonctionnels généraux.* – Amer. J. Math., 1952, v.74, p.168–186.

Guivarc'h Y.

1. *Généralisation d'un théorème de von Neumann* – Comptes Redus Acad. Sci., 1969, v.268, ser. A, B, No.18, p.1020.

Gurevich B.M.

1. *On one-side and two-side regularity of stationary processes.* – Doklady Akademii Nauk SSSR, 1973, v.210, p.763–767 (in Russian).
2. *Variational principle for one-dimensional Gibbsian random fields with countable number of states.* – Doklady Akademii Nauk SSSR, 1978, v.241, No.4, p.749–752 (in Russian).

Guzmán M.

1. *Differentiation of integrals in R^n.* Springer – Verlag, 1975.

Hahn F. – see Auslander L., Blum J.R.

Hahn L.S. – see Blum J.R.

Halmos P.R.

1. *Measure Theory.* New York, 1950.
2. *Lectures on Ergodic Theory.* The Mathematical Society of Japan, Tokyo, 1956.

Halperin I. – see Krotkov B.
Humphreys J.E.
Linear Algebraic Groups. Springer, 1975.
Hanson D.L., Pledger G. (see also Blum J.R.)
1. *On the mean ergodic theorem for weighted averages*. – Z. Wahrsheinlichkeitstheoric und verw. Geb., 1969, v.13, p.141–149.
Hattori I.
1. *On ergodic theorem in reflexive spaces*. – Mem. Fac. Sci. Kyusyu Univ., 1951, A. 6, p.17–19.
Helgason S.
1. *Differential Geometry and Symmetric Spaces*. Academic Press. New York and London, 1962.
Herz C.S.
1. *Fourier transforms related to convex sets*. – Ann. Math., 1968, v.75, No.1, p.81–92.
Hewitt E., Ross K.A.
1. *Abstract Harmonic Analysis*. Vol. I. Springer Verlag, Berlin, Gőttingen, Heidelberg, 1963.
Hille E., Phillips R.
1. *Functional Analysis and Semigroups*. American Math. Soc. Collog. Publications, v.31, Providence, 1957.
Hopf E.
1. *Ergoden Theorie*. – Ergebn. M., 5, Springer, 1937.
2. *Statistic der geodätischen Linien in Mannigfaltigkeiten negativer Krümmung*. – Leipzig: Ber. Verhandl. Sächs. Akad. Wiss., 1939, v.91, p. 261–304.
Howe R., Moore C.C.
1. *Asymptotic properties of unitary representations*. – J. Funct. Analysis, 1979, v.32, No.1, p.72–96.
Hulanicki A.
1. *Means and Følner condition on locally compact groups*. – Studia Mathematica, 1966, v.27, No.2, p.87–104.
Humphreys J.E.
Linear Algebraic Groups. Springer, 1975.
Ibragimov I.A., Rozanov Y.A.
Gaussian Random Processes. Springer; Appl. of Math. 9, 1978.

Ionescu Tulcea C.

1. *Ergodic theorems I.* – J. Math. Anal. and Appl., 1980, v.78, No.1, p.113–116.

Iwanik A.

1. *Weak convergence and weighted averages for groups of operators.* – Colloq. Math., 1979, v.42, p.241–254.

Jacobs K.

1. *Neuere Methoden und Ergebnisse der Ergodentheorie.* – Berlin – Göttingen – Heidelberg: Springer – Verlag, 1960.
2. *Ein Ergodensatz für beschränkte Gruppen im Hilbertschen Raum.* – Math. Ann., 1954, v.128, p.340–349.
3. *Periodizitätseigenschaften beschränkter Gruppen im Hilbertschen Raum.* – Math. Z., 1955, v.61, p.408–428.
4. *Ergodentheorie und fastperiodische Funktionen auf Halbgruppen.* – Math. Z., 1956, v.64, p.298–338.
5. *Fastperiodizitätseigenschaften allgemeiner Halbgruppen in Banachräumen.* – Math. Z., 1957, v.67, p.83–92.

Jenkins J.W.

1. *Growth of connected locally compact groups.* – J. Funct. Analysis, 1973, v.12, p.113–127.

Jones L.K., Kuftinec V.

1. *A note on Blum–Hanson theorem.* – Proc. Amer. Math. Soc., 1971, v.30, p.202–203.

Jones R.L (see also Bellow A.)

Jones R.L.

1. *Ergodic averages on spheres.* – To appear in "J. d'Analyse Mathématique".

Jones R.L., Olsen J.

1. *Multi-parameter moving averages.* – To appear in "Proc. of the

2nd International conference on almost everywhere convergence".
del Junco A. – see Akcoglu M.
Kaimanovich V.A. – see Vershik A.M.
Kakutani S. – see Yosida K.
Kampe de Feriet J. – see Birkhoff G.
Kantarovich L.V. – see Akilov G.P.
Karpelevich F.I., Tutubalin V.N., Schur M.G.
1. *Limit theorems for compositions of distributions in the Lobachevsky plane and space.* – Teoriya Veroyatnostei i ee Primenenia, 1959, v.4, No.4, p.432–436.
Katznelson Y., Weiss B.
1. *A simple proof of some ergodic theorems.* – Israel J. Math., 1982, v.42, p.291–296.
Keane M. – see Brunel A.
Kerstan J. – see Matthes K.
Kieffer I.C.
1. *A generalized Shannon – McMillan theorem for the action of an amenable group on a probability space.* – The Ann. of Probab., 1975, v.3, p.1031–1037.
Kirillov A.A.
1.*Elements of the Theory of Representations.* – Springer, 1976.
Kolmogorov A.N.
1. *A simplified proof of the Birkhoff- Khintchine ergodic theorem.* – Uspekhi Matemat. Nauk, 1938, No.5 (in Russian).
Kolmogorov A.N., Fomin S.V.
Introductory Real Analysis. Dover Publications, New York, 1975.
Kornfeld I.P., Sinai Ya.G., Fomin S.V.
1. *Ergodic Theory.* – Moscow: Nauka, 1980 (in Russian).
Kozlov O.K.
A Gibbsian discription of a system of random variables. – Problemy Peredachi Informacii, 1974, v.10, p.94–103(in Russian).
Krengel U.(see also Akcoglu M.A., Derriennic Y.)
1. *A local ergodic theorem.* – Invent. Math., 1969, v.6, p.329-333.
2. *Recent progress on ergodic theorems.* – Astèrisque. 1977, No.50, p.151–192.

3. *Ergodic theorems.* De Gruyter Studies in Math., v.6. Walter de Gruyter, Berlin–New York, 1985.

Krickeberg K.

1. *Convergence of martingales with a directed index set.* – Trans. Amer. Math. Soc., 1956, v.83, p.313–337.
2. *Stochatsche Derivierte.* Math. Nachr., 1958, v.18, p.203-217.

Krotkov V., Galperin I.

1. *The ergodic theorem for Banach spaces with convex compactness.* – Trans. Roy. Soc. Canada. Sect. III, 1953, v.47, p.17–20.

Krylov A.L. – see Arnold V.I.

Kubokava Y.

1. *A general local ergodic theorem.* – Proc. Japan Acad. Sci., 1972. v.48, p.461–465.
2. *A local ergodic theorem for semi-groups on L_p.* Tôhokn. Math. J. (2), 1974, v.26, p.411-422.
3. *Ergodic theorems for contraction semi-groups.* – J. Math. Soc. Japan, 1975, v.27, p.184-193.

Kuftinec V. – see Jones L.K.

Künsch H.

1. *Almost sure entropy and the variational principle for random fields with unbounded state space.* – Z.Wahrscheinlichkeitstheoric verw. Geb., 1981, v.58, p.69–81.
2. *Thermodynamics and statistical analysis of Gaussian random fields.* – Z.Wahrscheinlichkeitstheoric verw. Geb., 1981, v.58, p.407–421.

Lanford O.E., Ruelle D.

1. *Observables at infinity and states with short-range correlations in statistical mechanics.* – Comm. Math. Phys., 1968, v.13, No.3, p.194-215.

Lebowitz I.L., Presutti E.

1. *Statistical mechanics of systems of unbounded spins.* – Comm. Math. Phys., 1976, v.50, p.195-218.

Levitan B.M.

1. *Almost Periodic Functions.* – Moscow: Gostexizdat, 1953 (in Russian).

Leichtweiß K.

1. *Konvexe Mengen.* VEB Deutsher Verlag der Wissenschaften. Berlin, 1980.

Lin M. (see also Derriennic Y.)

1. *Mixing for Markov operators.* – Z.Wahrscheinlichkeitstheorie verw. Geb., 1971,v.19, p.231-242.
2. *Ergodic properties of an operator obtained from a continuous representation.* – Ann. Inst. H.Poincarè, 1977, v.13, No.4, p.321-331.

Loève M.

1. *Sur le fonctions alèatoires stationnaires de second ordre.* Rev. Sci., 1945, v.83, p.297–303.

Loomis L.

An introduction to abstract harmonic analysis. – Van Nostrand, New York, 1953.

Lorch E.R.

1. *Means of iterated transformations in reflexive vector spaces.* – Bull. Amer. Math. Soc., 1939, v.45, p.945-947.

Losert V. – see Bellow A.

Lyubarsky G.Ya.

1. *On integration in the mean of almost periodic functions on topological groups.* – Uspekhi Matem. Nauk, 1948, v.3, No.3(25), p.195–201.

Maak W.

1. *Fastperiodische invariante Vektormoduln in einen metrischen Vektorraum.* – Math. Ann., 1950, v. 122, p.157–166.
2. *Periodizitätseigenschaften unitärer Gruppen in Hilbertraümen.* – Math. Scand., 1954, v.2, p.334–344.

Mackey G.W.

1. *Ergodic transformation groups with a pure point spectrum.* – Illinois J. Math., 1964, v.8, p.593–600

Martin N.F.G., England J.W.

1. *Mathematical Theory of Entropy.* Addiso–Wesley Publishing Company, 1981.

Matthes K., Kerstan J., Mecke J.

1. *Infinitely Divisible Point Processes.* Akademie–Verlag, Berlin, Wiley and Sons, 1978.

Mautner F.I.

1. *Geodesic flows on symmetric Riemann spaces.* – Ann. Math., 1957, v.65, p.416–431.

McGrath S.A. (see also Chacon R. V.)

1. *An Abelian ergodic theorem for semigroups in L_p space.* – Proceedings of the American Math. Society, 1976, v.54, p.231–236.
2. *On the local ergodic theorems of Krengel, Kubokawa, and Terrill.* – Comm. Math. Univ. Carolinae, 1976, v.17, p.49–59.
3. *Some ergodic theorems for commuting L_1 contractions.* – Studia Math., 1981, v.LXX, No.3, p.153–160.

McMillan B.

1. *The basic theorem of information theory.* – Ann. Math. Stat., 1953, v.24, p.196–219.

Mecke J. – see Matthes K.

Michel P.

1. *Sur les moyennes ergodique et la desintegration de mésure.* – Université de Rennes, Faculté des Sci., 1967–1968.

Mirzahanian E.A. – see Aleksandrian R.A.

Moore C.C. (see also Howe R.E.)

1. *Ergodicity of flows on homogeneous spaces.* – Amer.J. Math., 1966, v.88, p.154–178.
2. *Restrictions of Unitary representations to subgroups and ergodic theory.* – In: Group Representations in Mathematics and Physics, Lecture Notes Phys., No.6, New York: Springer–Verlag, 1970.

Moore R.T.

1. *Measurable, continuous and smooth vectors for semigroups and group representations.* – Memoirs of the American Math. Soc., 1968, No78.

Moulin–Ollagnier J., Pinchon D.

1. *Groupes pavables et principe variationnel.* – Z. Wahrscheinlichkeitstheorie verw. Geb., 1979, v.48, p.671–679.

Mourier E. (see also Fortet R.)

1. *Lois des grands nombres et théorie ergodique.* – C.R. Acad. Sci. Paris, 1951, v.232, p.923–925.
2. *Éléments aléatoires dans un espace de Banach.* – Ann. Inst. H. Poincaré, 1953, v.13, p.161–244.

Naimark M.A.

1. *Linear Representations of the Lorentz Group.* A Pergamon Press Book. The Macmillan Co., New York, 1964.
2. *Normed Rings.* Groningen, 1959.

Naimark M.A., Fomin S.V.

1. *Continuous direct sums of Hilbert spaces and some applications.* – Uspekhi Matemat. Nauk, 1955, v.10, No.2 (64), p.111–142 (in Russian).

Naimark M.A., Štern A.I. (Shtern A.I.)

1. *Theory of Group Representations.* Fundamental Principles of Mathematical Science, 246. Springer–Verlag, New York–Berlin, 1982.

Nakamura M.

1. *Note on Banach spaces. II. An ergodic theorem for Abelian semi-groups.* – Proc. Imp. Acad. Tokyo, 1942, v.18, p.131.

Namioka I., Asplund E.

1. *A geometric proof of Ryll–Nardzewski's fixed point theorem.* – Bull. Amer. Math. Soc., 1967, v.73, No.3, p.443–445.

Natanson I.P.

1. *Theory of functions of a real variable.* – Constable & Co. Ltd., London, 1960.

Neumann J. von

1. *Proof of the quasi–ergodic hypothesis.* – Proc. Nat. Acad. Sci. USA, 1932, v.18, p.70–82.
2. *Almost periodic functions in a group.* – Trans. Amer. Math. Soc., 1934, v.36, p.445–492.

Neumann J. von, Wigner E.P.

1. *Minimally almost periodic groups.* – Ann. of Math. (2), 1940, v.41, p.746–750.

Neveu J.

1. *Bases Mathématiques du Calcul des Probabilities.* Masson et cie, Paris, 1964.

Nguyen X.X., Zessin H.

1. *Ergodic theorems for spatial processes.* – Z. Wahrsheinlichkeits-theorie verw. Geb., 1979, v.48, p.133–158.

2. *Ergodic theorems for subadditive spatial processes.* – Z. Wahrsheinlichkeitstheorie verw. Geb., 1979, v.48, p. 159–176.

Novikov C.P. see Doubrovine B.A.

Olsen J.H. (see also Jones R.L., Chacon R.V.)

1. *Multiple sequence ergodic theorem.* – Canadian Math. Bull., 1983, v.26 (4), p.493–497.
2. *Multi-parameter weighted ergodic theorems from their single parameter version.* – Almost everywhere convergence. Academic Press, 1989, p.297–303.
3. *Two consequences of Brunel's theorem.* – To appear in "Canadian Math. Bull."

Onishchik A.L.

1. *Classification of compact Lie groups.* Appendix to Adams [1a] p.132–144.

Ornstein D.S., Weiss B. (see also Chacon R.V.)

1. *Entropy and isomorphism theorems for actions of amenable groups.* J.d'Analyse Mathématique, 1987, v.48, p.1–141.
2. *The Shannon–McMillan–Breiman theorem for a class of amenable groups.* – Israel J. of Math., 1983, v.44, No.1, p.53–60.

Oseledec V.I.

1. *Markov chains, skew products and ergodic theorems for "general" dynamical systems.* – Teoriya veroyatnostei i ee primeneniya, 1965, v.10, No.3, p.551–557.

Parasyuk O.S.

1. *Oricyclic flows on surfaces of constant negative curvature.* – Uspekhi Matemat. Nauk, 1953, v.8, p.125–126 (in Russian).

Peck J.E.L.

1. *An ergodic theorem for a non-commutative semigroup of linear operators.* – Proc. Amer. Math. Soc., 1951, v.2, p.414–421.

Penrose O. – see Goldstein S.

Perez A.

1. *Notions généralisées d'incertitude, d'entropie et d'information du print de vue de la théorie des martingales.* – In: Trans. First Prague Conf., 1956, p.209–243.

Petersen K.

1. *Ergodic Theory.* Cambridge studies in advanced mathematics, 2.

Cambridge University Press, 1983.

Pinchon D. – see Moulin–Ollagnier J.

Pitckel B.S.

1. *Some remarks on the individual ergodic theorem of information theory.* – Matemat. Zametki, 1971, v.9, p.93–103 (in Russian).
2. *On informational future of amenable groups.* – Dokl. Akad. Nauk SSSR, 1975, v.223, p. 1067–1070 (in Russian).
3. *Entropy of random fields which are homogeneous with respect to a commutative group of representations.* – Matemat. Zametki, 1978, v.3, No.3, p.447–468 (in Russian).

Pitckel B.S., Stepin A.M.

1. *On the equidistribution property of commutative groups of metric automorphisms.* – Dokl. Akad. Nauk SSSR, 1971, v.198, No.5, p.1021–1024 (in Russian).

Pitt H.R.

1. *Some generalizations of the ergodic theorem.* – Proc. Cambridge Phil. Soc., 1942, v.38, p.325–343.

Pledger G. – see Hanson D.L.

Pontryagin L.S.

1. *Topological groups.* – New York: Gordon and Breach, 1966, p.520.

Preston Ch.I.

1. *Random Fields.* – Lectures Notes in Mathematics, Berlin–Heidelberg–New York: Springer, 1976, v.534.
2. *Gibbs states on countable sets.* Cambridge Tracts in Math., v.68, Cambridge University Press, 1974.

Presutti E. – see Lebowitz I.L.

Raikov D.A.

1. *Positive definite functions on commutative groups with an invariant measure.* – Dokl. Akad. Nauk SSSR, v.28, p.296–300 (in Russian).

Reich J. (see also Blum J.R.)

1. *Random ergodic sequences on local groups.* – Transactions of the American Math. Soc., 1981, v.265, No.1, p.59–68.

Renaud P.F.

1. *General ergodic theorems for locally compact groups.* – Amer. J. Math., 1971, v.93, No.1, p.52–64.

Riesz F.

1. *Some mean ergodic theorems.* – J. London Math. Soc., 1938, v.13, p.274–278.
2. *Sur la théorie ergodique des espaces abstraits.* – Acta Szeged, 1941, v.10, p.1–20.
3. *Another proof of the mean ergodic theorem.* – Acta Szeged, 1941, v.10, p.75–76.
4. *Rectification au travail "Sur la théorie ergodique des espaces abstraits."* – Acta Szeged, 1941, v.10, p.141.
5. *Sur quelques problèmes de la théorie ergodique.* – Mat. Fiz. Lapok, 1942, v.49, p.34–62.
6. *Sur la théorie ergodique.* – Comm. Math. Helv., 1945, v.17, p.221–239.
7. *On a recent generalization of G.D. Birkhoff's ergodic theorem.* – Acta Unv. Szeged. Sect. Math., 1948, v.11, p.193–200.

Riesz F., Sz.–Nagy B.

Über Kontraktionen des Hilbertschen Raumes. Acta Sci. Math. Szeged, 1943, v.10, p.202–205.

Rokhlin V.A.

1. *Lectures on the entropy theory of transformations with an invariant measure.* – Uspekhi Matemat. Nauk, 1967, v.22, No.5, p.3–56 (in Russian).

Rosen W.G.

1. *On invariant means over topological semigroups.* – Thesis, University of Illinois, Urbana, 1954.

Rosenblatt J. (see also Bellow A., Bergelson V.)

1. *Ergodic group actions.* Arch. Math., 1986, v.47, p.263–269.

Rosenfeld B.A.

1. *Non–euclidean spaces.* – Moscow, Nauka, 1969, p.548 (in Russian).

Rota G.C.

1. *An "Alternierende Verfahren" for general positive operators.* – Bull. Amer. Math. Soc., 1962, v.68, p.95–102.

Ruelle D.

1. *Statistical mechanics. Rigorous results.* – W.A. Benjamin, 1969.
2. *Thermodynamic formalism.* – London–Amsterdam, Eddison–Wesley publishing company, 1978, p.183.

Ryll–Nardzewski C.

1. *Generalized random ergodic theorems and weakly almost periodic functions.* – Bull. de l'Acad. Polonaise de Sciences Serie des Sci. Math., Astr., Phys., 1962, v.10, No.5, p.271–275.
2. *On fixed points of semigroups of endomorphisms of linear spaces.* – In: Proc. Fifth Berkeley Symposium on Math. Statist. and Probability, v.2, Berkeley, 1966, p.55–61.

Saks S.

Theory of the integral. 1939.

Sato R.

1. *On a local ergodic theorem.* – Studia Math., 1976, v.58, p.1–5.
2. *On abstract mean ergodic theorems.* – Tohoku Math. J, 1978, v.30, p.575–581.
3. *On local ergodic theorems for positive semigroups.* – Studia Math., 1978, v.63, p.45.
4. *On a mean ergodic theorem.* – Proc. Amer. Math. Soc., 1981, v.83, p.563–564.
5. *A note on operator convergence for semi-groups.* – Comment. Math. Univ. Carolinae, 1974, v.15, p.127–129.

Savichev A.O.

1. *The strong law of large numbers for homogeneous random fields on the Lobachevsky space.* – In: Application of Probability Theory and Math. Statistics, No.5. Vilnius: Institute of Math. and Cybernetics of Lithuanian Acad. Sci., 1983, p.161–163 (in Russian).

Savichev A.O., Tempelman A.A.

1. *Ergodic properties of homogeneous and isotropic random fields.* – In: III international Vilnius conference on probability theory and math. statistics. Summaries, v.3, Vilnius, 1981, p.308–309.
2. *Ergodic theorems on mixing homogeneous spaces.* Lietuvos Matematikos Rinkinys, 1984, v.24, No.4, p.167–175 (in Russian).

Scalora F.S.

1. *Abstract martingale convergence theorems.* – Pacific J. Math., 1961, v.11, p.347–374.

Schmidt K.

1. *Asymptotic properties of unitary representations and mixing.* – Proc. London Math. Soc. (3), 1984, v.48, p.445–460.

Schwartz J.T. – see Dunfard N.
Shilov G.E. – see Gel'fand I.M.
Shulman A.
1. *Maximal ergodic theorems on groups.* Dep. Lit. NIINTI, No.2184–Li(1988).
Shur M.G. – see Karpelevich F.I.
Simon B.
1. *The $P(\varphi)_2$ euclidian (quantum) field theory.* – Princeton University Press. Princeton, N.J., 1974.
Sinai Ya.G.
1. *On the notion of entropy of a dynamical system.* – Dokl. Akad. Nauk SSSR, 1959, v.124, No.4, p.768–771 (in Russian).
Smythe R.T.
1. *Multiparameter subadditive processes.* – Ann. of Probab., 1976, v.4, p.772–782.
Snell I.L. – see Föllmer H.
Spitzer F.
1. *A variational characterization of finite Markov chains.* – Ann. Math. Statist., 1971, v.93, No.1, p.1303–1307.
Stein E.M.
1. *Maximal functions: spherical means.* – Proc. Nat. Acad. Sci., USA, 1976, v.73, p.2174–2175.
Stein E.M., Wainger S.
1. *Problems in Harmonic analysis related to curvature.* – Bull. American Math. Soc., 1978, v.84, No.6, p.1239–1295.
Stepin A.M. (see also Pitchel B.S., Tagi–zade A.T.)
1. *Equidistribution of the entropy of transformations of amenable groups.* – Dokl. Akad. Nauk Azerb. SSR, 1978, v.34, p.3–7 (in Russian).
Stepin A.M., Tagi–zade A.T.
1. *Variational characterization of the topological pressure of amenable transformation groups.* – Dokl. Akad. Nauk SSSR, 1980, v.254, No.3, p.545–549 (in Russian).
Struble R.A.
1. *Almost periodic functions on locally compact groups.* – Proc. Nat. Acad. Sci. USA, 1953, v.39, No.2, p.122–126.

Sucheston L. (see also Akcoglu M., Fong H.).

1. *On one-parameter proofs of almost sure convergence of multiparameter processes.* – Z. Wahrscheinlichkeitsth. verw. Geb., 1983, v.63, p.43–49.

Sullivan W.G.

1. *Potentials for almost Markovian random fields.* – Commun. Math. Phys., 1973, v.33, p.61–74.

Sz.–Nagy B. – see Riesz F.

Tagi–zade A.T (see also Stepin A.M.)

1. *Entropy properties of actions of amenable groups.* – In: Functional Analysis and its Applications, No.3, Baku: ELM, 1978, p.169–183 (in Russian).
2. *The entropy of actions of amenable groups.* – Dokl. Akad. Nauk Azerb.SSR, 1978, v.34, No.6, p.18–22 (in Russian).
3. *On entropy characteristics of amenable groups.* – Dokl. Akad. Nauk Azerb.SSR, 1978, v.34, No.8, p.11–14 (in Russian).
4. *Some problems on the entropy theory of group actions.* – Candidate Thesis, Moscow, 1979 (in Russian).

Takahashi W.

1. *A nonlinear ergodic theorem for an amenable semigroup of nonexpansive mappings in a Hilbert space.* – Proc. Amer. Math. Soc., 1981, v.81, No.2, p.253–256.

Taylor K. – see Bagget L.

Tempelman A.A. (see also Garonas E., Gorbis A., Savichev A.)

1. *Ergodic properties of homogeneous random fields on groups.* – In: Trans. of the VI Soviet Union Symposium on Probability Theory and Math. Statistics, Vilnius, 1962, p.253–254 (in Russian).
2. *Some problems in ergodic theory of homogeneous random fields.* – Candidate Thesis, Vilnius, 1961, p.71 (in Russian).
3. *An ergodic theorem for wide sense homogeneous random fields.* – Dokl. Akad. Nauk SSSR, 1962, v.144, No.4, p.730–733. (English translation: Soviet Math. Dokl., 1962, p.817–820.)
4. *Ergodic theorems for wide sense homogeneous generalized random fields and homogeneous random fields on groups.* – Lietuvos Matematikos Rinkinys, 1962, v.2, No.1, p. 195–213 (in Russian).

5. *Ergodic theorems for general dynamic systems.* – Dokl. Akad. Nauk SSSR, 1967, v.176, No.4, p.790–793. (English translation: Soviet. Math. Dokl., 1967, v.8, No.5, p.1213–1216.)
6. *Criteria of metric transitivity of Gaussian homogeneous random functions.* – Lietuvos Matematikos Rinkinys, 1970, v.10, No.4, p.815–834 (in Russian).
7. *Ergodic theorems for general dynamic systems.* – Trudy Moskovskogo Matematicheskogo obshchestva, 1972, v.26, p.95–132. (English c/c translation: p.94–132.)
8. *Generalization of a theorem due to Hopf.* – Teoriya Veroyatnostei i ee Primeneniya, 1972, v.17, No.2, p.380–383 (in Russian).
9. *Consistent estimates of characteristics of homogeneous random fields. Admissibility estimates of a multidimensional shift.* – In: International Conference on Probability Theory and Math. Statistics. Summaries. Vilnius, 1973, p.271–274 (in Russian).
10. *On ergodicity of Gaussian homogeneous random fields on homogeneous spaces.* – Teoriya Veroyatnostei i ee Primeneniya, 1973, v.18, No.1, p.177–179 (in Russian).
11. *Consistent linear regression estimates.* – In: III International Symposium on Information Theory. Summaries. Tallin, 1973, p.129–132.
12. *Consistent and strongly consistent linear regression estimates.* – In: Proceedings of the Soviet Union Symposium on Statistics of Random Processes. Kiev, 1973, p.188–191.
13. *Ergodic theorems for amplitude-modulated homogeneous random fields.* – Lietuvos Matematikos Rinkinys, 1974, v.14, No.4, p.221–229 (in Russian).
14. *Convergence and consistency of regression estimates.* – Doctor Thesis, Vilnius, 1975, p.290 (in Russian).
15. *Ergodic theorems and their applications in statistics.* – Teoriya Veroyatnostei i ee Primeneniya. 1975, No.2, p.443–444 (in Russian).
16. *On averaging sequences of sets.* – Lietuvos Matematikos Rinkinys, 1976, v.16, No.2, p.216 (in Russian).
17. *Specific informational and thermodynamical characteristics of homogeneous random fields on groups.* – Lietuvos Matematikos

Rinkinys, 1977, v.17, No.3, p.199–200 (in Russian).

18. *Specific characteristics and the variational principle for homogeneous random fields on groups.* – Dokl. Akad. Nauk SSSR, 1980, v.254, No.2, p.297–302. (English translation: Soviet Math. Dokl., 1980, v.22, No.2, p.363–369.)
19. *Ergodic functions and averaging sequences.* – Dokl. Akad. Nauk SSSR, 1981, v.259, No.2, p.290–294. (English translation: Soviet Math. Dokl., 1981, v.24, No.1, p.78–82.)
20. *On linear regression estimates.* – II International Symp. Inform. Theory Proc. B.N. Petrov and T. Csaki, eds., Budapest, 1973, p.329-354; reproduced in "Reproducing Kernel Hilbert Spaces. Applications in Statistical Signal Processing", H.L. Weinert ed. Hutchinson Ross Publishing Company, Strassburg, 1982, p.301–326.
21. *Variational principle for homogeneous random fields on groups.* – In: Colloq. on "Random Fields: Rigorous Results in Statistical Mechanics and Quantum Field Theory". Abstracts. Estergom, June 24–30, 1979, Suppl. 1, p.6–7.
22. *Specific characteristics and variational principle for homogeneous random fields.* – Z.Wahrscheinlichkeitstheorie verw. Gebiete, 1984, v.65, p.341–365.
23. *Ergodic and mixing homogeneous spaces.* – Dokl. Akad. Nauk SSSR, 1982, v.269, No.5, p.1045–1049. (English translation: Soviet Math. Dokl., 1983, v.27, No.2, p.452–455.)
24. *Ergodic theorems on groups.* – Vilnius, Mokslas, 1986.

Terrell T.R.

1. *Local ergodic theorems for N-parameter semigroups of operators.* – Lecture Notes Math., 1970, v.160, p.53–63.

Thouvenot J.P.

1. *Convergence en moyenne de l' information pour l'action de Z^2.* – Z. Wahrsheinlichkeitstheorie verw. Geb., 1972, v.24, p.135–137.

Tortrat A. – see Blanc–Lapierre A.

Tutubalin V.N. – see Karpelevich F.I.

Ulenbek J., Ford J.

1. *Lectures On Statistical Mechanics.* 1965.

Urbanik K.

1. *Random processes realizations of which are generalized functions.* – Teoriya Veroyatnostei i ee Primeneniya, 1956, v.1, No.1, p.146–149 (in Russian).
2. *Generalized stochastic processes.* – Studia Math., 1958, v.16, p.268–334.

Verbickaja I.N.

1. *On conditions of validity of strong law members to wide-sense stationary processes.* – Teoriya Veroyatnostei i ee Primeneniya, 1966, v.11, No.4, p.715–720.

Vershik A.M.

1. *Countable groups which are close to finite ones.* – Appendix to Greenleaf [1a], p.112–135 (in Russian).

Vershik A.M., Kaimanovich V.A.

1. *Random walks in groups: the boarder, entropy, uniform distribution.* – Dokl. Akad. Nauk SSSR, 1979, v.249, p.15–18 (in Russian).

Vilenkin N.Ya. (see also Gel'fand I.M.)

1. *Special functions and the theory of group representations.* – Translations of Mathematical Monographs, v.22. American Math. Society, Providence, R.I., 1968.
2. *Editor's notes to Weil* [1a].

Wainger S. – see Stein E.M.

Walach N.R. – see Bord A.

Warren P. – see Beck A.

Weil A.

1. *L'integration Dans Les Groupes Topologiques Et Ses Applications.* – Publications de L'institut de mathématiques de Clepmont–Ferrand, Paris, 1940.

1a. *Integration in topological groups and its applications.* – Moscow: Inostrannaya Literatura, 1950, p.223 (Russian translation).

Weiss B. – see Katznelson Y., Ornstein D.S.

Wigner E.P. – see Neumann J.von

Yadrenko M.I.

1. *Ergodic theorems for isotropic random fields.* – In: Probability Theory and Math. Statistics, Kiev, 1970, No.1, p.249–251.

Yaglom A.M.

1. *Some classes of random fields in the n–dimensional space related to stationary random processes.* – Teoriya Veroyatnostei i ee Primeneniya, 1957, v.2, No.3, p.292–338 (in Russian).
2. *Positive definite functions and homogeneous random fields on groups and homogeneous spaces.* – Dokl. Akad. Nauk SSSR, 1960, v.135, No.6, p.1342–1345 (in Russian).
3. *Second–order homogeneous random fields.* – In: Proc. 4th Berkeley Symp., Math. Statist. and Probab., 1960, v.2, p.593–622, Berkeley–Los Angeles, 1961.

Yoshimoto T.

1. *An ergodic theorem for noncommutative operators.* – Proceedings of the Amer. Math. Society, 1976, v.54, p.125–129.

Yosida K.

1. *Functional Analysis.* Springer, 1965
2. *Mean ergodic theorem in Banach spaces.* – Proc. Imp. Acad. Tokyo, 1938, v.14, p.292–294.
3. *Ergodic theorems of Birkhoff–Khintchines type.* – Jap. J. Math., 1940, v.17, p.31–36.
4. *An abstract treatment of the individual ergodic theorem.* – Proc. Imp. Acad. Tokyo, 1940, v.16, p.280–284.

Yosida K., Kakutani S.

1. *Birkhoff's ergodic theorem and the maximal ergodic theorem.* – Proc. Imp. Acad. Tokyo, 1939, v.15, p.165–168.

Yurinsky V.V.

1. *A strong of large numbers for homogeneous random fields.* – Matemat. Zametki, 1974, v.16, p.141–148 (in Russian).

Zessin H. – see Nguyen X.X.

Zimmer R.J.

1. *Compact nilmanifold extensions of ergodic actions.* – Trans. Amer. Math. Soc., 1976, v.223, p.397–406.
2. *Orbit spaces of unitary representations, ergodic theory, and simple Lie groups.* – Ann. of Mathem., 1977, v.106, p.573–588.

Zygmund A.

1. *An individual ergodic theorem for noncommutative transformations.* – Acta Sci. Math. Szeged., 1951, v.14, p.103–110.

Index

absolutely continuous random measure – 249
absorbing set – 231
action of a semigroup – 300
– – – – completely continuous – 300
– – – – continuous – 300
– – – – left, right – 300
admissible reduction – 339
almost semidirect factorization (right, left) – 307
– simple group – 301
– periodic compactification – 335
– – element – 333
– – function – 334
amenable semigroup – 319
averageable function – 15
– element – 24
averaging net – 85,86
– operator – 20, 26, 41
– sequence – 11
background–free random field – 45
background of a field – 45
Bohr compactification – 335
canonical semidirect factorization – 31
character – 326
cofinally separable net – 357
commutative semigroup – 296
completely measurable random field – 348
condition (Γ_λ) – 253
– (Γ) – 259
– (A_0) – 112
conditional density – 266
– entropy – 278
– relative entropy – 291
configuration – 264
conservative representation – 232
consistent operators – 38
– representations – 38
continuous random field – 348
coordinate image – 340
– reduction – 340
correlation function – 339, 342
covariance function – 339, 342
covariancy with respect to a group action – 340
covariant random set function – 265
cross–section – 306
cyclic element – 328
density – 265, 320
direction (directed set) – 357
dissipative representation – 232
DOM (class of dominant nets) – 157
dominant function – 157
– net – 157
dominated ergodic theorem – 157

dynamical system – 337
– –, completely measurable – 346
– –, continuous – 345
– –, ergodic – 52
– –, G/K–relatively (quasi–)mixing – 57
– –, measurable – 345
– –, strongly (quasi–)mixing – 55
– –, translation – 341
– –, uniformly continuous – 345
– –, weakly (quasi–)mixing – 56
D–robust averaging net – 255
$\mathcal{D}$–sequence of sets – 250
elementary positive definite function – 326
energy – 265
entropy – 278
equivalent actions – 300
- dynamical systems – 337
- G–spaces – 310
 representations – 330
rgodic dynamical system – 54
 net – 320
 theorem – 86
ıclidean homogeneous space – 313
:tor group large – 315
 small – 315
epresentation – 89
:orization – 139
–continuous spectrum – 351
–discrete spectrum – 351, 352
l element – 328
t–function – 349
ier transform – 139
free energy – 286
full subsemigroup – 298
Gaussian random function – 338
– – vector – 338
Gibbsian free energy – 287
– measure – 266
– specification – 266
graph of a random σ–algebra – 353
group – 296
– almost simple – 301
–, – – of non–compact type – 301
–, amenable – 319
–, orderable – 266
–, perfect – 303
–, representable into a compact group – 336
–, simple – 301
–, almost simple – 301
group–type homogeneous space – 312
G/K–relative strong (weak) mixing – 57
– – – quasi–mixing – 57
G–space – 309
Haar measure – 304
HRM_λ–mean averaging net – 256
HRM_λ–pointwise averaging net – 256
homogeneous generalized random field – 342
– random field – 341
– – function – 340

– – measure – 246
– – order – 267
– – signed measure – 261
homogeneous space
(right, left)– 310
– –, Euclidean – 313
– –, group–type – 312
– –, Lobachevsky – 313
– –, simple – 315
– –, strongly quasi–mixing – 77
– –, weakly quasi–mixing – 69
HRM–averaging net – 256, 260
ideal (two–sided) – 319
identity – 296
image of a random variable
or function – 339
imbeddable semigroup – 304
IUR–function – 331
invariant measure
(right, left) – 304
irreducible group of
automorphisms – 316
interior, proper – 299
–, topological – 299
intransitive normal subgroup – 315
invariant element – 329
– function – 337
– linear order – 266
– mean – 318
– measure – 303
– set – 337
– subspace – 329
kernel of a semigroup – 319
K–uniform prototype – 318
large factor group – 315
lattice – 112, 115
left–averageable
function – 15
left–quasi–averageable
function – 11
left mean of a function – 11
left semigroup – 299
Lobachevsky homogeneous
space – 313
matrix coefficient – 330
MAX – 156
maximal ergodic
theorem – 157
maximal net – 156
mean averaging
net – 121, 126
mean of an element – 24, 41
– of a function – 15
–, invariant
(right, left) – 318
mean value theorem – 86
measurable random field – 348
metric homogeneous
space – 311
metrically transitive
dynamical system – 52
modular function
(right, left) – 303
motion – 310
multiplication – 296
–, jointly continuous – 296
–, – measurable – 296
–, seperately continuous – 296
–, – measurable – 296
net – 357
open semigroup – 299
orbit of an element – 328

orbital function – 24, 329, 338
orderable group – 266
ordering dynamical
system – 267
perfect group – 303
phase mean – 48
piecewise (transfer)
regular net – 174
point random measure – 248
pointwise averaging
net – 209, 241
– ergodic theorem – 210
Poisson point random
measure – 249
positive definite
function – 325, 326, 327
potential – 265
potential energy – 265
preimage (prototype)
of a measure – 317
– – – –, K–uniform – 318
projection of a measure – 317
proper interior: $\mathrm{Int}_p(X)$ – 299
q–conservative
representation – 232
q–dissipative
representation – 232
quasi–averagable function – 15
– element – 24
quasi–covering – 181
quasi–lattice – 112, 115
Q–ergodic net – 87
random measure (r.m.) – 246
– signed measure – 261
– set – 249, 267
– set function – 265
– variable, vector, function,
process, field – 338
– σ–algebra – 353
realization of an orbital
function – 338
– of a random function – 338
regular net – 174
– semigroup – 304
regularity index – 174
relative entropy – 276
– index – 162
relative strong mixing – 57
– weak mixing – 57
relatively invariant
measure – 303
– regular net 162
– semiinvariant measure – 303
representation (of a semigroup
or group) – 328
–, associated to a
dynamical system – 338
–, – – a positive definite
function – 330
–, bilaterally uniformly
bounded – 328
–, class–*1* – 329
–, completely mesurable – 347
–, continuous – 328, 343
–, contractive – 328
–, generalized isometric – 333
–, irreducible – 329
–, isometric – 328
–, measurable – 328, 343
–, regular – 329
– , strongly contractive – 328
–, strongly isometric – 328

–, translation – 330, 341
–, uniformly bounded – 328
–, uniformly continuous – 343
–, unitary – 328
– with FD–continuous spectrum – 352
– with FD–discrete spectrum – 352
reproducing kernel – 325
– – Hilbert space – 325
right–averageable function – 15
right–quasi–averageable function – 11
right mean of a function – 11
– transformation semigroup – 299
semidirect factorization – 315
– product (right, left) – 312, 314
semigroup – 296
semiinvariant measure (right, left) – 304
semisimple group – 301
semitopological semigroup – 296
semimeasurable semigroup – 296
separability set – 357
separable direction – 277
– net – 357
shift – 264
simple point random measure – 249
simple group – 301
– homogeneous space – 315
singular random measure – 249
small factor group – 315
smooth Banach space – 33
specific energy – 271
– entropy – 286
– –, lower – 282
– –, upper – 282
– free energy – 287
– Gibbsian free energy – 290
– – relative entropy – 291
– – – , lower – 291
– – –, upper – 291
spectral condition – 211
– decomposition – 326, 327, 332
– measure – 326, 332, 343
– system of a group – 331
sphere – 311
spherically symmetric measure – 317
– – set – 311
spherical positive definite function – 327
SM–(strongly quasi–mixing) group – 77
SM–homogeneous space – 77
standart modification – 354
stationary subgroup – 310
strict sense homogeneous random function – 340
strong Cesaro limit – 60
strong (quasi–)mixing – 55, 59
– – in the wide sense – 60
strongly contractive representation – 40
subrepresentation – 329
subsemigroup – 298
temporal mean – 47
topological semigroup – 296

transfer condition – 167
– principle – 167
– regular net – 174
transitive action – 310, 315
– normal subgroup – 315
translation – 264
uniform separability – 359
uniformly continuous function – 298
universally averaging net – 85
universal mean value
(ergodic) theorem – 86
vacuum state – 265
Van Hove net – 322
weak (quasi–)mixing – 56, 59
– – in the wide sense – 61
weakly almost periodic
compactification – 335
– – – element – 333
– – – function – 334
weakly ergodic net – 320
wide sense homogeneous
random function – 340
– – – – field – 341
– – – generalized
random field – 342
WM–(weakly quasi–mixing)
groups – 69
– – – homogeneous spaces – 69
α–contractive
representation – 40
γ–irreducible group of
automorphisms – 315
ε–subnet – 360
μ–density – 265
μ–entropy – 278
κ–equivalent operators – 37
– – representations – 37

FSC
www.fsc.org
MIX
Papier | Fördert
gute Waldnutzung
FSC® C083411

Zeitfracht Medien GmbH
Ferdinand-Jühlke-Straße 7
99095 Erfurt, Deutschland
produktsicherheit@kolibri360.de